Essentials of Modern Measurements and Final Elements in the Process Industry

A GUIDE TO DESIGN, CONFIGURATION, INSTALLATION, AND MAINTENANCE

Essentials of Modern Measurements and Final Elements in the Process Industry

A GUIDE TO DESIGN, CONFIGURATION, INSTALLATION, AND MAINTENANCE

By Gregory K. McMillan

ISA™

Notice

The information presented in this publication is for the general education of the reader. Because neither the author nor the publisher have any control over the use of the information by the reader, both the author and the publisher disclaim any and all liability of any kind arising out of such use. The reader is expected to exercise sound professional judgment in using any of the information presented in a particular application.

Additionally, neither the author nor the publisher have investigated or considered the affect of any patents on the ability of the reader to use any of the information in a particular application. The reader is responsible for reviewing any possible patents that may affect any particular use of the information presented.

Any references to commercial products in the work are cited as examples only. Neither the author nor the publisher endorse any referenced commercial product. Any trademarks or tradenames referenced belong to the respective owner of the mark or name. Neither the author nor the publisher make any representation regarding the availability of any referenced commercial product at any time. The manufacturer's instructions on use of any commercial product must be followed at all times, even if in conflict with the information in this publication.

Printed in the United States of America.
10 9 8 7 6 5 4 3 2

ISBN 978-1-936007-23-3

ISA
67 Alexander Drive, P.O. Box 12277
Research Triangle Park, NC 27709
www.isa.org

Library of Congress Cataloging-in-Publication Data

McMillan, Gregory K., 1946-

Essentials of modern measurements and final elements in the process industry : a guide to design, configuration, installation, and maintenance / by Gregory K. Mcmillan.

p. cm.

ISBN 978-1-936007-23-3 (pbk.)

1. Process control. 2. Engineering instruments. 3. Automatic control. 4. Measurement. I. Instrument Society of America. II. Title.

TS156.8.M3954 2009

629.8--dc22

2009049143

Dedication

This book is dedicated to the automation engineers and technicians who are responsible for the thousands of details in the specification, configuration, installation, and maintenance of measurements and control valves. The performance and integrity of field instrumentation set by their dedication to excellence is the foundation that is the basis for achievements at all levels of process control and optimization.

Acknowledgment

The author wishes to thank Carol McMillan, Mark Nixon, Ron Pozarski, Galen Wilke, Grant Wilson, and John Wright for their support in this endeavor; Ken Brandt for the cover graphics; and the following individuals who wrote sections for the chapters:

Chapter 1 - *Field Measurement Fundamentals* - Mike Boudreau (safety systems)

Chapter 2 - *Temperature Measurement* - Jim Cobb (thermocouples and resistance temperature detectors)

Chapter 3 - *Pressure Measurement* - Caitlin Thelemann (differential pressure and pressure) and Bill Zhou (diagnostics)

Chapter 4 - *Flow Measurement* - Trever Ball (magnetic), Michael Bechthold (multivariable), Kevin Green (differential head), John Herczeg (Coriolis), and Mike Schaefer (vortex)

Chapter 5 - *Level Measurement* - Christopher Chu and Michael Olivier (DP and remote seals), Lori Egbers (Radar) and Sarah Parker (Ultrasonic)

Chapter 6 - *pH Measurement* - Jim Gray

Chapter 7 - *Final Element Fundamentals* - Owen Campney (variable speed drives)

Chapter 8 - *Control Valves* - Bert Peterson

Chapter 9 - *Wireless* - Mark Nixon

In appreciation of the significant effort by Robert Montgomery and Frank Seibert in the use of wireless conductivity, flow, pH, pressure, and temperature measurements for carbon dioxide capture research in the Center for Energy and Environmental Resources at the University of Texas, the royalties from this book are donated to the University of Texas.

Table of Contents

Preface

This book gave me a chance to get back to my instrumentation roots. I spent the first seven years of my career in the field as an instrument design and construction engineer. I was responsible for the calibration, installation, and commissioning of instruments for a half-dozen plants, where I became painfully aware that the actual performance of the process measurements and control valves (the most prevalent type of final element) was largely unknown. These were the days before the advent of smart instrumentation. We didn't know the effects of stiction and backlash on control valves or the effects of impulse line, process, and ambient conditions on measurements. We didn't know the performance of measurements and control valves that were installed in industrial plants greatly differed from catalog information. We shifted set points and just shook our heads when the material and energy balances did not close. At the time the plants could sell everything produced. Since we were mostly interested in capacity we just pushed on to make more product. Operating efficiency and turndown were not a primary concern, which was fortunate because we didn't have the speed of response and accuracy of "today's" instrumentation needed for measuring and improving process performance and tracking down valve problems.

I then moved into engineering technology where I had the chance to develop an expertise in solving difficult problems in particularly challenging loops. I spent most of the 1980s working on pH, furnace pressure, and compressor surge loops, which were the ultimate test of measurement and valve performance.

My perspective on the importance of field devices was solidified in the 1990s, when I was part of a corporate-wide process control improvement program. Most of the opportunities involved tuning control loops and adding feedforward control for continuous processes and feedback loops for fed-batch operations. However, a lot of great ideas "fell by the wayside" because of missing or imprecise measurements and inadequately responsive valves.

An important point in process control improvement is that if you don't have a way to determine the capability and benefits of a proposed automation system, projects will tend toward the lowest-cost alternatives. The biggest mistakes I made as lead engineer in the 1970s were traced back to misguided attempts for project cost savings suggested by a super experienced instrument engineer at the contract design firm, who otherwise did a great job. My most memorable mistake was not putting positioners (valve position controllers) on all control valves to save about $500 per valve. Many of the valves without positioners didn't move until the controller output was greater than 25%. A more insidious example of capital cost taking priority over performance was the proliferation from the 1970s to the 1990s of relatively inexpensive rotary piping valves that were promoted and used as throttling valves by the addition of spool-type valve positioners on actuators, linkages, and shaft connections that had been designed for on-off service, not for throttling. The complete assembly including the high friction valve design was fundamentally wrong for regulating a flow. The leakage specs and price were attractive. Deadband (backlash) and resolution (stiction) were not considered. Since the valve specification didn't require the valve to actually move in response to the small changes in signal commonly typically incurred in a control loop, there was no valve shaft position feedback measurement, and field tests typically involved visual observation of valve shaft movement for changes in output of 25% or larger, the user did not know the real price paid for the use of this type of valve. Aggravated by noisy flow measurements with poor turndown, increased process variability was attributed to mysterious sources. Without online loop metrics, there was little recognition of the deterioration in loop performance resulting from the use of these piping valves.

Putting a smart positioner on a rotary piping valve with the feedback measurement at the actuator shaft rather than sensing ball or disk stem position only added to the confusion. The actuator shaft would move in response to the positioner but the ball or disk would not respond in proportion due to extensive seal friction, twisting (windup) of the ball or disk shaft, and backlash in the connections and linkages. It was only after actual tests in the flow labs of control valve manufacturers was the true cost of using piping valves recognized. The publication of the lab test results and the subsequent ISA standards developed on valve step testing, the availability of position feedback as a secondary process variable on digital signals, and the analysis of resolution and deadband led to increased awareness of valve response and dramatic improvement in valve dynamics.

The awareness of the importance of measurement and valve performance gained during these career experiences as a user in a major chemical and life science company led to a greater appreciation of recent advances.

Today we have smart transmitters and true control valves with rangeability, resolution, and sensitivity that are an order of magnitude better than the typical fare of the last century. A combination of embedded intelligence and new sensor, transmitter, valve, and positioner technology has resulted in dramatic improvements to measurements and valves. Combined with the ability to have additional process variables, diagnostics, and alerts reported to the control room by digital signals and the mobility afforded by wireless communication, we can increase the spectrum of applications and the utility of the field automation system including finding the optimum locations for process measurement and control. Doors are opening for online data analytics, process performance metrics (e.g., energy, quality, and yield) and new opportunities for basic and advanced control improvements to address the increasing need for greater efficiency, flexibility, and rangeability of processes.

It was recently realized that research and development could also greatly benefit from the advanced performance, intelligence, and historization capabilities of smart industrial automation systems. The future is best exemplified by the lab-optimized industrial distributed control systems with pH, dissolved oxygen, pressure, temperature, and mass flow measurements for benchtop and pilot plant bioreactors described in Chapter 1 that were pioneered by Broadley-James Corporation. The addition of wireless instrumentation to these systems increases the already significant advantages of moving advanced industrial automation capability upstream in the commercialization process.

This book makes no assumptions about prior knowledge beyond that the reader has some technical background. In Chapter 1, Modern Measurement Fundamentals, special care has been taken to explain technical terms and concepts relating to the use and performance of measurements in the process industry. There is special emphasis on the advances in smart instrumentation and wireless communication. Chapters 2 through 6 focus on the details needed for the best implementation of the specific types of measurements that would be used today on automation upgrade and new plant projects in the process industry. Chapter 7 on Final Element Fundamentals follows an approach similar to that of Chapter 1 in assuming no industrial experience; the material on control valves, dampers, guide vanes, and variable speed drives will be beneficial to students and new employees. Chapter 8 gets into the details of the types of control valves that are used in 95% of the applications in the chemical and petrochemical industry. The book concludes with the latest details on WirelessHART automation systems in Chapter 9. The questions at the end of each chapter are designed to stimulate the thought processes needed for a successful application.

The foundation of a process automation system is the measurements and final elements. If you don't get this right, not much else matters. Measurements provide the only window into the process and final elements provide the only means of affecting the process. The degree of process control improvement from increasingly advanced layers of process analysis and control depends upon the integrity and breadth of the foundation. The goal of this book is to create a foundation where imagination is the limit for automation.

1

Field Measurement Fundamentals

1-1. Learning Objectives

A. Appreciate the importance of measurements in process monitoring, analysis, control, and risk prevention.

B. Be introduced to the types of measurements and signals.

C. Understand the role of measurements in batch and continuous processes.

D. Learn how process dynamics and disturbances affect measurement requirements.

E. See how industrial instruments can be used in research and development.

F. Find out how the gap between universities and industry can be closed.

G. Learn terminology and issues to intelligently discuss industrial applications.

H. Recognize the advantages of wireless smart measurements.

I. See how measurements are used in safety systems.

J. Know the definition and relative importance of accuracy parameters.

K. Find out how to reduce measurement errors.

L. Distinguish the important aspects of process and measurement dynamics.

M. Gain insight into the relative importance of measurement speed.

N. Be able to set the transmitter damping to meet process requirements.

O. Appreciate the advantage of the electronic device description language in the setup, calibration, and interrogation of measurement devices.

P. Gain an overview of calibration methods and their effectiveness.

Q. Learn the basic types and advantages of process connections.

R. Understand the significance of sensor installation and location.

1-2. Introduction and Overview

Field measurements provide the only window into what is really happening in a process. Any technical job in the process industry depends upon measurements. Jobs in research, process development, process design, process technology, automation, operations, or maintenance depend upon the ability to see, trend, analyze, diagnose, and collect the information from measurements. Ultimately, what you want to know as an engineer or technician is "why." Modern instruments have made great progress in not only answering "why" but also offering a higher level of plant performance.

The performance and integrity of control systems and safety systems depend upon the accuracy, reliability, and speed of the measurements. You cannot control or protect something you can't measure.

The most common measurements in the process industry are flow, pressure, temperature, level, and pH. The measurement of these process variables is important for a plant to run safely and efficiently and provide products that meet customer requirements for quantity, quality, and price.

Advances in Measurements

Modern measurement instruments have made significant advances in addressing an industrial plant's requirements through:

1. Technological advances in sensing element technology
2. Indication and integration of multiple measurements

3. Compensation of application and installation effects
4. Online device alerts and diagnostics
5. Remote configuration and calibration
6. Digital signals with extensive embedded user-selected information
7. Wireless communication

Transmitters with features 1 through 6 are classified as "smart" or "intelligent."

There are 50 or more major types of instruments used to measure the most common process variables. The most prevalent type of instrument to measure flow, level, and pressure is the differential pressure transmitter, simply referred to as the DP. This chapter will use the DP as an example to dig into the details of accuracy, configuration, calibration, and installation. (Note: accuracy is the maximum deviation [error] of the measured value from the true value over a stated range of the measurement). Additional information on the DP and details on the other instruments with the best technology will be covered in the succeeding chapters.

Since the 1980s, the out-of-the-box accuracy of modern industrial instrumentation has improved by an order of magnitude. Consider the differential pressure transmitter. The 0.25% accuracy of an analog electronic DP has improved to 0.025% accuracy for a smart microprocessor based DP. Furthermore, analog DP accuracy often deteriorated to 2% when it was moved from a nice bench-top setting to service outdoors in a nasty process with all its non-ideal effects of installation, process, and ambient conditions [1][16]. A smart DP with its integrated compensation for non-ideal effects will stay close to its inherent 0.025% accuracy. Additionally, a smart DP takes 10 years to drift as much as an analog DP did in one year. (Note: drift is an error that increases with time).

Smart instruments offer the ability to report additional process variables, such as the local ambient and process temperature, and alerts based on a broad spectrum of device diagnostics. These variables and alerts are communicated digitally on the same signal as the primary process variable that is being measured. The visibility of this wealth of additional information has been greatly improved in the modern distributed control system (DCS) and the associated asset management systems (AMS). The use of the Electronic Device Description Language (EDDL) standard technology for the interface to smart instrumentation enables easier and more effective access to and visualization of the information. Instrument suppliers can readily provide an interface that optimizes the look and feel of the data items in their devices [2].

Wireless Opportunities

The wireless transmission of the signals from smart devices eliminates the time and cost of the design, installation, checkout, and troubleshooting of wiring. Devices can be easily replaced and relocated. In plants, this facilitates the movement of measurement instruments for troubleshooting and for finding optimum operating points; for example, the optimum location for temperature control in a fractionation column is the point where the greatest and most symmetrical change in temperature occurs with a change in reflux or steam flow. Also, the temperature profiles in certain sections of the column are more informative than those in others. While some simulation models can provide an idea of the best locations for temperature measurements, the locations could change with load and feed composition. To facilitate the online identification and use of the best locations, additional column connections for temperature sensors have long been provided. However, signal wiring and conduit runs have to be replaced frequently whenever a measurement point is moved [3].

Wireless smart instrument measurements create the opportunity to readily relocate instruments for better analysis and control [3]. Some of the possibilities:

- Wireless temperatures and differential pressures for packed absorber and distillation column hot spot and flow distribution analysis and control
- Wireless temperatures and differential pressures for fluidized bed reactor hot spot and flow distribution analysis and control
- Wireless pressures to debottleneck piping systems, monitor process filter operation, and track down the direction and source of pressure disturbances
- Wireless temperatures and flows to debottleneck coolant systems

- Wireless instrumentation to increase the mobility, flexibility, and maintainability of skids for process equipment service such as cleaning and sterilization
- Wireless instrumentations to increase the mobility, flexibility, and maintainability of skids for lab and pilot plant unit operations. (Note: skids are platforms of pre-assembled equipment, piping, and automation to perform unit operations)

The traditional concerns about using wireless devices in a plant for process control have been addressed in the WirelessHART network [3, 4]. The WirelessHART (Highway Addressable Remote Transducer) network has a sophisticated inherent security mechanism design. Specific keys are required at both ends for a device to join the wireless mesh network. A rolling encryption of messaging protects devices as they communicate. The self-organizing and optimizing capability of the wireless network makes the setup and commissioning of a device fast, easy, and efficient. Simple modifications to the control algorithm in the DCS and exception reporting by the device remove concerns about battery life and control system performance. Wireless smart devices will become the norm for most new plant installations [3].

Smart instruments offer an improvement in installed accuracy, a reduction in drift, an increase in reliability, and the integration of device diagnostics. The latest advances in intelligent features, such as electrode impedance diagnostics for pH, have been put into wireless transmitters. Wireless transmitters eliminate the installation and maintenance cost of wiring and offer the mobility to optimize the measurement location. The significantly lower life-cycle cost and faster speed of installation, checkout, and commissioning offers the opportunity to install more measurements to provide additional process variables. Applications of smart transmitters can be developed in labs, such as inferential measurement of concentrations from conductivity and pH measurements, and readily connected to the control system. When the application has been successfully demonstrated in the lab, the wireless transmitters can be directly moved to the field. These advantages can yield a dramatic improvement in the speed and scope of process information and a reduction in the troubleshooting for real and imagined problems.

Industrial R&D Opportunities

In the past, most process and automation system analysis was done through pattern recognition of trends and calculations performed in custom spreadsheets developed by process engineers. Offline statistical and empirical modeling was done by specialists in these areas. Tools for online advanced data analysis (e.g., multivariate statistical process control), process dynamics identification (e.g., adaptive tuning), and advanced process control (e.g., model predictive control), are now integrated into the modern DCS. These tools take advantage of the increased accuracy, integrity, and scope of smart instrumentation.

The greater accuracy of smart measurements, the greater portability of wireless measurements, and the integrated advanced analysis and control capability of the DCS open up the door for industrial instrumentation and automation systems in the pilot plant and the industrial research lab.

In the past, the measurements used in the lab and plant were quite different. The dramatic improvement in accuracy, self-diagnostics, and wireless communication of smart industrial instrumentation and the advanced capability of the modern DCS make a strong case for better data and control early in the product and process research and development (R&D) phase. The use of industrial systems in R&D offers faster and more effective specification and configuration in the production facility.

The new approach, pioneered by the biopharmaceutical industry, is to use the best smart measurement and a lab optimized industrial DCS [5]. For example, bench-top bioreactors use industrial online measurements of pH, dissolved oxygen, temperature, turbidity (cell density), capacitance (cell viability), and NIR (precursors to cell death) [5, 6]. New at-line analyzers with automated sampling systems provide measurements of the concentration of glucose, amino acids, nutrients, and byproducts. These additional online measurements (analyzers with probes in the process) and at-line measurements (analyzers with automated sample systems connected to the process) are essential for unlocking the secret profiles of these complex batch processes [7]. New basic control systems for glucose and amino acid concentration can set the stage for model predictive control of the batch profiles. Studies have shown a Model Predictive Control (MPC) system that sustains maximum cell growth and product formation rates can provide a more repeatable batch profile, a shorter batch cycle time, and a higher yield [7, 8].

Figure 1-1 shows the integration of wireless smart industrial instrumentation and a lab optimized DCS for a portable single-use bioreactor (SUB). The portable SUB has a disposable liner that eliminates the cost and time required for cleaning and sterilization in place. The sizes of SUBs have increased to the point where they can be used for industrial production of high value-added pharmaceuticals [3][5].

Figure 1-1. Portable Bioreactor with Wireless Industrial Instrumentation and a Lab Optimized DCS (*Courtesy of Broadley-James Corporation*)

Some of the more advanced lab and plant opportunities offered by a modern DCS and wireless smart measurements are as follows:

- Process experimentation
- Process development
- Plant design
- Plant debottlenecking
- Root cause analysis
- Predictive maintenance
- Abnormal situation prevention
- Plant efficiency improvement
- Plant capacity improvement

- Online data analytics
- Adaptive controller tuning
- Process control improvement
- Advanced control

University R&D Opportunities

Universities teach the theoretical basis of process research, development, design, and control. The equations presented and first-principle models developed are essential for developing a conceptual understanding of processes. However, equations and models have parameters that are unknown and relationships that are incomplete. Process measurements are essential for fitting the parameters. There is a great opportunity for universities to work more closely with industry [9].

The DCS is the most prevalent means of implementing batch and continuous control in the process industry. The DCS platform may also be effectively used for the evaluation and demonstration of innovative algorithms developed at a university. The integrated online advanced controls, diagnostics, data analytics, and process modeling capability of a DCS facilitate university research. This working environment also prepares students for careers in industry. Industrial standards adopted by the DCS allow students and researchers to concentrate on the application of new control capabilities. External applications such as Matlab can be used with the DCS to explore control and process modeling.

In this environment, students and researchers may take advantage of preconfigured and automated features of a DCS. Process simulations may be integrated into the DCS to create a virtual plant capability that is capable of running faster than real-time [5][8]. The DCS supports wireless and Internet access to the virtual plant, which increases ease of use in a lab or pilot plant. The DCS enables rapid exploration, discovery, and prototyping of process modeling and control technologies in a university environment and deployment in industry [9].

High technology companies are realizing the synergy between university and industry R&D for new process technology, advanced data analytics, and advanced control. These companies promote the use of the DCS in the labs of the universities that they support. Examples of university labs and pilot plants with modern industrial measurements and a DCS are the following:

- Purdue Engineering Research Center for Structured Organic Particle Systems
- Rose-Holman Institute of Technology Chemical Engineering Lab
- University of Texas Chemical Engineering Unit Operations Lab
- University of Texas Environmental Energy and Resources Pilot Plant
- Washington University in Saint Louis Computer Control Lab

We now move from getting a basic understanding of the capability and opportunities of modern instrumentation to a conceptual view of how measurement signals are created by sensing elements, generated by transmitters, communicated by wiring and networks, and received by control systems. The basic classification and requirements of enclosures for weather protection and safety are briefly summarized. Common measurements are used as examples.

Perspective

This book focuses on modern measurements and final elements with the highest performance, reliability, and maintainability. There are many older technology devices and valves still offered that are not described in this book. The addition of smart features doesn't overcome the inherent limitations of older, more primitive technologies. The life-cycle cost (cost of hardware, design, calibration, installation, commissioning, maintenance, and process variability) of the best technologies is generally less. Some project proposals unfortunately only consider the hardware cost.

The life-cycle cost of even the best technology can be excessive if there is a mistake in selection, specification, sizing, calibration, installation, or diagnosis. A common mistake is to use 1960s and 1970s installation practices [1].

To summarize, the user of smart instrumentation should not be deceived into thinking that smart features in themselves guarantee fool-proof performance. Smart instruments can be handicapped by dumb technologies, dumb applications, and dumb installations. The goal of this book is to provide guidance for getting the most out of the inherent capability of modern measurement instruments.

Sensing Elements

A sensing element converts a process variable into a quantifiable output that can be measured by another sensing element, a safety or control system input card, or a transmitter. The output of the sensing element is the **measurand** [10]. The term "sensing element" is preferred to "primary element" or "primary detecting element" [10]. The term "primary element" should be reserved for elements placed in the process to create a process variable (PV) that can be detected by the sensing element.

Differential Head Flow Measurement Example

The most common example of a measurement with a primary element is a differential head flowmeter with a capacitance type differential pressure transmitter (DP). This flowmeter is widely used because of the low cost of its primary element, which is simply a plate with a hole in it (an orifice plate). The volumetric flow is proportional to the square root of the pressure drop across the orifice. The primary element creates an intermediate PV (differential pressure), the diaphragm sensing element converts the PV to a proportional change in measurand (capacitance), and the transmitter computes the flow from the measurand with square root extraction and communicates it as an analog or digital signal to an automation system. In instruments, such as the DP, where the sensing element is integrated with the transmitter, the instrument is simply referred to as a transmitter with no distinction made to the sensing element component.

The density and hence the mass flow can be computed from the volumetric flow for a known composition by the addition of pressure and temperature measurements. Figures 1-2a and 1-2b show the installation of pressure, temperature, and DP measurements as part of a multivariable transmitter to compute and communicate mass flow instead of volumetric flow.

The DP can be installed at a more accessible location by the use of sensing lines called "impulse lines" as shown in Figure 1-2a. The trade-off for more convenient accessibility of the transmitter is the extra installation and winterization costs, and more importantly, the increased uncertainties and maintenance created by impulse lines. Plants have moved toward the direct coupling of the transmitter to the process connection as shown in Figure 1-2b to eliminate impulse line issues with the realization that a smart DP rarely needs to be removed or calibrated. In the section on installation, some of the many problems with keeping these lines at a known process composition or state will be discussed.

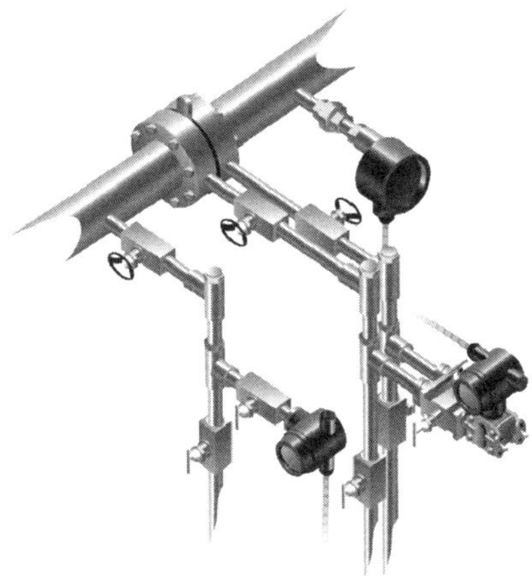

Figure 1-2a. Multivariable Differential Head Flowmeter with Sensing (Impulse) Lines

In Figure 1-2b, the isolation valves are shown as part of a manifold connected to the DP. This manifold would be left in place for removal of the DP. In many plants, piping isolation valves at the orifice taps are used, which provides isolation per the piping spec and enables the manifold to be removed with the DP.

Figure 1-2b. Multivariable Differential Head Flowmeter Mounted on Orifice Taps

The differential pressure (DP) transmitter has an integral sensing element that detects the pressure drop from the resistance to flow imposed by the reduced area of the orifice. The DP transmitter's process connections are piped into the isolation valves at the orifices upstream and downstream pressure taps. The installation in Figures 1-2a and 1-2b also shows an integral valve manifold that allows the user to isolate or equalize DP transmitter's high and low sides. The upstream pressure and downstream pressures that are connected to the high and low side of the isolating diaphragm respectively create a minute displacement of the internal sensing diaphragm. In DP transmitters that use capacitance as a measurement principle, the sensing element diaphragm is inside a chamber enclosed by capacitor plates that is filled with a dielectric. The deflection of the sensing element and consequential change in capacitance is proportional to the difference in pressures on the high and low side of the transmitter. The transmitter extracts the square root of the differential pressure to provide an output signal proportional to flow.

Examples of sensing elements that are often separately mounted from the transmitter are thermocouples, resistance temperature detectors (RTD), remote seals for differential pressure transmitters, pH electrodes, and conductivity sensors. Thermocouples and pH electrodes generate a millivolt measurand that is carried by lead wires from the sensing element to the transmitter, which computes and communicates a temperature and pH signal, respectively. The transmitter is installed nearby or in the case of the thermocouple, can be mounted on the sensing element as shown in Figures 1-2a and 1-2b, which eliminates the need for thermocouple extension lead wires.

While the pH transmitter stays in place, the pH electrode is transient because of its life expectancy and need for periodic removal for calibration and possibly cleaning. The cost for calibration, cleaning, and replacement of electrodes makes the maintenance costs for pH much higher than for other instruments. Modern pH smart electrodes have a micro-chip with their model and serial number and factory calibration. Smart pH electrodes also have a microprocessor and preamplifier to offer the collection and communication of the probe's calibration history in the field. These features significantly improve the effectiveness of the calibration, cleaning, and replacement of electrodes.

Transmitters

Modern systems have progressed from sensing elements, such as thermocouples being directly wired to the input card of a safety system or control system, to using smart field transmitters that can digitally and wirelessly communicate individual diagnostics, scaling, and multivariable compensation for process conditions.

The transmitter converts and scales the sensing element output into a standard signal that becomes the input to a safety system or control system. Since a transmitter converts a measurand into an output signal it is sometimes referred to as a transducer. For the thermocouple and pH electrode example, the millivolt measurand from the sensing element is converted to the process variables of temperature and pH, respectively, and communicated as an analog (milliamps direct current), frequency (cycles per second), or digital signal (bits per second). It is important to realize that implied in these signals is always the engineering units and scale of the process variable to be displayed at the transmitter output's destination.

The transmitter may be integrated with the sensing element (e.g., DP transmitter) or mounted on the sensing element or primary element. For temperature measurements, the transmitter may be mounted on the thermowell that contains the thermocouple or RTD. For flow measurements, the combination of a sensing element and transmitter is often called a "flowmeter" or "meter" with a descriptor in front. Examples are Coriolis mass flowmeters, differential head flowmeters, magnetic flowmeters (magmeters), positive displacement meters, thermal mass flowmeters, turbine meters, and vortex meters.

Early practices that used zero current to represent 0% output were abandoned to use a positive current for 0% to distinguish a failed signal from 0% output and indicate negative measurement errors at 0% output. In the 1960s, the analog output predominantly used a 10- to 50-mA DC signal. In the 1970s, it was changed to a 4- to 20-mA DC signal, which remains the standard today for analog electronic signals. The 4- and 20-mA correspond to 0 and 100%, respectively, of the process variable range specified in the setup of the transmitter. The nonzero or "live zero" low end of the signal range facilitates the detection of sensing element and device failures and the observation of negative errors. The 4-mA signal is also essential for providing power to the microprocessors of smart transmitters to keep the devices awake and alert to problems. Each electronic signal is transmitted from the device by a single pair of wires that is twisted and shielded to reduce electro-magnetic interference (EMI). Single pairs of wire are often terminated in a junction box and/or interface panel to become part of a multi-pair "home run" cable that carries the signals of multiple transmitters to a safety system or control system. For transmitters that require more power than can be provided by the DC voltage on the signal pair of wires, a second separate pair of wires is used to provide power. These transmitters are termed "four-wire."

HART (Highway Addressable Remote Transducer)

In the 1980s, the HART Communication Foundation (HCF), an independent nonprofit organization, developed a communication protocol for digital devices. HART devices transmit the primary process variable via the customary 4–20 mA analog output but also provide the ability to communicate additional process variables, device identification, commands, status and diagnostic alerts, and calibration information via a digital signal imposed on the analog signal. HART devices can communicate one primary process variable and several additional process variables in a single message. The number of additional process variables for simultaneous communication has increased from three to eight in HART version 7. The user has a choice of up to 256 process variables in a HART device [11]. The availability of additional process variables eliminates the need for installation of additional measurements and wiring. For example, a Coriolis mass flowmeter can simultaneously transmit fluid density, percent solids, and temperature besides mass flow on a single pair of wires.

The HART standard provides for the communication of events (occurrence of abnormal conditions) by the setting of bits. The device manufacturer can choose to communicate a signal bit as an alert or a series of bits to provide detailed diagnostics on the event.

Fieldbus

In the mid 1990s, Fieldbus systems first appeared based on International Society of Automation (ISA) and International Electrotechnical Commission (IEC) standards. The Fieldbus Foundation (FF) was set up as an independent nonprofit organization to support the commercialization of the ISA/IEC Fieldbus standard. Fieldbus devices use the ISA/IEC 1158 physical layer and stack for communication and can embed function blocks per the architectural concepts and terminology of ISA/IEC 61804. The function blocks enable the user to readily configure process computations and process control loops in field devices.

Table 1-1. Example of Data Available for Communication between a HART Device and a Host ("Unleash the Power." *Control* Supplement, Aug. 2001)

Process Variable Values	Commands from Host	Status and Diagnostic Alerts	Device Identification	Calibration Information for 4-20 madc Transmission of Primary Process Variable
Primary Process Variable (PV) - analog current signal transmitted to host	Set Primary Variable Units	Device Malfunction - Indicates device self-diagnostic has detected a problem in device	Instrument Tag - User defined, up to 8 characters	Date - date of last calibration
Primary Process Variable (PV) - digital value in engineering units	Set Upper Range	Configuration Changed - Indicates device configuration has changed	Descriptor - User defined, up to 16 characters	Upper Range Value - PV value in engineering units for 20 mA set by user
Percent Range - primary process variable in percent of calibration range	Set Lower Range	Cold Start - Indicates device has gone through power cycle	Manufacturer Name (Code) - code established by HCF and set by manufacturer	Lower Range Value - PV value in engineering units for 4 mA set by user
Loop Current - loop current value in milliamps	Set Damping Value	More Status Available - Indicates additional devices status data available	Device Type and Revision - set by manufacturer	Upper Sensor Limit - set by manufacturer
Second Process Variable (SV) - digital value in engineering units	Set Message	Primary Variable Analog Output Fixed - Indicates device in fixed current mode	Device Serial Number - set by manufacturer	Lower Sensor Limit - set by manufacturer
Third Process Variable (TV) - digital value in engineering units	Set Tag	Primary Variable Analog Output Saturated - Indicates 4-20 mA signal is saturated	Sensor Serial Number - set by manufacturer	Sensor Minimum Span - set by manufacturer
Fourth Process Variable (FV) - digital value in engineering units	Set Date	Secondary Variable Out of Limits - Indicates secondary process variable value outside the sensor limits		PV Damping - Primary process variable damping factor set by user
	Set Descriptor	Primary Variable Out of Limits - Indicates secondary process variable value outside the sensor limits		Message - Scratch pad message area (32 characters) set by user
	Perform Loop Test			Loop Current Transfer Function - Relationship between primary value digital value and 4-20 mA signal
	Initiate Self Test			Loop Current Alarm Action - Loop current action on device failure (upscale or downscale)
	Get More Status Available Information			Write Protect Status - Device write-protect indicator

In Fieldbus devices, signal wiring between the device and the control room is replaced by a spur to a Fieldbus segment that is connected to a controller or control system bridge via a home run trunk as shown in Figure 1-3. Up to 32 devices can be connected to an individual segment and 12 of these devices can be powered from the segment. Existing twisted pair wire for 4–20 mA signals can be used for spurs and segments. Fieldbus reduces the wiring and termination cost of new installations by 30–40% [13]. Fieldbus also enables local high-speed control and can communicate more descriptive and quantitative diagnostics.

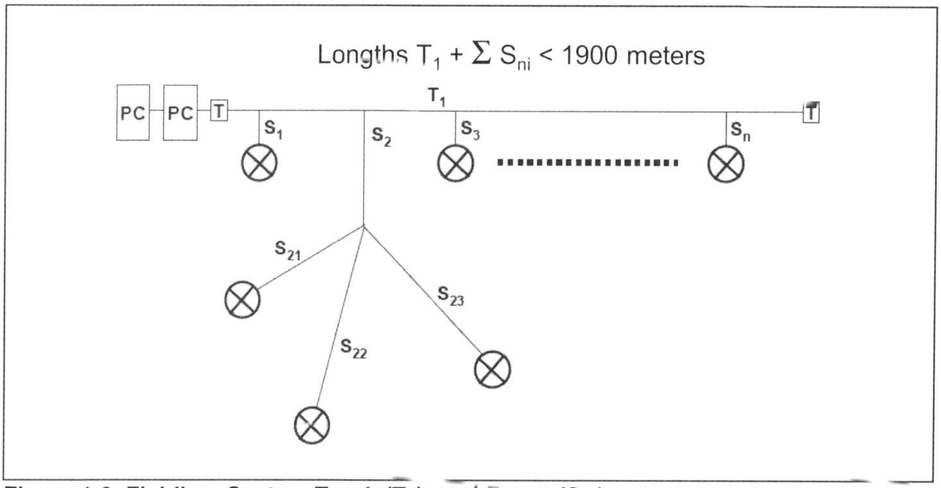

Figure 1-3. Fieldbus System Trunk (T_1), and Spurs (S_{ni})

The life-cycle savings potential for WirelessHART is more than double what is achieved by Fieldbus. WirelessHART eliminates the entire cost of wiring (design, installation, modification, and troubleshooting) and associated EMI noise problems. WirelessHART can meet the speed requirements of all but the fastest loops. Wireless periodic reporting of the primary process variable can be as fast as once per second. To extend battery life, significantly longer update time intervals are used for slow loops.

WirelessHART

The advent of WirelessHART measurements in 2008 eliminated the need for wiring between the transmitter, final elements, and the control system. WirelessHART devices broadcast to a self-organized network (Figure 1-4) that is optimized for security and reliability. WirelessHART transmitters integrally mounted with the sensing element directly to a process connection, such as the DP shown in Figure 1-2b, can save up to 70% of the life-cycle cost of a measurement by the elimination of the design, installation, checkout, and troubleshooting of wiring and sensing (impulse) lines [3].

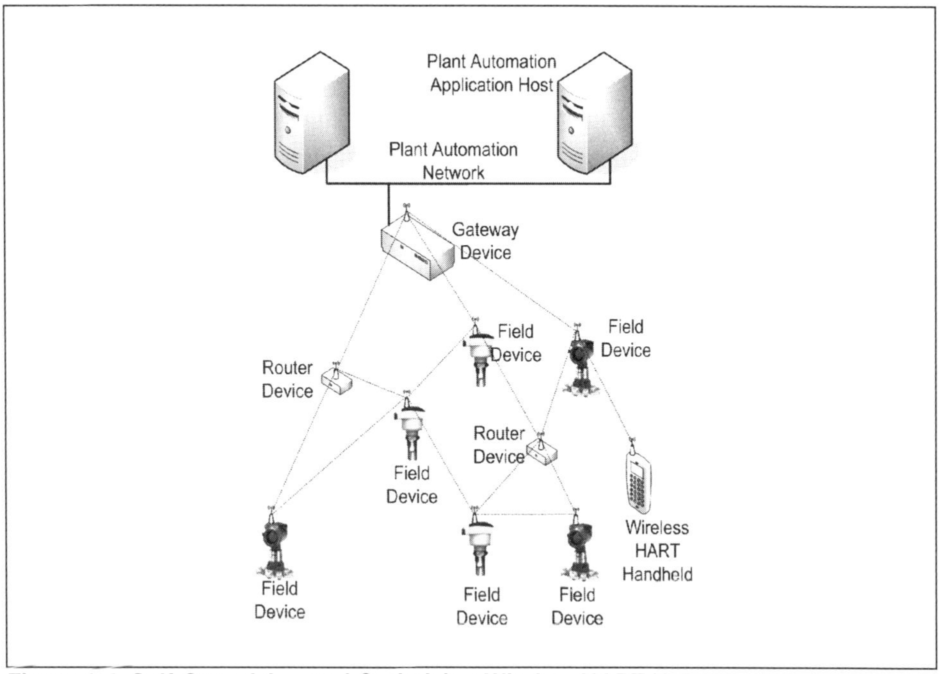

Figure 1-4. Self-Organizing and Optimizing WirelessHART Network

A major disadvantage of wired systems is that the spikes and noise from wiring problems are often intermittent, difficult to track down, and extremely disruptive [12]. The cost in terms of process variability and maintenance is usually underestimated. Whether wiring problems truly exist or not, there is always uncertainty about the integrity of the wiring and grounding system [3].

Enclosures

In the process industry, the enclosure for the transmitter typically has a NEMA 4 (National Electrical Manufacturers Association) high pressure hose proof rating. In solids process areas, the enclosures have a NEMA 5 dust proof rating. In the petrochemical industry, explosion-proof enclosures are frequently specified suitable for Class 1 Div II areas, which means ignitable concentrations of flammable gases, vapors, or liquids can exist but are not present during normal operating conditions. The Class 1 Div I ratings are for ignitable concentrations that are normally present. Plants are designed so that ignitable concentrations are an abnormal event, so this classification is rare. Explosion-proof enclosures have a NEMA 4 or NEMA 5 rating.

The most common field measurement devices are differential pressure (DP) transmitters and thermocouple or resistance temperature detector (RTD) temperature transmitters. Figures 1-2a and 1-2b show a DP flow transmitter with an integral diaphragm sensing element and temperature transmitter with an integral thermocouple sensing element for a Class I Div II electrical area classification.

Types of Measurements

There are more than a thousand different types (model numbers) of transmitters. This book will cover general classes of transmitters that account for 95% of the measurements in modern industrial process plants. Older technologies or devices specific to a particular product that represent less than 5% of instruments purchased for a new facility are omitted to facilitate a focus on the future of automation systems and measurements common to most batch and continuous processes. The most prevalent measurement technologies will be covered in the greatest detail.

A typical production plant has hundreds of process flows. Each piece of process equipment has process flows that enter and exit the equipment, which are then manipulated to control the amount of material, and in many cases the relative concentration of the materials, in the equipment. In addition, to control the temperature within the process equipment, utility flows for heating and cooling go through coils inside the equipment or through a jacket that surrounds the outside of the equipment.

Figure 1-5 shows a process vessel with common field measurements. There are flow measurements of the liquid and gas streams entering and exiting the vessel. There is a level measurement for the amount of liquid in the vessel, a pressure measurement of the overhead vapor space in the vessel, a temperature measurement of the vessel liquid, and a pH probe in the recirculation stream. You can see from this example that there are many more flow measurements than any other type of measurement. A continuous plant with 1000 total online measurements would typically have 600 flow measurements, 200 temperature measurements, 100 pressure measurements, 80 level measurements, and 20 composition or physical property measurements. Batch plants tend to have a higher percentage of flow measurements.

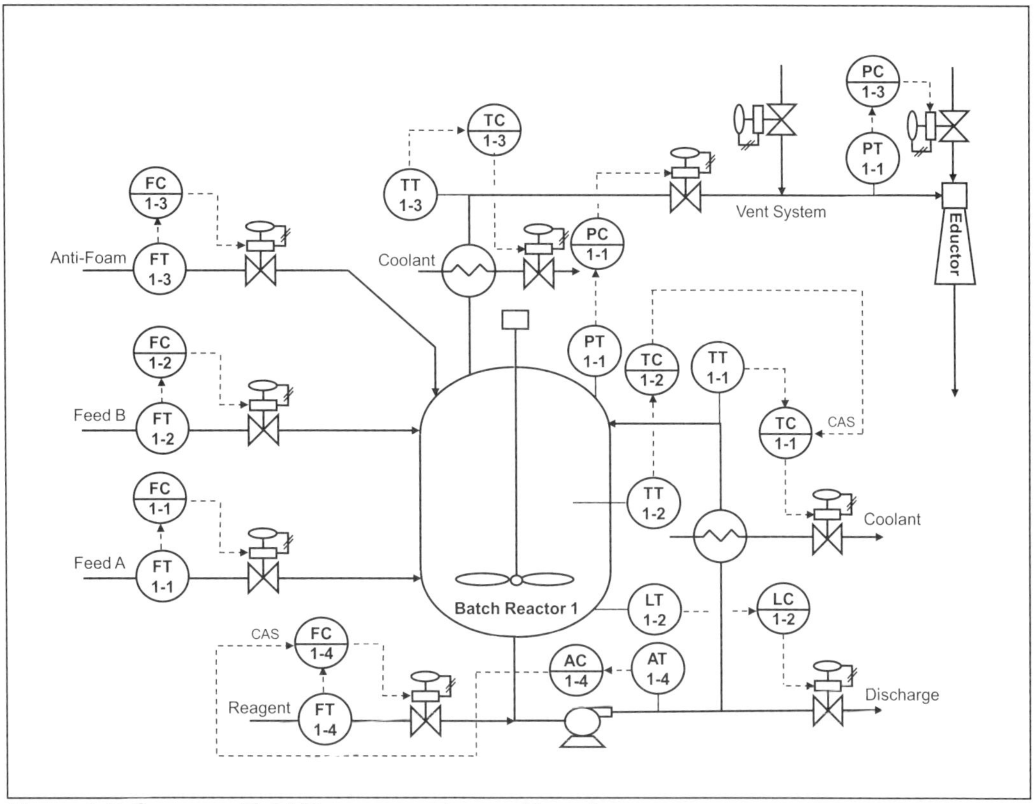

Figure 1-5. Common Field Measurements on a Vessel

Transmitters provide an analog output that is a quantized numerical value of a process variable. The fineness of the quantization is set by the sensing element sensitivity and by micro-processor and I/O resolution.

Distributed Control Systems

Figure 1-6 shows some of the functionality in a manufacturing automation network that results from measurement inputs. The following applications rely on timely and accurate measurement inputs from the field and the laboratory:

- Basic Process Control System (e.g., PID loops)[1]
- Dedicated Unit Operations Control (e.g., Batch Manager)[1]
- Advanced Process Control (e.g., Model Predictive Control)[2]
- Loop Tuning and Performance (e.g., Adaptive Control)[2]
- Data Analytics Systems (e.g., Multivariate Statistical Process Control)[2]

1. These applications are normally in the DCS.
2. These applications are integrated into the more advanced DCS.

- Laboratory Information Management Systems (LIMS)
- Process Information Management Systems (PIMS)
- Supervisory Control Systems (e.g., Real-Time Optimization)
- Manufacturing Logistics (e.g., SAPP)

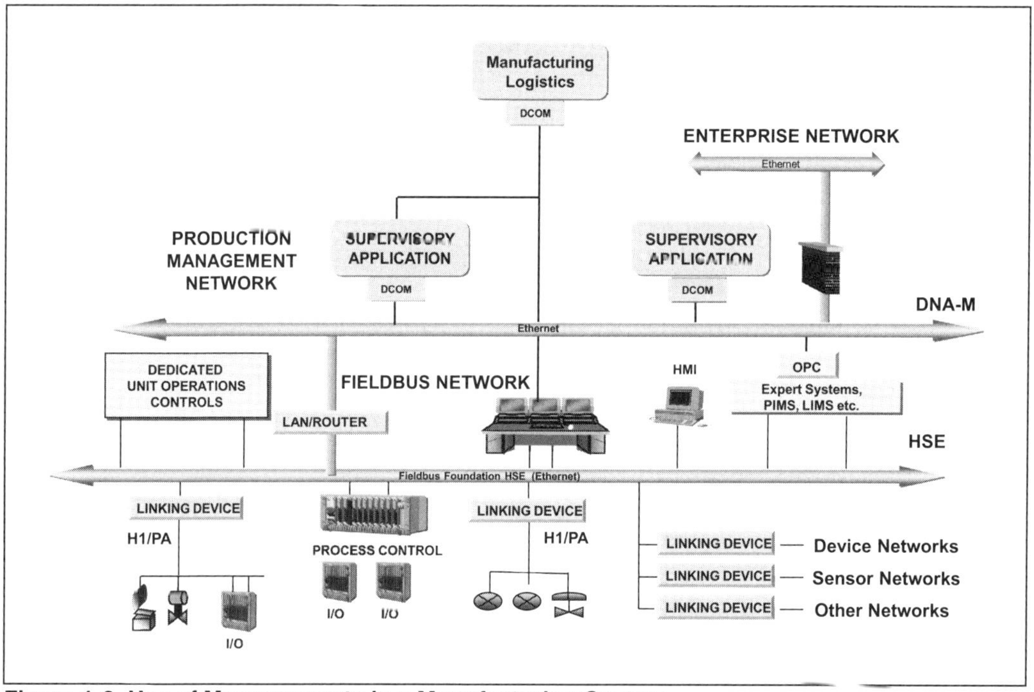

Figure 1-6. Use of Measurements in a Manufacturing System

The DCS has conventional 4–20 mA, HART 4–20 mA, Fieldbus digital, and WirelessHART analog inputs and outputs. An input card and Analog Input (AI) block converts the input signal to a numerical value. Similarly an Analog Output (AO) block and output card converts a numerical value in the DCS to an analog or digital output signal. The analog-to-digital (A/D) input cards and digital-to-analog (D/A) output cards have improved resolution as compared to earlier versions. For example, the 12-bit I/O cards of the 1980s were increased to 16 bit in the 1990s. Numerical values are used within the DCS for monitoring, analysis, prediction, basic process control, and advanced process control. The analog outputs are typically used to manipulate flows for the throttling of control valves or setting the speed of motors for agitators, blowers, conveyors, extruders, fans, pumps, and sheet lines.

The DCS has discrete inputs and outputs with 120 VAC or 24 VDC dry contact or isolated contact closures. Discrete values with a software state of a 1 or 0 (true or false) in the DCS are used in conjunction with measurements in sequential function charts and batch managers for startup, shutdown, grade transitions, and batch process control. The discrete outputs are used to turn motors on and off and to open and close valves. Discrete outputs have a pulse duration option where the width of the pulse in the discrete output for a specified cycle time is proportional to an analog value.

1-3. Process Automation

In order to better understand the role and importance of various measurement features, it is helpful to learn the basics of how processes respond and how measurements are used in automation systems for basic process control and risk prevention.

Control Systems

Important process variables become the controlled variables for feedback control systems. Nearly all these feedback control systems are control loops as shown in the block diagram in Figure 1-7. The control system is called a loop because a signal travels in the loop from point 1, the process output, to point 9, the process input. The process output at point 1 is detected by a sensing element and is transmitted as an input to an analog input (AI) block at point 2 in the DCS as the process variable (PV) in engineering units. The term analog input is used even though the transmitter output may not be an analog signal but may be a digital Fieldbus, an analog signal with a superimposed digital signal (HART), or a frequency-derived value of a WirelessHART signal. The AI block provides scaling, filtering, alarms, and other options. The PV output of the AI block is generally in the same engineering units as the AI block input. At point 4a, the PV is scaled to 0-100% to become a controlled variable (CV) and at point 3a, the set point (SP) is scaled to 0-100% for the Proportional-Integral-Derivative (PID) controller. The CV is subtracted from the SP to create an error in percent at point 5 that is the input to the PID controller. The negative sign applied to the measurement provides "negative feedback," which may be a familiar term from amplifier and servomechanism design that is essential for self-regulation and stability.

When in automatic mode, the controller computes a 0-100% output at point 6 from its input error that is proportional to the error, the integral of the error, and the derivative of the error. (In most loops, the derivative

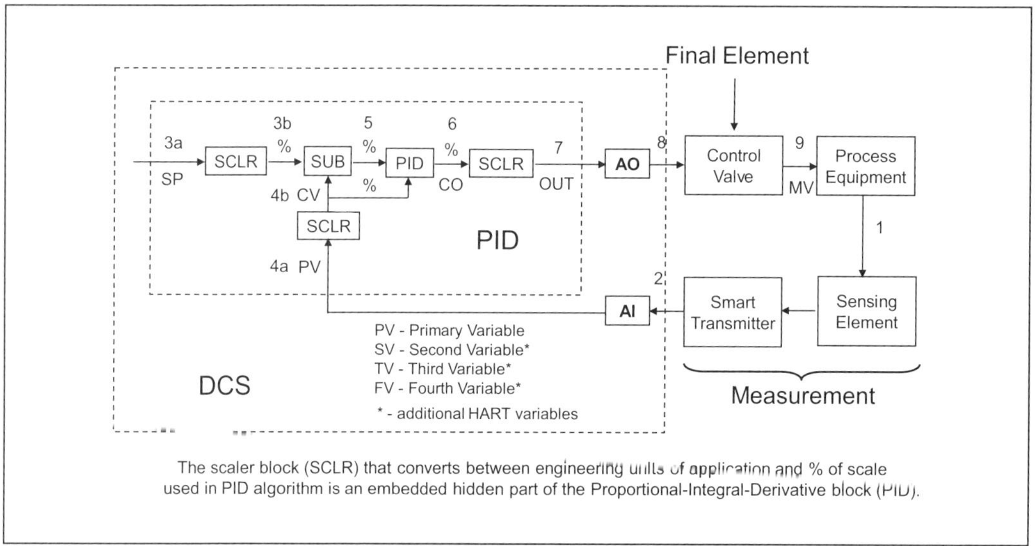

Figure 1-7. Block Diagram of Typical Feedback Process Control Loop

mode is not used). At point 6, the sign of the input error is retained for "direct control action" or reversed for "reverse control action." The output is scaled to match the scale of its manipulated variable (MV) and becomes the set point for an analog output (AO) block at point 7. The field output signal at point 8 is used by the control valve to manipulate a flow that goes into the process at point 9.

It is an important but not well-recognized fact that the PID algorithm processes signals in percent and that the conversion from engineering units to percent affects the controller gain calculation. What users see in the operator interface, trend charts, and configuration is the PV, SP, and MV in the specified engineering units.

The AO block offers scaling and a reversal of the signal for reverse acting control valves, also known as "increase-to-close" and "fail-open" valves. The output of the AO block typically becomes the signal to a control valve where it is converted from an analog current or digital Fieldbus signal to a pneumatic actuator pressure via an electronic or digital valve positioner. The valve positioner provides air flow to the actuator until the air pressure in the actuator provides the desired valve position. For older valves without an electronic or digital valve positioner, a current to pneumatic (I/P) converter is used to provide the pressure to a pneumatic valve positioner or directly to the valve actuator. In Chapters 7 and 8, we will see that all valves should have a smart electronic or digital positioner.

The control valve converts the position to a flow at point 9 for a given pipeline pressure upstream and downstream of the valve. While a control valve is considered here, in some cases a variable speed drive is used. Each process has input and output flows called streams. These flows can be liquids, gases, or solids. In 99% of the plant applications in the process industry, control loops ultimately affect the process by manipulation of flows (e.g., liters per minute, gallons per minute, pounds per hour, and kilograms per hour). There may be intermediate loops that control process variables such as stream pressure and temperature, but these loops eventually cascade to a loop that manipulates a final element that changes a stream or mixing (circulation) flow. The flow manipulated is not always measured. In some cases, the control loop output will manipulate the power to an electrical heater or an electrode but these applications predominantly occur in benchtop or pilot plant scale equipment or in specialized equipment such as silicon crystal pullers and electrolytic reactors and furnaces.

Continuous Processes

A process is continuous if the principal discharge flow is non-zero during normal operation. In batch processes, the principal discharge flow is zero until the batch is done. This simple distinction, and its significant implications as to the dynamic response of these processes, is somehow lost in the complexity of automation systems. Most books and courses on process control are based on continuous processes. This section and the following sections are designed to introduce the reader to the significant distinctions and implications of continuous, pure batch, and fed-batch processes.

Most of the older and higher volume processes (e.g., commodity chemicals) are continuous operations, which are characterized by steady state values of the process variables. The dynamic behavior of the process is classified as self-regulating. There can be long periods of quiet operation where the process variables are drawing straight lines (not changing). However, continuous processes may have long lines (trains) of equipment and parallel lines of equipment. The flows of the parallel lines of equipment may diverge from and converge to common points in the process, which are often surge tanks. Consequently, the startup, shutdown, production rate change, and product grade transition of any piece of equipment can spread to other lines and ripple through and upset the whole system.

Recycle streams are process output streams that are returned as process input streams to recover solvent or raw materials that are not completely converted into product. The use of recycle streams makes processes more efficient by a decrease in makeup solvent, an increase in conversion, and a reduction in waste treatment. However, recycle streams can cause the dynamic response of concentrations to be integrating, i.e., if the controller is put into manual, the process variable will ramp up or down at a constant rate provided there are no disturbances or load changes.

In continuous processes that are running, the discharge flows from vessels are open. A level controller manipulates an inlet or outlet flow to keep the level constant except during startup, shutdown, and transitions. In surge tanks, an error squared algorithm or notch gain may be used to reduce the correction for levels well within the tank low and high level alarm limits. This strategy reduces the variability in feed to downstream units. If the level controller is in manual or is not keeping the level constant, the composition and temperature responses move from self-regulating to integrating to a degree dependent upon the mismatch between inlet and outlet flows.

Flow and liquid pressure processes are inherently continuous and self-regulating when in service because the flow out is equal to the flow into the control valve whenever the valve is open. There is no change in mass inventory between the entrance and exit of the system. Gas pressure processes are usually integrating, because there is a change in mass inventory if the loop is in manual. An increase in pressure will not increase the flow out of the volume enough to match the flow going into the volume and vice versa.

When there are several diverging and converging streams from parallel unit operations, the process can be difficult to analyze and control. Complicating factors such as catalyst activity deterioration, coating or fouling of heat transfer surfaces, equipment being pushed well beyond its nameplate capacity, and undersized surge tanks can lead to a system that is in a perpetual state of disruption.

Some people view a continuous process as a long batch process because like a batch, the continuous process has a beginning and an end, and the startup and shutdown sequences are similar to batch sequences. This view is interesting but is misleading because during normal operation, the level of a continuous process is constant and the response is self-regulating. The most important period, which is when the product is being produced, is a continuous process.

Self-Regulating Processes

The dynamic response of a temperature, concentration, or pH measurement as shown in Figure 1-8 is self-regulating for a continuous process. If the controller is put in manual, each process variable will line out at final value provided there are no disturbances or load changes or recycle streams. The observed dead time (θ_o) is the time of no process response after a change in the controller output. The process time constant is the time to reach 63% of the final change after the process breaks out of the noise band with a recognizable response. While the terms "process dead time" and "process time constant" are normally used, more correct terms would be "total loop dead time" (θ_x) and "open loop time constant" (τ_o), respectively. The descriptor "open loop" is an important distinction for tuning methods that use a "closed loop time constant" (τ_c) called Lambda (λ), which is the time to reach 63% of a set point change. Lambda is used to specify the controller's speed of response in the automatic, cascade, or remote cascade mode which is important for the coordination and performance of loops. In contrast, the "open loop time constant" does not depend upon controller tuning but describes the response of the rest of the automation system. When a controller is in manual or remote output modes, there is no feedback control. The loop is open at the controller in the Figure 1-9 block diagram [8][18].

The open loop time constant (τ_0) is the largest time constant in the loop (largest time constant in the Figure 1-9 block diagram). Ideally, the largest time constant is in the process so it slows down process disturbances.

The largest time constant can occur in the measurement with an excessively large sensing element time constant, transmitter damping setting, or DCS filter time setting.

The largest time constant can occur in the final element if a control valve has a large actuator or a positioner with poor sensitivity or a large prime mover has an undersized motor. These scenarios are particularly troublesome for flow and liquid pressure loops because these processes are so fast. Poor temperature sensing elements and fan or vent damper installations can be problematic for furnace and dryer temperature and pressure loops even though their process dynamics are slower. In these situations, the measurement is providing an attenuated and slow view of the actual

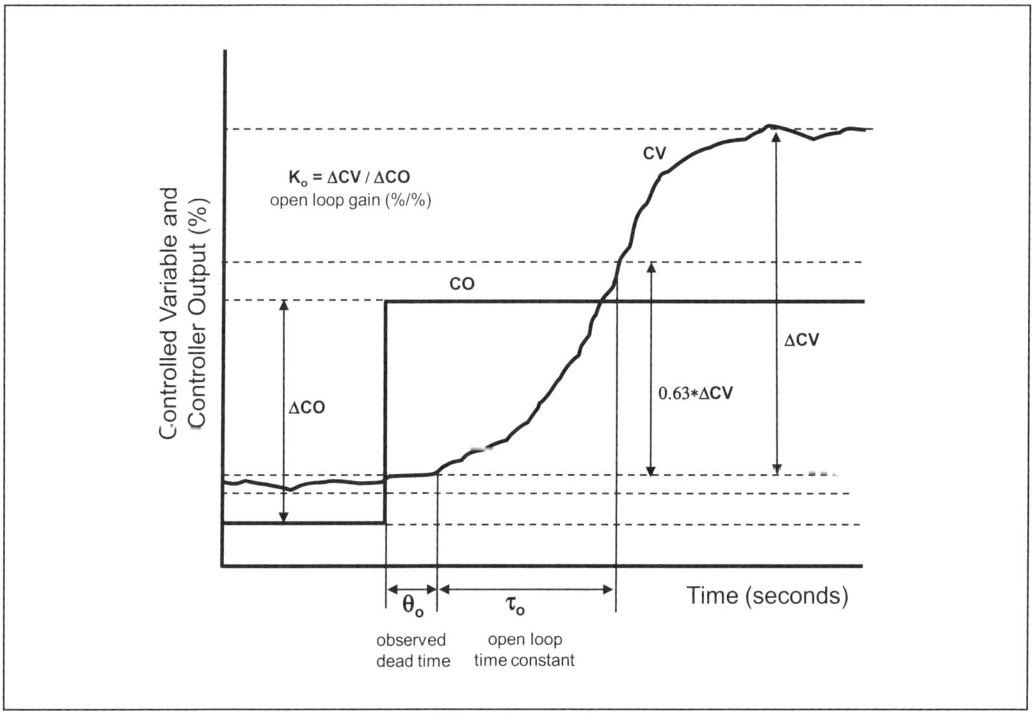

Figure 1-8. Self-Regulating Response of Continuous Process Measurements

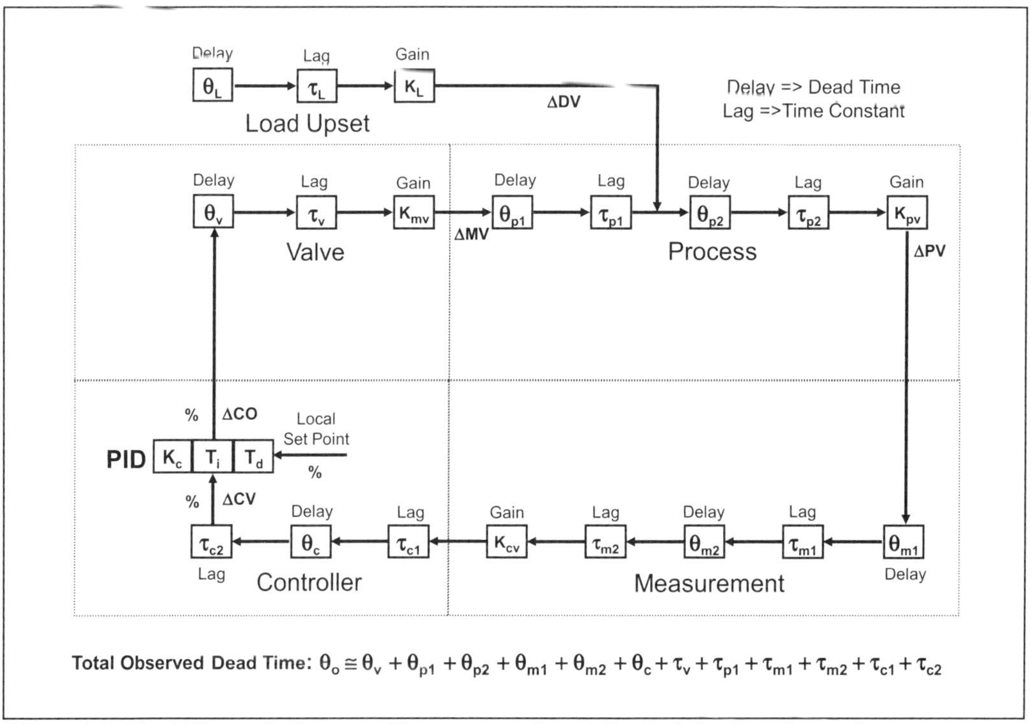

Figure 1-9. Final Element, Process, Measurement, and Control System Dynamics

process variable or the final element is providing a slow correction of the process variable [8][18].

The total loop dead time (θ_0) is the sum of the dead times and small lag times in the loop. Dead time is a problem no matter where it occurs because it delays either the observation of or the reaction to a disturbance. The minimum integrated absolute error (IAE) for loops tuned for maximum load rejection (e.g., flow disturbance correction) is proportional to the square of the total loop dead time. The peak error is proportional to the loop dead time — open loop time constant ratio [8][18].

The process gain is the change in the measurement output in percent divided by the change in controller output in percent. The process gain is more properly known as the open loop gain. The open loop gain depends upon the dynamics of the final elements and measurements, as well as the dynamics of the process, as denoted in Equation 1-1a for the typical case where the final element regulates a flow into the process. Equation 1-1b shows the measurement gain is inversely proportional to the measurement span, an important consideration in the effect of measurement setup on controller tuning.

The literature rarely breaks down dynamics in a process control loop into contributing parts of an automation system as shown in Figure 1-9, leading to misunderstandings. In the chapters on final elements, we will see how the response of a control valve depends upon size, installed characteristic, deadband, hysteresis, resolution, and the size and direction of the change. In the chapters on measurements, we will explore how the response of a measurement depends upon the speed of the sensing element and transmitter. The good news is the proper design, setup, and installation of control valves, variable speed drives, and inline measurement devices makes the speed of response of these components negligible for all but the fastest processes.

Some process responses have a time constant so slow that their response in the region of interest around the set point looks like a ramp. These processes are called "near-integrating" or "pseudo-integrating" processes and can be analyzed and tuned as if they had an integrating response (as shown in Figure 1-10) by converting the self-regulating process gain to an integrating process gain per Equation 1-1c. Most integrating processes would eventually reach a steady state if there were no process or measurement range limits. Also, some auto tuning methods do not handle integrating processes very well. For these and other reasons, Equation 1-1c is useful for converting a measured self-regulating process gain to an estimated integrating process gain and vice versa.

$$K_o = K_f * K_p * K_m \tag{1-1a}$$

$$K_m = \frac{100\%}{S_m} \tag{1-1b}$$

$$K_i = \frac{K_o}{\tau_o} \tag{1-1c}$$

Where

K_o = open loop gain for self-regulating processes (CV %/CO%)

K_i = open loop gain for near-integrating processes (CV %/CO%)

K_f = final element gain (Flow e.u./CO %) (e.u.= engineering units)

K_p = process gain (PV e.u./Flow e.u.)

K_m = measurement gain (CV %/PV e.u.)

S_m = span of measurement scale in PID (PV e.u.)

(CV % is percent of controlled variable scale and CO % is percent of controller output)

The span of the measurement scale is the difference between the low and high values of the process variable (PV) scale of the PID controller. The span of the PV scale in the PID is generally but not necessarily the calibration span of the measurement.

Integrating Response

The following process variables and processes tend to have an integrating response:

- Level in vessels with pumps on discharge
- Gas pressure in columns, dryers, and vessels
- Composition in continuous processes with no level control
- Composition in continuous processes with recycle streams

- Composition, pH, and temperature in batch operations

An integrating response has no steady state. When the controller is in manual, the process variable will ramp at a rate dependent upon the unbalance. The easiest integrating process to visualize is level. If the main discharge flow is pumped out, for example, an increase or decrease in level will not appreciably increase or decrease the discharge flow, respectively. The level will ramp up or down whenever the total of the flows coming into the vessel does not exactly match the total of flows going out of the vessel. Any exact match of incoming and outgoing flows is temporary at best. The ramp rate depends on the unbalance in these flows and the size of the vessel. Integrating processes are more difficult to control but fortunately tend to be much slower, giving the controller a chance to catch up with a disturbance provided it is properly tuned to take advantage of the slow ramp rate.

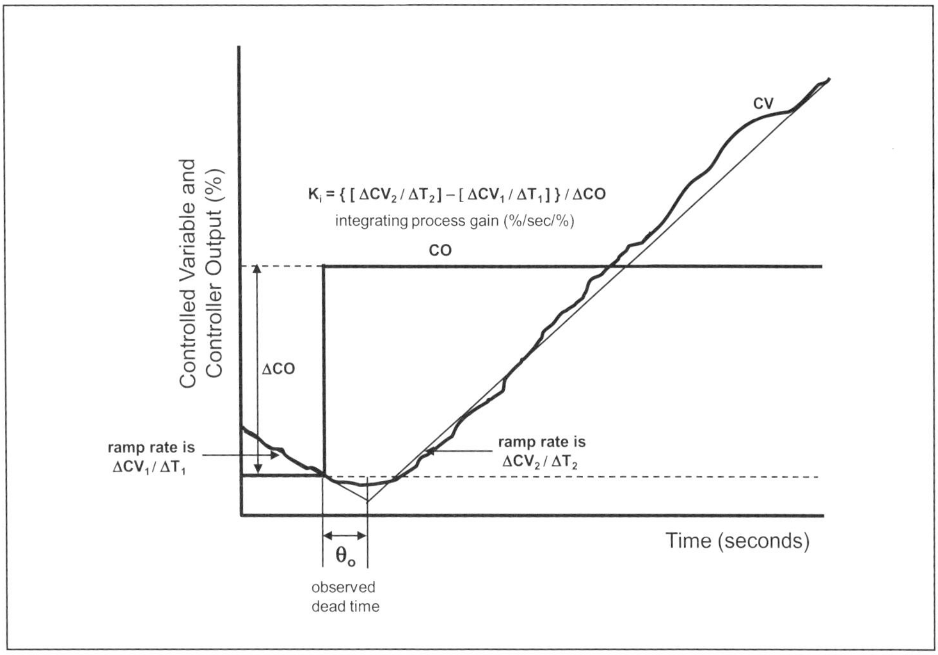

Figure 1-10. Integrating Response of Batch Process Measurements

Batch Processes

Most ingested or injected products (foods, beverages, and pharmaceuticals) use batch processes because the quality is predictable if the recipe is carefully followed and good and consistent quality raw materials are used. There is no retention of old material or buildup of impurities or contaminants as associated with continuous processes since the vessels are completely emptied and can be cleaned and sterilized between batches. New processes for specialty chemicals also use batch operations because a batch system is easier to develop and design since it essentially follows a cookbook approach. However, for a given size of equipment, a batch process will not produce as much as a continuous process. As products mature and production capacity requirements increase, production tends to move from batch to continuous mode.

In batch processes, the liquid or solids discharge flow does not match the total flow coming into a vessel. Consequently, the liquid or solids level is continually rising or falling. In pure batch processes, the discharge flow is shut off until the end of the batch cycle when the vessel is drained. The rise or fall in level from the mismatch of flow going in and out changes the dynamic response of vessel temperature, concentration, and pH from self-regulating to integrating, which has important implications for process control.

To understand the distinctive difference between a batch and continuous operation consider the preparation of a meal. The making of a pot of chili is a pure batch operation as shown in Figure 1-11. There is a definitive sequential recipe where tomato sauce, beans, spices, and secret ingredients are added in precise quantities until the pot is full. The contents of the pot are cooked until the mixture reaches a temperature, simmers, and melds. When the peak flavor is reached, the pot is emptied. The pot is stirred and sampled at a frequency and loss in volume that depend on the taste and hunger of the stirrer and whether your spouse is looking.

A continuous chili process might start out like a pure batch process but once the chili reached peak flavor, a drain valve would be opened and a simultaneous addition of ingredients would be started whose total flow matches the discharge flow of chili as shown in Figure 1-12. Getting to the balance condition would be difficult but when you reached it, the flow and taste should be consistent. For a given size pot, a continuous process can produce more chili because you don't have the addition and drain times and the flow rate can be increased to a point commensurate with the heating capacity. The quality or taste might not be as good as a pure batch process with a correct batch time because a fraction of the ingredients added are going out without enough simmer time or sufficient mixing, particularly as the production rate is increased.

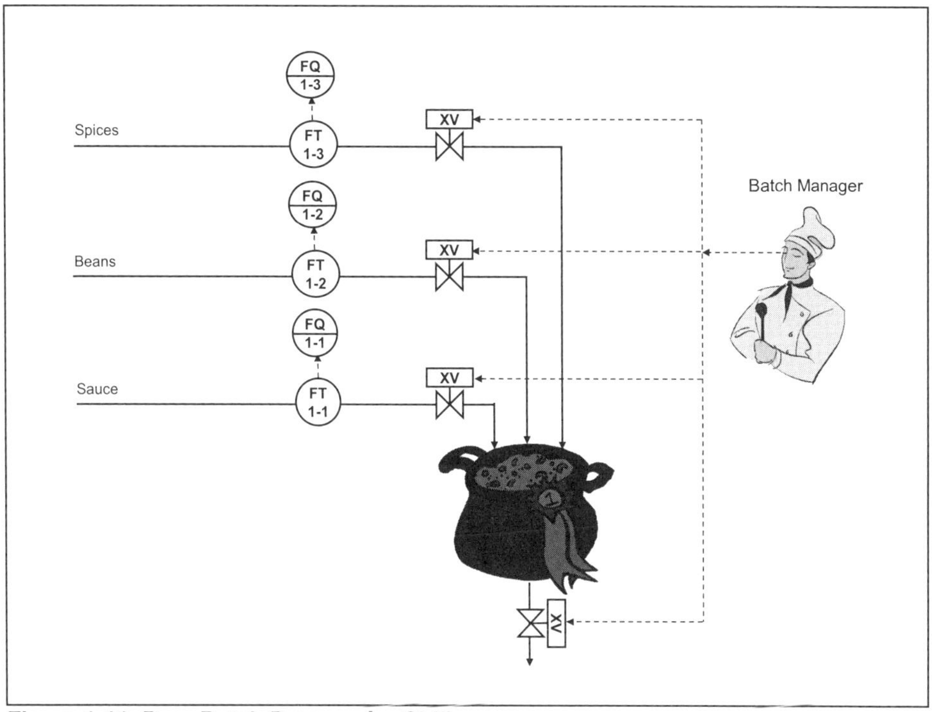

Figure 1-11. Pure Batch Process for Chili

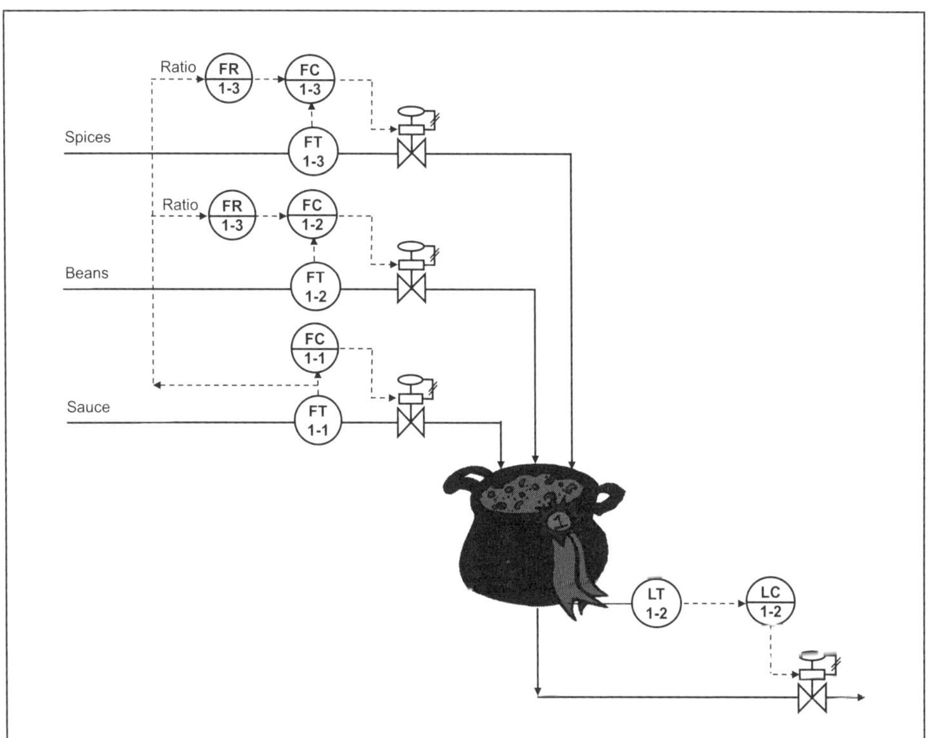

Figure 1-12. Continuous Process for Chili

Fed-batch Processes

Fed-batch operation is also termed semi-continuous operation because ingredients are simultaneously added under a ratio control scheme where the flow of each ingredient is kept in the proper proportion to the others as shown in Figure 1-13. This usually occurs after a pre-charge of some ingredients and heating of the mixture to operating temperature. The simultaneous addition continues until the pot is full, at which time it goes into a simmer mode. In some fed-batch processes, the ingredients do not flow simultaneous but flow sequentially. In either case, the flows are often manipulated by composition, pressure, or temperature control loops. The manipulation of flows may be by the direct adjustment of a flow set point or by the correction of a ratio factor. While this process is termed semi-continuous, it does not have the dynamic characteristics of a continuous process because the level is rising.

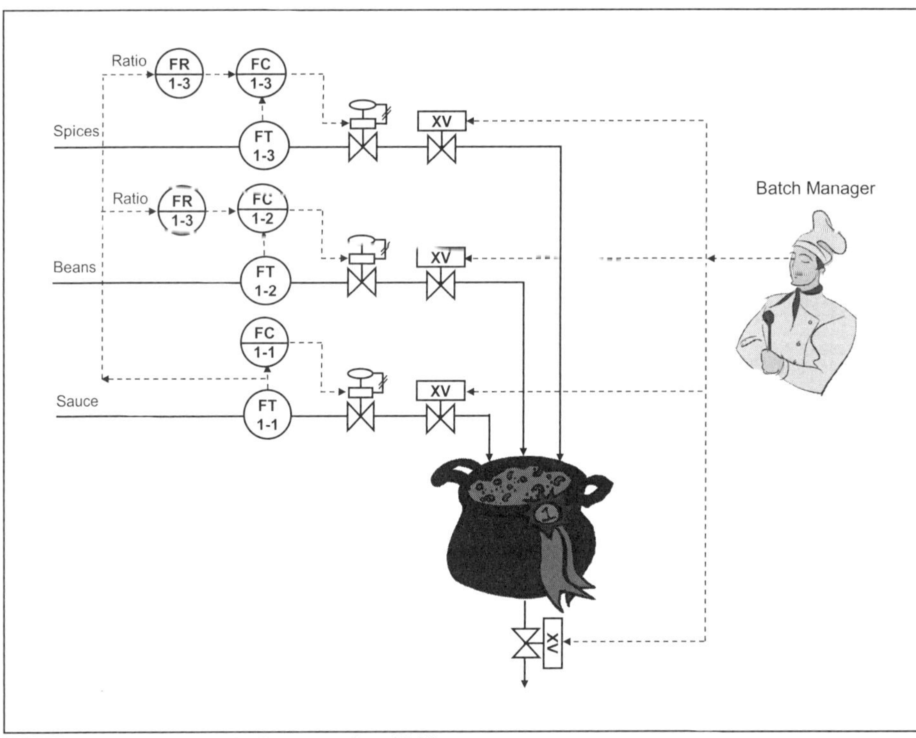

Figure 1-13. Fed-batch Process for Chili

The control systems for pure batch chemical processes tend to not to have as many control loops as do continuous control systems. The feeds are often charged based on a weight or flow total basis without control loops. Online concentration measurement and control is rare in chemical batch processes. The profiles of intermediate and product concentrations are largely secret. Since the product quality depends upon these concentrations, the performance of a batch process is largely unknown until after the batch is done and lab results are available.

Batch processes typically have pressure and temperature loops. The pressure is often controlled by manipulation of a gas or vent flow. Temperature is usually controlled by manipulation of cooling or heating. We should not generalize that batch processes don't have as much process control as continuous processes. While batch chemical processes generally do not have as many field analyzers as continuous chemical processes, there are often more flow control loops in a fed-batch process. Biochemical and biopharmaceutical processes (predominantly batch to date) have more online measurements than chemical processes.

Fed-batch control systems have feed flow loops and pH or concentration control loops to correct the ratio of ingredient flows. In the case of a gas phase reaction, the gas reactant feed flow is often manipulated by a pressure loop that corrects for the consumption of the gas reactant.

Fed-batch cell culture bioreactors have dissolved oxygen, viable cell, biomass concentration, and pH loops besides temperature and pressure loops, and in the near future will have glucose and glutamine concentration control loops as shown in Figure 1-14. This, combined with air, oxygen, inerts (e.g., argon), carbon dioxide, glucose, and glutamine flow loops, add up to a complexity seldom seen in continuous chemical processes.

Many of the batch variables follow a batch profile (curve) that usually starts at a low value and reaches a peak that corresponds to a significant decrease in product formation rate.

Profile Control

The degree of process control determines how many of the measurements of process outputs (batch variables) follow a batch profile (curve). Process control loops make process outputs draw straight lines by transferring variability in process outputs to process inputs.

A simple bioreactor example, where the media (glucose and nutrients) and seed cells have been pre-charged and the batch has reached operating temperature, illustrates the principle of transfer of variability, which has

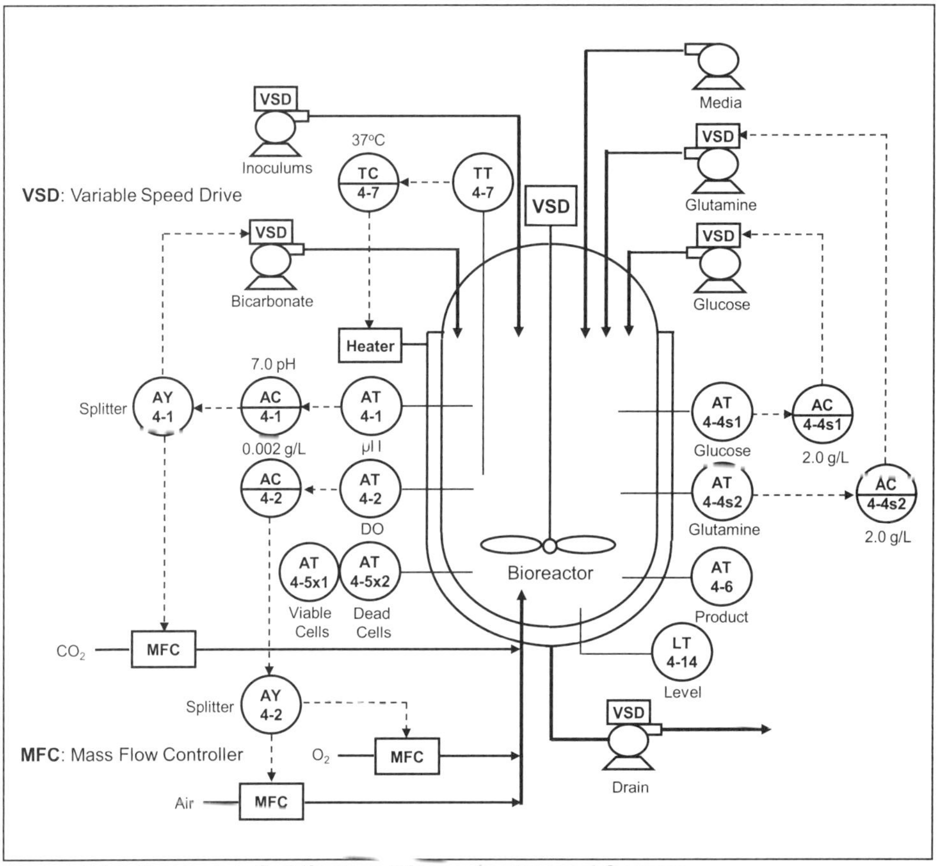

Figure 1-14. Fed-batch Cell Culture Bioreactor Control System

profound implications not commonly recognized in the literature on process analysis, efficiency, and consistency.

The mindset of process research, development, and design in universities and industry is to determine process outputs by fixing process inputs per a script. This mindset, while performance limiting, is difficult to change in practice. The experimentation in research and process development sets the path of fixing the values and timing of additions and heating and cooling. The lack of process control loops in research and process development translates to process definitions that are deficient in basic and advanced process control. If there is variability in the process outputs, there is a more stringent effort to keep most process inputs fixed and to improve the role of a few inputs viewed as key. Process engineers who have achieved particular expertise in orchestrating the process inputs are reluctant to give up personal control of these inputs to a control loop and a process control engineer. The solution is to put a lab-optimized distributed control system with control loops and advanced control tools into the research lab, similar to what has been done for benchtop and pilot plant bioreactors [5].

In Figure 1-15, there are no process control loops on the process outputs. The feeds are sequenced in a batch operation. All the process variables follow a batch profile. The feed rates and timing are experimentally derived for a given set of feed conditions and starting points.

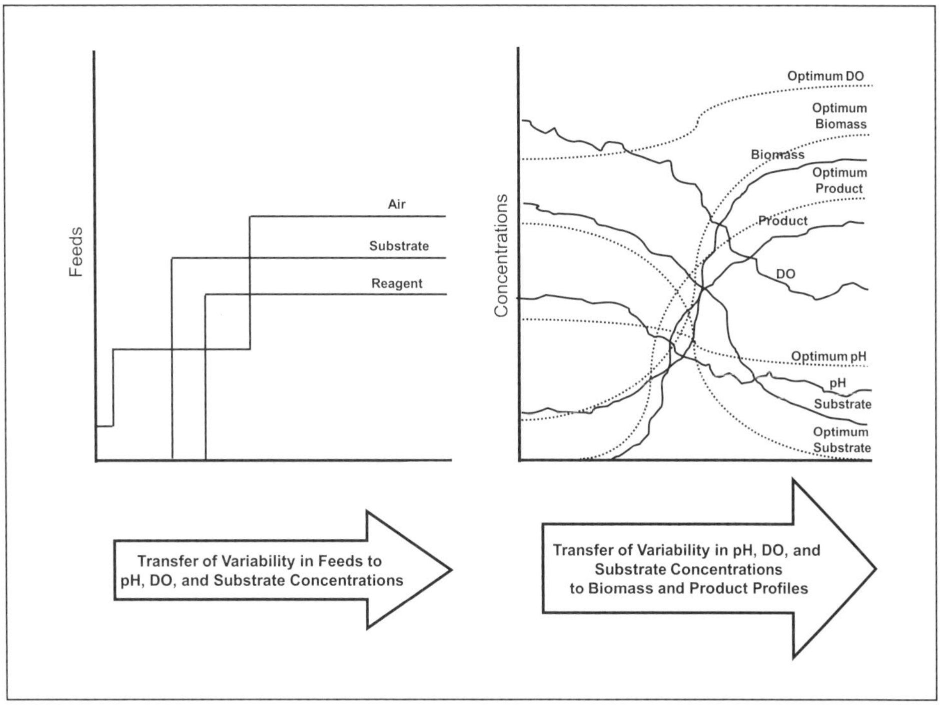

Figure 1-15. Transfer of Variability for Pure Batch Control of a Bioreactor

In Figure 1-16, dissolved oxygen, pH, and glucose are drawing relatively straight lines by the use of process control loops that have transferred variability in these process outputs to the feed rates. These loops automatically take care of disturbances and unknowns in the media charge and seed cells and mutations and alterations in the metabolic pathways of the growing cells.

In Figure 1-17, the biomass and product concentration profiles achieve a more proficient curve shape and end point by the use of model predictive control (MPC) to automatically optimize the dissolved oxygen, pH, and glucose set points. MPC transfers variability from the bottom line process outputs (e.g., product concentration) to the less important process outputs (pH and dissolved oxygen), which in turn via control loops, transfer the

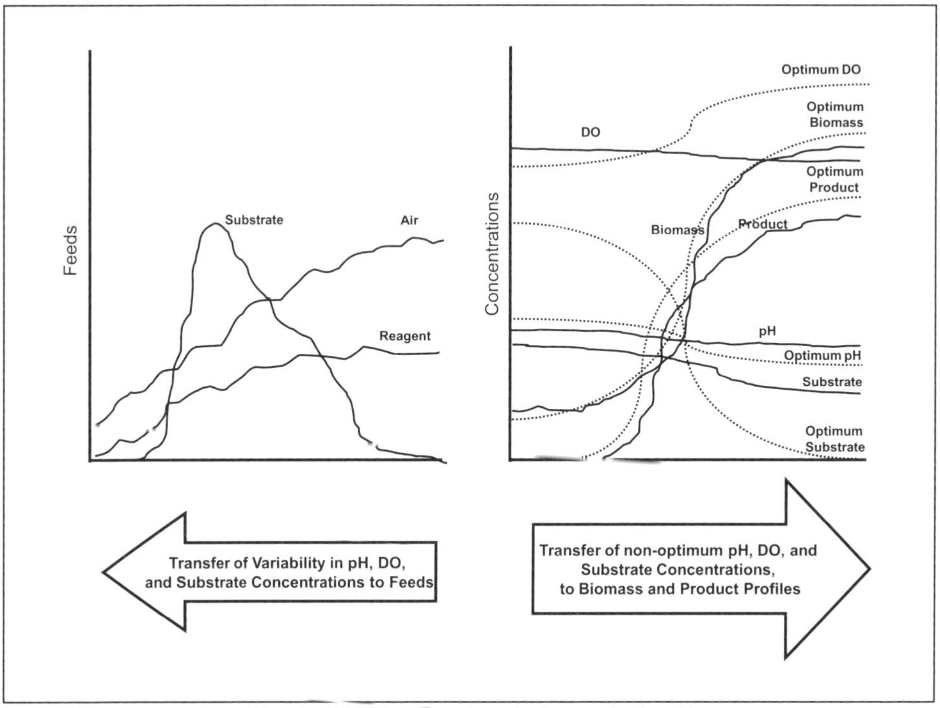

Figure 1-16. Transfer of Variability for Fed-batch Control of a Bioreactor

deviations from the optimum set point to the process inputs (e.g., reagent and air flows).

Real Processes

Regardless of whether a process is batch or continuous the feeds are subject to disturbances in terms of pressure, temperature, concentration, physical properties such as heat capacity, density, and viscosity, and quality parameters such as color. The variation in bulk material shipments, recycle streams, and intermediate products from other processes in the plant and the degree of smoothing in storage tanks is usually not tracked or modeled.

Most university chemical and petrochemical engineering programs exclusively study continuous processes during normal operation and use steady-state simulation software to generate process flow diagrams (PFD). The flows, temperatures, pressures, concentrations, and physical properties do not change with time but are constant for a given set of feed conditions.

Nearly all control theory classes and textbooks on process control focus on continuous processes. Students are not prepared for the dynamics and

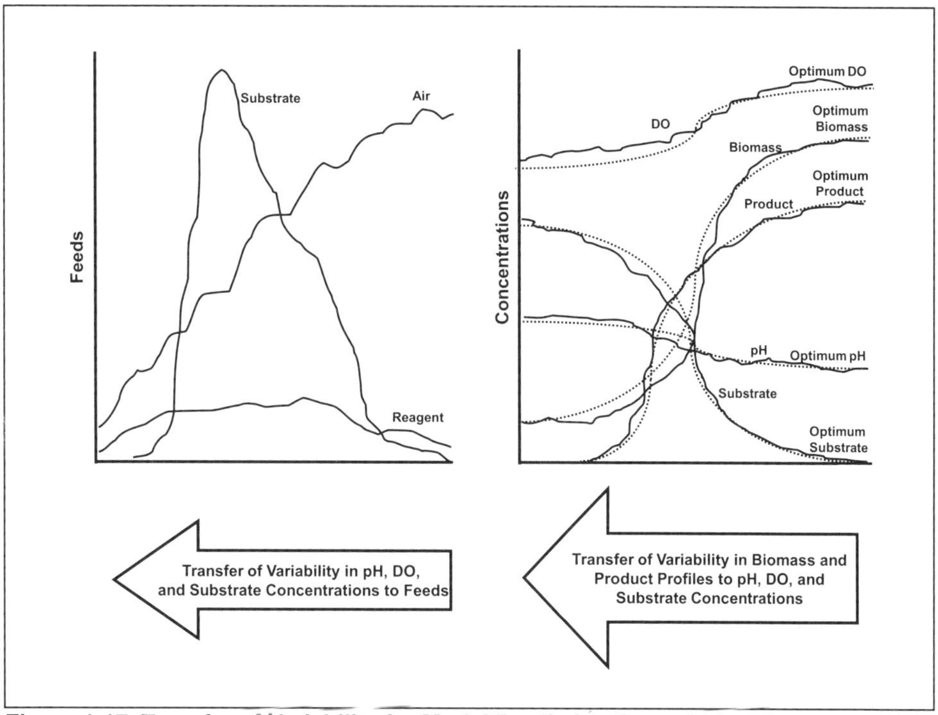

Figure 1-17. Transfer of Variability for Model Predictive Control of a Bioreactor

measurement and control considerations of batch processes or the startup, shutdown, and transitions of continuous processes. Students may even think the streams in a plant are at the conditions depicted in a PFD.

Continuous processes may operate for long times at one steady state and can appear to be quite boring, but assuming the concentrations not measured online are at PFD values is wishful thinking. The real thrill of process control is moving the process to reach a more efficient or productive operating point. In a batch operation, reaching an operating point that creates saleable product is a natural course of action, because the batch manager and process control system are pushing the batch to reach its end point. Continuous processes are more dependent on the operator or an advanced control system to move the process operating point. On the other hand, there is generally more opportunity to increase batch production capacity by reducing batch cycle times by adding process control loops or model predictive control. Increases of 25% capacity from additional process control are common for batch process, whereas increases of 5% capacity in continuous processes is considered exceptional. For both batch and continuous processes improved process efficiency, in terms of providing higher yields, lower utility use, or producing higher grade product are process control opportunities.

Many plants have a combination of batch and continuous processes. A common configuration consists of batch reaction and crystallization processes followed by continuous separation (e.g., distillation and evaporation) and purification processes.

Measurements and process control systems also have the important function of property and personnel protection. These often go hand in hand because damaged or malfunctioning process equipment usually poses a safety risk to personnel.

Safety Systems

Most plants have safety instrumented systems (SIS) as automated protective layers to reduce the risk in process operations, as shown in Figure 1-18. A safety integrity level (SIL) rates the severity of the potential hazard. A higher SIL number indicates a higher severity and greater safety system requirements. The American National Standards Institute (ANSI) and the International Society of Automation (ISA) have developed standards for the Functional Safety: Safety Instrumented Systems for the Process Industry Sector (ANSI/ISA-84.00.01-2004 Parts 1, 2, and 3). These standards provide the framework, requirements, and guidelines for the application of SIS and guidance for the determination of SIL.

Many processes include flammable, explosive, harmful, or even lethal chemicals. A safety hazard can be the result of contact with the chemicals or from the effects of damaged or malfunctioning process equipment. The SIS will initiate actions to prevent or mitigate a hazardous event if process variables get too high or low for safe operation. The discrete status of a process variable being out of safe range is set by field process switches, transmitter signal switches, or software value switches as shown in Figures 1-19a through 1-19c.

Field process switches have a process connection that puts the process fluid in contact with the trigger mechanism or sensing element. Older mechanical examples are bellows or bourdon tube pressure switches, bimetallic or filled capillary temperature switches, paddle type flow switches, and float or displacer level switches. Newer electronic examples include capacitance, ultrasonic, and microwave level switches. Field pressure and temperature switches, despite technical improvements, are not recommended by the author. Field process switches don't provide an online measurement of the process variable (e.g., pressure, temperature, flow, or level). Consequently, it is difficult to monitor their sensing ability

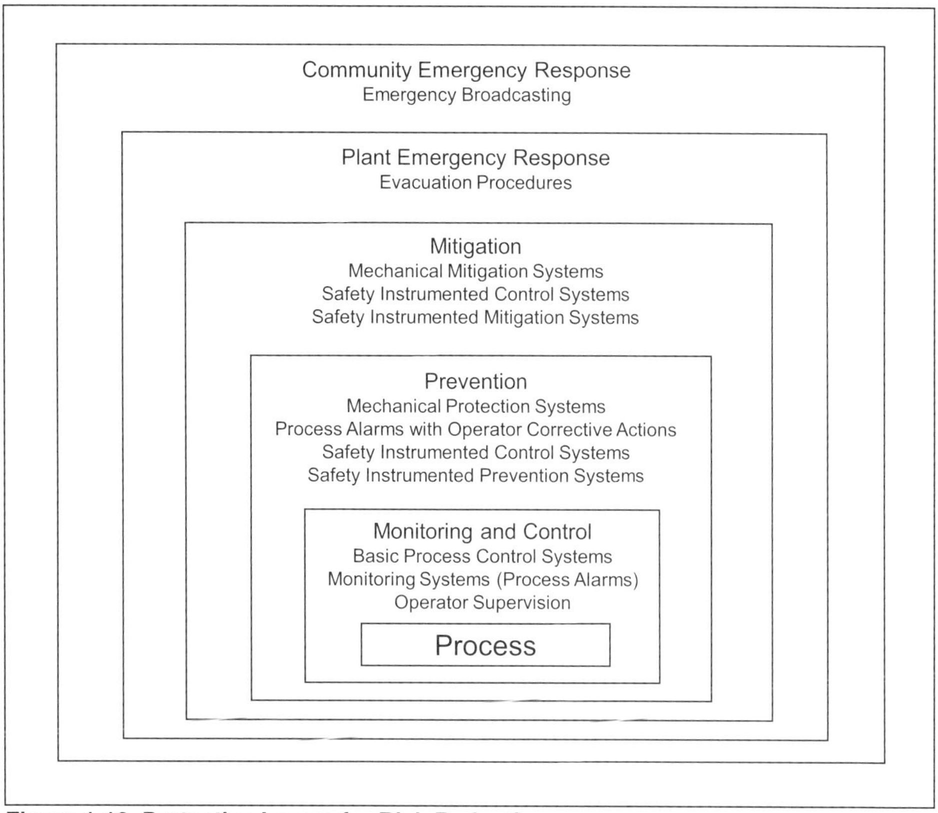

Figure 1-18. Protective Layers for Risk Reduction

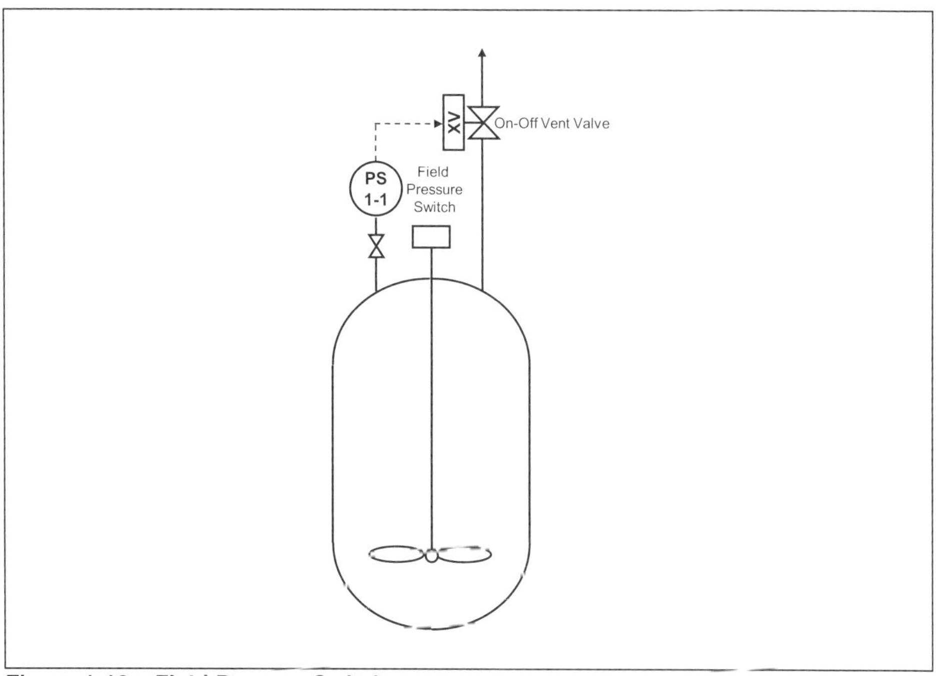

Figure 1-19a. Field Process Switch

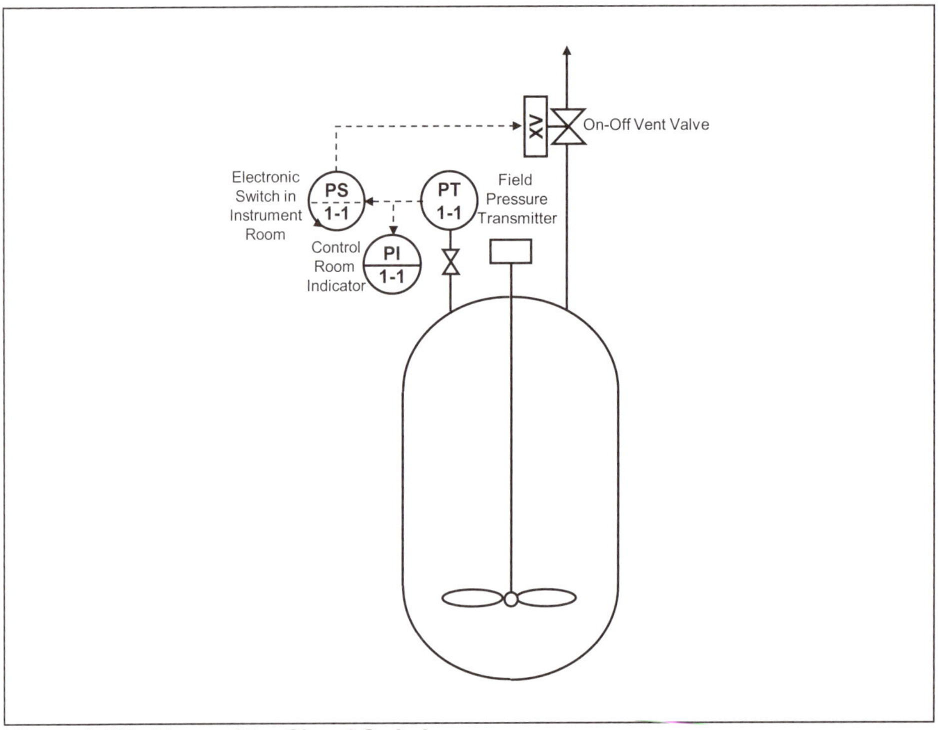

Figure 1-19b. Transmitter Signal Switch

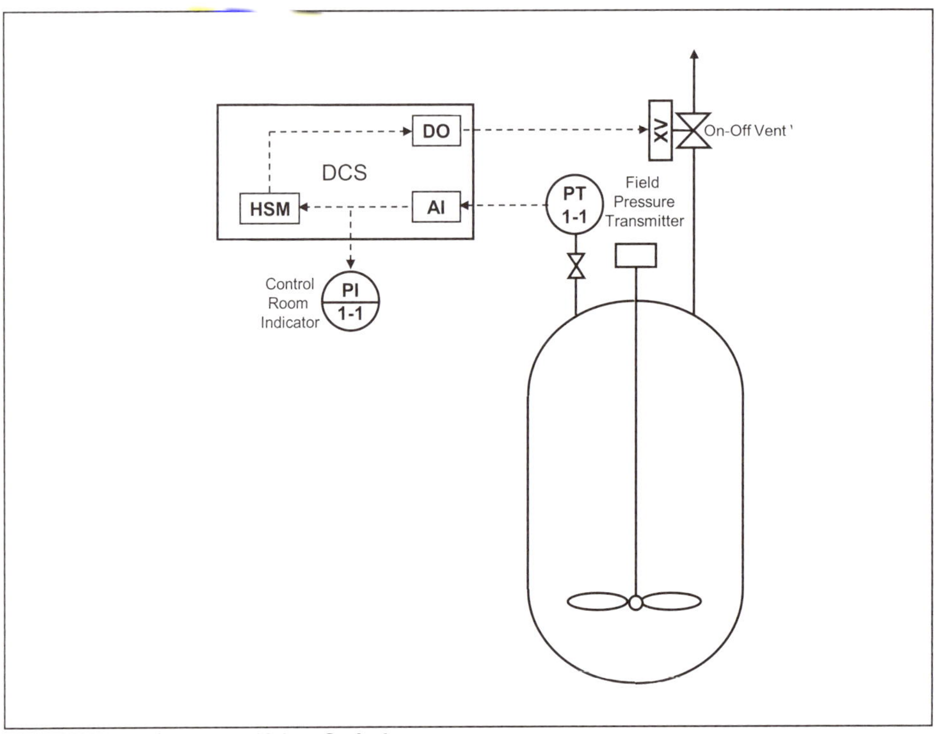

Figure 1-19c. Software Value Switch

or integrity online. Older mechanical switches are particularly problematic because of problems that include hysteresis and repeatability and wear and tear from process exposure and age. These field process switches were predominantly used in packaged equipment and utility systems to reduce the cost of the instrumentation.

Experienced users realized that the reliability and life-cycle cost of safety systems is improved by the use of continuous measurements and switches on the transmitter output signal. In the early days of process control (e.g., 1960s), the pneumatic transmitter output (3-15 psig) went to rack mounted pneumatic pressure switches. With the advent of electronic instrumentation (in the 1970s), the electronic transmitter output (e.g., 4–20 mA or 1-5 VDC) went to rack-mounted current or voltage switches.

The discrete outputs of field process switches and transmitter signal switches were typically contact closures that would de-energize relays, motor control, or solenoid valves for the prevention or mitigation of unsafe conditions. De-energization is normally used as the discrete output state to ensure an electrical failure puts the process in a safe state. The solenoid valve (SV) is a 3-way valve, whose operating and trip positions are set by the energization and de-energization, respectively, of an electrical solenoid. Solenoid valves take less than a second to switch position. The wetted materials for the body and switching element are not extensive and the flow path is not smooth enough for installing an SV on a process line. Some SVs are used on air, water, and other utility streams, particularly on packaged equipment. In most process applications, the SV is in the air line of a control valve, to exhaust air upon de-energization and allow a spring return actuator to open or close the control valve. The spring puts the valve in the fail-safe position. For double acting piston actuators, dual SVs are used to fill and exhaust the appropriate sides of the piston to open or close the valves. Local air tanks are used to ensure a fail-safe position.

Safety instrumented systems are often activated by high temperature or pressure measurements but many types of online measurements are used. These measurements need to be fast and reliable.

A high SIL requires instrumentation and logic solver redundancy for reliability. NooM (N out of M) input voting architectures are commonly used with redundant measurements to trigger a trip, where N indicates the number of votes required to trip and M is the total number of devices used. The typical architectures employed are 1oo1, 1oo2, and 2oo3. A 1oo1 architecture can be used for simple applications, but undetected device failures can cause loss of protection since only one input device is used. A 1oo2 voting system is a common method of improving the reliability of a safety system. If one measurement reaches the safety setting, a trip is initiated. However, this makes the plant more susceptible to false trips due to sensing element or transmitter problems. False trips are quite costly for large continuous processes. The startup after a trip can be difficult, require significant time, and necessitate a transition through process conditions that are more hazardous than during normal operation. Thus, minimization of trips is both an economic and a safety issue.

Consequently, a 2oo3 voting system is sometimes used, where two measurements have to indicate an unsafe process condition to initiate a trip. This voting system eliminates a false trip or disabling of the SIS due to a single measurement problem. A 3oo5 voting system requires three measurements out of five measurements to vote for a trip. These systems are used in nuclear reactors and aerospace to protect against two measurement problems causing either a false trip or the inability to trip on a true unsafe condition.

Large continuous chemical and petrochemical plants with a high production loss from downtime use a "middle signal" system particularly when the startup requires significant operator expertise and time to clean out the equipment and piping and/or get the process back to operating conditions. The selection of the middle signal not only provides protection against a single measurement failure similar to a 2oo3 voting system for safety systems, but also protects an associated control loop against a single measurement failure. The continued operation of the loop may prevent the process from reaching the trip point. Middle signals are used in polymerization lines that might solidify on shutdown and catalyst bed reactors that require careful fluidization and reintroduction of explosive feeds. Here, three transmitters are installed, and the middle signal is selected for both safety systems and process control. This middle signal selection protects against a false trip and prevents a true trip as the result of an unsafe condition created by a disabled process control system from a single measurement failure.

The process control system and its measurements and final elements should be designed so that the safety instrumented system is never activated. The process control system is the primary automatic protective layer. The process control system should be able to handle disturbances and keep the measurements that are inputs to the SIS well within the settings that trigger SIS actions.

These safety systems may set coolant valves or vent valves wide open or close feed valves. The shutdown and startup of a process are usually the most dangerous times because the process is going though large and rapid transients, particularly during a trip. Most shutdowns are completely automated and generally very fast for safety. Solenoid valves described in the chapters on Final Element Fundamentals and Control Valves are electrically de-energized to rapidly move air in or out of the control valve actuator to put the valve in the trip position as fast as possible.

1-4. Dynamics

Measurement devices do not instantly respond to changes in process variables. All sensing elements have a time delay (period of no response) and time lag (time duration after the delay to reach 63% of the complete response). What the control system sees is old information. However, in most cases, the response speed of the sensing element is fast enough to enable the control system to correct for the change. Flow, pressure, and level sensors have a time delay and time lag that are much less than one second. Temperature sensors have multiple time lags due to the time it takes for enough heat to transfer to make the internal sensor temperature equal to the process temperature. These time lags are usually a few seconds but can be much longer when the sensor is in a thermowell. If the thermowell has a large metal mass, the tip of the sensor is not touching the bottom of the well, or the fit is not tight, creating an air gap, the time lag can increase from seconds to minutes (see Chapter 2).

The transmitter can also introduce a time lag through a damping adjustment. The purpose of the adjustment is to smooth out measurement noise, which is any fluctuation in the sensor output that is not representative of changes in the process variable. Some manufacturers have non-zero default damping settings that could be too slow for some processes.

Measurement Response to a Step Change

Figure 1-20 is a trend plot of actual (dotted line) and theoretical (solid line) dynamic response of the transmitter output to a step change in the process variable.

As mentioned, the measurement gain is the change in percent transmitter output divided by the percent change in process input in engineering units. This gain is mostly just a reflection of the measurement scale. The time delay, also known as the dead time, is the period of time before there is any distinguishable change in the output of the transmitter after a change in a controller output. Once the transmitter output gets out of the noise band, it goes through an S-shaped response. While a better fit to the curve would be to use several time constants plus the time delay, a single time constant plus a time delay (dead time) called a first order plus dead time (FODT) approximation is commonly used in industry. The additional complexity of a higher order fit is generally a wasted effort because the improvement in fit is less than the uncertainty in these parameters.

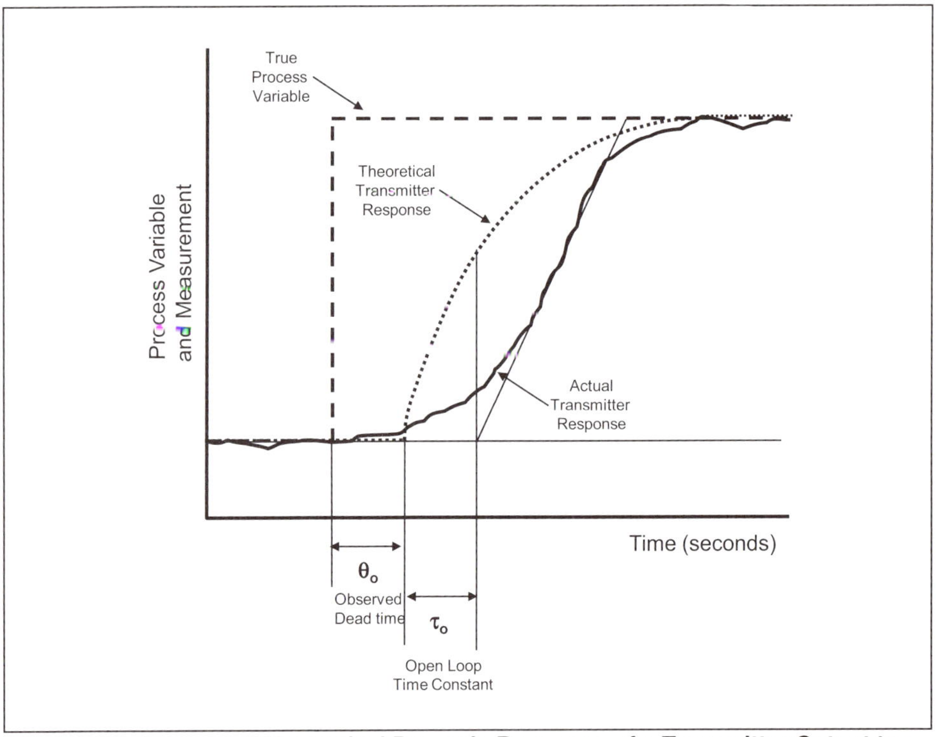

Figure 1-20. Actual and Theoretical Dynamic Response of a Transmitter Output to a Step

Time delays and time constants are not constant but depend upon the size and direction of the change as well as process conditions. This is dramatically the case for pH electrodes and temperature sensing elements. The FODT time constant is the time from the start of recognizable response to the time the output reaches 63% of its final value. The term *lag time* (as opposed to "time lag") is sometimes used in practice to denote a time constant but lag time is defined in *ISA's Automation, Systems, and Instrumentation Dictionary, 4th Edition*, as the transportation delay through an analyzer sample system. The terms *lag* and *lag time* are also used to define how late a changing (ramping) measurement is compared to a changing (ramping) actual value, which could be the sum of the time delay and time constant as noted in Figure 1-21. In this book, "time constant" will be used interchangeably with "time lag" and "dead time" with "time delay," which is the common practice in process control.

The use of the terms "time response" and "response time" is inconsistent in the discussion and specification of the dynamic response of measurements. The consequences can be drastically different assessments of measurement speed. "Time response" may or may not include the time delay and may be used to simply mean "time constant." The "response time" should always have a specified amount of the final response in its subscript for a change in the process variable as seen in the measurand.

A response time is the total time from the step change in the process variable to the time the transmitter output reaches a specified percent of its final change in output. For example, the response time to 95% of the change in output (T_{95}) would take 4 time constants plus the time delay in a FODT approximation. This book will use the 95% response time (T_{95}) for measurements and 86% response time (T_{86}) for valves instead of the 98% response time (T_{98}), because the last part of the response time is often lost in the noise and disturbances and can yield drastically different results for nonlinear systems.

Often there is more than one time constant, which gives the response its curvature as it moves out of the noise band. The small time constants can be summed with the pure time delays from digital device execution and scan times and sample transportation delays as an approximation of the total dead time.

Disturbances from safety systems, switches (e.g., on-off level control), and batch sequences are fast enough to be steps unless smoothed by an intervening volume. Steps are the most disruptive because of the abruptness of the change in the process input. Disturbances that originate from changes in set points, loads, and ambient conditions tend to be ramps. Disturbances from process and sensor noise, sticking control valves, aggressive controller tuning, and interacting (fighting) loops tend to be oscillations. Oscillatory disturbances are particularly problematic if the frequency is near the natural frequency of the affected loops causing amplification of the amplitude (resonance).

Measurement Response to Ramping Disturbance

Whether a measurement response is fast enough depends upon the size and speed of the change in the process variable. Fortunately, process variables tend to ramp rather than make step changes. Figure 1-21 is a trend plot of a transmitter's response to a ramp change in the process variable. The ramp in the transmitter output is behind the process variable ramp by a time that is approximately the measurement time delay plus time lag. Figure 1-21 also shows how much the actual process variable is above the corresponding transmitted value. If the ramp rate is very slow compared to the sensing element time lag, the transmitter digital delay, and the damping setting time lag, the difference between process variable and transmitter output will be negligible.

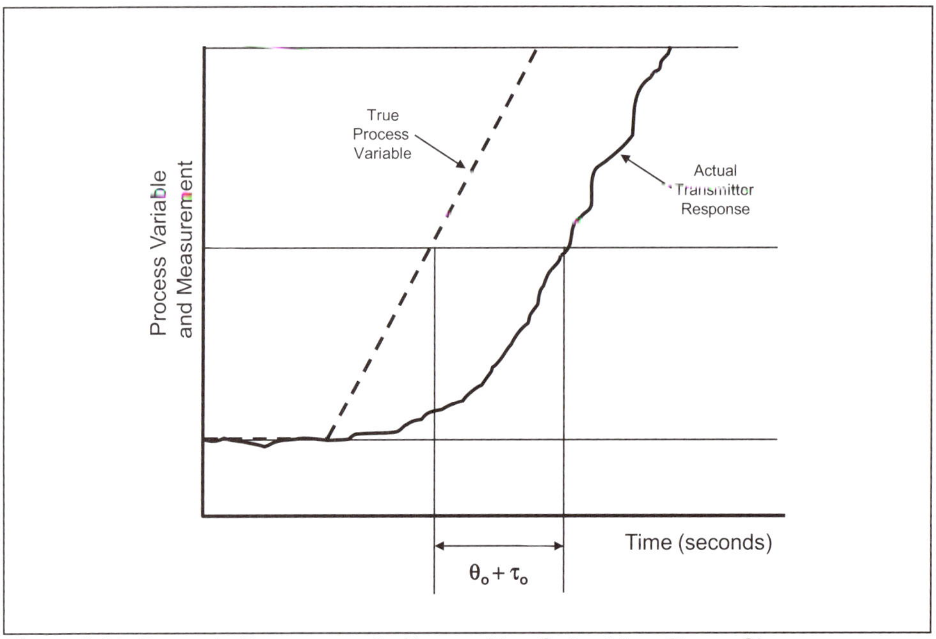

Figure 1-21. Dynamic Response of a Transmitter Output to a Ramp Change

The dynamic response of a transmitter output to a sinusoidal input is an attenuated and delayed sinusoidal output. This section will describe how the attenuation of the oscillation can be estimated if the measurement time lag is much larger than the oscillation period, which is the general case for measurement noise. Oscillatory process disturbances originating from control loops tend have a period much longer than the measurement time constant. For these disturbances, there is no attenuation of the oscillation and the measured oscillation is delayed (shifted) by the sum of the measurement time delay plus the time lag.

Analyzers

In this book, we are only covering "online" measurements which use sensors mounted directly in a process vessel or pipeline. There are also "at-line" measurements, which are typically analyzers that receive a sample that is automatically withdrawn and transported via a sample line to a sample conditioning system and the analyzer in the field. The transportation delays through the sample line, sample system, and analyzer plus the analyzer cycle time usually adds up to a time delay that is orders of magnitude larger than exhibited by other measurements. While the effect of time delay on loop performance is beyond the scope of this book, it is important to know that the maximum error from a disturbance is proportional to the total time delay. The integrated error (area between the set point and measurement on a trend chart) is proportional to the total time delay squared for a well-tuned controller.

"Off-line" measurements are typically done on manually withdrawn samples by analyzers in a laboratory. The time between when the sample is withdrawn and the lab analysis is made is often large and variable. The lab results should be time stamped with the time the sample was withdrawn. Some laboratory information systems automatically communicate the lab results to the process control system. In many plants, the lab results are entered manually. Regardless of whether there is manual or automatic entry of the lab results, the data historian should have the ability to show the results for the point in time that the sample was taken. This historization of the lab sample at the correct process time is essential for visually correlating results and for process analysis and data analytics. The time stamping of when the sample was taken helps deal with issues with the sample results of the raw material or end point being available after the batch has gone downstream. Time stamping is also essential for the correlation of the samples taken from tanks used for raw materials, intermediates, and products with the online measurements from batch and continuous operations in series and parallel.

Disturbances

Process variables are not constants. Any flow, temperature, pressure, or composition value that matches what is shown on a process flow diagram

(PFD) is a momentary coincidence. A production facility is a dynamic environment in a state of flux.

The biggest transients occur during the shutdown, startup, and batch operation of equipment. Some shutdowns are planned for maintenance or product changeovers, but many shutdowns resulting from trips are due to equipment or instrument failures, malfunctioning safety systems, equipment degradation (e.g., fouling or frosting), or operator errors. Even when equipment is operating continuously, there are changes in the composition of feeds from raw material shipments, intermediate chemicals produced elsewhere in the plant, and recycle streams where a portion of a process output flow is returned as a process input feed. Recycle streams can cause disturbances to perpetuate and even grow in magnitude because the upsets are returned as feed disturbances.

The temperature of process streams and the performance of equipment change with ambient temperature and utility stream temperatures. Often, cooling towers and refrigerant units have not kept up with increases in production unit capacity from process control improvement and debottlenecking projects. During hot weather and peak demand, high cooling water and refrigerant temperatures cause process upsets. Similarly, undersized steam headers and boilers cause dips in steam pressure during the coldest months and periods of peak demand. Heat recovery operations, where process output streams are passed through heat exchangers to preheat feeds, cause changes in the feed temperature based on process performance and production rate.

Process control loops can be the source of oscillations in process variables even if there are no process disturbances. The output of a controller in automatic is always moving because an exact match of the process variable to the set point is rare. Too much gain will cause controllers to oscillate whenever there is a load or set point change. Even if the controller is tuned properly and there are no changes, stick-slip in control valves creates limit cycles (sustained oscillations with constant amplitude) and interactions can cause loops to fight each other.

The most frequently encountered source of oscillations in a production unit is a level loop on a surge tank that is tuned with too much reset action (too small of a reset time). For a waste treatment system the use of level switches to pump out sumps is the most disruptive upset. On-off control of any type creates fast and large disturbances.

The constant state of change from disturbances and sequential or manual operations, non-ideal behavior, and imperfect knowledge of the process mean that timely and accurate measurements are needed to meet quality and production rate requirements. Measurements provide the only window into what is actually happening in the process. While each application has specific requirements, there are some concepts about what is most important in measurement performance in terms of measurement accuracy and speed.

Process automation systems rely on PID controllers to use measurements to improve production rate, quality, and efficiency. The performance of a PID depends on whether the PID has measurements of non-desirable process behavior that are sensitive and fast enough. During the time a PID doesn't see a change in a process variable, the process can move to an increasingly poor operating point. Sensing element time lags and transmitter damping adjustments for most slow processes (e.g., level, vessel pressure, temperature, and vessel pH) are effective sources of time delay that increase the total loop dead time. For well-tuned controllers, the total of the measurement time lags should be less than 1/5 the process dead time. In reality most controllers have a controller gain that is less than ½ the gain for maximum disturbance rejection. A loop detuned by a factor of 2 gives about the same performance as an aggressively tuned loop with twice the dead time. Measurement time lags totaling less than the process dead time have a negligible effect on the performance of these loops if these measurement lags are small compared to the process time constant.

1-5. Accuracy

Part of the difficulty in comparing the quality and speed of measurements is the number of parameters used to specify quality and speed and their misunderstanding and misuse. The functional definition and value of a multitude of terms are offered in this section to help practitioners intelligently discuss and understand the effects of device selection, configuration, calibration, installation, and operating conditions for an application.

Accuracy is often used as a simple qualitative term to describe how well a measurement represents the true value of the process variable. The quantitative accuracy of a measurement is the maximum deviation (error) of the measured value from the true value over a stated range of the measurement (e.g., 1000 to 4000 lb/hr flow [453 to 906 kg/hr] for differential head meters), measurand (e.g., 5 to 100 inches [12.7 to 254 cm] orifice differential pressure for differential head flowmeters), or process condition (e.g., 0.1 to 10 fps [3.05 to 305 cm/sec] fluid velocity for magmeters and vortex meters). The maximum error is the worst case of the combined systematic and random errors in the measurement. Systematic errors are errors that can be predicted from scientific principles. The most common systematic error is a bias as shown in Figures 1-22a through 1-22d, and discussed in more detail below. The bias error can be due to an offset (intercept) error or a span (slope) error (see Section 1.6 on setup and calibration). A bias can also originate from a fixed nonlinearity error that is a function of the true value.

Measurement *span* is the difference between the lowest and highest values of a measurement range. For example, a temperature transmitter calibrated for a temperature range of 50 to 150°C (122 to 302°F) has a span of 100°C (180°F). For an offset error, the bias is constant. For a span error, the bias increases with measurement.

An instrument's accuracy specification is often expressed in terms of a *tolerance,* which is the allowable deviation (error) from a true value (more on this later in the chapter). Sensing element error can result from randomness in its manufacturing process, with the result that a given sensor will often display a bias error, which is a systematic error, in the measurement. Systematic errors can be largely eliminated by linearization and calibration, and by set point changes done manually or automatically by higher level primary cascade or supervisory loops.

Random (as opposed to systematic) measurement errors show up as false variability in the process and cannot be eliminated by calibration or changes in the set point of the measurement loop. However, the effect of random errors on important process variables downstream can be attenuated by a "back mixed" volume. A higher level loop on a downstream volume can control the average of the fluctuations so the key variables are closer to the desired value. For example, the effect on vessel pH of random errors in a reagent flow measurement can be minimized by the cascade of the output of a primary pH loop that automatically adjusts the secondary reagent flow set point. In the section on repeatability, the ability of back mixed volumes to attenuate random errors is discussed.

The *precision* of a measurement is the standard deviation of the total random error. The principal sources of random errors are repeatability errors and noise that have a statistical distribution as shown in Figures 1-22a through 1-22d. Sensitivity and resolution limits and hysteresis and deadband create errors that appear to be random due to changes in the start and stop points and direction of the change in the true value of a measurement. For the modern measurement instruments detailed in this book, the random errors from resolution, sensitivity, hysteresis, and deadband are generally negligible.

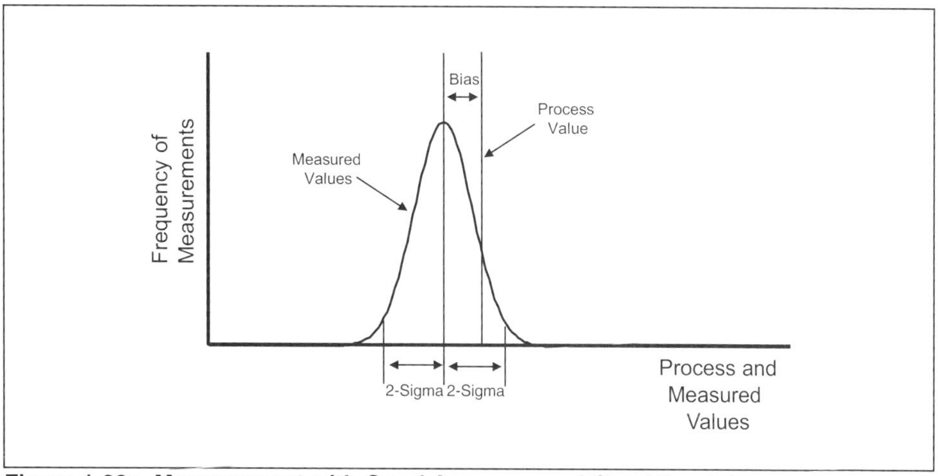

Figure 1-22a. Measurement with Good Accuracy and Good Precision

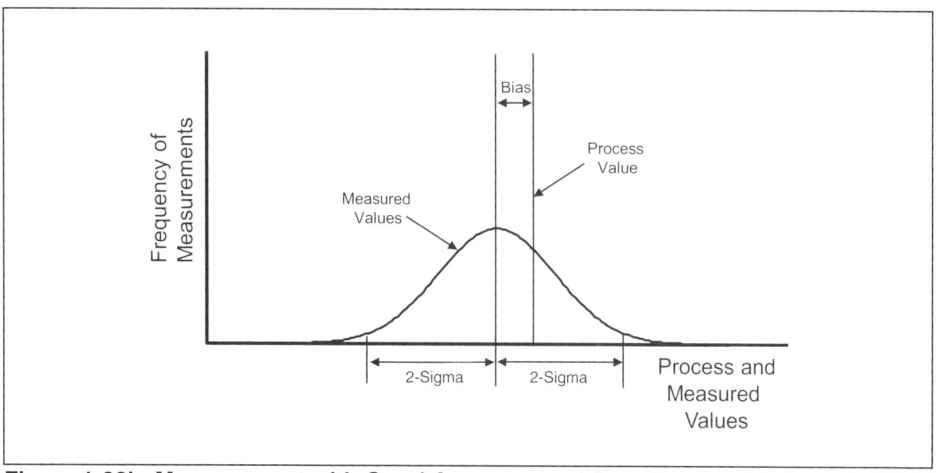

Figure 1-22b. Measurement with Good Accuracy and Poor Precision

Turndown (Rangeability) and Span

Also known as rangeability, turndown is the ratio of the maximum to the minimum measured value of the process variable for which the device's performance is within the specified limits for the device or final element. The maximum value is the high end of the selected device's range rather

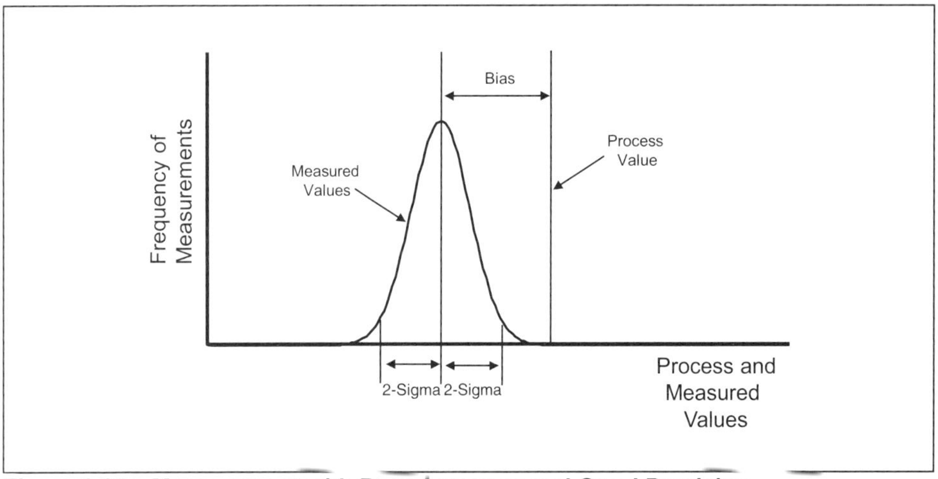

Figure 1-22c. Measurement with Poor Accuracy and Good Precision

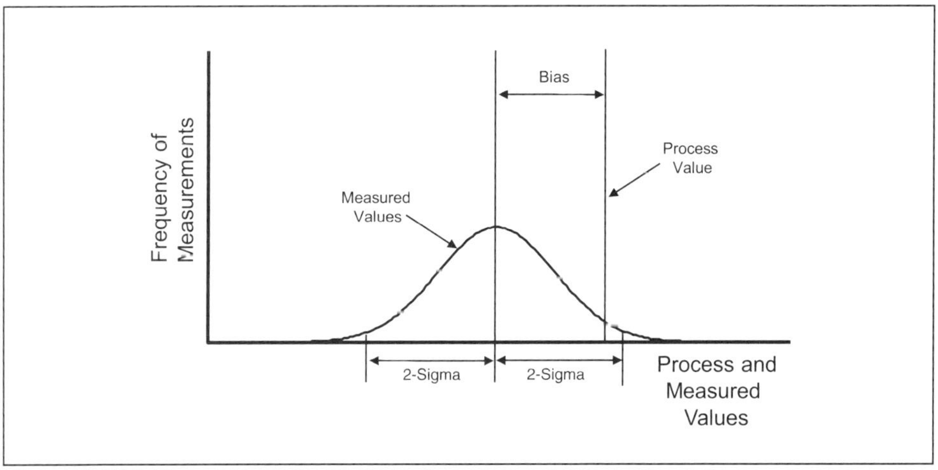

Figure 1-22d. Measurement with Poor Accuracy and Poor Precision

than the high scale value of the calibration. To maximize turndown, the device must be selected to provide as close a match of the device's maximum to the application maximum. For flowmeters, this corresponds to choosing a meter size whose maximum flow is as close as possible to the maximum flow in the process.

Errors are normally expressed in terms of a percent of span (see Section 1.6 on Calibration) but in some cases it is a percent of the reading of the measurand and the transmitter output. The difference in performance can be dramatic, particularly in terms of turndown (rangeability). For example, a 0.1% of span accuracy transmitter for a linear process relationship would have a maximum error of 0.5% at 20% output (5 times larger than a 0.1% of reading accuracy at 20% output).

For a differential pressure flowmeter, the deterioration in accuracy with turndown is much more drastic because the flow is proportional to the square root of the differential pressure measurement. A 0.1% of span accuracy transmitter would have a maximum error of 25% at 20% output (25 times larger than the error at 100% output).

Dramatic Improvements in Modern Measurement Instruments

The quality of measurement instruments has improved dramatically since the advent of smart transmitters. These transmitters have exceptional stability (resistance to drift) and correct for extraneous effects whether sensing element nonlinearity, ambient temperature, or process conditions. Many of these instruments no longer require periodic process calibration. For example, coplanar pressure and differential pressure transmitters have an accuracy of 0.025% of span and 10 year stability. Furthermore, the installed accuracy remains essentially the specification accuracy, because of the compensation for ambient and process effects. In the 1970s, a typical electronic pressure transmitter had a benchtop accuracy of 0.25%, an installed accuracy of 1 to 2%, and a stability specification of 1 to 2 years. Smart microprocessor based measurements such as Coriolis mass flowmeters, coplanar pressure and differential pressure transmitters, and radar level meters have an installed accuracy that is 50 times better than what existed in the 1970s. The accuracy of these devices approaches the uncertainty in process calibration standards and methods. Furthermore, the drift is so small and the stability is so good the average instrument engineer will have changed jobs or retired before the device needs a calibration check. Additionally, dynamic errors are negligible. Their response is so fast that the measurement time constant is too small to be measured in a DCS.

Estimation of Total Error

Since errors can be in opposite directions, where a portion of the errors cancel out, the total measurement error can be approximated per Equation 1-2 as the square root of the sum of the individual errors squared. Equation 1-2 includes the error in the reference value used for calibration (E_c), which is small for properly used and maintained calibration standards. While it is not customary to consider noise in accuracy calculations, noise is a term in Equation 1-2 because noise is often the largest source of measurement uncertainty. However, it is critical to note that noise can be attenuated by a proper setting of the transmitter damping or signal filter in the control system. The most important terms for control loop performance and process variability are the repeatability error (E_r) and measurement noise (E_n). For accuracy statements, the largest source of error is typically a bias error (E_b). For thermocouples and RTDs, the sensing element manufacturing tolerance (E_t) is a key term for temperature measurement accuracy. Equation 1-2 is the magnitude of a vector and does not include any dynamic errors from a measurement lagging behind a true change in the process variable.

$$E_m = \sqrt{E_b^2 + E_l^2 + E_q^2 + E_s^2 + E_r^2 + E_h^2 + E_d^2 + E_n^2 + E_t^2 + E_c^2} \qquad (1\text{-}2)$$

Where:

E_m = total measurement error (%)

E_b = bias error (%)

E_l = linearity error (%)

E_q = resolution (quantization) error (%)

E_s = sensitivity error (%)

E_r = repeatability error (%)

E_h = hysteresis error (%)

E_d = deadband error (%)

E_n = measurement noise (%)

E_t = sensing element tolerance (%)

E_c = calibration error (%)

Uncertainty

Uncertainty is the interval within which the true value of a measured quantity is expected to lie with a stated probability [10]. Thus, uncertainty statements imply a quantifiable probability in confidence interval. Uncertainties with a 95% and 99% probability are often used. For a normal distribution, the probabilities that the measurements lie within one, two, and three standard deviations are 68.26%, 95.46%, and 99.73%, respectively. There is also an uncertainty in the true value of the measurement. The term "uncertainty" plays an important role in the analysis of metrological determinations of the ability of calibration standards to provide true values. The term has limited general practical utility for describing the accuracy of measurements in industrial applications but is used for some types of measurements, particularly when there are some unknowns in the measurand. For example, in discussions of the accuracy of differential head flow measurements an uncertainty of the discharge coefficient of the orifice plate, venturi tube, or flow nozzle is stated. In some cases accuracy is viewed as the total uncertainty with a probability that 99%+ of the measurements lie within the stated accuracy range.

Tolerance

Tolerance is the allowable deviation from a specified or true value [10]. It is generally used to quantify the allowable variation in accuracy of a sensing element during the manufacturing process and is usually given in engineering units. A common use is in specification of the manufacturing tolerances of thermocouple (TC) and resistance temperature detector (RTD) elements. Most of the tolerance is a systematic error that can be eliminated by calibration.

The tolerance of a sensing element is indicative of the interchangeability of the element, with a narrower tolerance enhancing interchangeability. Consider the case where in the middle of the night a thermocouple fails. The loop needs to be put back in service immediately so calibration of the transmitter by testing the new thermocouple at a known temperature is not done. If there is a redundant temperature measurement, a one point calibration of the transmitters can be done online. If calibration is not possible, an error is introduced into the measurement that is some fraction of the thermocouple tolerance. A conservative estimation approach would be to take the full tolerance of the sensing element for the term in Equation 1-2.

There are many other uses of the term tolerance. Voltage tolerance is specified as the permissible percentage deviation of the actual voltage (DC or root mean square for AC) from the nominal voltage. Dimensional tolerances are used in manufacturing to specify the permissible variation in size [10].

Resolution

Resolution error (E_r) is the minimum change in measurement output that is most commonly a quantization error determined by digital processor and I/O bits. A large resolution limit is clearly identifiable as a staircase response where the step size is the resolution limit or a multiple of the resolution limit as depicted in Figure 1-23. A resolution limit will cause a limit cycle in a standard PID loop with integral (reset) action. The limit cycle period is much longer than the period of oscillations from steps triggered by sensor noise, which is slower than the oscillations of pure noise with no resolution limits. Consequently, the oscillations from resolution limits are not as readily attenuated as pure noise by signal filters in the control system. The attenuation by a signal filter of resolution error or noise can be approximated by Equation 1-3.

The resolution error in most modern sensors is less than 0.001% because there are no mechanical movements and there are 18 bits or more in the transmitter's processor and digital to analog convertor (D/A). The resolution error from a digital device can be estimated by Equation 1-4 where one bit is a sign bit. In older devices and control systems, the processor of the digital-to-analog convertor and analog-to-digital convertor was 12 bits, which provided a resolution of 0.05%. For large span thermocouple input cards of 500°C (900°F) span or more, resolution error was thus 0.25°C (0.45°F) for older systems which precluded the use of derivative and high gain action in temperature controllers even though these tuning settings were beneficial in terms of load rejection and set point response.

In most temperature processes the true change in the temperature is slower than 0.025°C (0.045°F) per minute. The result is a poor signal to noise ratio. In these temperature loops the use of signal filters with relatively large filter times (e.g., 30 seconds) helps attenuate the resolution error triggered by thermal noise to the point where higher controller gains can be used, but this adds an undesirable time lag that could be detrimental to the ability of the loop to recognize and correct fast disturbances.

$$A_f = A_o * \frac{T_o}{2\pi * \tau_f} \tag{1-3}$$

$$E_r = \frac{100\%}{2^{n-1}} \tag{1-4}$$

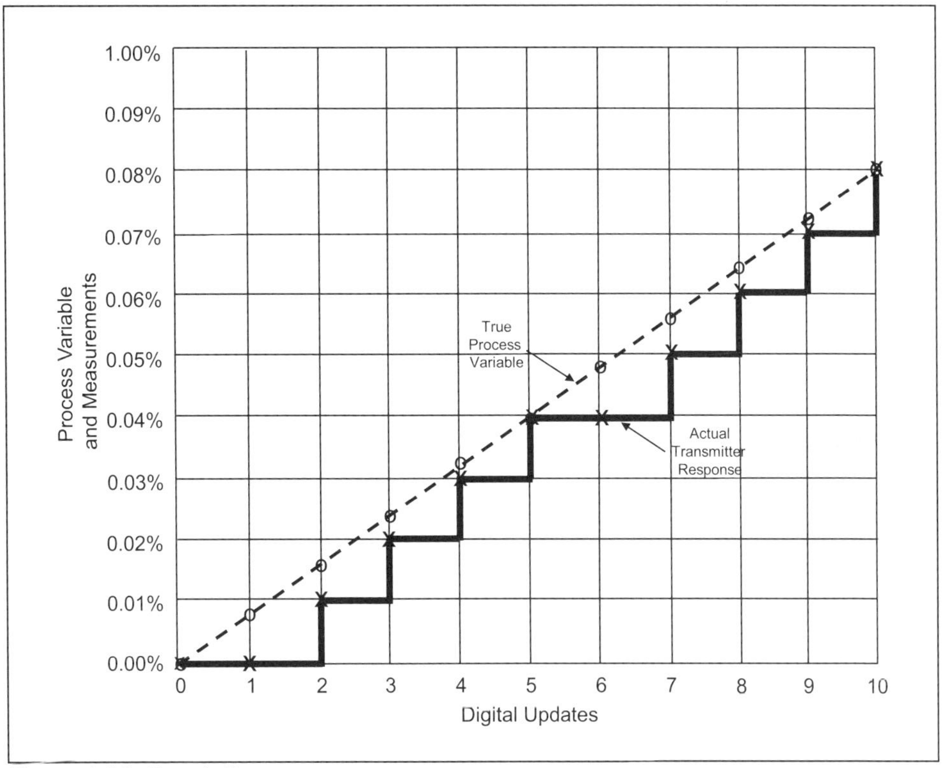

Figure 1-23. Resolution

Where:

A_f = amplitude of the filtered oscillations (e.u.)

A_o = amplitude of the original oscillations (e.u.)

T_o = period of the original oscillations (sec)

τ_f = measurement time lag, damping setting, or signal filter time constant (sec)

Resolution is the top consideration in terms of a loop being able to see a change in the process because there are no intermediate values between the steps. Instead of steps, trend charts may show a slope to the change in the measurement due to smoothing by the historian algorithm and attenuation by signal filters DCS. The presence of the steps, whether seen or hidden, means it would be extremely rare for the process variable to come to rest exactly at the set point. As a result of the continual control error, reset action in the controller will continually ramp the controller output up and down. The result is a sustained oscillation with a period that depends on the tuning of the controller. Detuning (decreasing the controller gain and increasing the reset time) makes the period longer. The amplitude of the oscillations remains constant for a given damping or filter time setting. The oscillations are difficult to filter out since the period is relatively large in Equation 1-3. Fortunately, the resolution of most modern transmitters is extremely fine. However, for wireless control we will see that a resolution (deadband) setting is used to transmit signals on an exception basis to extend battery life. The resolution setting is chosen to be more than the noise after damping, less than the allowable control band (maximum peak to peak amplitude of fluctuations about set point), and less than the resolution (stick-slip) of the control valve quantified as a step in flow multiplied by the process gain. Fortunately a wireless PID algorithm has been developed that eliminates the limit cycle from the resolution setting. Equation 1-3 can also be used to predict the smoothing by continuous back mixed volumes on the loop oscillations from resolution limits. In this case, the time lag in Equation 1-3 is the volume's residence time (volume divided by total throughput flow).

Sensitivity

Sensitivity is the smallest detectable change in the process variable (see Figure 1-24). If the step in the process variable is larger than the sensitivity, then the measurement shows a step size that matches the step in the process variable. Thus a sensitivity error will show step sizes that are in between multiples of the sensitivity limit. Normally sensitivity is determined by the measurement principle and type of sensing element. For example, a resistance temperature detector (RTD) has a sensitivity (0.001°C; 0.0018°F) that is 50 times greater than thermocouple sensitivity (0.05°C; 0.09°F). It is important to realize that sensitivity is a term commonly used in electronics for amplifier gain and automation systems for process gain, where sensitivity is the change in output divided by the change in input to the component or process.

Repeatability

The repeatability error (E_r) is the difference between indications for traverses upscale and downscale as depicted in Figure 1-25a. There is also a practical definition of repeatability which is the maximum difference between changes in the indicated value for the same magnitude changes in the true value as shown in Figure 1-25b. The deviations around the average could be due to random error, more commonly experienced as sensor noise.

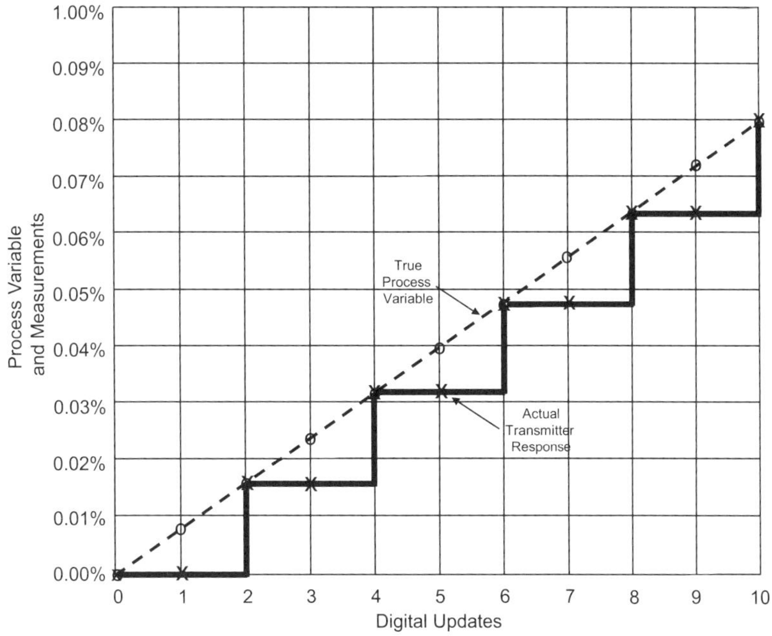

Figure 1-24. Sensitivity

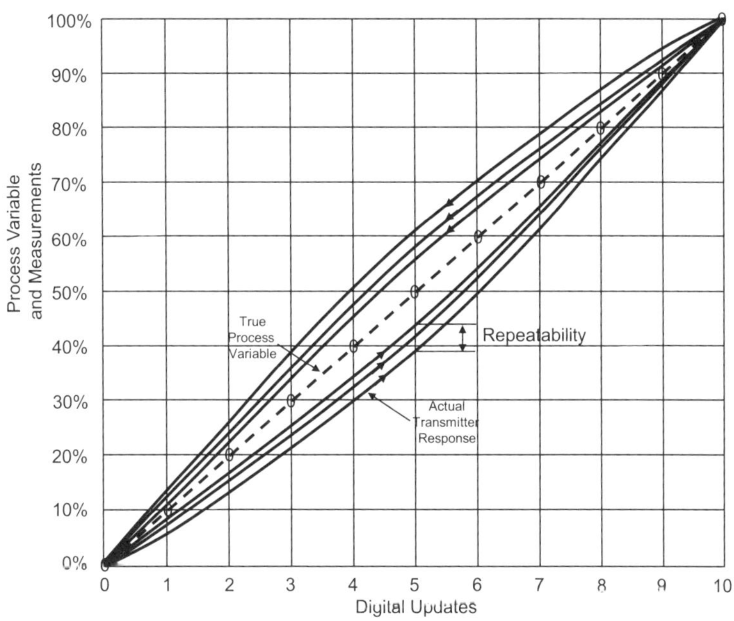

Figure 1-25a. Technical Definition of Repeatability

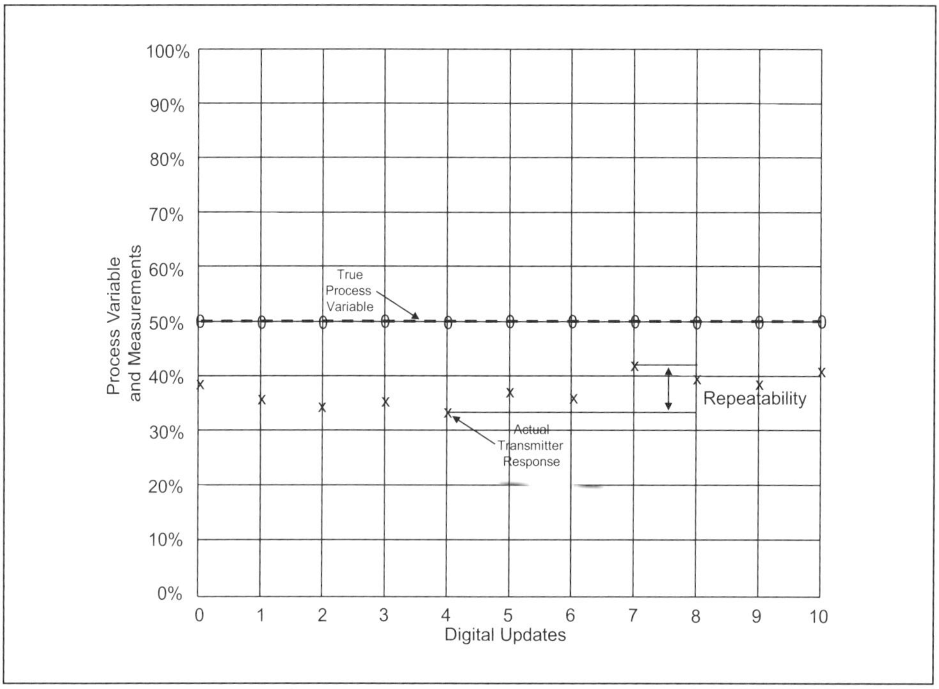

Figure 1-25b. Practical Definition of Repeatability (maximum difference between repeated measurements of same value)

Hierarchy

There is a hierarchical relationship between some of the terms in Equation 1-2 in determining accuracy in that the error level of one term depends upon the error level of the previous one. However, a low error level in one term does not guarantee a low error level in the other term. For example, the repeatability error cannot be less than the sensitivity error but a very sensitive measurement may not be repeatable. Additionally, the sensitivity error cannot be less than the resolution error but a high degree of measurement resolution does not imply a sensitive instrument [14]. In this example, the resolution (quantization) error (E_q) must be less than or equal to the sensitivity error (E_s) which in turn must be less than or equal to the repeatability error (E_r)

$$E_q \le E_s \le E_r \tag{1-5}$$

Noise

Sensor noise (E_n) is the deviation between the indicated value and a constant true value as shown in Figure 1-26. The indicated value can take on intermediate values. If the period between fluctuations in the measurement is much shorter than the period of the control loop, the random error is termed noise and can be filtered out by a damping adjustment in the transmitter or a signal filter in the analog input or PID block of the control system.

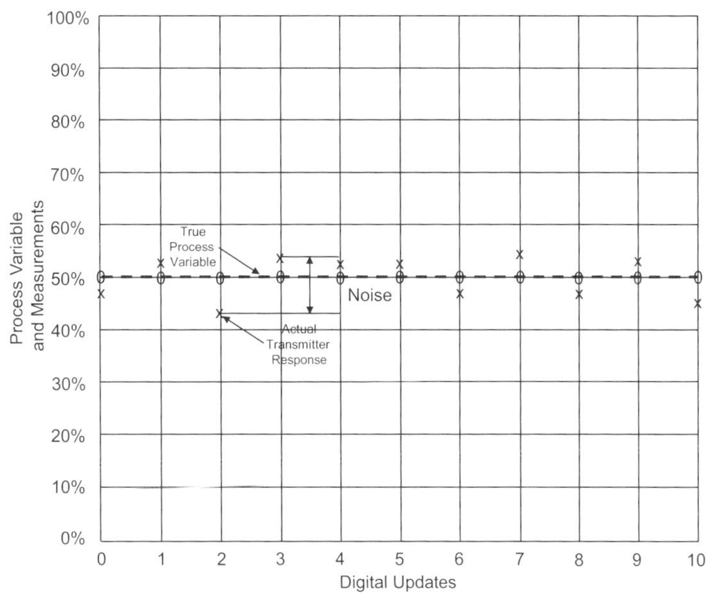

Figure 1-26. Noise (peak to peak)

Hysteresis and Deadband

The maximum difference between the upscale and the downscale traverses is hysteresis error (E_h) if there is an immediate reversal of output for a change in direction of the input. The trend chart shows a football shaped response to full scale traverses as illustrated in Figure 1-27. The amount an input has to change direction before the output changes is deadband error (E_d). Hysteresis error is frequently accompanied by a deadband error. A plot of the measured value versus the true value of the process variable shows a slanted and bowed box shaped response to full scale traverses similar to what is depicted in Figure 7-40b for a control valve. Technically, these parameters are defined for a full traverse of the span but in reality both errors occur with any reversal in the direction of the process variable.

Hysteresis and deadband are usually associated with the movement of mechanical elements. In this book we will concentrate on the more modern measurement devices that primarily involve a fluid or electromagnetic sensing element response so that hysteresis and deadband errors are negligible. However, we will see for control valves, these errors can be considerable, resulting from actuator response and shaft backlash, particularly for rotary valves that originated as on-off or isolation valves.

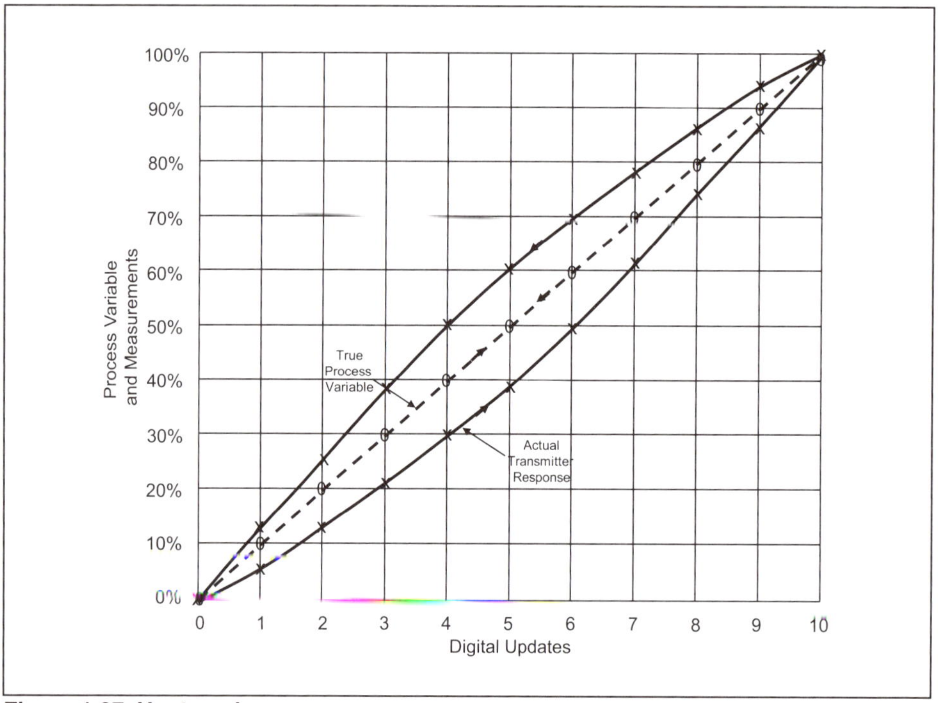

Figure 1-27. Hysteresis

Nonlinearity

Nonlinearity error as shown in Figure 1-28 is the largest deviation of the indicated value's curve from the true value's straight line across the entire measurement range. Well known nonlinearities are usually compensated in the transmitter. Thermocouple (TC) and resistance temperature detector (RTD) sensing elements are important examples of a known nonlinearity that can be corrected by equations in the transmitter.

Bias

Bias error is the offset of the indicated value's average value from the true value's straight line as depicted as the difference in the peak of random values from the true value in Figures 1-22a through 1-22d. A constant bias for all true values can originate from an offset (intercept) error as shown in Figure 1-29. A bias that increases with true value can be due to nonlinear-

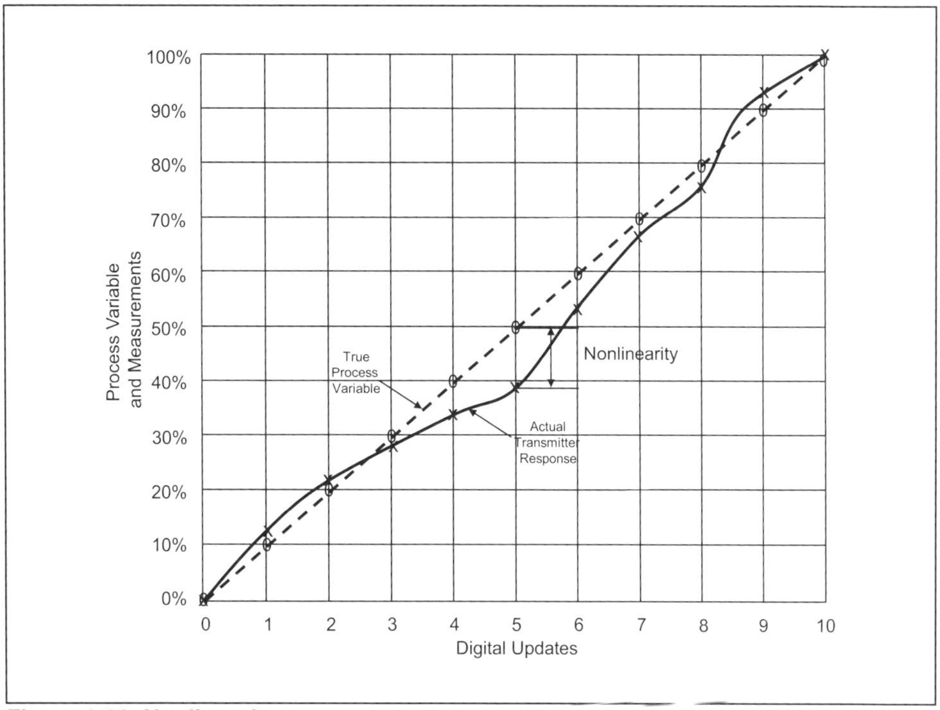

Figure 1-28. Nonlinearity

ity or a span (slope) error. Bias errors are readily corrected by calibration and by manual and automatic adjustment of the control loop's set point. The manual adjustment of process set points is very common but not usually recognized as being due to a measurement bias error until process performance changes when a measurement device is replaced. The best temperature found for running a column, crystallizer, dryer, or reactor (or example) is not exactly at the true value of the optimum but at a value that corrects for the offset in the measurement from the true value. For example, if a column temperature measurement is reading 1°C (1.8°F) higher than the true value, the best column temperature set point is 1°C (1.8°F) lower than the true optimum. Many columns use thermocouples, which have a significant bias error.

Stability

Stability is the period of time during which the accuracy of the measurement stays within some specified bound. Stability is often defined in terms of **drift** which can be visualized as an offset that increases with time. The drift is expressed as % per month for pH and % per year for most other measurement devices. In a trend that covers months of the measurement output where the true value is constant as shown in Figure 1-30 the drift is

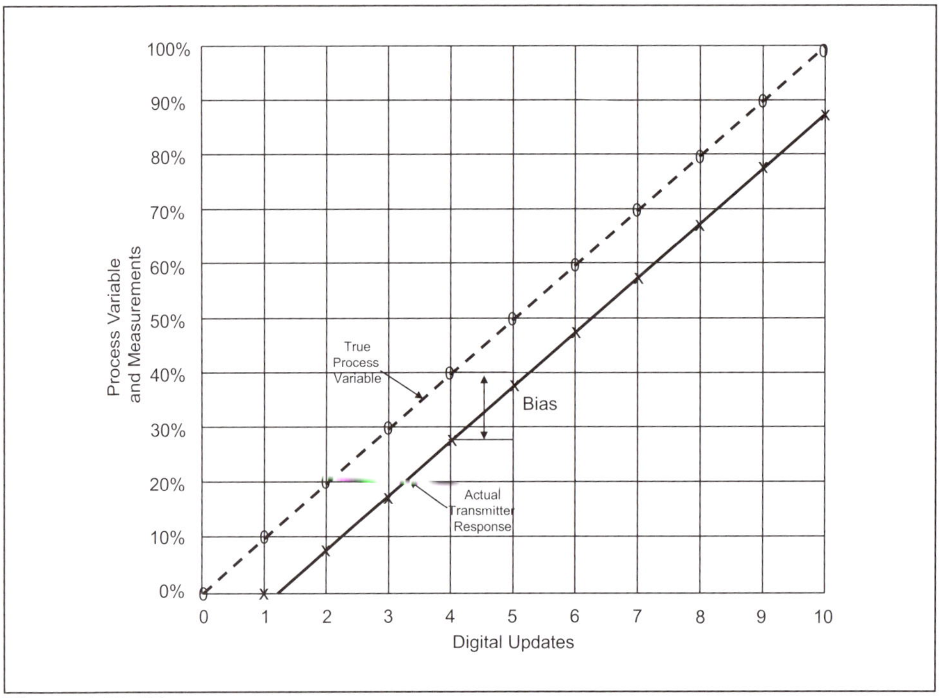

Figure 1-29. Bias

the slope of the line. In practice, drift is difficult to ascertain because the true value is not constant for the months and the measurement may be the process variable of a control loop. For a PID controller with a constant set point and load, drift is seen in the slope of the controller output since the control loop is working to keep the measurement at set point.

Reproducibility

ASTM International, formerly the American Society of Testing and Materials (ASTM), uses the term reproducibility to quantify the variation in calibration results for different testers who use the same test method and test equipment. For temperature measurements, it is used to specify the possible variation in temperature measurements for different sensing elements for the same temperature with the same transmitter [17]. The International Society of Automation (ISA) defines reproducibility as the closeness of agreement among repeated measurements of the output for the same value of input made under the same operating conditions over a period of time, approaching from both directions. Reproducibility includes the effect of hysteresis, deadband, drift, and repeatability. Reproducibility is usually measured as non-reproducibility but is termed reproducibility and quantified as a percent of measurement span over a specified time

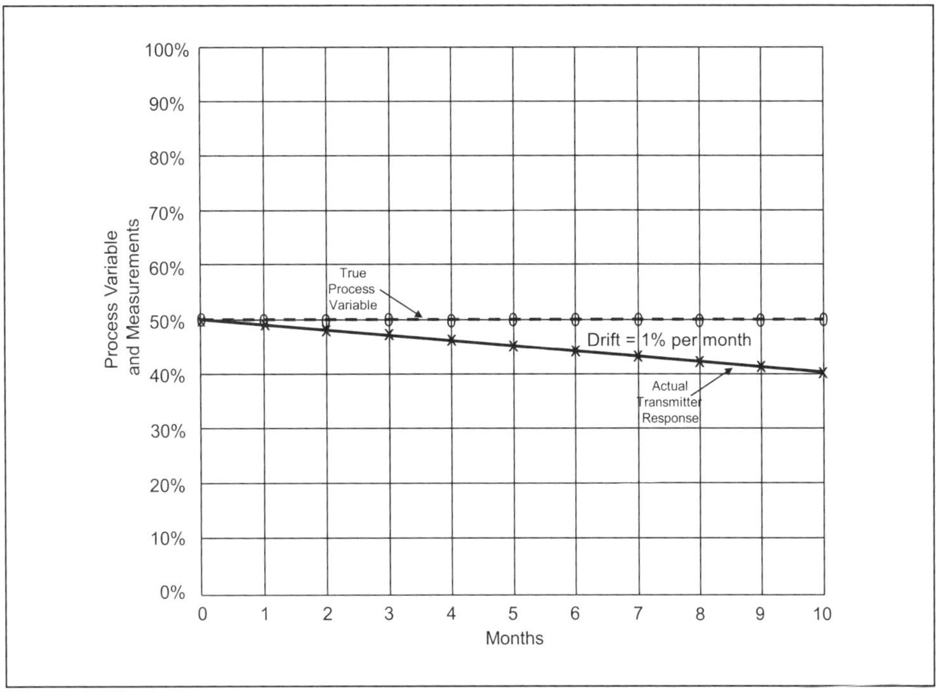

Figure 1-30. Drift

period. If drift is omitted, reproducibility can be evaluated over a much shorter period (hours instead of months).

1-6. Setup (Configuration) and Calibration

Range and Span

Measurement devices have a range and span. The range is the low and high values of the process variable that correspond to 0% and 100%, respectively of the transmitter output. The low end of the range in engineering units is usually (but not always) zero, particularly for flow measurements. The span is the difference between the low and high ends of the range. For measurement devices that have a separate sensing element such as differential pressure flow and level measurements, TC and RTD temperature measurements, and pH measurements, there is a measurand range that corresponds to the process variable range. Some typical examples of PV measurement ranges with the type of sensing measurand and measurand range (in parentheses) are:

- Differential Pressure Flow: 0 to 1000 liters per minute [0 to 264 gal/min] (DP: 0 to 25 kPa [0 to 3.6 psi])

- Differential Pressure Level: 0 to 100% (DP: 10 to 110 kPa [14.5 to 16.0 psi])
- Absolute Pressure: 50 to 250 kPa (7.2 to 36.2 psi)
- Temperature: 50 to 150°C (122 to 302°F) (RTD: 119.4 to 157.3 ohms)
- pH: 2 to 12 pH (glass electrode: 295 to -295 millivolts)

As mentioned, the low end of the measurand range is not necessarily zero. A positive low end of the DP measurand range occurs for a high side connection of a DP transmitter that is below the desired 0% level measurement, as will be illustrated in the section on installation. The negative low end of the pH measurand range for a glass electrode occurs for a high concentration (> 1 normality) of a strong acid.

The transmitter signal is converted at the AI block in a DCS to the engineering units with the same range used by the transmitter so that the measurements seen in the data historian have the same engineering units used in the setup and calibration of the transmitter. The scale of a PID controller usually matches the range in engineering units but most DCS offer the option of a different process variable scale for the PID. This is important because this scale determines the conversion back to percent in the PID block and thus affects the process gain per Equation 1-1.

Setup (configuration) determines the device functionality and specifies the measurement range and span in engineering units. Calibration involves tests and adjustments to match the transmitter output in engineering units to the process variable or the measurand.

Since most device accuracies are given as a percent of span, a practice developed decades ago was to minimize the measurement span to reduce the measurement error. Advances in sensing element technology, the compensation of nonlinearities and non-ideal effects, and the high resolution of the computations and I/O make this practice not as important. Some smart instrumentation specifications will calculate the accuracy in percent of span based on the actual upper and low range limits, eliminating most of the supposed improvement in accuracy from narrowing the span.

Historical View

Before the advent of microprocessor based measurement instruments, instrument setup and calibration was rudimentary. Setup was largely done structurally and preset at the factory, based on options specified in the model number. Calibration was mostly done by screwdriver adjustments of potentiometers. There was usually a zero, span, and damping adjustment. Basically, the output of the transmitter was set to match the measurand range and it was often up to the user to figure out the relationship between the measurand and the process variable. There were few if any alerts to measurement problems. Troubleshooting relied upon sorting through lists of symptoms and possible causes and making circuit measurements, or the comparison of related or redundant measurements.

You would think that not many of these old instruments still exist until you realize that most refineries, chemical plants, and steam plants in the United States were built in the 1960s and 1970s. Some of these plants have the original instrumentation and possibly a DCS that does not support smart instrumentation. The original measurement devices in these plants have installed errors and drift an order of magnitude larger than modern instrumentation. The poor performance of these original devices is due to older sensing element technology, missing compensation for non ideal and non-benchtop conditions, no multivariable calculations, and poor visibility of measurement integrity.

Smart Instrumentation Capability

Microprocessors opened the door to smart measurements that offered:

- Compensation of effects of ambient conditions on sensing elements
- Compensation of effects of process conditions on sensing elements
- Compensation of nonlinearities of sensing elements
- Measurement of secondary variables such as ambient and process temperature
- Multivariable measurement outputs such as density and mass flow

- Computation of abnormal process conditions such as pressure over range
- Detection of sensing element problems such as fixed output or abnormal value
- Detection of device malfunction and possible cause
- Identification of device such as model and serial numbers
- Calibration history
- Installation configuration for sophisticated devices such as radar level

The question then became, how does the supplier develop and the customer use an interface to provide access to this wealth of capability and data? A medium-size chemical plant may have a hundred or more different model numbers of measurement devices. Each smart device may provide a hundred or more data items. The plant could have ten thousand or more different pieces of useful information on the performance and integrity of the measurement system and the process. Obviously a huge need materialized for the standardization of the communication and interface to make it easier for the supplier to develop and the customer to use programs and displays for a wide spectrum of devices, functions, and data items.

Smart Instrumentation Interface

Fortunately, the International Electrotechnical Commission (IEC) developed the IEC 61804 standard that defined the Electronic Device Description Language (EDDL). The ISA104 committee adopted the standard and continues to promote and improve the use of EDDL. The standard was cooperatively designed by industrial authorities to ensure EDDL meets the needs of the end user. The standard specifies a generic language to describe the properties of automation system components. The specific language is capable of describing:

- Device parameters and their dependencies
- Device identification and history
- Device functions
- Historian functions
- Graphical representations, such as charts, plots, and diagrams
- Displays with conditional images

- Interactions with control devices

EDDL is a text-based language that enables an interface to smart instrumentation, such as measurement devices with microprocessor based transmitters and final elements with microprocessor based positioners or digital valve controllers. EDDL is used to provide the user friendly and complete access to smart instrumentation for the following:

- Setup (configuration)
- Calibration
- Testing
- Commissioning
- Tuning
- Identification (model and serial numbers)
- History
- Investigation
- Simulation
- Monitoring
- Analysis

EDDL has several unique characteristics not found in other device integration technologies. EDDL provides a well-defined structure for supporting the most simple to the very complex field device with a common look and feel across applications, which reduces the learning curve and supports multiple host applications. Also, this enables field device additions to be incorporated without affecting the runtime stability of the control system. Taken together, these technology advantages of EDDL provide significant benefits over other approaches in terms of the longevity and stability of the instrumentation and control system over the plant life cycle [2].

One of the greatest benefits that EDDL provides the process industry is interoperability across multiple hosts, devices, and technologies. This flexibility allows the end user to choose the best combination of price and performance for devices and software. EDDL allows instrument suppliers to define their products in a single, open, and consistent format. This format is readable by handheld communicators, control systems, and other process interface devices that are EDDL-enabled. This interoperability simplifies the control system integration process, because all a systems integrator or end user has to specify in the bid process is that all instrumentation and control equipment must conform to the International Standard, EDDL. Competitive bids are easier to evaluate, because the end user does not have to worry if specific software packages are available to support various devices; with EDDL, all comparable EDDL devices can be supported by all control systems. Finally, interoperability means that operators and maintenance personnel can easily find the calibration and diagnostic information needed for a particular device, and all EDDL-compatible devices will provide the necessary information in the "look and feel" of the host system [2].

EDDL is a declarative technology, not a software program. Because it is a text-based description of the field device and its properties, EDDL offers significant advantages over technologies that rely on any code from the device manufacturer to enable data to be displayed or printed. EDDL is independent of operating systems and control platforms. Operating system and platform independency, along with backward compatibility, are some of the biggest advantages of the technology. Through the use of EDDL, it is possible to avoid problems caused by operating system upgrades, control system revisions, and new versions of the device software from multiple suppliers different than that of the host system [2].

The EDDL files can reside in any device that can see and access the parameters, functions, and data in smart devices. EDDL files are used in host devices that have either direct plug-in access or HART or Fieldbus communication access to the smart instrumentation. The smart instrumentation determines what will be displayed while the host determines the "look and feel" via EDDL. Examples of hosts with direct plug-in access are hand-held communicators and laptops. The primary example of a host with HART or Fieldbus access is the DCS and multi-user device management software, such as asset management systems (AMS). These management systems get their information through the DCS if the DCS has "pass through" capability where information can be passed to and from the smart instrumentation through the DCS. If there is no "pass through" capability, the HART and Fieldbus signals of the smart instrumentation have to be connected to the I/O of the asset management system.

DCS and asset management systems provide better visualization with graphics and faster navigation by tabs. The secondary screens may have a subset of tabs so not all screens are directly accessible from the main screen. However, navigation is much easier and intuitive, particularly going back and forth between widely separated screens. This better visualization and navigation is particularly important for the setup of more sophisticated devices, such a radar level measurement, and the monitoring of sensing elements more susceptible to deterioration from process exposure, such as pH electrodes. Figure 1-31a shows a radar level setup screen with basic and more advanced probe settings. Figure 1-31b shows the radar level diagnostic screen details for an invalid measurement due to missing volume configuration. Figure 1-32a shows a pH setup screen for the range, impedance, standard electrode temperature compensation, and custom solution temperature compensation. Figure 1-32b shows the dashboard for a diagnostics screen with dials for the measurand, process temperature, electrode impedances, and RTD resistance.

Handheld communicators rely on a tree menu and navigation via branches from screen to screen. Because of the small display window, graphics and plots are not generally used and only a portion of the information and its interdependency can be seen at the same time. Figure 1-31c shows a partial menu of communicator screens for a radar device. Figure 1-32c shows a communicator menu of screens for the configuration of a pH measurement device.

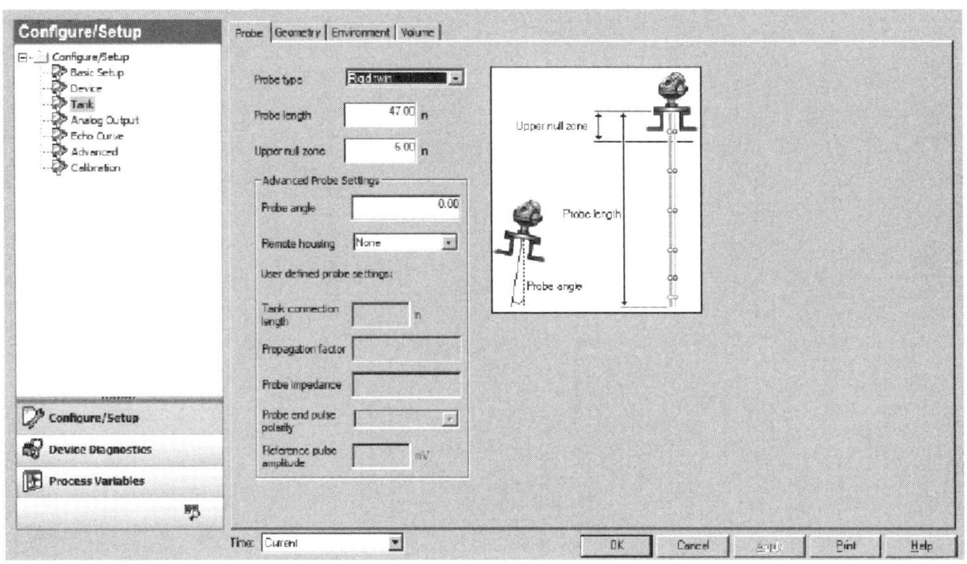

Figure 1-31a. Radar Level AMS Screen for Basic and Advanced Probe Settings

Calibration Adjustments

Calibration basically consists of setting a zero (offset) and span (slope) adjustment to provide a better match between the transmitter output and the true value of the process variable. Zero and span adjustments require

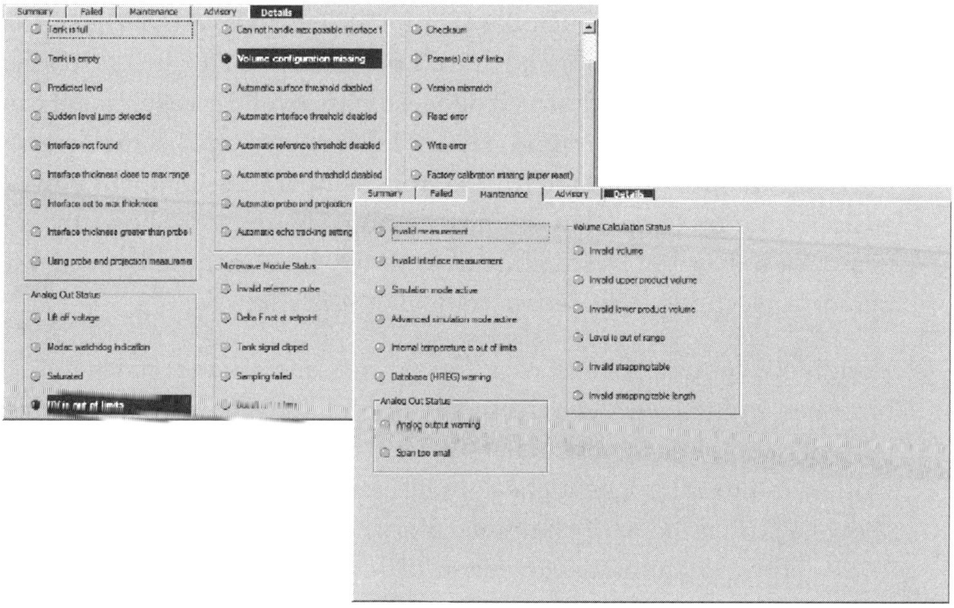

Figure 1-31b. Radar Level AMS Diagnostic Screen

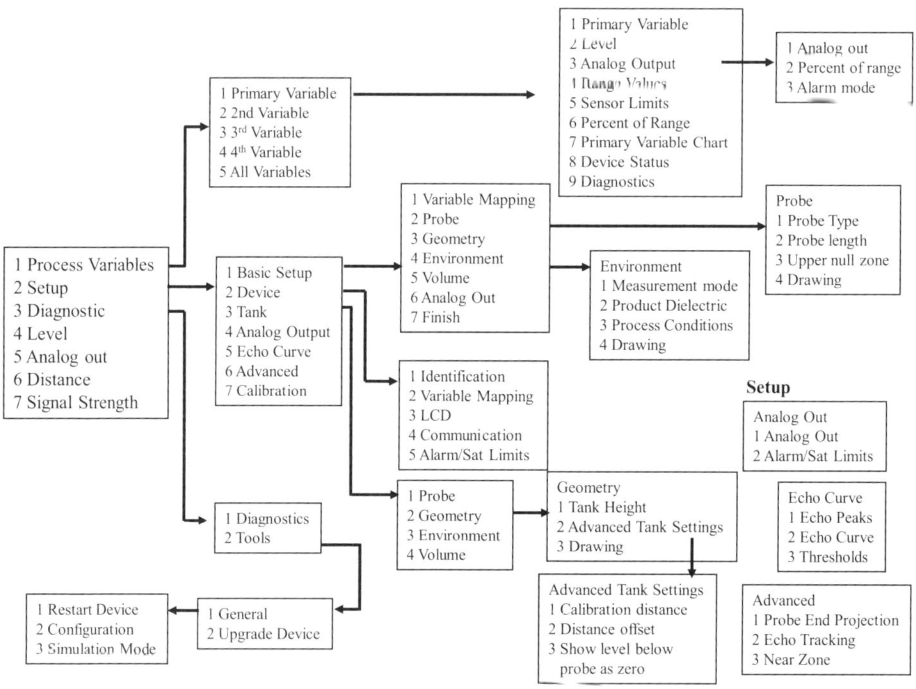

Figure 1-31c. Radar Level Handheld Communicator Menu

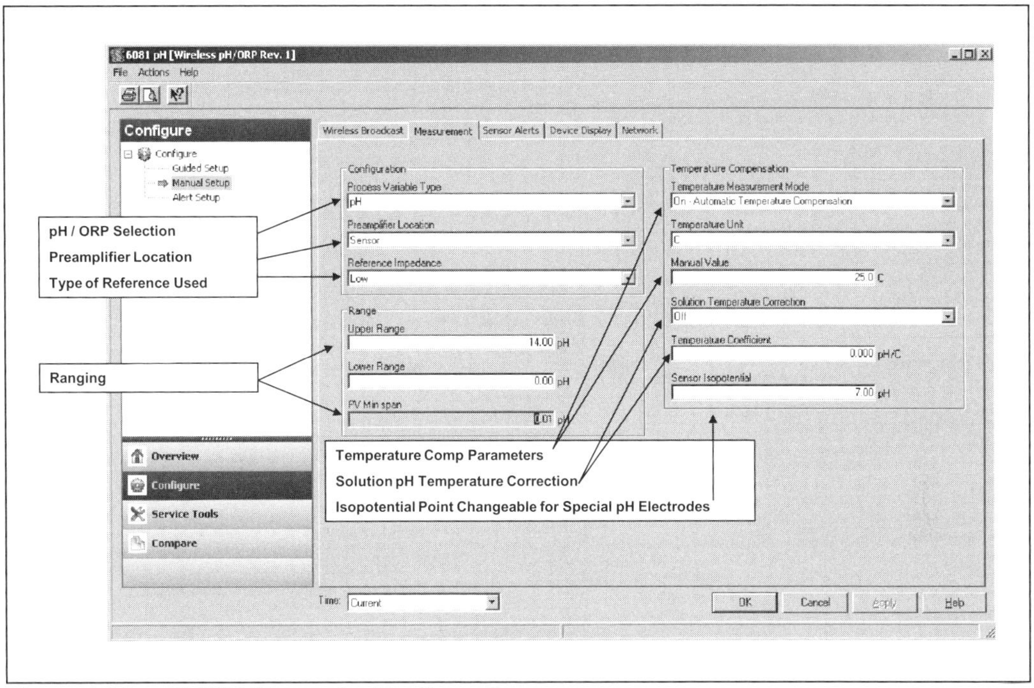

Figure 1-32a. pH AMS Screen for Configuration

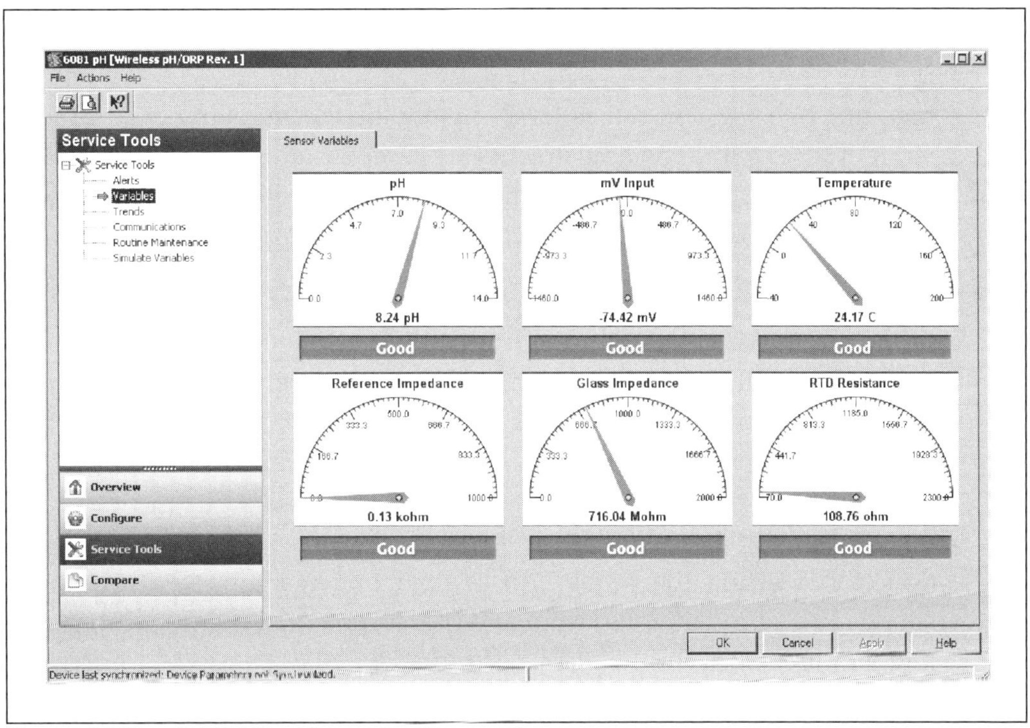

Figure 1-32b. pH AMS Dashboard for Diagnostics

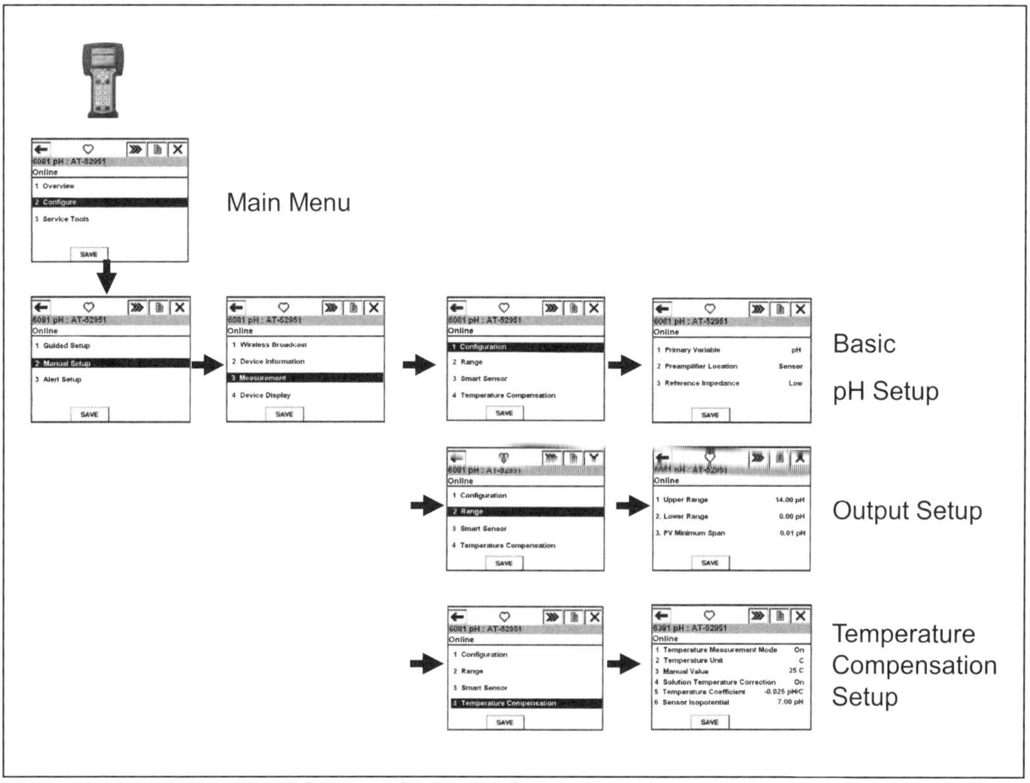

Figure 1-32c. pH Handheld Communicator for Configuration

one and two known values, respectively, of the measurand or process value. These calibration adjustments are compensating for errors or uncertainties in the sensing element and transmitter. If the calibration is done online it also compensates for uncertainties in the process and installation.

The zero adjustment provides a parallel shift to compensate for an offset of the measured value from the true value as shown in Figure 1-33. This zero adjustment is done by comparing the transmitter output with the true value of the measurand or process variable at one point. This adjustment is most typically done at the low range limit (0% transmitter output). For pH electrodes, the zero adjustment is termed an offset adjustment. For initial benchtop pH transmitter calibrations the offset adjustment is often done by inserting the probe into a 7 pH buffer solution, which corresponds to a zero millivolt output and the iso-potential point (point of zero temperature effect) for the glass electrode.

The span adjustment rotates the calibration line to compensate for a mismatch in the slope between the measured and the true value at two points

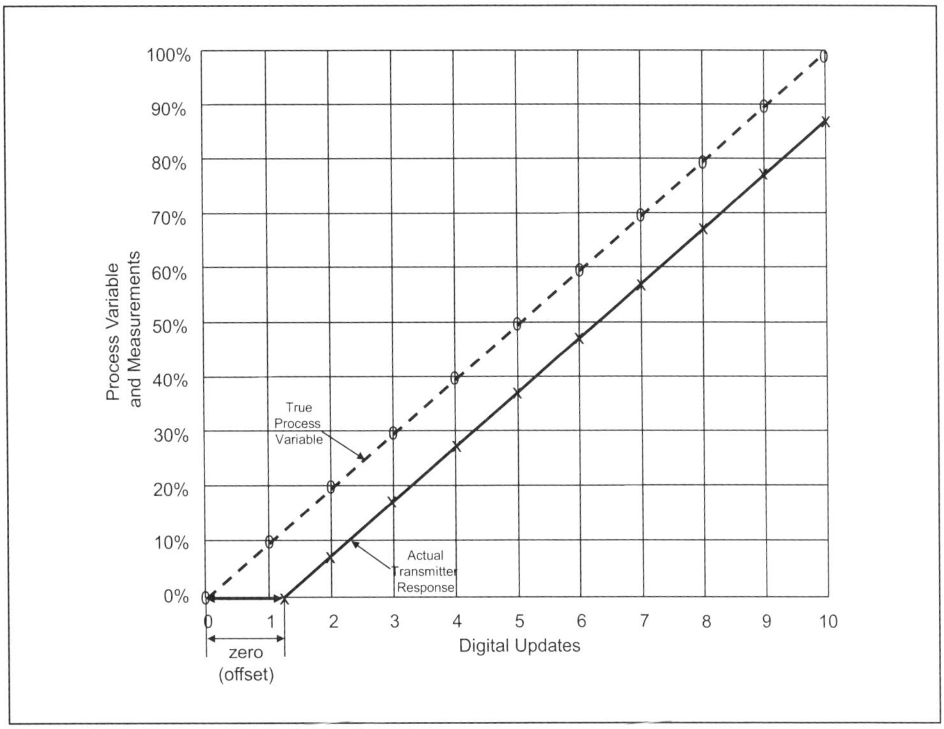

Figure 1-33. Zero (offset) Adjustment

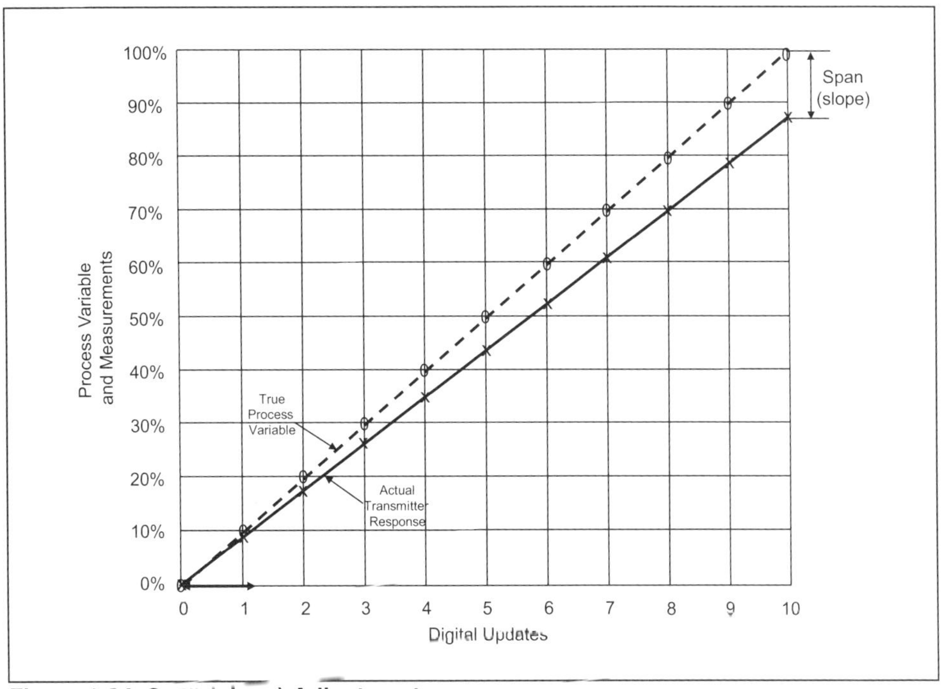

Figure 1-34. Span (slope) Adjustment

as shown in Figure 1-34. It is best to make the zero adjustment first to pin the calibration line at the zero percent output.

In some smart instrumentation, the zero and span adjustments are done automatically when two known values of a process variable or measurand are provided.

Calibration Methods

There are six basic calibration methods. Calibration methods 4 and 5 generally yield the best results because they take into account non-ideal installation and process effects. Calibration methods 1 and 2 are generally done at the supplier and are redone or at least checked in the plant before commissioning. The calibration methods used by the supplier may be proprietary. Suppliers will also perform calibration method 3 for level and flowmeters to establish the accuracy specifications of each model number. For method 3, the supplier installs the flow or level measurement device in a pipe and tank system at ambient temperature with a circulation of water and/or air. For flow, the installation is commonly called a "Meter Prover." Suppliers rarely offer calibrations at process conditions primarily due to the practical difficulties and safety issues with the transporting, handling, venting, and disposal of process chemicals.

1. Offline calibration using a simulator for the measurand
2. Offline calibration using a calibration standard for the process variable
3. Online calibration using redundant measurement(s) at standard conditions
4. Online calibration for zero value of process variable at process conditions
5. Online calibration using redundant measurement(s) at process conditions
6. Online calibration using a grab sample at process conditions

Offline calibration is either done on a benchtop or at the installation point by effectively removing or disconnecting the sensing element from the process. Offline calibration can be done with a simulator for the measurand (e.g., millivolts for thermocouples and pH electrodes) or a calibration standard (pressure, temperature, or pH). Simulators and calibration standards are both commonly referred to as "calibrators."

Consider the calibration of a typical differential pressure flow measurement. A sizing program for the primary element provides the differential pressure for the upper end of the flow measurement range at the process conditions and pipe size. The program also provides the primary element size, such as the orifice beta (ratio of orifice diameter to inside pipe diameter). With both sides of the DP open to atmosphere on the benchtop, a zero adjustment is made to provide 0% transmitter output. An adjustable pressure standard (calibrator) is connected to the high side and the pressure is set to match the upper range value from the orifice sizing program. A span (slope) adjustment is then made to provide 100% output.

If a controller is automatic, when an instrument is removed from service or an online calibration is done, the final element could be driven quickly to its output limit, upsetting the process and possibly shutting down the process unit. It is extremely critical that all plant procedures be followed before any maintenance or calibration is done on an instrument in service. Plant procedures require written details such as the duration of the calibration, approval by an operations supervisor, approval of the control room operator, placement of the affected controllers in manual, and verification of the proper commissioning of the instrument when the calibration is complete.

Online calibration for zero value of the process variable at process conditions is typically done for pressure, DP, flow, and level transmitters. After the instrument is installed and the line or vessel is at operating conditions, a zero adjustment is made when the pressure, differential pressure, flow, or level are known to be zero. For an inline flowmeter, particularly a magnetic or Coriolis flowmeter, the meter should be completely full of process fluid for the zero flow adjustment.

For a differential pressure transmitter (DP) it is recommended that a zero adjustment be done with the process pressure loaded to both sides of the DP. Figure 1-35 shows a DP with a manifold bolted to the high and low side connections. The manifold has an integral set of valves. The valves H1 and L1 enable the high and low side of the DP to be connected or isolated from the high and low process pressure connections, respectively. To provide a zero adjustment at process pressure, the low side valve L1 is closed, the high side valve H1 is opened, and the equalization valve E1 is then opened to load process fluid and pressure on both sides of the DP making the differential pressure zero.

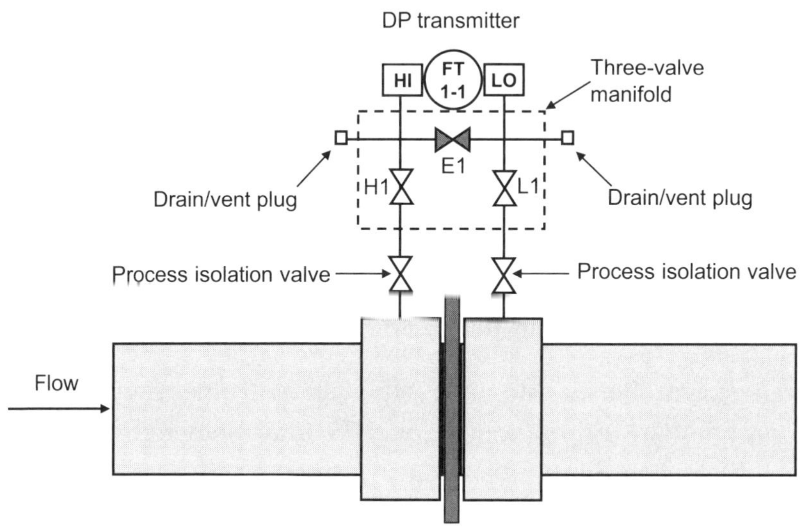

Figure 1-35. Differential Pressure Transmitter (DP) with Integral Valve Manifold

Online calibration using redundant online measurements is done by matching the subject transmitter's output to a more accurate online measurement or to the middle or median value of multiple online measurements of similar accuracy. This online calibration with redundant measurements can be done at standard conditions or at process conditions. See Chapter 6 on pH for a discussion of what middle signal selection offers for online calibration, noise reduction, inherent universal failure protection, and diagnostics.

A process sample is taken (grabbed) and used to make an online zero adjustment to correct for an offset of composition measurement such as conductivity and pH. For pH measurements this online grab sample calibration method is commonly termed a "standardization" adjustment. The sample is usually measured by inserting a fast and accurate lab sensor into the process sample bottle or beaker at the installation site or in the lab. The sample should be stirred and at process temperature.

The calibration methods used in the plants vary considerably in degree of sophistication based on practical considerations such as application requirements, lab and shop resources, and technician staff. The practices for nuclear power plants and pharmaceutical plants are generally the strictest. Metrological labs would probably consider most industrial plant practices as being not up to lab standards. In industry, non-ideal installation and process effects can easily obliterate the benefits from more exact lab calibrations, making the discussion of metrological standards academic.

Online Calibration Examples

The following descriptions are meant to be examples for major types of measurements and do not necessarily represent best practices. The best calibration method for an application depends upon the features and capability of the device, the installation, the process conditions, and the accuracy requirements.

It is assumed for each example that the zero and span adjustments for these devices have already been done on a benchtop in the instrument shop.

Online Level Calibration

An online calibration adjustment of a level device for a zero level can often be done for batch processes and before the startup and filling of vessels in continuous processes. An online adjustment for a specific level requires another level indication. Sight level gauge glasses that provide an outside visual indication of level are not used much any more, particularly in chemical plants, because of safety issues with the glass. During water batching, the water level can be varied, the water level manually measured, and online calibration adjustments made. For level devices that are affected by the physical properties of the fluid (e.g., density and dielectric constant), the calibration adjustment must take the difference between water and the process fluid into account.

A differential pressure transmitter is commonly used for level measurement. The high and low sides of the DP are connected to the bottom and top of the vessel, respectively. The high side measures the pressure from the height of the liquid level plus the pressure in the vapor space in the top of the vessel. If the vessel is open to atmosphere, the pressure in the top of the vessel is atmospheric and the low side connection of the DP can be simply vented. However, for environmental and safety reasons, the top of most process vessels is closed. During the filling and emptying of the vessel, the vessel pressure will rise and drop from the compression and decompression of the gases and vapors in the top of the tank. The vessel pressure set point may also be changed by a control loop. For closed vessels, the low side of the DP is connected to the top of the vessel. The net differential pressure between the high and low side of the DP is then proportional to liquid head (level). The DP zero adjustment for zero level should be made with the vessel pressurized to make sure the low side sensing line is properly installed and functioning. Figures 1-39a through 1-39d show the installation of major types of DP level measurements

Relative Accuracy of Inventory Measurements

Sometimes online level calibration is done by computing the level from a total of feed flow(s) measured by Coriolis mass flowmeter(s) or the change in feed tank weight measured by load cells. The accuracy of the total mass added to the vessel provided by these measurements is better than the accuracy of most level measurements except for radar.

In the author's assessment, the accuracies of properly set up and calibrated inventory measurements for changing process fluid compositions are listed below where 1 is the highest and 10 is the lowest accuracy. For the level devices, random errors are most important because we are looking for a change in level that would cancel out fixed (offset) errors. Note that even the lowest accuracy device might be perfectly adequate for many applications where knowing the exact level or change in level is not necessary, such as surge tank level. For applications where residence time and material balance control are important, such as continuous reactor and column overhead receiver level, measurement noise and sensitivity are more important than overall accuracy.

1. Radar level measurements (provided dielectric constant effects are negligible)
2. Coriolis flowmeter total (provided pipeline inventory effects are negligible)
3. Load cells (provided structural support and wind effects are negligible)
4. Ultrasonic level (provided vessel vapor effects are negligible)
5. Multivariable DP level (provided DP sensing line effects are negligible)

6. Magnetic flowmeter total (provided velocity and conductivity are above low limit)
7. Vortex flowmeter total (provided velocity and Reynolds number are within limits)
8. Multivariable DP flowmeter total (provided flow and Reynolds number are within limits)
9. Single direct connect DP level (provided DP sensing line effects are negligible)
10. Bubbler DP level (provided DP sensing line and bubbler effects are negligible)

A measurement device with higher installed accuracy can be used for the online calibration of devices with lower installed accuracy. For example, radar level and Coriolis flowmeter totals could be used to calibrate the other level and flow measurement devices on the list.

Online Temperature Calibration

The use of dual temperature sensing elements permits an online check of calibration. The millivolt outputs of dual thermocouple elements and the ohms outputs of dual resistance temperature detector (RTD) elements are measured and compared to quantify sensor manufacturing tolerance and reproducibility. The best sensing element indication is selected based on other online measurements or previous test results in the shop using a calibration standard, provided the current sensing element output is reasonable. A one-point zero (offset) calibration adjustment is made to make both transmitter outputs agree with the temperature that corresponds to the best sensing element.

Online pH Calibration

The use of three pH measurements and middle signal selection enables a simple online calibration where an offset adjustment is made to match the high and low pH measurements to the middle pH measurement. More commonly there are only two pH measurements and a judgment is made based on past history and current operating conditions as to which one is most representative. If one electrode is always lagging behind the other electrode, it may be fouled or be losing efficiency due to old age.

A more exact online calibration would involve taking a grab sample. It is preferable to measure the sample immediately in the field using a portable lab pH meter before the temperature or composition changes in the sample or in the process. If the sample is taken to the lab, the time the sample was drawn from the process and the corresponding field pH indication must be noted so they can be correlated in time. The sample in the lab must be kept at the same temperature and the sample container must prevent the evolution, evaporation, or absorption of components that would change the pH. Pure water samples, for example, readily absorb ions from a glass container and carbon dioxide from the air.

1-7. Installation

Process Connections

Figures 1-36 through 1-38 are block diagrams for a generic measurement device with an integrated sensing element and transmitter for three types of process connections. The process fluid can flow through the sensing element. Alternatively, the sensing element can be mounted flush or inserted into an opening into the process. The connection of the sensing element to the process can be flanged, threaded, or welded. Food, beverage, and pharmaceutical applications have special sanitary connection requirements. Figure 1-36 shows a sensing element in which the process flows through the body of the sensing element. This flow-through body is called a "meter." The meter has inlet and outlet flanges that match up with flanges on a pipe. Meters are often used to measure flow.

Figure 1-37 shows a sensing element inserted into a flanged process nozzle on a pipe or vessel wall. This sensing element is called a "probe" and is typically used to measure temperature and pH. The tip of the probe should extend through the nozzle and into the pipe or vessel. Since the fluid is relatively stagnant, the fluid in the nozzle may be at a different temperature and pH than the fluid in the vessel or pipe volume. The measurement does not see changes in the process or the true value of the process variable. pH probes are almost always separate from the transmitter. A pH probe may have an integrated preamplifier to magnify the low level millivolt signal sent to the transmitter. Temperature probes tend to have the transmitter mounted on top of the sensor unless the transmitter is designed to accept multiple temperature sensing elements.

Figure 1-38 shows a sensing element mounted flush on a flanged process nozzle on a pipe or vessel wall. This sensing element typically has a flat but rippled surface called a *diaphragm* that is used in measuring pressure.

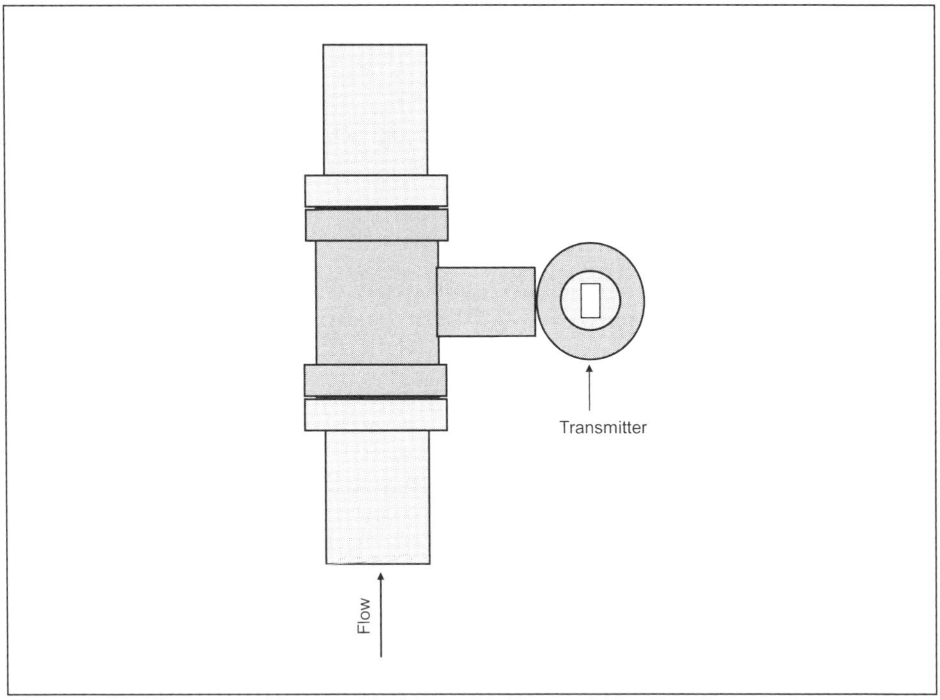

Figure 1-36. Flowmeter Mounted In-Line with Flanged Pipeline Connections

Differential Pressure Level Example

Figures 1-39a through 1-39d show some of the many installation options for the differential pressure measurement of level. The high side of the DP could have a flush or extended diaphragm mounted on the vessel flange (direct mounted). These figures show the use of direct-mounted transmitters, sensing lines, and a filled capillary system and diaphragm seal.

The sensing line to the high side in Figure 1-39a is filled with process fluid. The sensing line to the low side, termed the "equalization line," is empty or in some cases filled with a known liquid such as water. In practice, it is difficult to keep the "equalization line" at its assumed state because of evaporation and condensation. There are often valves at the top and bottom of the "equalization line" to drain, flush, and fill this line.

Figure 1-39b shows the use of a second DP that measures the pressure at the top of the vessel. The bottom DP now measures the total of the liquid

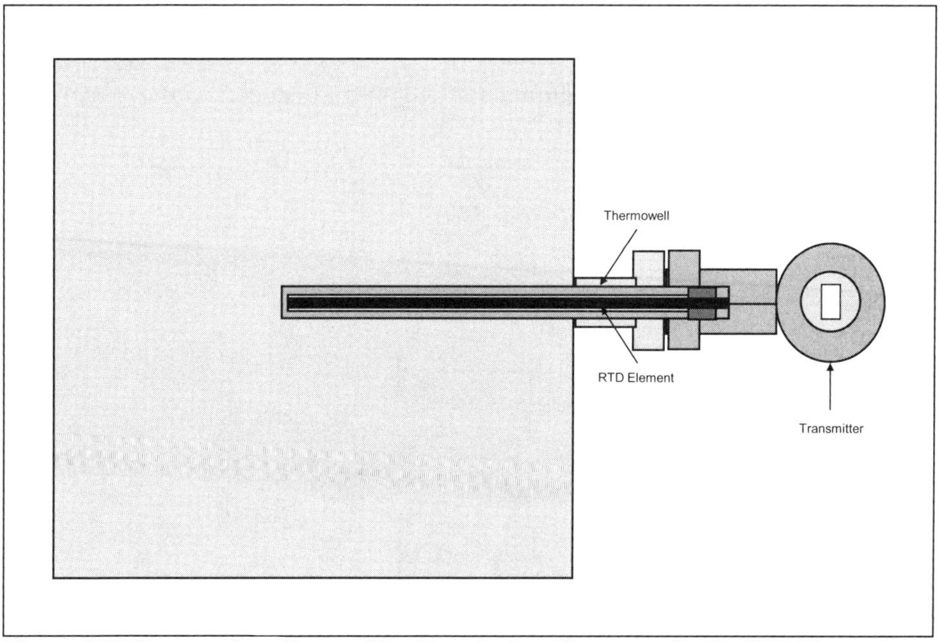

Figure 1-37. Temperature Transmitter Mounted on Thermowell with RTD Element on a Flanged Nozzle

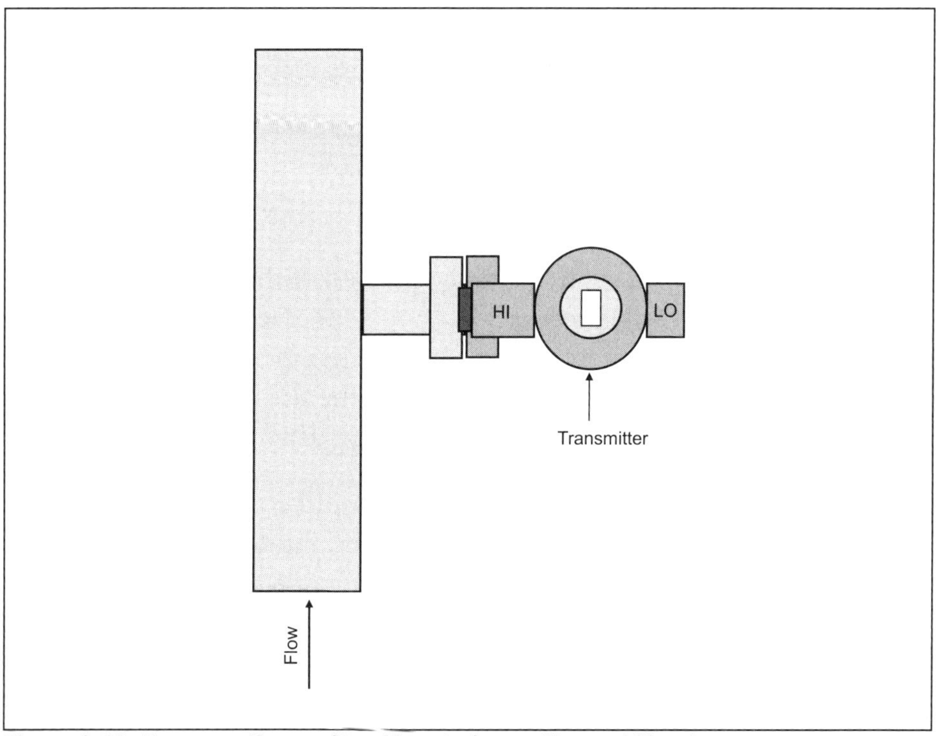

Figure 1-38. Pressure Transmitter Mounted on a Flanged Nozzle

head and the vessel pressure. The signal from the top DP is subtracted from the bottom DP to give the differential pressure from just the liquid head. The dual direct-mounted DP installation eliminates the need for sensing lines. However, if the vessel pressure is large compared to the liquid head, the error introduced from the vessel pressure measurement can be excessive.

Purges may be used to keep sensing lines from plugging, trapping process fluids, or collecting vapors or condensate. These purges can cause problems as well. Liquid purges can adversely affect the process composition so gas purges are often used as shown in Figure 1-39c. The tube extending into the liquid at the vessel bottom is called a "bubbler" because gas bubbles come out of the tube's tip. Sometimes the tube comes down from the top instead of through the side of the vessel and is called a "dip tube." However, a sudden increase in process pressure can cause the process to backfill the tube and sensing line with process fluid. The result is a compression of the gas in the tube and line and a bump in measurement indication. The pressure and flow or level indication does not return to normal until the gas purge flushes out the process fluid. The size and duration of a bump in the measurement indication makes the 200-millisecond response time and 0.02% repeatability of the differential pressure transmitter a wasted opportunity.

The solution would seem to be to just crank up the purge rate but a high purge rate can create an appreciable pressure drop in the tube and sensing lines and can appreciably increase the cost of the purge gas, which in many cases is nitrogen. High purge rates can also increase the cost of the vent system in large projects.

A separate problem with bubblers is that a drying out of the process at the tube tip can cause solidification and restriction or even plugging of the tip. In some of these applications, the addition of a small water purge to make a wet gas has helped.

The last example of a DP level installation uses a filled capillary system and diaphragm seal connected to the low side of the DP, as illustrated in Figure 1-39d, to eliminate the need for an equalization line. The diaphragm seal is usually mounted flush at the process connections. The large connection and surface area reduces concerns about plugging and coatings. The capillary line is filled under vacuum conditions to ensure the line has essentially no gas. However, temperature changes cause a slight expansion or contraction of the liquid fill that is sensed as a pressure change. To reduce extraneous effects the capillary line is as short as possible, and insulated in some cases. The DP is direct mounted on the lower nozzle of the vessel so that a capillary system is not required for the high side for the DP. This asymmetrical design provides some compensation of temperature effects and a faster response to changes in level (see Chapter 5).

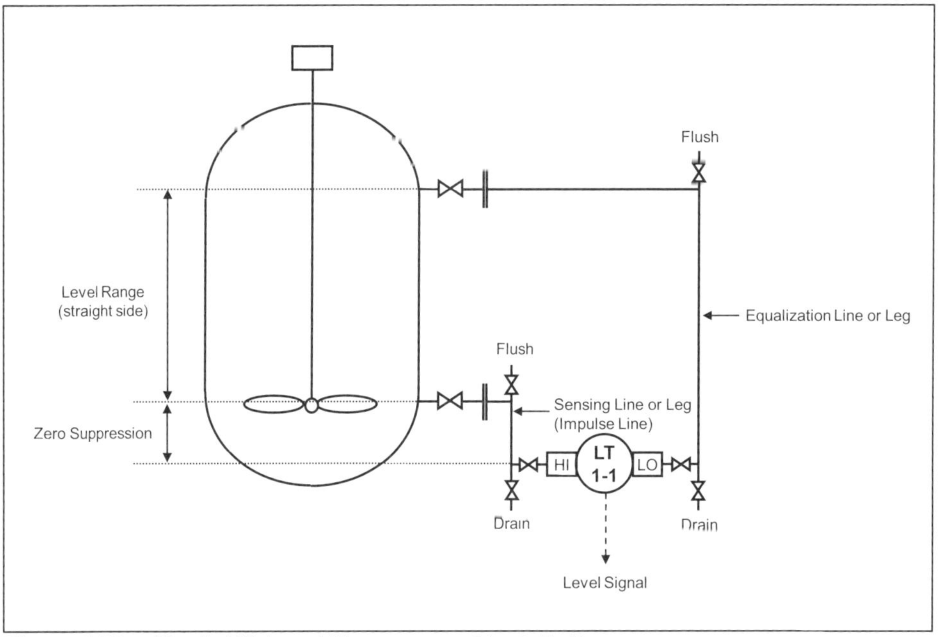

Figure 1-39a. Single DP Level Measurement Installation

Poor or outdated installation practices can waste the exceptional accuracy, stability, and reliability of modern smart wireless measurements. Yet installation problems are rarely discussed in articles and books on instrumentation. The best source of installation "watch-outs" is usually the instruction manual for the specific device even though these may be sketchy. The sensing element installation method and location should be designed to:

1. Prevent the presence of a non-representative process variable at the sensing element(s)
2. Prevent extraneous effects at the sensing element(s)
3. Minimize plugging, coating, and fouling of sensing elements
4. Minimize time lag(s) of sensing elements

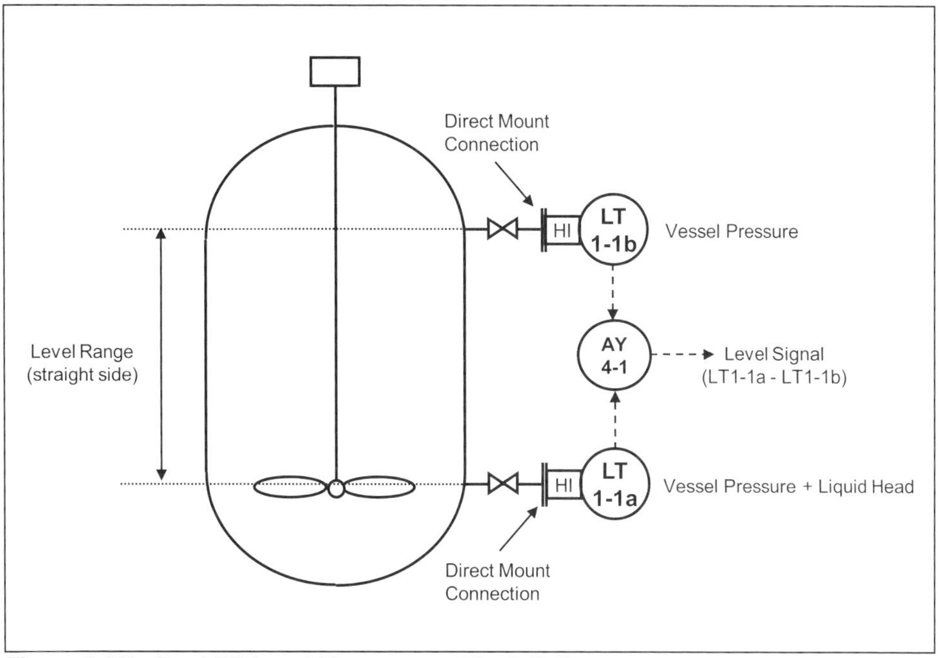

Figure 1-39b. Dual DP Level Measurement Installation

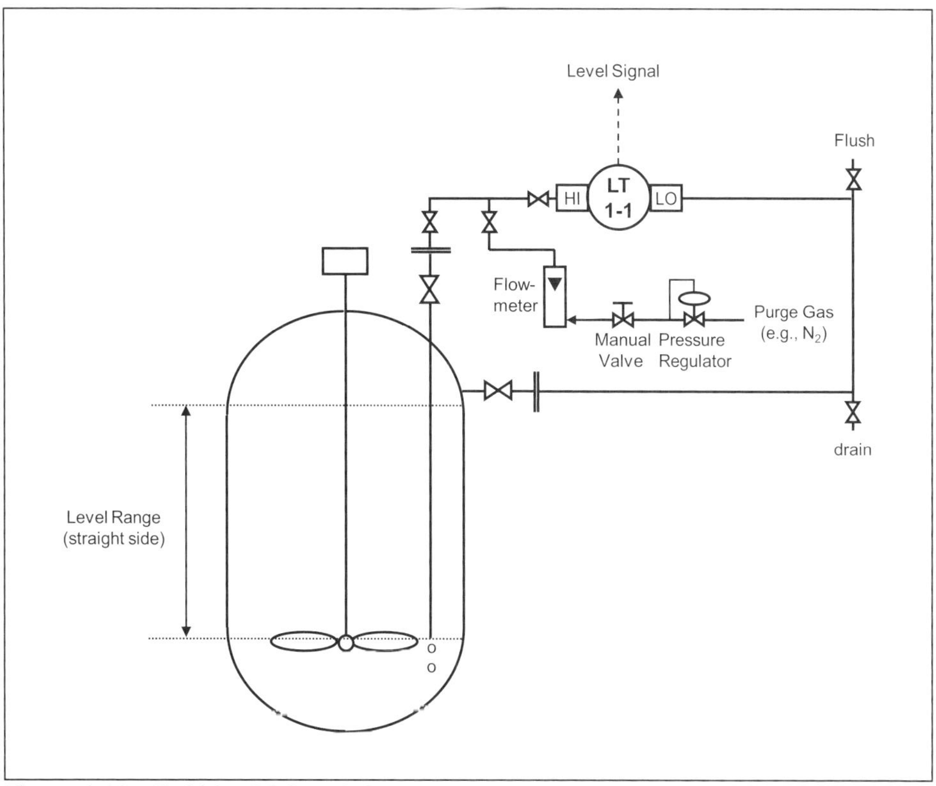

Figure 1-39c. Bubbler DP Level Measurement Installation

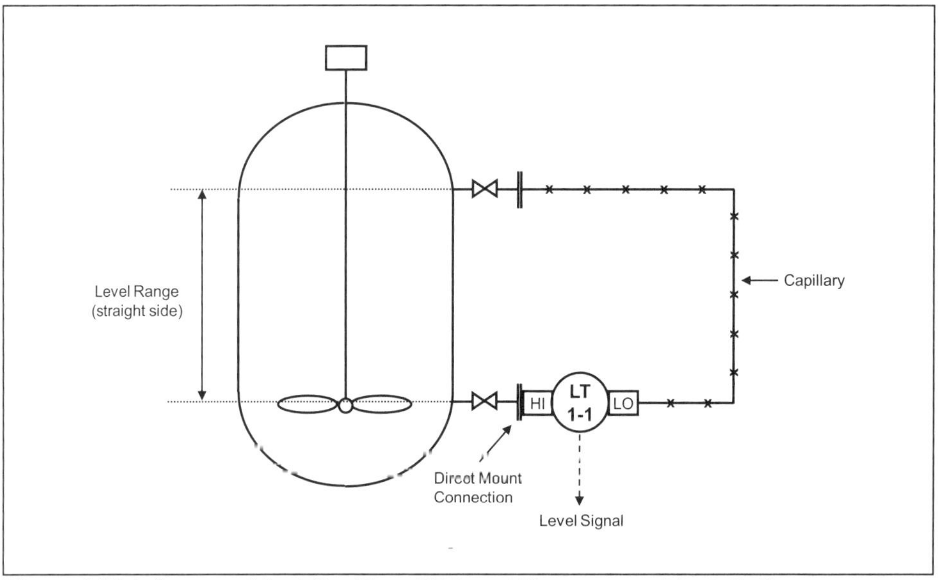

Figure 1-39d. Asymmetrical Capillary DP Level Measurement Installation

5. Minimize measurement noise
6. Minimize transportation time delay to sensing elements

The maximum transportation delay can be readily estimated for temperature and pH sensing elements. For pipelines, estimate the distance between the process equipment and the sensing element and multiply it by the minimum pipeline velocity. For more complex piping systems and sample lines, estimate the volume of the respective section of line or equipment and divide it by the minimum flow through the line or equipment.

The sensing time lag(s) of temperature and pH sensing elements increase as the fluid velocity decreases, from the decrease in the heat transfer coefficient and mass transfer coefficient, respectively, at the tips of these elements. The fouling and coating rate also increases as the velocity decreases, which greatly increases the sensing time lag(s), especially for pH glass electrodes. Stagnant process fluid areas can also result in a non-representative process variable due to a lack of mixing.

There are many examples of installation problems. The following is a list of installation mistakes the author has seen and the problems created. The list should clue the reader into what to look for as potential installation considerations in the succeeding chapters. The reader may want to review this list after reading the respective measurement chapters to gain a greater appreciation and understanding of the potential problems.

1. A differential head flowmeter or vortex meter installed downstream of a partially open valve (could be a manual valve throttled in the field or a control valve) – causes variations from the assumed flow profile for these volumetric flowmeters. Bubbles formed by sudden decompression in the vena contracta of a control valve (flashing) cause inaccurate and noisy flows. The implosion of bubbles (cavitation) downstream can damage flowmeters.

2. Insufficient straight runs from two non-planar elbows to a differential pressure flowmeter or vortex meter – swirling causes erratic and inaccurate flow measurements.

3. Condensation or surge of process fluid into sensing lines for DP measurements designed to be empty – accumulation of liquid head in the sensing lines causes a variable bias in the DP.

4. Evaporation or suction of liquid out of sensing lines for DP measurement designed to be full – loss of liquid head in the sensing lines causes a variable bias in the DP.

5. Variability in the inside diameter of pipelines for differential pressure flow measurements – allowable tolerance of inside diameter for industrial piping introduces excessive uncertainty in the beta ratio of the orifice element.

6. Temperature sensing elements located too close to the outlet of desuperheater – water droplets that haven't vaporized cause an inaccurate and noisy temperature signal.

7. Temperature and pH sensing elements located too close to the outlet of a static mixer or heat exchanger – liquid streams divided by the mixer or exchanger have not combined or bubbles that have not dissolved completely cause inaccurate and noisy temperature and pH signals.

8. Temperature sensing elements with insufficient immersion in the process (immersion length less than 10 times the thermowell diameter or bare sensing element diameter) – conductive heat transfer along the wall of the thermowell or

bare element to the process connection causes an inaccurate temperature that is affected by jacket and ambient temperature. For polymer lines with hot jacketing, the viscous effect at the wall and the higher temperature of the jacket can cause a higher than actual temperature reading that is affected by flow rate and jacket temperature.

9. Temperature sensing elements installed in glass line baffles – the insulating effect of glass slows heat transfer and response.
10. Temperature sensing elements inserted into thermowells with construction debris or sand from sandblasting – the insulating effect of debris and sand slows heat transfer and response.
11. Temperature and pH sensing elements that do not protrude past baffles on a vessel wall – low mixing and low velocity cause non-representative and slow response of temperature and pH measurements.
12. Temperature element mounted on an extruder barrel wall – heat capacity of barrel wall slows down the response.

Readers are encouraged to send installation mistakes to the author via ISA. The list will be periodically updated on the ISA website.

1-8. Questions

1. What are the typical installed accuracies of a DP in the 1980s and a smart DP with advanced sensor technology?
2. How can wireless temperature transmitters be used to find the optimum location for temperature control of a distillation column?
3. Why do wireless transmitters offer the potential for more process information?
4. For pharmaceuticals, at what phase in the commercialization process are improvements in measurement and control best demonstrated?
5. How does the use of industrial automation systems help students and professors at universities teaching process control?
6. What are the advantages of eliminating sensing (impulse lines)?
7. What are the advantages of eliminating wiring?

8. What information is transmitted in a HART or Fieldbus signal?
9. What is the most frequently measured process variable in the process industry?
10. How does a control loop ultimately affect the process in nearly all process control applications in the process industry?
11. How does the principle discharge flow of a vessel determine whether the vessel is operating in a continuous or batch mode?
12. How does the dynamic open loop response of temperature and composition differ for continuous and batch processes?
13. What are the sources of deadtime in a control loop?
14. Does a continuous or batch process tend to have more automation?
15. How does the transfer of variability differ for pure batch, fed-batch, and batch profile control?
16. How do industrial processes differ from the processes taught at universities?
17. Why are field pressure and temperature switches not recommended?
18. How does a discrete 2oo3 voting logic or analog middle signal selection prevent a false trip from a single instrument failure?
19. Why are false trips highly undesirable?
20. What types of disturbances are most disruptive?
21. What are the relative dynamic consequences of online, at-line, and off-line (lab) analyzers?
22. What are some of the major causes of disturbances to temperature loops?
23. For an equivalent size, are measurement systematic or random errors more detrimental to loop performance?
24. Why does an error expressed in percent of reading rather than in percent of span usually translate to better measurement performance?
25. For a control loop, what types of errors are most problematic?
26. What are some of the advantages offered by the Electronic Device Description Language (EDDL) for interface to smart instruments?

27. What are the advantages of an asset management system compared to a handheld communicator for smart measurements?
28. If the measure change is larger than the true change in the process variable, what calibration adjustment is needed?
29. What is the principal problem with making online calibration adjustments at process conditions?
30. What are the primary advantages of dual direct-mounted differential pressure transmitters for level measurement?

1-9. References

1. McMillan, G. K., and Weiner, S. "Sensible Sensor Speed - Part 2." *Control Talk*, *Control*, Jun. 2009.
2. Diedrich, C.; Winkel, L.; Berge, J., and Blevins, T. L. "Keeping Systems and Communicators Up-to-date using EDDL." Paper presented at ISA EXPO 2007.
3. McMillan, G. K. "Is Wireless Process Control Ready for Prime Time?" *Control,* May 2009.
4. McMillan, G. K. and Weiner, S. "Secure Answers to a Risky Business." *Control Talk*, *Control*, Nov. 2008.
5. McMillan, G. K., Benton, T., Zhang, Y., and Boudreau, M. "PAT Tools for Accelerated Process Development and Improvement." *Process Design Supplement* to *Bioprocess International*, Mar. 2008.
6. Boudreau, M. and Benton, T. "Bioprocess Control: What the Next 15 years Will Bring." *Pharmaceutical Manufacturing*, May 2008.
7. McMillan, G. K. "Unlocking the Secret Profiles of Batch Reactors." *Control*, Jul. 2008.
8. Boudreau, M. and McMillan, G. K. *New Directions in Bioprocess Modeling and Control.* Research Triangle Park: ISA, 2006.
9. Blevins, T. L. and McMillan, G. K. "Bridging the Gap Between Universities and Industry." Automatic Control Conference (ACC) Proceedings, June 2009
10. ISA's *Automation, Systems, and Instrumentation Dictionary, 4th Edition.* Research Triangle Park: ISA, 2003 (ISBN: 1-55617-941-3).
11. Helson, R., "Unleash the Power." *Control* Supplement, Aug. 2001.

12. Sanders, Fred. F. "Watch Out for Instrument Errors." *Chemical Engineering Progress*, Jul.1995.

13. Blevins, T. L. *An Overview of the ISA/IEC Fieldbus, Process/Industrial Instruments and Controls Handbook,* 5th Edition. New York: McGraw-Hill, 1999.

14. Benedict, R. P. *Fundamentals of Temperature, Pressure, and Flow Measurements, 3rd Edition*, Hoboken: Wiley-Interscience, 1984.

15. Girard, M. and Visser, J. L. "Fundamental Differential Pressure Calibrations." ISA *Transactions*, Vol. 23, No. 1, 1984.

16. Szani, R., Ratermaan, M., and Eryurek, E. "Diagnostic Capabilities of Foundation Fieldbus Pressure Transmitters." *Hydrocarbon Processing* Apr. 2003.

17. Kerlin, T. and Shepard, R. *Industrial Temperature Measurement.* Research Triangle Park: ISA, 1982.

18. Blevins, T. L., McMillan, G. K., Wojsznis, W. K., and Brown, M. W. *Advanced Control Unleashed.* Research Triangle Park: ISA, 2004.

19. McMillan, G. K. and Weiner, S. "Sensible Sensor Speed – Part 1." *Control Talk* in *Control* May, 2009.

20. Stichweh, L. "Gas Purged DP Transmitters for Liquid Level and Flow." *InTech,* Research Triangle Park: ISA, Nov. 1992.

2

Temperature Measurement

2-1. Learning Objectives

A. Appreciate the importance of temperature measurements.

B. Understand the relative performance of thermocouples and resistance temperature detectors.

C. Gain an overview of optical pyrometer performance.

D. Learn about the latest technological advances in transmitters and sensors.

E. Find out how to select specific sensor and thermowell designs.

F. Understand important aspects of extension wire effects.

G. Gain some insight into best transmitter and communication options.

H. Learn the basic thermowell location and installation requirements.

I. Learn terminology and issues to intelligently discuss industrial applications.

2-2. Introduction and Overview

Temperature is a measure of a material's internal molecular activity. As the level of molecular activity rises, the temperature of the material increases. *Hot* and *cold* are subjective, qualitative descriptions of a change in molecular activity.

Temperature is often the most important of the common measurements because it is an indicator of process stream composition and product quality. It would be nice if we had online analyzers throughout the process but the fact of the matter is that most plants have infrequent, offline lab analysis at best. In the chemical industry, nearly all loops that are controlling the composition of a unit operation use temperature as the primary controlled variable. Even when online or at-line analyzers exist, these are usually relegated to monitoring and the manual trim or optimization of temperature set points by supervisory or model predictive control. Temperature measurements are also essential for equipment protection and performance monitoring. Some examples of common unit operations that have a critical dependence upon tight (minimum variability) temperature control are:

- Bioreactors and fermentors
- Calciners and kilns
- Chemical reactors
- Columns (e.g., absorber, distillation, stripper)
- Condensers
- Crystallizers
- Dryers
- Evaporators
- Extruders
- Furnaces
- Superheaters and desuperheaters
- Vaporizers

Over the years, the need in the process industry for more consistent and accurate ways to describe temperature led to the invention of temperature-measuring devices, or *sensors*. Sensors use standard, universally recognized temperature scales. Because these scales rely on fixed points in nature (e.g., freezing point of water), they provide a way to describe temperature that is both objective and quantitative. The four temperature scales in use today are Fahrenheit, Celsius (also called Centigrade), Kelvin, and Rankine. In commercial applications, Fahrenheit and Celsius are the most commonly used scales.

In industrial environments, high process temperatures, pressures, and vibration make it necessary to have a robust temperature sensor. Fast response time, accuracy, and stability are also needed. While several types of temperature sensors are available, such as thermistors, infrared pyrometers, fiber optic, and others, the two most commonly used in the process measurement industry are resistance temperature detectors (RTDs) and thermocouples (TCs).

Comparison of Thermocouples and Resistance Temperature Detectors

In the process industry as a whole, 99% or more of the temperature loops use thermocouples or resistance temperature detectors (RTD). The RTD provides sensitivity (minimum detectable change in temperature), repeatability, and drift that are an order of magnitude better than the thermocouple, as shown in Table 2-1 [1, 2]. Sensitivity and repeatability are 2 of the 3 most important components of accuracy as discussed in Chapter 1. The other most important component, resolution, is set by the transmitter. Drift is important for extending the time between calibrations. The data in this table dates back to the 1970s and consequently doesn't include the improvements made in thermocouple sensing element technology and premium versus standard grades. However, the differences are so dramatic that the message is still the same.

Table 2-1 includes data on thermistors, which have seen limited use in the process industry despite their extreme sensitivity and fast (millisecond) response, primarily because of their lack of chemical and electrical stability. Thermistors are also highly nonlinear but this can be addressed by smart instrumentation.

Table 2-1. Accuracy, Range, and Size of Temperature Sensing Elements [1, 2]

Criteria	Thermocouple	Platinum RTD	Thermistor
Repeatability (°C)	1 - 8	0.02 - 0.5	0.1 - 1
Drift (°C)	1 - 20	0.01 - 0.1	0.01 - 0.1
Sensitivity (°C)	0.05	0.001	0.0001
Temperature Range (°C)	–200 - 2000	–200 - 850	–100 - 300
Signal Output (volts)	0 - 0.06	1 - 6	1 - 3
Power (watts at 100 ohm)	1.6×10^{-7}	4×10^{-2}	8×10^{-1}
Minimum Diameter (mm)	0.4	2	0.4

For bare sensing elements, thermistors are much faster-responding than thermocouples, which are slightly faster than RTDs. This point rarely comes into play because for most industrial processes a 1 or 2 second additional lag time in a temperature loop is well within the uncertainty of the loop's dynamics. The secondary process time lags can easily change by 10 to 20 seconds for slight changes in operating conditions. Also, once these sensing elements are put inside a thermowell or protection tube (a closed-end metal tube that encapsulates and protects a temperature sensor from process flow, pressure, vibration, and corrosion), the fit, fill, material, and construction of the thermowell have the biggest impact on temperature measurement time lags, as noted in Tables 2-2a and 2-2b [1][3, 4]. Protection tubes like thermowells provide isolation of the element from the process but unlike thermowells, protection tubes do not necessarily provide a pressure tight attachment to a vessel, a tapered or stepped wall, or a tight fit of the element. Protection tubes may be ceramic for high temperature applications. The measurement lags from protection tubes are generally larger than for thermowells.

Table 2-2a. Dynamics of Bare Sensing Elements [1][3, 4]

Bare Sensing Element Type	Time Constant (seconds)
Thermocouple 1/8-inch sheathed and grounded	0.3
Thermocouple 1/4-inch sheathed and grounded	1.7
Thermocouple 1/4-inch sheathed and insulated	4.5
Single Element RTD 1/8 inch	1.2
Single Element RTD 1/4 inch	5.5
Dual Element RTD 1/4 inch	8.0

Table 2-2b. Dynamics of Thermowells [1][3, 4]

Process Fluid Type	Fluid Velocity (feet per second)	Annular Clearance (inches)	Annular Fill Type	Time Constants (seconds)
Gas	5	0.04	Air	107 and 49
Gas	50	0.04	Air	93 and 14
Gas	150	0.04	Air	92 and 8
Gas	150	0.04	Oil	22 and 7
Gas	150	0.02	Air	52 and 9
Gas	150	0.005	Air	17 and 8
Liquid	0.01	0.01	Air	62 and 17
Liquid	0.1	0.01	Air	32 and 10
Liquid	1	0.01	Air	26 and 4
Liquid	10	0.01	Air	25 and 2
Liquid	10	0.01	Oil	7 and 2
Liquid	10	0.055	Air	228 and 1
Liquid	10	0.005	Air	4 and 1

There are many stated advantages for thermocouples, but if you examine them more closely you realize they are not as important as perceived for industrial processes. Thermocouples are more rugged than RTDs. However, the use of good thermowell or protection tube design and installation methods makes an RTD sturdy enough for even high-velocity stream and nuclear applications [5, 6]. Thermocouples appear to be less expensive until you start to include the cost of extension lead wire and the cost of additional process variability from less sensor sensitivity and repeatability.

The minimum size of a thermocouple is much smaller. While a tiny sensor size is important for biomedical applications, miniature sensors are rarely useful for industrial processes. The thermocouple millivolt output is a result of the temperature along the entire length of the sensing element, where the ohms output of the RTD is the result of the temperature at its tip. This may be or may not be an advantage for thermally inhomogeneous fluids. As users we often want to think of a point measurement, particularly if we are looking to locate the most representative points in the process, or problems such as hot spots.

The main reason to go to a thermocouple is if the temperature range is beyond what is reasonable for an RTD or you don't need the accuracy of an RTD. Thus, for temperatures above 850°C (1500°F), the clear choice is a thermocouple for a contacting temperature measurement. For temperatures within the range of the RTD the decision comes often down to whether the temperature is used for process control or just the monitoring of trends. If you have lots of temperatures for trending where errors of several degrees are unimportant, you could save money by going to thermocouples with transmitters mounted on the thermowell (integral mount) or nearby. If you are using temperature for process control, data analytics, statistical or neural network predictions, process modeling, or in safety systems, a properly protected and installed RTD is frequently the best choice for temperatures lower than 850°C (1500°F).

Optical Pyrometers

While thermocouples can be used at high temperatures in furnaces and wireless transmitters have expanded their use in rotating equipment, such as kilns and calciners, there is still a significant niche market for optical pyrometers in the process industry.

Non-contacting temperature measurements are needed for solid products that are moving, such as plastic webs, fiber spin lines, and paper sheets. Non-contacting temperature measurements are also useful for furnaces where the installation integrity and reliability of thermocouples is questionable.

While the market is small for optical pyrometers, an overview of the technology is merited to better understand some of the limitations and difficulties not commonly discussed in product catalogs. Equations are used to clarify the assumptions and quantify the effects of a non-ideal target, installation, intervening space, and viewing window.

For details on selection, configuration, installation, and calibration, the user should use information and application specialists from the manufacturer.

Pyrometers infer temperature from the optical radiation intensity at one or more wavelengths from a target. The most prevalent non-contacting temperature measurements devices in industry are single-color and ratio (2-color) pyrometers [1][7, 8, 9, 10]. The concept of blackbodies, graybodies, and non-graybodies is necessary to understand how the radiation seen by the pyrometer for a particular temperature varies with target and process conditions.

Blackbodies

A blackbody is a theoretical surface that absorbs all energy (does not reflect or transmit energy). A blackbody is also a perfect radiating body. The total net energy radiated by a blackbody is proportional to the difference between the target temperature (T_t) and the sensor temperature (T_s) each raised to the 4th power as detailed in Equation 2-1a (Stephan-Boltzmann Equation). The total energy radiated is the area under the curve in Figure 2-1 at the given temperature. The parameter K takes into account the areas of the target and sensor and the distance between them. For high temperatures, the sensor temperature is considered negligible so this temperature is dropped in Equation 2-1b but the K factor remains to take into account the characteristics of the sensor [1][7, 8, 9, 10].

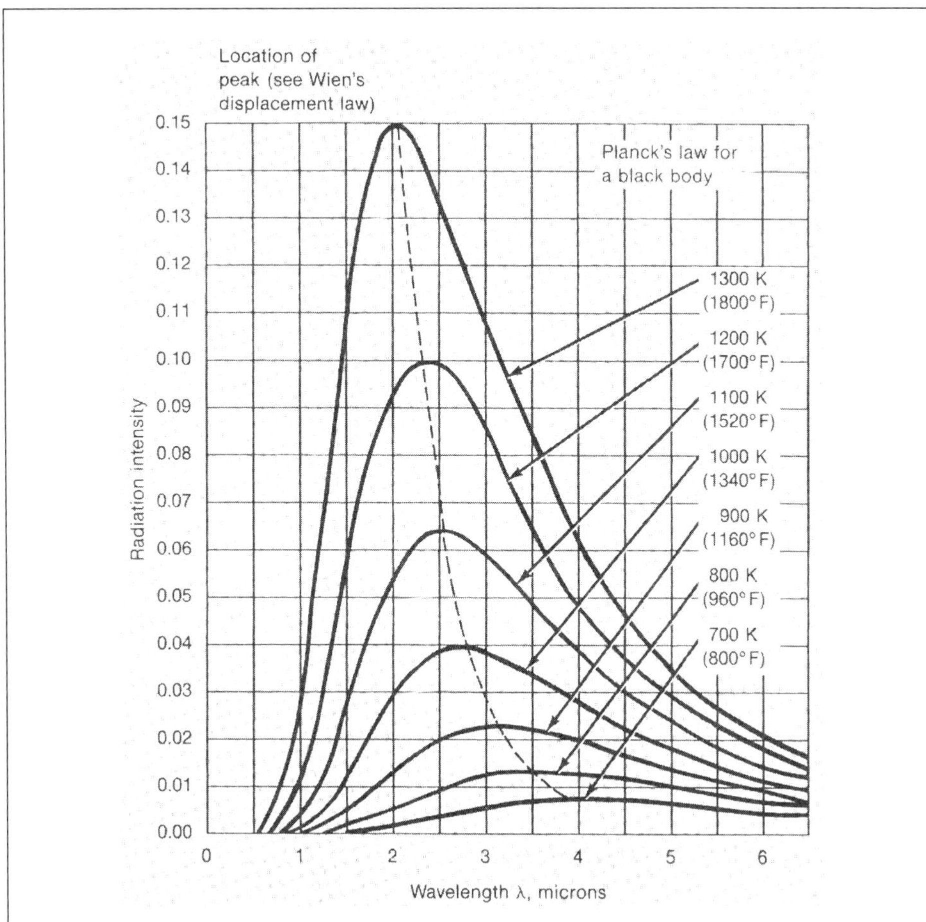

Figure 2-1 Blackbody Radiation Intensity as a Function of Wavelength and Temperature *(Source: Figure 8-3. Temperature Measurement in Industry [ISA, 1990])*

Emissivity and Emittance

Process temperature targets do not behave as blackbodies. In Equation 2-1b, the total emissivity term is introduced as a multiplier to take into account the non-ideal radiating behavior of materials (e.g., aluminum, carbon, Monel, steel, titanium, zirconium, plastic, paper). The total emissivity is the ratio of the total energy radiated by a material to the total energy radiated by a blackbody. The total emissivities of most materials encountered in process applications at various temperatures can be found in tables in books with chapters on pyrometers.

The emissivity is a physical property of the material and is specified for a sample with a highly polished surface. The targets in industrial equipment (e.g., furnaces, calciners, and kilns) have different roughness, geometry, and sizes and are often covered with oxides and corrosion products. To take into account the effects of installation and operating conditions, a factor called *emittance* is used instead of emissivity. The total emittance is the ratio of total radiation for the physical and operating conditions of the target to the total radiation from a blackbody at a given temperature. In Equation 2-1c total emissivity is replaced with total emittance. Tables are available to give some typical values of total emittance but the actual emittance depends upon application conditions and changes to the surface from coatings, corrosion, oxidation, and erosion during a production run.

Sensitivity

In Figure 2-1 for blackbodies the peak in the curve for each temperature shifts to the left as the temperature increases. The maximum change in radiation intensity for a change in temperature occurs near the peaks. To the left of the peak, the radiation intensity drops off suddenly. The change in intensity with temperature is appreciable until the intensity becomes close to zero. To the right of the peak, the radiation intensity tails off and the curves start to converge. The change in intensity with temperature becomes smaller and the pyrometer loses sensitivity. Figure 2-1 is for a blackbody [1][7, 8, 9, 10].

Graybodies

For graybodies, a fraction (e.g., 0.65 to 0.95) of the radiation intensity radiated by a blackbody is emitted. The fraction is fixed for all wavelengths for a given set of operating conditions. The shape of the graybody curve is similar to the blackbody curve and the peak would occur at the same wavelength but the graybody curve would be shifted below the blackbody curve as illustrated in Figure 2-2 by an amount that corresponds to the emittance of the graybody. The emittance can be a function of temperature and can vary with time but these changes are the same for all wavelengths. The graybody curve is simply the blackbody curve's values multiplied by the same emittance factor for all wavelengths. Equation 2-1c gives the total energy radiated for a graybody [1][7, 8, 9, 10].

Non-Graybodies

For non-graybodies, the fraction of the radiation emitted is variability and generally lower. The variation in emittance with wavelength and can result in multiple peaks as shown in Figure 2-2. The area under each of the peaks must be integrated to get the total radiation. We can get back to a simple general equation if we compute the spectral radiation (radiation at a specific wavelength) rather than the total radiation from all wavelengths. Equation 2-1d shows that a further modification of the Stephan-Boltzmann equation (that simply consists of raising the target temperature to the power N as a function of wavelength and temperature as defined by Equation 2-1e) provides a prediction of the spectral radiation for graybodies. If we make the emittance a function of wavelength besides temperature, we end up with Equation 2-1f, which provides a prediction of spectral radiation for non-graybodies.

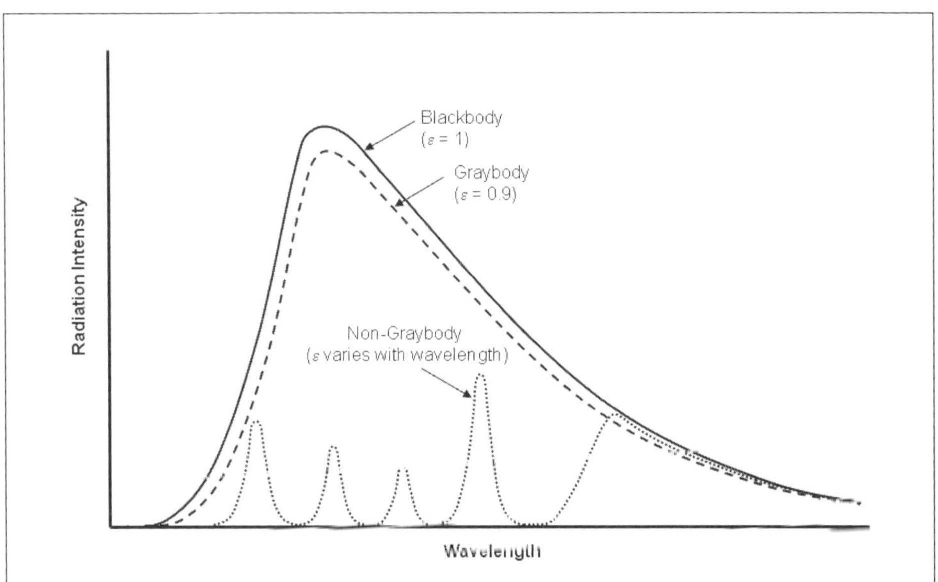

Figure 2-2. Radiation Intensity for Blackbody, Graybody, and Non-graybody [10]

Single and Two-color Pyrometers

In single-color and two-color pyrometers, wavelengths are chosen to provide a maximum change in radiation intensity with temperature but with minimum absorption by anything in the space between the target and the sensor. There is a peak in the absorption for the various gases, vapors, particles, and steam in the process equipment and the condition and material of the window into the process equipment. There is also a peak in the detector output. Ideally, the peak in radiation intensity and detector output coincide and are sufficiently separated from the peaks in absorption. As the maximum temperature increases, smaller wavelengths are desired because of the shift of the peak to the left in Figure 2-1 but this corresponds to peaks in absorption by gases, notably carbon dioxide.

Single-color or narrow band pyrometers use filters to narrow the view to a particular wavelength. In general a single-color pyrometer is more accurate than total radiation or broadband pyrometers (e.g., hand-held pyrometers) because smart instrumentation can take into account the known effects at a particular wavelength and a color can be chosen that is a better match for the application. However, there are often many unknown and time-varying effects, such as changes in the surface, the intervening vapors, and the viewing window, that introduce errors into the readings from a single-color pyrometer [1][7, 8, 9, 10].

Two-color or ratio pyrometers measure the radiation at two wavelengths. If the change in emittance at each wavelength with temperature is identical (graybodies), the effect of emittance can be cancelled out by ratio calculations. In reality, the change in emittance with temperature varies with wavelength (non-graybodies). Additionally, the change in emittance with changes in surface, operating conditions, and the composition of the intervening space may vary with wavelength.

In one comparison test on a blackbody, single-color and two-color pyrometers exhibited errors of 2 and 30°C (3.6 and 54°F), respectively [10]. Equal changes in emittance due to surface and operating conditions and intervening gases, particles, and vapors may make a two-color ratio pyrometer more accurate than a single-color pyrometer but it puts into question any accuracy statements for two-color pyrometers that are much better than 30°C. As a surface becomes closer to a blackbody and the absorption of radiation in the intervening space decreases, a single-color pyrometer may be best. The proper selection of the pyrometer type and wavelengths requires application experience or field trials.

Equations for Total and Spectral Radiation

Total radiation intensity from a ***blackbody*** target can be expressed by the Stephan-Boltzmann Equation with the factor K to account for differences in target and sensor area and the distance between them:

$$E(T_t) = K * \sigma * (T_t^4 - T_s^4) \quad \text{(2-1a)}$$

For a negligible sensor temperature and a ***graybody*** with a target material emissivity that is a function of temperature, the equation for total radiation becomes:

$$E(T_t) = \varepsilon(T_t) * K * \sigma * (T_t^4) \quad \text{(2-1b)}$$

The substitution of target emittance for material emissivity takes into account installation and application effects on a ***graybody*** target on the total radiation:

$$E(T_t) = \kappa(T_t) * K * \sigma * (T_t^4) \quad \text{(2-1c)}$$

The substitution of N for 4 for the exponent of the ***graybody*** target temperature, gives the radiation intensity at a particular wavelength (spectral radiation):

$$E(\lambda_i, T_s) = \kappa(T_t) * K * \sigma * (T_t^N) \quad \text{(2-1d)}$$

$$N = \frac{14388}{\lambda_i * T_t} \quad \text{(2-1e)}$$

The spectral radiation intensities of ***non-graybodies*** have emittances that depend upon both wavelength and temperature:

$$E(\lambda_i, T_s) = \kappa(\lambda_i, T_t) * K * \sigma * (T_t^N) \quad \text{(2-1f)}$$

Where:

$E(T_t)$ = total radiation at a given temperature (Watts/cm^2)

$E(\lambda_l, T_t)$ = spectral radiation at a given wavelength and temperature (Watts/cm^2)

$\varepsilon(T_s)$ = emissivity of a surface at a given temperature

$\kappa(T_s)$ = emittance of a surface at a given temperature

$\kappa(\lambda_l, T_t)$ = emittance of a surface for a given wavelength and temperature

K = factor for distance and size of target and type of detector

σ = Stephan-Boltzmann constant (5.669×10^{-12} W/cm^2°K^4)

T_t = target temperature (°K)

T_s = sensor temperature (°K)

λ_l = wavelength i (cm)

2-3. Thermocouples and RTDs

Sensing Element

As has been mentioned, the sensing element is located at the tip of the temperature sensor on an RTD, while the sensing element on a TC consists of the entire length of the TC's wires. The sensing element, which is constructed of metal, responds to the process temperature by generating a measurable resistance (RTDs) or voltage (TCs) signal.

Sensor Sheath

Most temperature sensors used in the process industry today have a sheath to prevent damage during handling and installation and from debris and solids. Furthermore, they are usually installed in a thermowell for additional protection and to meet the various material of construction requirements.

The sensor sheath, or cable housing, is constructed of metal and holds most of the component parts of the temperature sensor as shown in Figures 2-3a and 2-3b for an RTD. Typically magnesium oxide (MgO) sensor packing (also known as minerally insulated [MI]) surrounds the sensing element and is contained within the sensor sheath. The sensor packing decreases the impact of process vibration on the sensing element and thus ensures a more accurate measurement over time. The end of the sensor sheath is sealed with a fill (e.g., epoxy) that keeps moisture out of the sheath and away from the sensing element.

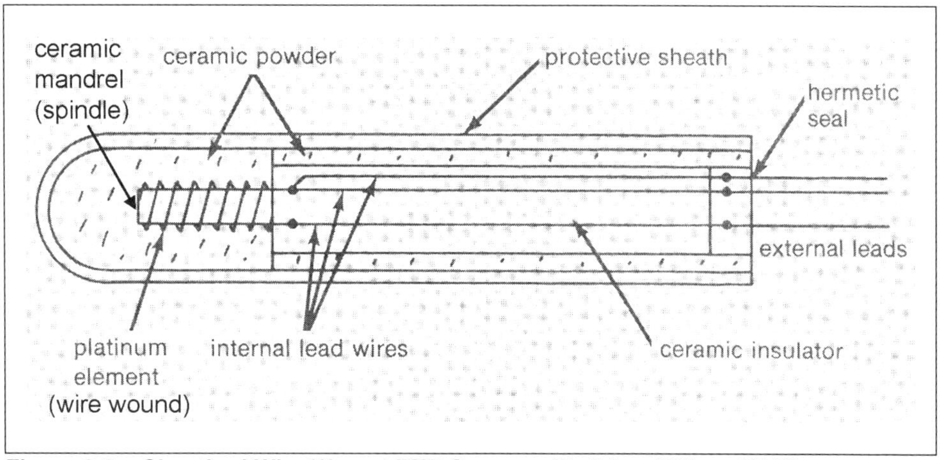

Figure 2-3a. Sheathed Wire Wound RTD Sensor with 3 Lead Wires [8] (*Source: Figure 6-1. Temperature Measurement In Industry – [ISA, 1990]*)

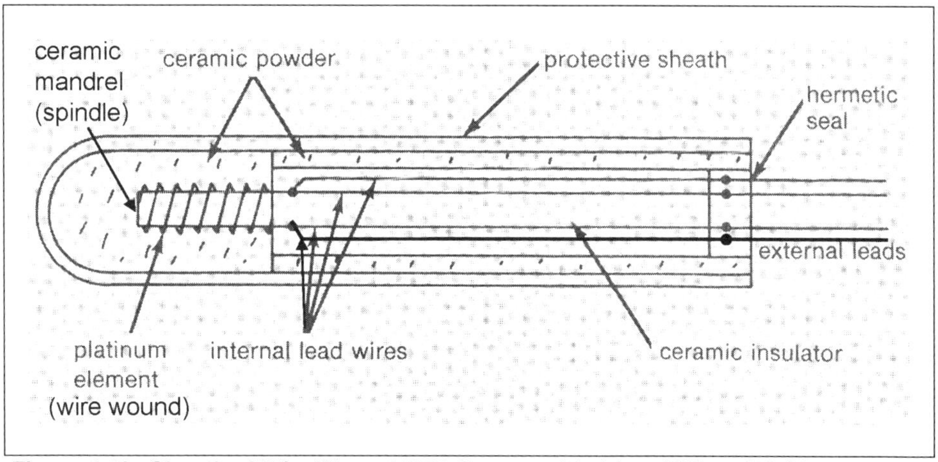

Figure 2-3b. Sheathed Wire Wound RTD Sensor with 4 Lead Wires [8] (*Source: Figure 6-1. Temperature Measurement In Industry – [ISA, 1990]*)

Resistance Temperature Detectors (RTD)

RTDs operate on the principle that the electrical resistance of a metal increases as temperature increases, a phenomenon known as *thermoresistivity*. A temperature measurement can be inferred by measuring the resistance of the RTD element. The thermoresistive characteristics of RTD sensing elements vary depending on the metal or alloy from which they are made.

Platinum

Platinum RTD elements are the most common type used in process industries. Platinum elements have high accuracy, high repeatability, and a high resistance change per degree of temperature change. In addition, the output of platinum RTD elements is highly linear throughout their temperature range.

Copper

Copper RTD elements are highly linear throughout their temperature range, but have limited accuracy and a narrower temperature range than platinum elements. Copper elements, which are less expensive, are most often used for measuring temperature in bearings and motor windings—applications in which accuracy is not critical.

Wire-Wound RTD

Wire-wound RTD sensing elements are constructed by coiling a platinum (or other resistance metal) wire inside (internally wound) or around (externally wound) a ceramic *mandrel* (spindle). Most RTD sensors for the process industry are internally wound and sheathed for protection as shown in Figures 2-3a and 2-3b.

A dual-element wire-wound RTD can be created by coiling a second set of wires inside or outside the ceramic mandrel. If connected to a second transmitter, a transmitter with dual-sensor capabilities, or to another distributed control system (DCS) card, a *dual-element sensor* increases the reliability of the temperature measurement.

Wire-wound RTD elements are very sturdy and reliable. Compared to thin-film RTD elements, their accuracy tends to be higher, and their time response (how quickly the output reflects the temperature change) is several seconds faster than thin-film RTD elements. Wire-wound RTD elements work well for a wide variety of applications, although they may fail in high-vibration applications.

Thin-Film RTD

Thin-film RTD sensing elements are constructed by depositing a thin film of resistance metal onto a ceramic substrate (base piece) and trimming the metal to specifications. Sensing elements of thin-film construction are typically less expensive than those of wire-wound construction because less resistance metal is required for construction. However, thin-film RTDs tend to be less stable over time, typically have a more limited temperature range, and may be more susceptible to damage from rough handling.

Extension Lead Wires

To get an accurate temperature reading from an RTD, the resistance of the RTD sensing element must be measured. Each copper lead wire that connects the RTD sensing element to the resistance measuring device adds a small amount of resistance to the measurement. If this added resistance is ignored, an error is introduced and an inaccurate temperature measurement results. The error is referred to as the *lead wire effect*. The longer the wire run, the greater the error, or lead wire effect, reflected in the temperature measurement. To compensate for lead wire effect, three-wire and four-wire RTDs are used instead of two-wire RTDs. Three-wire RTDs are created by connecting one additional copper wire to one of the lead wires. Four-wire RTDs are created by connecting one additional copper lead wire to each of the existing lead wires. These additional wires are used by the transmitter to compensate for lead wire resistances.

The third wire compensates for the resistance of the lead wires based on the assumption that each wire has exactly the same resistance. In fact, there is a tolerance of 10% in the resistance of standard wires. The fourth wire compensates for the uncertainty in the resistance of wires. For example, 500 feet of 20 gauge cable would add 10 ohms which would cause a measurement error of 26°C (47°F) for a two-wire RTD. The 10% tolerance of the cable could create an error as large as 2.6°C (4.7°F) for a three-wire RTD [11]. For high accuracy applications or long extension wire runs, a four-wire RTD or a transmitter mounted on the thermowell (integral mount) should be used. The increased accuracy, stability, and reliability of microprocessor based transmitters, the redundancy of dual sensing elements, and the advent of secure and reliable wireless networks makes integral mounted transmitters an attractive option. Accessibility is less of an issue because maintenance requirements are drastically reduced. The transmitters rarely need recalibration, wiring problems are gone, and a failure of a sensing element does not create an emergency.

Once the resistance value is determined, it is converted to a temperature measurement. One of two conversion methods may be used by the transmitter:

- RTD standard (i.e., IEC 751 standard)
- Callendar-Van Dusen equation

IEC 751 standard

The IEC 751 standard describes an ideal relationship between the resistance of a platinum RTD and the temperature to which the RTD is subjected. For example, at 212°F (100°C), the IEC 751 standard shows that an ideal platinum RTD (one that exactly matches the IEC 751 standard) would have a resistance value of 138.5 ohms.

When a transmitter or control system accepts a resistance signal from a platinum RTD, the IEC 751 standard curve is often used to translate that resistance signal into a temperature reading. However, since actual RTDs are never ideal, they do not match the resistance versus temperature relationship as described in the IEC 751 standard. The difference between the actual RTD curve and the ideal RTD curve results in a measurement error, which is referred to as a *sensor interchangeability error.*

The maximum allowable sensor interchangeability error at a given process temperature is defined by two IEC 751 standard classifications—Class A and Class B. These classifications are used to identify platinum RTDs. Figure 2-4 compares these two classes to the IEC 751 standard ideal.

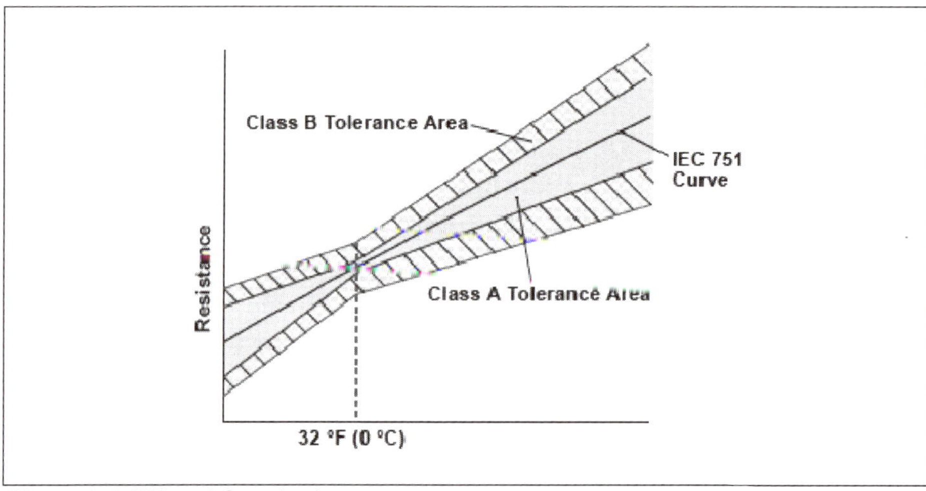

Figure 2-4. IEC 751 Standard

Callendar-Van Dusen Equation

The Callendar-Van Dusen equation offers an alternative to the IEC 751 standard. The equation, used in "Transmitter-Sensor Matching," can create a curve that approximates an RTD's resistance-temperature relationship. The Callendar-Van Dusen equation is:

$$R_t = R_O + R_O\alpha[(t-\delta)(0.01t-1)(0.01t) - \beta(0.01t-1)(0.01t)^3]$$

Where:

t = Temperature in °C

R_t = Resistance of the RTD at t

R_O = Resistance of the RTD at t = 0°C (a Callendar-Van Dusen constant)

α, β and δ = Callendar-Van Dusen constants

The Callendar-Van Dusen equation can be programmed into a transmitter so that the transmitter can use the actual RTD curve rather than an ideal curve (e.g., IEC 751 standard) to translate the sensor's resistance signal into a temperature value. The Callendar-Van Dusen equation provides a significant improvement in measurement accuracy, even when compared to Class A RTDs.

Thermocouples (TC)

A thermocouple (TC) consists of two wires of dissimilar metals (e.g., iron and constantan) that are joined at one end to form a *hot junction* (or sensing element). The temperature measurement is made at the hot junction, which is in contact with the process. The other end of the TC lead wires, when attached to a transmitter or volt meter, forms a *cold* or *reference* junction.

Thermocouple Types

Several types of TCs are available, each differing by the metals used to construct the element. While accuracies are better for type T and E compared to J, the type selected in industry often comes down to the plant standards and the application temperature range. Figure 2-5 shows the milliVolts generated over the approximately linear range of the following types of thermocouples:

- Type E—Chromel and constantan
- Type J—Iron and constantan
- Type K—Chromel and alumel
- Types R and S—Platinum and rhodium (differing in the % of platinum)
- Type T—Copper and constantan

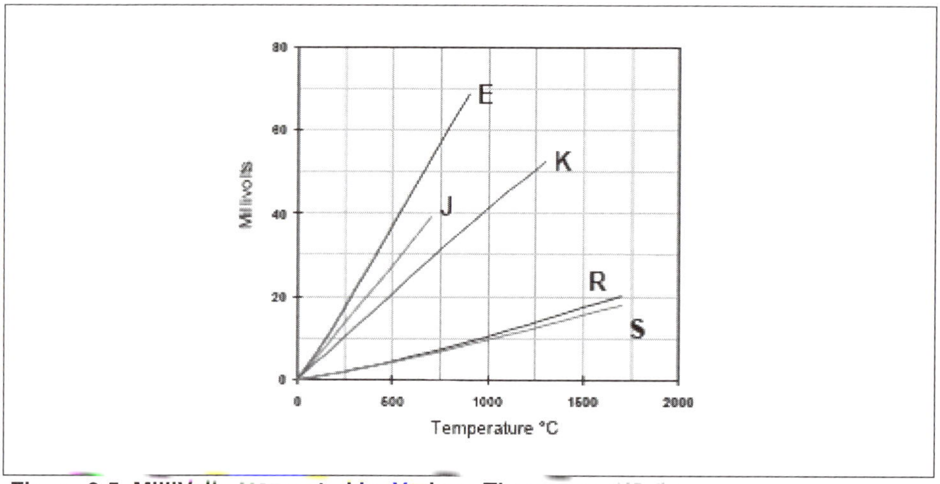

Figure 2-5. MilliVolts Generated by Various Thermocouple Types

Hot Junction Configurations

Junctions can be grounded or ungrounded to the sensor sheath. With dual-element TCs (two TCs in one sheath), the elements can be isolated or connected (unisolated). Each configuration offers benefits and limitations:

- *Grounded*—Grounding creates improved thermal conductivity, which in turn gives the quickest response time. However, grounding also makes TC circuits more susceptible to electrical noise (which can corrupt the TC voltage signal) and may cause more susceptibility to poisoning (contamination) over time.
- *Ungrounded*—Ungrounded junctions have a slightly slower response time than grounded junctions, but are not susceptible to electrical noise.
- *Unisolated*—Unisolated junctions are at the same temperature, but both junctions will typically fail at the same time.
- *Isolated*—Isolated junctions may or may not be at the same temperature. The reliability of each junction is increased, because failure of one junction does not necessarily cause a failure in the second junction.

Voltage Measurement and the Seebeck Effect

TCs use a phenomenon known as the Seebeck effect to determine process temperature. According to the Seebeck effect, a voltage measured at the cold junction of a TC is proportional to the difference in temperature between the hot junction and the cold junction. The voltage measured at the cold junction is commonly referred to as the Seebeck voltage, the thermoelectric voltage, or the thermoelectric EMF. As the temperature of the hot junction (or process fluid) increases, the observed voltage at the cold junction also increases by an amount nearly linear to the temperature increase.

If the hot junction temperature is held constant, an increasing cold junction temperature will produce a decreasing voltage, because the temperature difference between the hot and cold junction is decreasing. When the cold and hot junctions reach the same temperature, the observed voltage will be 0 V. The magnitude of the voltage signal produced at the cold junction also depends on the type of metals used to form the TC. Different combinations of metals, or different TC types, have different thermoelectric voltages at the same temperatures.

Cold Junction Compensation

As with RTDs, each type of TC has a standard curve. The standard curve describes a TC's voltage versus temperature relationship when the cold junction temperature is 32°F (0°C). As mentioned, the cold junction is where the TC lead wires attach to a transmitter or volt meter. Because the voltage measured at the cold junction is proportional to the difference in temperature between the hot and cold junctions, the cold junction temperature must be known before the voltage signal can be translated into a temperature reading. The process of factoring in the actual cold junction temperature (rather than assuming it is at 32°F [0°C]) is referred to as cold junction compensation (CJC).

2-4. Selection of a Temperature Sensor

Why Use an RTD Rather Than a TC?

The main reasons for selecting RTDs rather than TCs are as follows:

- RTDs have better sensitivity, repeatability, and stability.
- RTD signals are less susceptible to noise (higher signal-to-noise ratio).
- RTDs have better linearity over temperature ranges.

- RTDs can use the Callendar-Van Dusen equation to eliminate sensor interchangeability error.
- Cold junction compensation and related errors are not associated with RTDs.
- RTD drift is predictable, while TC drift is erratic and unpredictable. In addition, TC drift errors can be large as a result of element poisoning and element oxidation at high temperatures.
- The changes that affect the output of an RTD or TC occur over time due to mechanical shock, poisoning, and temperature cycling. These changes can be eliminated by an in-line RTD calibration, an option not available for a TC.
- RTDs do not need special extension wire.

Why Use a TC Rather Than an RTD?

In summary, the main reasons for selecting TCs rather than RTDs are as follows:

- TCs function at higher temperatures than RTDs (above 1500°F [850°C]).
- TCs are typically less expensive than RTDs.
- TCs have a faster response time than RTDs but this is only a consideration for bare elements (direct immersion of the element without a thermowell) in very fast temperature change processes (process time lag < 10 seconds). Bare elements are not used in industrial chemicals or elevated temperatures due to safety concerns.

2-5. Specifications

The accuracy of a temperature measurement is based on the combination of the sensor used and the performance specifications of the transmitter. The accuracy ranges shown in Table 2-3 are what can be expected for various sensor/transmitter combinations.

Table 2-3. Range and Accuracy Specifications

RTD Sensor Type	RTD Sensor Reference	Sensor Range		Accuracy over Range	
		°C	°F	°C	°F
Pt 100 (_ = 0.00385)	IEC 751; _ = 0.00385, 1995	–200 to 850	–328 to 1562	± 0.10 to ± 0.30	± 0.18 to ± 0.54
Pt 100 (_ = 0.003916)	JIS 1604, 1981	–200 to 645	–328 to 1193	± 0.10 to ± 0.30	± 0.18 to ± 0.54
Pt 200	IEC 751; _ = 0.00385, 1995	–200 to 850	–328 to 1562	± 0.22 to ± 0.54	± 0.40 to ± 0.98
Pt 500	IEC 751; _ = 0.00385, 1995	–200 to 850	–328 to 1562	± 0.14 to ± 0.38	± 0.25 to ± 0.68
Pt 1000	IEC 751; _ = 0.00385, 1995	–200 to 300	–328 to 572	± 0.10 to ± 0.40	± 0.18 to ± 0.72
Ni 120	Edison Curve No. 7	–70 to 300	–94 to 572	± 0.10 to ± 0.30	± 0.18 to ± 0.54
Cu 10	Edison Copper Winding No. 15	–50 to 250	–58 to 482	± 1.00 to ± 3.20	± 1.80 to ± 5.76
Cu 100 (a=428)	GOST 6651-94	–185 to 200	–365 to 392	± 0.48	± 0.86
Cu 50 (a=428)	GOST 6651-94	–185 to 200	–365 to 392	± 0.96	± 1.73
Cu 100 (a=426)	GOST 6651-94	–50 to 200	–122 to 392	± 0.48	± 0.86
Cu 50 (a=426)	GOST 6651-94	–50 to 200	–122 to 392	± 0.96	± 1.73

TC Sensor Type	TC Sensor Reference	Sensor Range		Accuracy over Range	
		°C	°F	°C	°F
NIST Type B	NIST Monograph 175	100 to 300	212 to 572	± 3.00 to ± 6.00	± 5.40 ±10.80
(varies by input range)		301 to 1820	573 to 3308	± 0.75 to ± 1.54	± 1.35 to ± 2.78
NIST Type E	NIST Monograph 175	–50 to 1000	–58 to 1832	± 0.20 to ± 0.40	± 0.36 to ± 0.72
NIST Type J	NIST Monograph 175	–180 to 760	–292 to 1400	± 0.25 to ± 0.70	± 0.45 to ± 1.26
NIST Type K	NIST Monograph 175	–180 to 1372	–292 to 2502	± 0.25 to ± 1.00	± 0.45 to ± 1.80
NIST Type N	NIST Monograph 175	–200 to 1300	–328 to 2372	± 0.40 to ± 1.00	± 0.72 to ± 1.80
NIST Type R	NIST Monograph 175	0 to 1768	32 to 3214	± 0.60 to ± 1.50	± 1.08 to ± 2.70
NIST Type S	NIST Monograph 175	0 to 1768	32 to 3214	± 0.50 to ± 1.40	± 0.90 to ± 2.52
NIST Type T	NIST Monograph 175	–200 to 400	–328 to 752	± 0.25 to ± 0.70	± 0.45 to ± 1.26
DIN L	DIN 43710	–200 to 900	–328 to 1652	± 0.35 to ± 0.70	± 0.63 to ± 1.26
DIN U	DIN 43710	–200 to 600	–328 to 1112	± 0.35 to ± 0.70	± 0.63 to ± 1.26
w5Re26	ASTME 988-96	0 to 2000	32 to 3632	± 0.70 to ± 1.60	± 1.26 to ± 2.88
GOST Type L	GOST R 8.585-2001	–200 to 800	–392 to 1472	± 0.71	±1.28

If supported by the transmitter, sensor matching using the Callendar-van Dusen constants for an RTD sensor will improve the accuracy by 75%. Ambient temperature effect: 0.001% of span or < 0.007°C per 1°C (0.0013°F per 1°F) of temperature change for most sensor types. Stability will range from ±0.25% of reading for 5 years for RTDs to ± 0.5% if reading for 1 year for thermocouples.

2-6. Setup and Calibration

Calibration is a means to verify the proper operation of the sensor and the electronics receiving the sensor signal.

Sensors can be calibrated by exposing them to one or more known temperatures and measuring the sensor output. The output is then compared to that of a certified sensor probe to determine the calibration error. It is also possible to calibrate the combined measurement system by measuring the resulting reading of the transmitter or other measurement electronics.

Transmitters and other electronics measurement equipment such as system I/O cards can be calibrated by applying an electronically generated signal to simulate the voltage or resistance of a sensor. A wide variety of calibrators are available for this purpose.

Most suppliers of temperature sensors and transmitters will provide factory calibration for their products with associated documentation of the results.

2-7. Installation

The best practice for making a temperature measurement is to keep the length of the sensor wiring as short as possible to minimize the effect of electromagnetic interference (EMI) and other interference on the low level sensor signal. The temperature transmitter should be mounted as close to the process connection as possible.

To minimize conduction error (error from heat loss along the sensor sheath or thermowell wall from tip to flange or coupling), the immersion length should be at least 10 times the diameter of the thermowell or sensor sheath for a bare element. Thus, for a thermowell with a 1 inch (2.54 cm) outside diameter, the immersion length should be 10 inches (25.4 cm). For a bare element with a ¼ inch (6.35 mm) outside diameter sensor sheath, the immersion length should be at least 2.5 inches (63.5 mm). This is just a rule of thumb. Computer programs can compute the error and do a fatigue analysis for various immersion lengths and process conditions.

For high velocity stream and bare element installations, it is important to do a fatigue analysis because the potential for failure from vibration increases with immersion length.

The process temperature will vary with process fluid location in a vessel or pipe due to imperfect mixing and wall effects. For highly viscous fluids such as polymers and melts flowing in pipes and extruders, the fluid temperature near the wall can be significantly different than at the centerline (e.g., 10 to 30°C; 50 to 86°F) [12]. Often the pipelines for specialty polymers are less than 4 inches (101.6 mm) in diameter, presenting a problem for getting sufficient immersion length and a centerline temperature measurement. The best way to get a representative centerline measurement is by inserting the thermowell in an elbow facing into the flow (position 1 in Figure 2-6). If the thermowell is facing away from the flow, swirling and separation from the elbow as can create a noisier and less representative measurement (position 2 in Figure 2-6). An angled insertion (position 3 in Figure 2-6) can increase the immersion length over a perpendicular insertion (position 4 in Figure 2-6) but the insertion lengths shown for both are too short unless the tip extends past the centerline. A swaged or stepped thermowell can reduce the immersion length requirement by reducing the diameter near the tip.

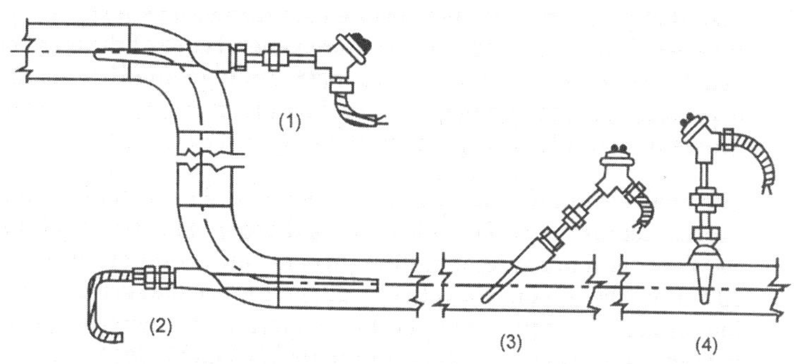

Figure 2-6. Ranking of Installations to Reduce Heat Conduction and Profile Errors
(Source: 614. Advanced Control Unleashed [ISA, 2003])

The distance of the thermowell in a pipeline from a heat exchanger, static mixer, or desuperheater outlet should be optimized to reduce the transportation delay but minimize noise from poor mixing or two phase flow. As shown in Figure 2-7, generally 25 pipe diameters is sufficient to ensure sufficient mixing after the recombination of divided flows from heat exchanger tubes or static mixer elements.

For desuperheaters, the distance from the outlet to the thermowell depends upon the performance of the desuperheater, process conditions,

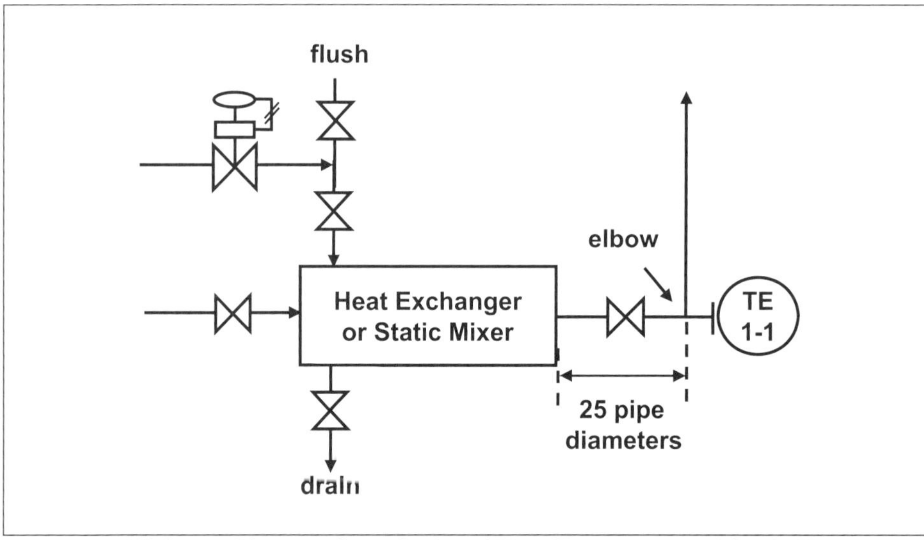

Figure 2-7. Minimum Distance from Outlet of Heat Exchanger and Static Mixer

and the steam velocity. To give a feel for the situation there are some simple rules of thumb for the straight piping length (SPL) to the first elbow and the total sensor length (TSL). Actual SPL and TSL values depend on the quantity of water required with respect to the steam flow rate, the temperature differential between water and steam, the water temperature, pipe diameter, steam velocity, model, type, etc. and are computed by software programs [13].

SPL (feet) = Inlet steam velocity (ft/s) * 0.1 (s)
SPL (m) = Inlet steam velocity (m/s) * 0.1 (s)

TSL (feet) = Inlet steam velocity (ft/s) * 0.2 (s)
TSL (m) = Inlet steam velocity (m/s) * 0.2 (s)

Typical values for the inlet steam velocity, upstream of the desuperheater range from 25–350 ft/s (7.6 to 107 m/sec). Below 25 ft/s there is not enough motive force to keep the water suspended in the steam flow. Water tends to fall out and run down the pipe to a drain. When this happens the water no longer cools the steam and the system thinks it needs to add more water, which compounds the problem. Problems can also include pipe wall erosion and high thermal stress gradients in the pipe wall (i.e., a hot top and cold bottom, which can crack welds or warp the pipe to an egg-shaped cross section). Current technology has an inlet velocity limitation of 350 ft/s (107 m/sec). Velocities higher than 350 ft/s cause the desuperheater to vibrate and damage the unit to the point where it breaks apart [13].

Thermowells

As mentioned, a *thermowell* is a closed-end metal tube that encapsulates and protects a sensor from process flow, pressure, vibration, and corrosion. Thermowells allow for the installation and removal of sensors without having to shut down the process. They are mounted in various ways to a process pipe or vessel.

Thermowells are available in several different materials, mounting methods, and stem types. The variety of design features renders thermowells suitable for various applications and environmental conditions. Three factors affect the choice of material:

- Type of corrosive environment the thermowell will be exposed to
- Temperature and pressure limits of the material
- Compatibility with the process piping or vessel material to ensure solid, non-corroding welds and junctions

Mounting Methods

Thermowells and sensing elements can be assembled and mounted as shown in Figures 2-8 through 2-9. Thermowells can be threaded, welded, or bolted (flanged style) onto the process pipe or vessel wall process connection.

Thermowells are most commonly threaded onto the process piping or vessel, which enables them to be easily installed and removed. Threaded thermowells are the weakest type of thermowell.

Welded thermowells are permanently welded onto the process pipe or vessel. Thus, removal is very difficult and requires cutting the thermowell out of the system. Welded thermowells are the strongest type of thermowell and are used with fluids of high velocity, high temperature, or high pressure. Welded thermowells are necessary for applications that require a leak-proof seal.

Flanged thermowells are bolted onto a pipe or vessel and can be easily removed or installed. Flanged thermowells are used in corrosive environments, as well as in high-velocity, high-temperature, or high-pressure applications. Flanged thermowells are the most expensive type of thermowell.

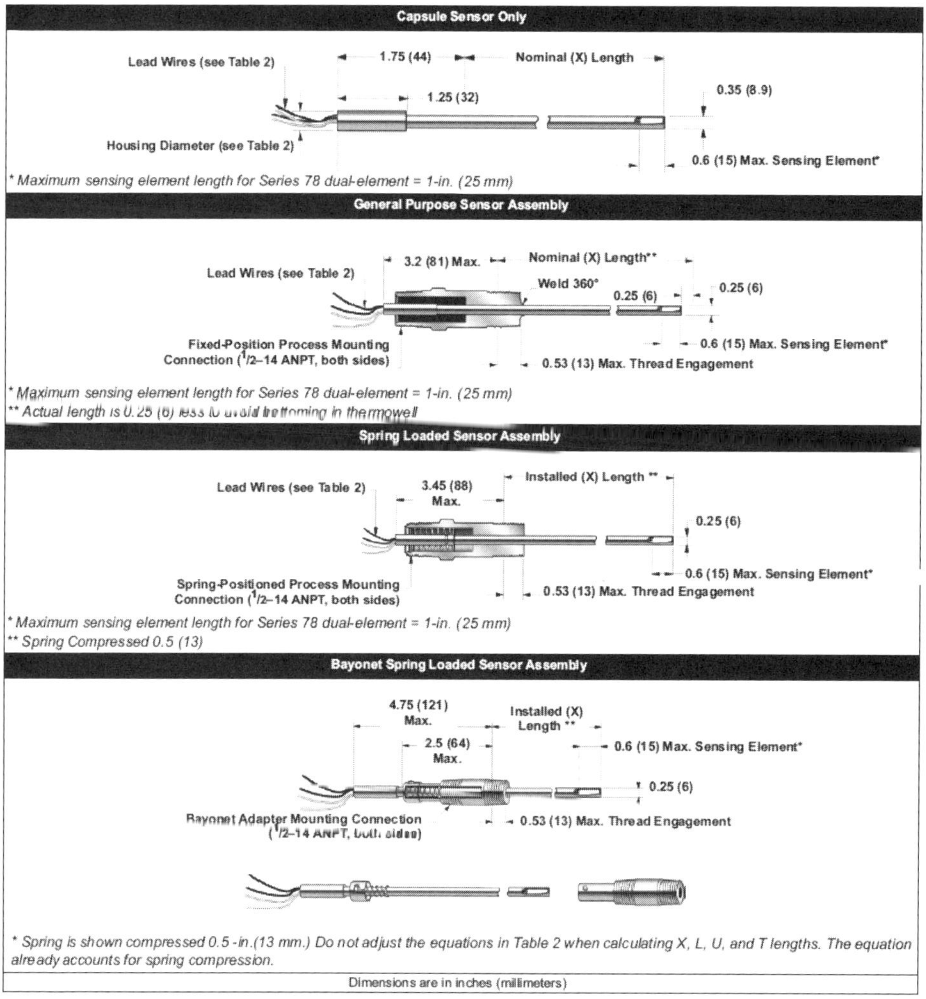

Figure 2-8. RTD Sensor Assembly Types

The mounting of the transmitter on the thermowell as shown in Figure 2-10 eliminates the need for cables between the element and transmitter and offers mobility for wireless transmitters important for finding the optimum temperature measurement location.

Stem Designs

The *stem* of a thermowell is the part that is inserted into the process stream. Stems can be tapered, straight, or stepped. The performance of a thermowell varies with its stem design. In general, a tapered or stepped stem provides a faster response, creates less pressure drop, and is less susceptible to conduction error and vibration failure. The choice of stem design is based on:

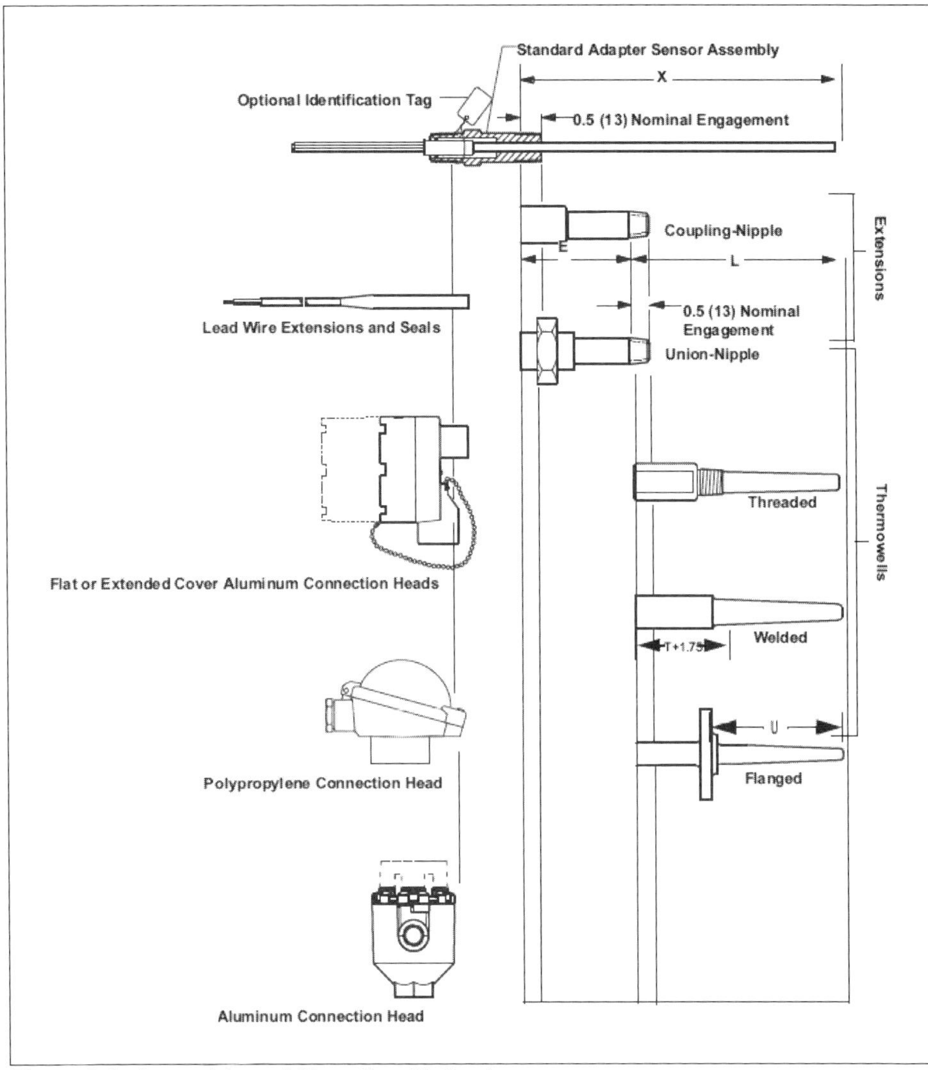

Figure 2-9. Assembly and Mounting Methods

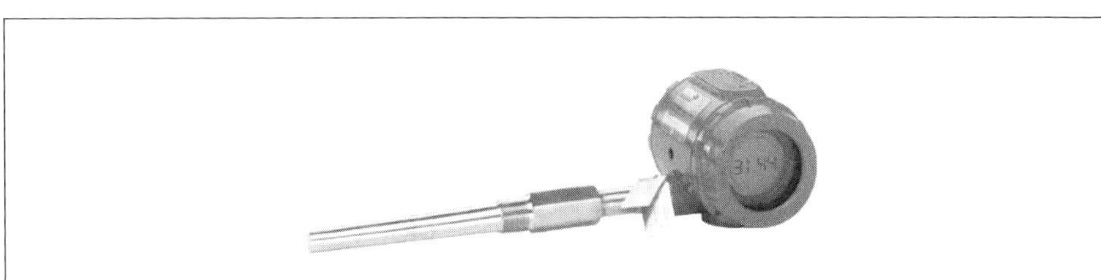

Figure 2-10. Transmitter Mounted on Thermowell

- Process pressure
- Time response required

- Permissible conduction error
- Wake frequency—frequency of alternating side-to-side movement of a fluid; depends on properties of the fluid
- Drag force—resistance to motion of a solid shape through a body of fluid
- Price

All stem designs perform equally well at equal process pressures. If the thicknesses of the thermowell walls are identical, thermowells with straight stems have the slowest time response because they possess the most material at the tip (largest diameter). Thermowells with stepped stems have the fastest time response because they possess the least material at the tip (smallest diameter). A small diameter also results in the least amount of drag force.

Thermowells with stepped stems also provide the maximum separation between the wake frequency (vortex shedding) and the natural frequency (oscillation rate determined by the properties of the thermowell itself). If the wake frequency is 80% or more of the thermowell natural frequency, resonance and probably damage can occur. Generally, thermowells with tapered stems are slightly more expensive as a result of a more complicated manufacturing process.

Communication

To be useful for control, safety, or monitoring applications, a temperature measurement signal must be communicated from the point of measurement to the control system of the process. The two most common ways are:

- *Transmitter*—the sensor is wired a short distance to the transmitter or connected directly to the transmitter, where its signal is converted to a digital, 4–20 mA, or wireless signal. The converted signal output is then communicated to the control system through transmitter wire or a wireless network.
- *Wired direct*—the sensor's lead wires are wired the entire distance to the control system. No signal conversion takes place along the route.

Three benefits of using a temperature transmitter over wiring directly to thermocouple and RTD input cards of control system are:

- *A more robust signal is delivered*—the 4–20 mA or digital signal output from the transmitter is much more robust than a sensor signal that is wired direct. Noise interference has less impact on 4–20 mA or digital signals.
- *Measurement accuracy is optimal*—Transmitters offer improved measurement accuracy over wiring direct. For example, sensors can be matched to transmitters (transmitter sensor matching), which improves the accuracy of the temperature measurement. The temperature span can be narrowed to match the process operating range (significant for older DCS with 12-bit input cards).
- *Time and money are saved*—Transmitter installation is often less expensive than wiring a sensor direct because of savings from cabling costs and installation (sensor wire, especially TC wire, is relatively expensive). Also, a robust signal and accurate measurements produce time and money savings through increased functionality and diagnostic capabilities of the transmitter.

Typically, the analog signal is linear with a process temperature measurement. Most transmitters now incorporate a microprocessor, which has improved their performance compared to analog designs. These smart (also called "intelligent") transmitters are able to compensate for ambient temperature variations and EMI, and provide cold junction compensation for thermocouples. Some also enable the transmitter to be matched to the characteristics of a specific sensor, providing very high accuracy. They are easily configured for a specific sensor type, providing an output that is linear over the temperature range. Intelligent transmitters with digital communications are also able to communicate diagnostic information about the health of the sensor and electronics.

Some transmitters provide the ability to connect more than one sensor. The most common transmitter of this type will support two sensors but has a single analog output. Coupled with dual sensor elements, this feature can be used to provide a more reliable measurement by comparing two measurements or the ability to switch to the second sensor in the event of the failure of the first.

The use of digital communications allows the additional flexibility of using a single transmitter to make more than one temperature measurement and communicate these back to the control system. There are temperature devices designed to specifically take advantage of this capability, providing the ability to measure four, or eight, or potentially more individual temperatures. The most common communication techniques are the HART® (including WirelessHART™), FOUNDATION™ Fieldbus and Profibus PA standard protocols.

The reliability, security, and ease of setup of WirelessHART (Highway Addressable Remote Transducer) networks combined with increased battery life from new communication rules and PID enhancements have made wireless communication an excellent option [14]. Since temperature changes in most processes are quite slow, the refresh time can be set longer than for other types of loops, extending battery life. Also, the noise amplitude and period in temperature loops is usually quite small compared to other loops unless there are two phases (e.g., liquid and gas) or poor mixing (e.g., poor uniformity — increased variability due to insufficient agitation), decreasing the number of exception updates triggered by noise, which also extends battery life.

Field-Mount Transmitters

Field-mount transmitters are the most rugged of all transmitter styles. Their robust housings protect against corrosion and humidity. Some field-mount transmitters house the electronics in dual-compartment housings, which completely isolates them from the effects of humidity. Dual-compartment transmitters are the best design for use in harsh environments. Field-mount transmitters can be integrally or remotely mounted.

- *Integral mount*—the transmitter is threaded onto the sensor directly (mounts directly to U.S. style [1/2-inch NPT] sensor or DIN plate).
- *Remote mount*—the transmitter is mounted on a pipe stand or other support near the sensor. Remote mount is preferred when the measurement point is inaccessible or when the process environment is too harsh for the transmitter to be installed directly on top of the sensor.

Head-Mount Transmitters

Head-mount transmitters are small, puck-shaped transmitters. They are typically housed in a protective enclosure—a *connection head* for direct mounting or a *junction box* for remote mounting.

Rail-Mount Transmitters

Rail-mount transmitters are designed to be attached to a DIN-rail (G-rail or top-hat rail) or directly screwed onto a wall. Rail-mount transmitters are also designed for compact mounting, which allows for a number of transmitters to be mounted very closely together

Wiring Direct

As mentioned, *wiring direct* refers to wiring the sensor's lead wires back to the control system. Because the sensor's lead wire (and original signal) is traveling the entire distance from the point of measurement to the control system, care must be taken to avoid two key problems:

- *Noise*—TCs are especially sensitive to noise interference and extension wires must be routed around such sources as generators and motors.
- *Heat sources*—a large change in the ambient temperature can affect the sensor's signal as it travels to the control system.

Extension Wire

TC extension wires are often used to wire a TC back to a control system or to a remote transmitter, which may be anywhere from 2–2000 ft (0.61–610 m) away. TC extension wire, with a few rare exceptions, must be of the same type of metal as the TC lead wires. If the metals do not match, the cold junction will be created prematurely. TC wire is relatively costly, particularly for platinum TC Types R and S. It is often not economically feasible to make continuous runs of TC wire (perhaps hundreds of feet long) from the hot to the cold junctions. To compensate for this problem, special lead wire is used that closely approximates the thermoelectric properties of TC wire. The special wire with less expensive material allows the user to minimize cost without sacrificing performance.

2-8. Maintenance

Table 2-4 lists checks that can be made to manually track down the source of a high, erratic, or low transmitter output.

Table 2-4. Temperature Measurement Troubleshooting Guide Example

High Output	Sensor Input	• Check for a sensor open circuit.
		• Check if the process variable is out of range.
	Loop Wiring	• Check for dirty or defective terminals, interconnecting pins, or receptacles.
	Power Supply	• Check the output voltage of the power supply at the transmitter terminals. It should be 12.0 to 42.4 V DC (over entire 3.5- to 23.0-mA operating range).
	Electronics Module	• Connect a Field Communicator and enter the transmitter test mode to isolate module failure.
		• Connect a Field Communicator and check the sensor limits to ensure calibration adjustments are within the sensor range.
Erratic Output	Loop wiring	• Check for adequate voltage to the transmitter. It should be 12.0 to 42.4 V DC at the transmitter terminals (over entire 3.5- to 23.0-mA operating range).
		• Check for intermittent shorts, open circuits, and multiple grounds. • Connect a 375 Field Communicator and enter the loop test mode to generate signals of 4 mA, 20 mA, and user-selected values.
	Electronics Module	• Connect a Field Communicator and enter the transmitter test mode to isolate module failure.
Low Output or No Output	Sensor Element	• Check if the process variable is out of range.
	Loop Wiring	• Check for adequate voltage to the transmitter. It should be 12.0 to 42.4 V DC (over entire 3.5- to 23.0-mA operating range).
		• Check for shorts and multiple grounds.
		• Check for proper polarity at the signal terminal.
		• Check the loop impedance.
		Connect a Field Communicator and enter the loop test mode.
		• Check wire insulation to detect possible shorts to ground.
	Electronics Module	• Connect a Field Communicator and check the sensor limits to ensure calibration adjustments are within the sensor range.
		• Connect a Field Communicator and enter the transmitter test mode to isolate an electronics module failure.

The use of intelligent transmitters can simplify the process of diagnosing problems, since they are continuously monitoring both the sensor and the electronics. Transmitter diagnostics typically include indication of an open or shorted sensor. They can also detect if the measurement is outside of the sensor range or beyond the range limits of the analog output.

As mentioned, more advanced transmitters add diagnostic capabilities such as the ability to have a second sensor input back up the first in the event of failure, and also to compare the inputs from two sensors to determine if one is drifting. Advanced diagnostics are available that can detect if a sensor's performance is deteriorating even before the condition affects the measurement, allowing the sensor to be replaced or the wiring corrected before the process is impacted.

2-9. Questions

1. Why do color pyrometers perform better in the lab than in industrial applications?
2. What type of sensor is needed to control reactor or column temperature to within 0.2°C of a 100°C set point?
3. How is the wake and natural frequency of a thermowell determined?
4. What is the most common type of thermocouple used in industrial processes?
5. Is the faster response of the thermocouple element important?
6. How do I ensure the reliability of my temperature measurement?

2-10. References

1. McMillan, G. K. and Toarmina, C. M. *Advanced Temperature Measurement and Control.* Research Triangle Park: ISA, 1995.
2. Loeffler, R. E. "Thermocouples, Resistance Temperature Detectors, Thermistors - Which?" *Instruments and Control Systems,* May 1973.
3. Kardos, P. W. "Response of Temperature Measuring Elements." *Chemical Engineering,* Aug. 1977.
4. Buckley, P. S. "Dynamics of Temperature Measurements", paper from Process Annual Symposium, Instrumentation Process Industry, Wilmington, Vol. 34, 1979.
5. Stroik, E. R. "RTDs are Sturdier Than You Think." *Instruments & Control Systems,* Jun. 1980.
6. Umrath, E. "Pt RTDs – What are They? Who Needs Them?" *Power,* Jun. 1970.
7. Kerlin, T. W. and Shepard, R. L. *Industrial Temperature Measurements.* Instructional Resource Package (IRP), Student Text. Research Triangle Park: ISA, 1982.
8. Magison, E. C. *Temperature Measurement in Industry.* Research Triangle Park: ISA, 1990.
9. Omega Corp. "Non-Contact Temperature Measurement." *Transactions in Measurement and Control, Vol. 1, 2nd Edition,* Omega Corporation, 1998.

10. Benedict, R. P. *Fundamentals of Temperature, Pressure, and Flow Measurement, 3rd Edition*. John Wiley & Sons, 1984.

11. Trietley, H. L. "Avoiding Error Sources in Platinum Resitance Temperature Detectors." *InTech*, Feb. 1982.

12. Yazbak, G. and Diraddo, R. W. "An Inside Look at Extrusion Melt Temperatures." *Plastics Technology*, Jun. 1993.

13. McMillan, G. K. and Weiner, S. "Straight Talk." *Control Talk*, *Control*, Jan. 2008.

14. McMillan, G. K. "Is Wireless Process Control Ready for Prime Time?", *Control*, May 2009.

Common Reference Information, Conversions, and Tables may be found at http://www.emersonprocess.com/rosemount/document/tds/3061b00q.pdf

3

Pressure Measurement

3-1. Learning Objectives

A. Appreciate the importance of pressure in process safety and performance.

B. Understand the implications of pressure dynamics and control band.

C. Gain an overview of the role of pressure in pipelines and processes.

D. Learn about the latest technological advances in transmitters and sensors.

E. Find out how line pressure and temperature affect transmitter accuracy.

F. Understand important aspects of transmitter installation.

G. Gain some insight into online diagnostics.

H. Learn basic location and installation requirements.

I. Learn terminology and issues to intelligently discuss industrial applications.

3-2. Introduction and Overview

Requirements

In process industries, pressure measurements play an important role in maximizing efficiency, capacity, and quality while protecting people, equipment, and the environment. In chemical, pharmaceutical, and petroleum plants, the process equipment is closed rather than open to the atmosphere to isolate the contents for personnel and environmental protection and to maintain product purity; food and pharmaceutical equipment is often closed to prevent contamination. The pressure in closed equipment should be monitored and regulated for safe and efficient operation. In many processes, there are optimum pressures. For columns and reactors, the control band (allowable peak to peak fluctuation around set point) may need to be as small as 0.1% for best process performance.

The process dead time and time constant in a liquid pressure control loop are often less than a second. For gas pressure, the process dead time is usually only slightly larger. The gas pressure time constant may be 6–60 seconds but a high process gain and/or tight control requirements can result in the process pressure zipping through the control band in less than a second.

Most of the dead time and time lag in a loop is in the control valve and the control system unless a transmitter damping setting is mistakenly set too high. For a controller tuned for maximum disturbance rejection, the peak error is proportional to the dead time, and the integrated absolute error (IAE) is proportional to the dead time squared [1]. Distributed control system (DCS) scan and execution times are reduced to less than a second, and variable speed drives are used in liquid pressure and in some gas pressure control loops to prevent peak errors from triggering trips or integrated errors from causing poor performance (see Appendix C) [2, 3].

The speed of the pressure measurement is important for safety instrumented systems, which is another reason for the elimination of sensing lines (impulse lines) by integral mounting of the pressure transmitter on the pipeline or equipment as discussed in Chapter 1.

The speed and accuracy of modern pressure transmitters are up to the task. Today's best transmitters have an accuracy of 0.02% and a time lag of 100 milliseconds.

Combustion Applications

In calciners, dryers, furnaces, and kilns, the pressure is kept slightly negative to prevent combustion products and dust from blowing out of the seams and joints. This equipment is inherently leaky due to rotating parts and ductwork. If the pressure is too negative, too much air is sucked into the equipment, which increases the heat requirements. The ideal pressure difference is often a fraction of an inch of water column.

Bio-Pharmaceutical Applications

In bio-pharmaceutical processes, positive pressures are used to prevent contamination by the infiltration of microorganisms. An increase in pressure in a bioreactor increases the dissolved oxygen, which can reduce the amount of oxygen that needs to be added. However, higher pressures in cell cultures cause an increase in the dissolved carbon dioxide concentration. The resulting decrease in pH causes an increase in sodium bio-carbonate that can lead to an increase in cell pressure from higher sodium concentrations.

Pipelines

Most pipelines in a production facility are pressurized. Gravity flow is rare. Compressors, fans, and pumps are used to provide the pressure that is the primary motive force for moving liquids and gases between equipment. Optimization strategies can save energy costs by lowering the discharge pressure set point of a prime mover until a control valve that supplies flow to a user goes below a minimum throttle position.

The pressure in headers is controlled to provide a constant pressure source. However, the pressure profile downstream changes with flow and time. The pressure drops from system resistances depend upon the positions of manual and automatic valves and the plugging, fouling, and frosting of equipment. For example, in the first field test of a new adaptive controller, a dramatic reduction in the process gain of the secondary temperature loop for exchanger "A" was diagnosed as a flattening of the upper installed characteristic (flow rate vs. degree of opening) of the coolant valve. The suspected cause of the loss of sensitivity was an increase in the system resistance. It was found that the manual coolant block valves had been partially closed because poor temperature control in the winter caused frosting of the heater exchanger [4]. The opening and closing of manual valves in the field can be a common occurrence. Usually, these valves have no limit switches or position feedback so their status depends upon a manual log entry. Exact positions are seldom known.

Pressure measurements are essential for tracking down changes to the pressure drop across a control valve and a system resistance. If the pressure changes in a pipeline, the flow through a control valve will change even though the valve position has not changed. Pressure disturbances at control valves frequently occur, unless there is a pressure loop or regulator immediately upstream of the control valve and the downstream pressure is constant (a rare case). To make things more interesting, there are usually no upstream or downstream pressure measurements at the control valve. Without these pressure measurements and actual position feedback or a flowmeter, disturbances to the flow and changes in the system resistances are largely unknown.

Column and Reactor Performance

The pressure is controlled in separation equipment (usually columns) to set equilibrium conditions and in reactors to control reaction rates. Control at a lower pressure in distillation columns lowers the boiling point and increases the efficiency of the separation. Pressure drop measurements are used to detect flooding and weeping. Higher pressures in gas phase reactors increase the rate of the reactions. However, too high a pressure can trigger undesirable side reactions.

Wireless Profiles

There is an opportunity for wireless pressure transmitters integrally mounted on pipelines and equipment to monitor and analyze pressure profiles. If project planners have the foresight to provide pressure connections in a variety of potential problem areas, wireless transmitters can be moved as needed to track down and diagnose increases in system resistances [5]. For example, the plugging and fouling of columns, filters, and catalyst beds can be monitored and the time until cleaning or catalyst regeneration predicted. In crystallization processes, the monitoring of pressures can be used to schedule the defrosting of the equipment. Since fouling rates tend to vary with application and accelerate late in the run since the last cleaning, regeneration, or defrosting, the use of integrally mounted wireless transmitters enables pressure measurements to be moved for the most opportune location and time.

Safety

Nearly all process equipment has a pressure limit. Pressure plays an important role in plant safety. Pressure loops in the basic process control system are designed to prevent a trip on high pressure. Pressures are used in safety instrumented systems to provide the next layer of defense to prevent the opening of relief valves or the bursting of rupture discs.

3-3. Principles

Electronic pressure transmitters convert process pressure into a digital or analog electrical signal. Electronic transmitters have two basic parts:

- Sensor
- Electronics

The sensor of an electronic pressure transmitter physically responds to changes in input pressure. The sensor converts the physical movement into an electrical property, such as capacitance, voltage, or inductance. The electronics part of the transmitter changes the output of the sensor into a standard electrical signal.

The most widely used electronic signal in the process control industry is the 4–20 mA signal. The 4- to 20-mA signal of an electronic transmitter is calibrated to relate the lowest process pressure point to an output of 4 mA from the transmitter. The maximum process pressure is calibrated to the 20 mA output of the transmitter. Transmitter outputs between 4 mA and 20 mA correspond to process pressures within the operating range.

There are several types of sensors used with electronic pressure transmitters that include:

- Variable capacitance
- Piezoresistive
- Piezoelectric
- Variable inductance
- Variable reluctance
- Vibrating wire
- Strain gauge

Only the two most common types of sensors, variable capacitance and piezoresistive, will be discussed in detail.

Variable Capacitance

Pressure transmitters with a variable capacitance sensor are widely used in the process control industry.

Capacitance is the ability of a substance to hold an electrical charge. A capacitor consists of two conductive plates aligned with one another but not making contact. The space between the plates is filled with an insulating medium known as a *dielectric*. The variable capacitance sensor for most pressure transmitters uses oil as the dielectric medium.

Three factors affect the capacitance of a variable capacitance sensor:

- Surface area of the plates
- Insulating properties of the dielectric
- Distance between the plates

Considering the three factors that affect capacitance, only the distance between the capacitor plates is allowed to vary. The surface area and dielectric oil do not change.

The variable capacitor consists of a sensing diaphragm located between two capacitor plates with dielectric oil between the plates and diaphragm. Because the sensing diaphragm needs to be surrounded by the dielectric, it cannot be directly exposed to process pressure or fluid. Exposed directly to the process is the isolating diaphragm (Figure 3-1).

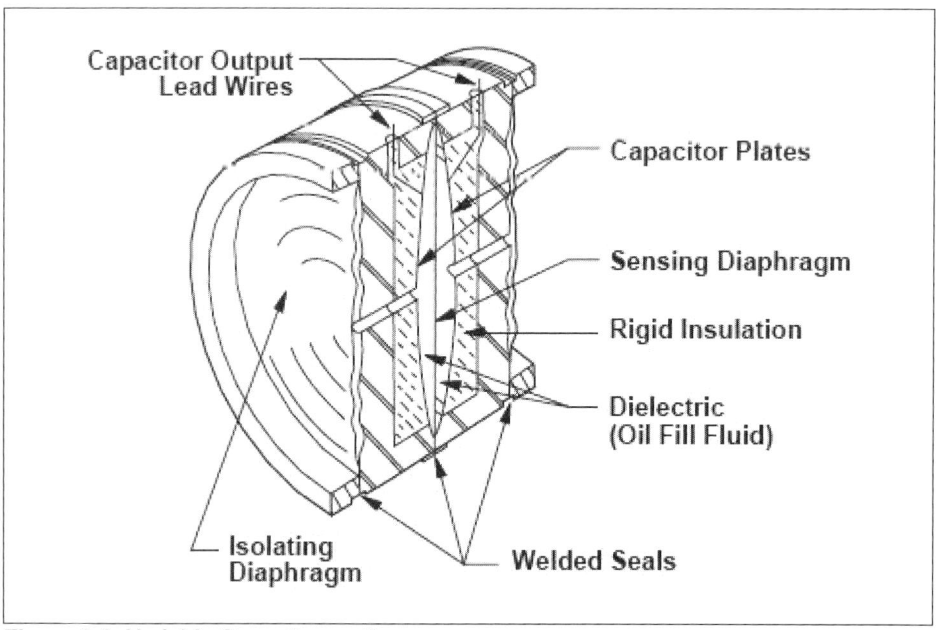

Figure 3-1. Variable Capacitance Sensor

During operation, the isolating diaphragms and dielectric fluid on the process and reference sides of the sensor transmit the process and reference pressure to the dielectric fluid. The fluid in turn transmits the process and reference pressure to the sensing diaphragm in the center of the sensor.

The sensing diaphragm deflects in response to the applied differential pressure. In gauge pressure transmitters, atmospheric pressure is transmitted to the low side of the sensing diaphragm. In absolute pressure transmitters, a reference pressure is maintained on the low side. The displacement of the sensing diaphragm is proportional to the differential pressure. Capacitor plates on both sides detect the position of the sensing diaphragm. The differential capacitance between the sensing diaphragm and the capacitor plates is converted electronically to an appropriate current, voltage, or digital output signal.

Variable capacitance devices are very durable, accurate, and vibration resistant. They can be calibrated to measure both large and small amounts of pressure. The design reduces the potential damage due to overpressure conditions. Variable capacitance devices can, however, be sensitive to changes in temperature.

Piezoresistive

Resistance is the amount of opposition to the flow of electricity exhibited by a particular conductor. A piezoresistor is a piece of metal or a semiconductor that exhibits a change in resistance when bent or stretched. In a piezoresistive transmitter, pressure is transmitted through a fill fluid from an isolating diaphragm to a piezoresistor (Figure 3-2). As the piezoresistor is flexed by pressure, its resistance changes. Typically, the piezoresistor is part of a Wheatstone bridge, a configuration of electrical conductors and resistors measuring changes in resistance. The change in resistance is then converted by the electronic components of the transmitter into a standard control signal, commonly a current, voltage, or digital control signal.

Piezoresistive devices are very responsive to pressure changes and are highly accurate. As with variable capacitance devices, piezoresistive devices can be sensitive to changes in temperature.

3-4. Selection

As has been mentioned, accurate measurement of liquid, gas, vapor, or steam pressure is a basic requirement for many industrial processes to operate safely, efficiently, and with optimum quality control. If differential pressure (DP) transmitters are used for level and flow besides pressure measurement, the number of DP transmitters used in a plant may be more than the total number of other types of transmitters.

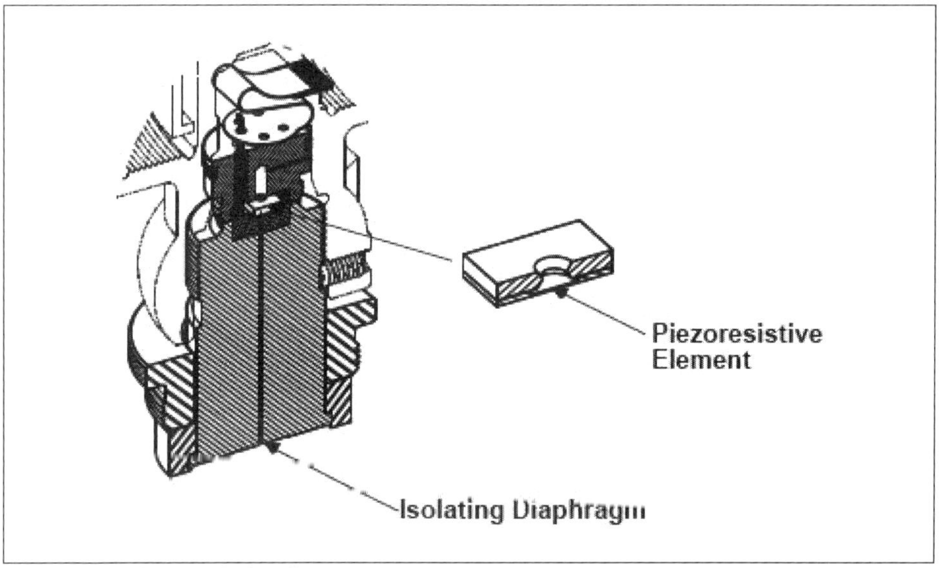

Figure 3-2. Piezoresistive Sensor

Pressure-measurement readings can be used to infer process variables. Examples include calculating flow rate, liquid level, and fluid density. Because a known relationship exists between pressure and density, pressure and level, and pressure and fluid flow through a pipe, these non-pressure process variables can be calculated from the pressure measurements.

For more information on inferring flow or level from pressure measurements, see their respective chapters.

3-5. Specification

Pressure-measurement devices can be categorized according to the measured reference pressure. *Reference pressure* is the pressure measurement that is compared to the measured process pressure that is of interest. The three reference pressures are:

- Absolute
- Gauge
- Differential

Absolute and gauge devices measure the difference between the pressure of the process fluid and a reference pressure—in gauge pressure, the reference pressure is atmospheric; in absolute pressure, the reference pressure is some lower pressure, often a specified vacuum. Differential devices take two pressure measurements of the process fluid at different points and measure the difference.

Differential pressure transmitters are most frequently used to measure level or flow. DP transmitters can be configured to measure differential, gauge or absolute pressure types. Pressure transmitters directly mounted on a pipeline (inline) or a vessel nozzle are most frequently used for measuring operating pressures. These transmitters can measure absolute or gauge pressure.

Performance

The reference accuracy of a pressure transmitter is generally specified as a percent of calibrated span. Even though reference accuracy is widely used for specifying transmitters, there's a fundamental problem: reference accuracy alone does not predict operating performance. Reference accuracy is measured under controlled laboratory conditions that bear little or no resemblance to real process conditions. For a true measure of performance, transmitters must be evaluated under more realistic conditions.

In a real process, a transmitter will not only exhibit error due to reference accuracy, but will also be affected by temperature changes and static pressure. Static pressure is the operating pressure for zero flow. For a DP with the high and low sides of the diaphragm connected to the diaphragm, the static pressure is on both sides. Line pressure is the operating pressure for a given flow. All manufacturers of transmitters publish errors (TPE) that occur due to reference accuracy, temperature, and static or line pressure in their product data sheets. These errors must all be taken into account to determine the actual error at operating conditions. This can be done with the following formula:

$$TPE = \sqrt{(reference\ accuracy)^2 + (line\ pressure)^2 + (temperature\ effects)^2} \quad (3\text{-}1)$$

A typical pressure transmitter includes a span turndown specification, which quantifies the magnitude the span can be decreased by and remain within the accuracy specification. For instance, if a transmitter has a 0 to 250 psi (0 to 1724 kPa) measurement range and a 100:1 turndown ratio, it can be calibrated to measure as low as 0 to 2.5 psi (0 to 17.24 kPa). The higher the turndown ratio, the more versatile a transmitter is. You can stock one transmitter for many different applications.

The response time of a measurement device is a simple concept that can get quite complex as you look into the details of how it is measured and specified. Response time for pressure measurement devices is a measure of how rapidly a transmitter's output responds to changing input pressures. A fast-responding transmitter is able to measure a dynamically changing process more accurately, enabling a user to control the process more tightly and improve plant efficiency. These plant efficiency gains are due to reduced process variation resulting in improved product quality and increased throughput, and in many cases, plant safety is also increased. A time constant (the time it takes the transmitter to respond to 63.2% of a step change in sensor output) is used to quantify the speed of a transmitter. A typical pressure transmitter's time constant is around 100 ms. A time constant may be inadvertently referred to as a response time (see Chapter 1 for the distinction between time constant and response time).

The stability, or drift, of a pressure transmitter is the percent of span error that is introduced over time. A typical value is 0.125% of span over three years. Obviously, a longer period of time and a lower drift value are desirable. This value helps determine the calibration period required for the desired performance.

There are other factors that affect a pressure transmitter's performance, such as linearity and hysteresis; they too should be taken into account when developing performance expectations.

Process Operating Range

Pressure transmitters often have published pressure and temperature operating ranges, and exceeding either of these metrics risks damage to the device. Most pressure transmitters have a process temperature operating range from -40°C to 80°C (-40°F to 176°F). For DP level, remote seals can be used for process temperatures above 80°C (176°F). For DP flow, impulse lines can be used to insulate the transmitter from temperatures exceeding 80°C (176°F). See the DP Flow and Level sections in Chapters 4 and 5, respectively, for more information.

Every pressure transmitter also has a published overpressure limit. This value varies with sensor technology and transmitter type and range, but the value is typically well above the upper range value of the device.

Wetted Materials of Construction

Certain parts of the pressure transmitter come in contact with process fluid. These materials are defined as wetted materials and the choice of the wetted materials for a pressure measurement device is driven by the application. Most manufacturers supply a wide range of materials, one of which should be suitable for the process conditions.

Typical wetted materials of construction include:

- Stainless Steel (SST)
- Alloy C-276
- Alloy K-500
- Plated Carbon Steel
- Tantalum
- Gold Plated SST

Process Connection Sizes and Types

Process connections vary by application and industry; they include threaded, flanged, and hygienic style connections among a variety of others.

Pressure-measurement devices come in two major model types:

- Differential pressure transmitters (Figure 3-3)
- Pressure transmitters (Figure 3-4)

Figure 3-3. Differential Pressure Transmitter

Figure 3-4. Pressure Transmitter

3-6. Setup and Calibration

The factory calibration of transmitters is generally a closely guarded secret with each manufacturer. How transmitters are calibrated at the factory determines how they will behave in the field and also their performance specifications (such as reference accuracy). That information is typically not made public, and varies between manufacturers.

Calibrating instruments in the field is needed for commissioning a device and for required maintenance. A desired outcome can often be obtained by either trimming or ranging the transmitter. Trimming is a calibration adjustment used to compensate for bias, drift, nonlinearity, and span measurement errors (see Chapter 1 for types of errors). The terms used for calibration adjustments vary with manufacturer. The more general terms are zero or offset and span or slope adjustment instead of various trim adjustments. Ranging is generally used to ensure the transmitter range covers the lowest and highest possible process pressures. For transmitters where the accuracy is truly in terms of percent of span, decreasing the span (the difference between the upper and lower ends of the range), can reduce measurement error. For many smart transmitters, the calculation used for the error in percent of span changes with range so the improvement is not proportional to the reduction in span.

"Trimming" is a minor adjustment to the transmitter to improve system accuracy. For instance, trimming would compensate for mounting position effects. There are two types of trims that can be performed:

- Sensor Trim
- Analog Output (AO) Trim

Sensor trim alters the transmitter so that the reported pressure matches the applied pressure. There are two sensor trims:

- Lower sensor trim (offset adjustment)
- Upper sensor trim (slope adjustment)

Lower trim adds or subtracts a value from the original digital pressure value. Lower trim is often done at a pressure lower than the upper sensor trim, but it can be higher. Zero Trim is a special form of lower trim. Used at 0 for applied pressure, it is the best way to eliminate mounting position errors.

Upper trim changes the slope of the value of the original digital pressure value. Upper trim is often done at a pressure higher than the lower sensor trim, but it can be lower. The upper trim is best used for high line pressure correction.

AO trim will adjust the actual transmitter output. The error to correct might be in the measurement system, external to the transmitter. There are two AO trims:

- AO Zero trim (zero adjustment)
- AO Gain trim (span adjustment)

"Ranging" is setting the 4 mA and 20 mA points to represent the application range of pressures.

3-7. Installation

Installation procedures vary with the type of desired measurement. Examples for fluid flow applications can be found in the DP flow section. Examples for level applications can be found in the DP level section.

Differential pressure transmitters are available today in two main configurations:

- Biplanar (Figure 3-5)
- Coplanar™ (Figure 3-6)

The biplanar design orients the process isolating diaphragms parallel to one another on either side of a sensor module, thus requiring two separate process flanges. Various process connections and instrument manifolds then bolt up to these two process flanges.

In the Coplanar design, the process isolating diaphragms are oriented in the same plane on the bottom of the sensor modules, creating a flat interface that requires only a single process flange. This orientation allows the sensor module to connect directly to numerous types of process connections. By eliminating the need for process flanges, the combined size and weight of the transmitter and manifold can be reduced, providing a weight reduction of over 10 pounds (4.5 kg).

Installation costs can also be reduced when the Coplanar instrument has an integrated manifold. The compact design also allows the transmitter to be mounted directly to the process, improving measurement performance by eliminating the need for impulse lines and reducing leak points by at least 50% when compared with biplanar design installations.

Figure 3-5. Biplanar Transmitter

Two typical mounting configurations for both differential and pressure transmitters are panel and pipe mounting (Figures 3-7 and 3-8).

3-8. Maintenance

The output signal (4–20 mA) discussed in this section is characteristic of analog transmitters. "Smart" transmitters were introduced late in the

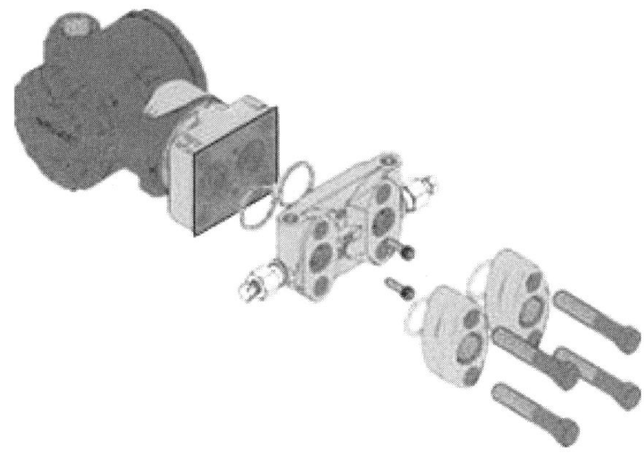

Figure 3-6. Coplanar™ Transmitter

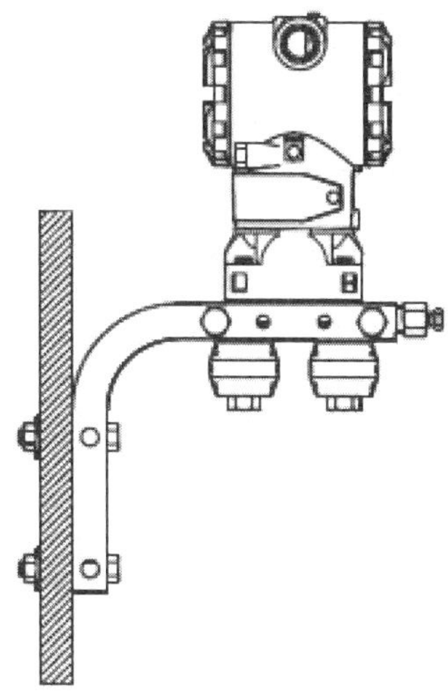

Figure 3-7. Panel Mount Transmitter

twentieth century and the market has shifted. HART transmitters are microprocessor based 4- to 20-mA transmitters that have a digital signal imposed on the analog output. The microprocessor has the intelligence to drive the mA output to certain values in advisory and failure conditions; the digital signal carries the on-line diagnostics to make troubleshooting easier. The diagnostics vary by manufacturer and by the generation of the pressure transmitter and the digital protocol, regardless of the application.

Pressure transmitter diagnostics can be classified into two groups: standard and advanced. Examples of standard diagnostics are diagnostics that

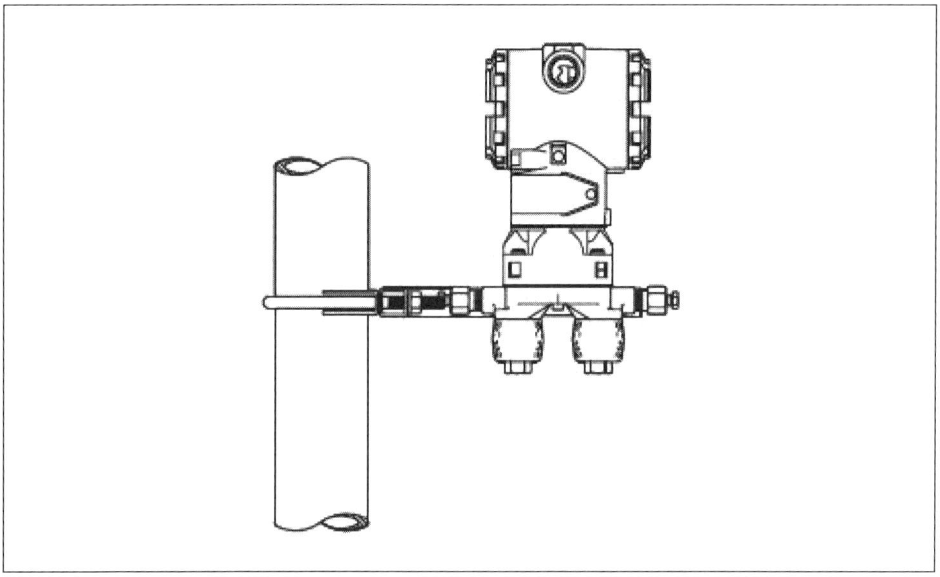

Figure 3-8. Pipe Mount Transmitter

check for sensor updates, as well as microprocessor and memory errors. The number of diagnostics and the details and thoroughness of the diagnostics typically vary by generation. Advanced diagnostics are typically diagnostics that are monitoring things outside the envelope of the transmitter. An example of advanced diagnostics is Statistical Processing Monitoring, a diagnostic that calculates the mean and standard deviation process noise. This diagnostic can be used for detecting a plugged impulse line or a problem such as entrained air or solids.

The amount and usefulness of diagnostics are protocol dependent. Pressure transmitters benefitted from the industry boom in diagnostics that started with the invention of HART (Highway Addressable Remote Transducer) smart transmitters. As with all other HART transmitters, pressure transmitters were originally limited to standard diagnostics as the amount of power or microprocessor bandwidth available was limited. With the introduction of low power consumption components, advanced diagnostics are now available in some HART pressure transmitters. Diagnostics included in these transmitters are statistical process monitoring (mentioned previously), variable logging, and diagnostic logging. Fieldbus transmitters are available with standard and advanced diagnostics, as there are no power constraints. More advanced Fieldbus transmitters have many function blocks that can be configured to be diagnostic in nature and detect anomalies outside the envelope of the transmitter.

The missing diagnostic for all pressure transmitters, regardless of protocol or measurement type, continues to be the oil fill system surrounding the pressure sensor. To date, there are no pressure transmitters that can detect the absence of, or damage to, the diaphragm or oil fill system. Applying pressure manually and/or calibrating the pressure transmitter are the only ways to verify it is working correctly.

3-9. Questions

1. Why are pressure measurements important in the process industry?
2. What unit operations typically require the tightest pressure control?
3. What other type of process measurements use DP transmitters?
4. What are some of the advantages of wireless pressure transmitters?
5. What can be done to reduce maintenance and improve the speed of response?
6. What advantages does a Coplanar™ transmitter offer?
7. What types of problems can be detected with a smart DP?

3-10. References

1. Blevins, T. L., McMillan, G. K., Wojsznis, W. K., and Brown, M. W. *Advanced Control Unleashed – Plant Performance Management for Optimum Benefits.* Research Triangle Park: ISA, 2003.
2. McMillan, G. K. *The Funny and Not so Funny Side of Life as an Automation Engineer* (CD-ROM), Research Triangle Park: ISA, 2009.
3. McMillan, G. K., and Weiner, S. "Sensible Sensor Speed - Part 1." *Control Talk, Control,* May 2009.
4. McMillan, G. K., Sowell, M. S., and Wojsznis, P.W. "The Next Generation – Adaptive Control Takes a Leap Forward." *Chemical Processing,* Sep. 2004.
5. McMillan, G. "Is Wireless Process Control Ready for Prime Time?" *Control,* May 2009.

4

Flow Measurement

4-1. Learning Objectives

A. Appreciate the importance of flow measurement in the process industry.

B. Understand the relative accuracy requirements of different operations.

C. Gain an overview of flowmeter capability and market.

D. Learn about the basic operating principles of modern flowmeter designs: differential pressure, multivariable differential pressure, vortex, magnetic, and Coriolis mass flow.

E. Find out the latest features.

F. Understand important aspects of selection and specification.

G. Gain an overview of performance capability.

H. Be introduced to the basic configuration, installation, and maintenance requirements.

I. Learn terminology and issues to be able to intelligently discuss industrial applications.

4-2. Introduction and Overview

The crucial role of flow measurements for the monitoring, control, and analysis of processes is best understood through the concept of streams. Streams are used on a process flow diagram (PFD) to define the quantitative performance of an industrial process. Streams (fluid flowing in pipes and ducts) convey the pressure, temperature, and composition of fluids.

How much of a stream that enters or exits equipment depends upon the flow rate of the stream.

Process equipment produces output streams that depend on input streams and utility streams. For example, the flows and concentrations of reactant feed streams and the flow and temperature of a coolant stream to the jacket determine the temperature and concentration of the discharge stream for a chemical reactor.

The measurement and regulation of the flow rates (often simply called *flows*) of streams are essential for individual equipment performance and for a plant to be able to efficiently and safely run to meet production requirements. These streams and their flows appear on the graphics for the operator interface to a control system. Sometimes temperatures are also measured and displayed. Unfortunately, compositions are rarely available except in the case where Coriolis meters are used to provide inferential measurements (see Section 4-7). The operator relies heavily upon flow measurements to run the plant. A typical plant will have hundreds of flow measurements.

Process engineers depend on accurate flow measurements for closing the material balance (making sure all flows are accounted for) and calculating process metrics. Flow measurements of raw materials and utility streams are important for computing the cost of goods. These flow measurements combined with the flow measurements of byproduct, intermediate, final product, and waste streams are important for determining process efficiency.

Process control depends on the control system's ability to manipulate flows through final elements. The predominant final element is a control valve. The use of a flow measurement for every control valve improves process control, process analysis, and process modeling. The opportunities offered by flow measurement can be best understood by looking at how we determine and characterize process stream flow through a valve.

Installed Valve Characteristics

The flow rate through a control valve depends upon the pressures upstream and downstream of the valve, fluid density and viscosity, valve position, and valve trim. The pressures upstream and downstream depend upon complex interrelationships between pressure sources (e.g., headers), prime movers (e.g., compressors, fans, and pumps), system resistances to flow (e.g., piping, heat exchangers, static mixers, filters, meters, manual valves, automated valves, and nozzles), and operating pressures of the destinations. There may be multiple sources and destinations, creating branches in the flow system. The pressures at the control valve may change rapidly with changes in source and destination pressures, prime mover operating speed, flow demand of users, and valve positions.

These pressures may also change more gradually with time due to changes in stream density and viscosity and the buildup of coatings or particles that increase system resistances. As a result, the pressures at the valve and consequently the flow through the valve are a time-varying nonlinear function of valve position. The total effect is captured by the installed characteristic of the control valve, which is a plot of the valve flow (flow rate through the valve) versus valve position. The curve can be based on the best process system information, but there is considerable uncertainty in the system parameters and operating conditions. The installed valve characteristic is only a best guess for a snapshot in time. Before position readback smart valve positioners, there was also considerable uncertainty as to actual valve position due to stick-slip and deadband.

The equations for flow through a control valve drastically change between turbulent and laminar flow (more on these terms later in the chapter). For the low flow rate of streams with high viscosities, such as 98% sulfuric acid for pH control, the flow can go from being turbulent to laminar at low valve positions (e.g., positions < 20%). Flows in the transition region between turbulent and laminar are extremely unpredictable. Valve sizes and types need to be chosen to ensure the flow stays laminar or turbulent with the full range of flow rate. Without a consistent flow measurement not affected by the transition (e.g., Coriolis), there is not enough awareness of the situation to solve the problem.

The slope of the installed valve characteristic curve at the operating point is the valve gain. In Equation 4-1, we see how the valve gain changes the open loop gain (commonly referred to as the *process gain*). The maximum controller gain is inversely proportional to the open loop gain. Controller gains higher than this maximum will result in excessive oscillation. Unless there is an adaptive controller that can identify and schedule the controller gain as a function of valve position, the controller must be tuned for the maximum slope of the installed characteristic for a given set of process conditions and measurement span. Operation at other points on the valve curve can be quite sluggish at low and high valve positions due to flattening of the installed characteristic. To keep the changes in valve gain less 4:1, the valve position typically needs to be kept within 5 to 75% for sliding stem (globe) valves and within 25 to 45 degrees rotation for butterfly valves from closed to open [1]. Figure 4-1 shows a snapshot of an installed valve characteristic curve and valve gain (slope) for a butterfly valve. In plants operating at higher than nameplate capacity, butterfly valves show a greater flattening of the curve and a corresponding drop in the gain (slope) below the low specification limit at a lower rotation (e.g., 30 degrees) than in Figure 4-1 because the pressure drop available for the control valve decreases more drastically with flow due to the higher flow resistance of undersized piping.

$$K_o = K_v * K_p * K_m \qquad (4\text{-}1)$$

Where:

K_m = measurement gain (CV % per kg/sec)

K_o = open loop gain for self-regulating processes (CV % per CO %)

K_p = process gain (1.0)

K_v = control valve gain (slope of installed characteristic) (kg/sec [lbf/sec] per CO %)

(CV % is percent of controlled variable scale and CO % is percent of controller output)

Cascade Control Opportunity

A flow control loop can be closed around the control valve to provide a correction of valve position to maintain a flow set point regardless of the shape of the installed valve characteristic. If the valve is manipulated by process control loops, the effect of the non-linearity and uncertainty of the installed characteristic is moved from the primary process variable (e.g., composition, level, pH, pressure, or temperature) to a secondary variable of flow. The result, as shown in the Figure 4-2a block diagram, is a cascade control system where the primary process controller output is the cascade set point of a secondary flow controller. Since flow controllers compared to other process controllers use much more reset action rather than gain

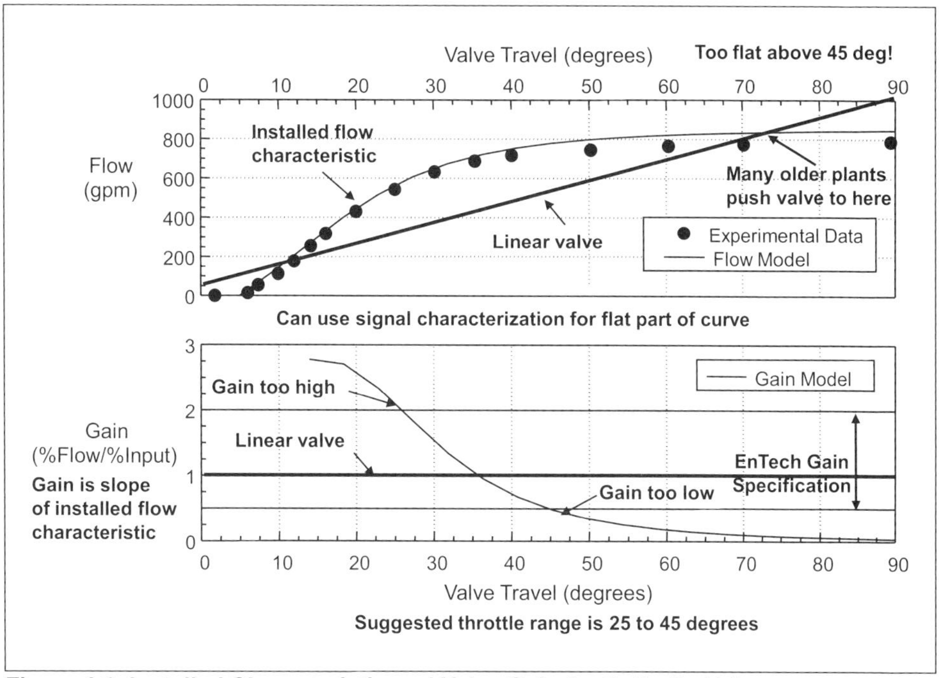

Figure 4-1. Installed Characteristic and Valve Gain for Butterfly Valve

action, the nonlinearity of the valve is less of a concern. The flow controller gain is well below the maximum for stability. Reset action is used to provide a gradual compensation of the nonlinearity. A flow loop also deals with the friction in the valve stem packing and trim seal (stick-slip), the backlash from linkages and connections (dead band), pressure disturbances, and changes in fluid density much faster than a primary process variable controller because the flow loop sees the effect within one second. The flow loop can correct for a time-variant, non-linear, and non-ideal valve response and for process upsets (e.g., pressure and density) before they appreciably affect the primary process variable. The correction is particularly effective if the rule for cascade control is followed in which the controllers are tuned to ensure the secondary flow loop is five times faster than the primary process variable loop [1]. When this rule is followed, the period of the oscillation in the flow loop is fast enough to be filtered out by the slow primary loop per Equation 1-3 in Chapter 1, where the filter time constant is the process time constant of the primary loop. Also, this rule guarantees there will not be an interaction and resonance between the primary and secondary loops.

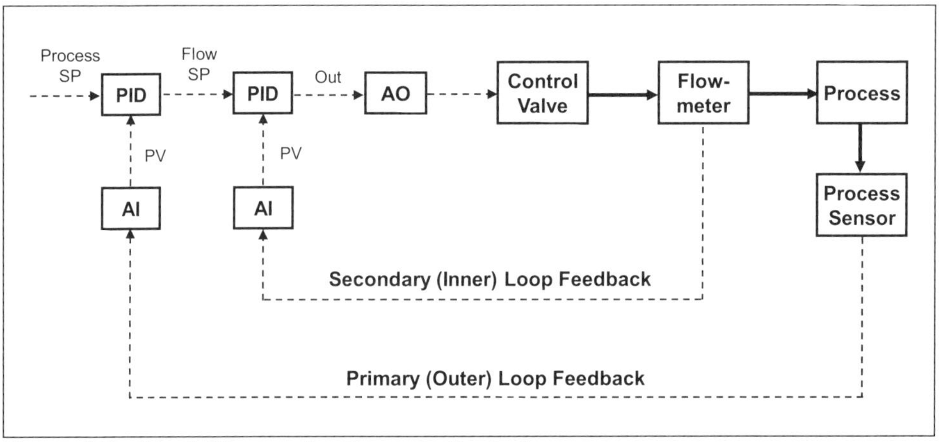

Figure 4-2a. Block Diagram for Cascade Control

Feedforward Control Opportunity

There are composition and temperature disturbances in the manipulated stream and in the other streams that affect the primary process variable. There are also flow disturbances from the other streams going into and out of the process. If a disturbance can be measured and the effect of the disturbance on the primary process variable can be calculated or identified, the measurement of the disturbance can be used to set the manipulated flow in anticipation of the disturbance's effect on the primary process variable. This preemptive action, taken before the feedback measurement of the primary process variable fully sees the disturbance, is called *feedforward control*. The disturbance is often simply multiplied by a feedforward gain and passed through dynamic blocks (deadtime, filter, and/or lead-lag blocks) before being added to the primary controller output. The dynamic blocks are set to help the correction arrive at the same point in the process and at the same time as the disturbance.

If there were no secondary flow loop and primary controller output went directly to set the valve position, the unknowns in the installed valve characteristic would translate to poor feedforward control. The feedforward gain would have to be set based on the maximum possible slope of the curve to avoid overcorrection. The result would be severe under-correction by feedforward action for all other operating conditions.

A secondary flow loop is important for feedforward control of all types of disturbances and all final elements that manipulate a flow. While the relationship between speed and flow is more linear for a variable speed drive than for a control valve, the flow through a compressor, fan, or non-positive-displacement pump also depends upon the slip of fluid between vanes or rotors and process conditions such as viscosity, density, and pressure.

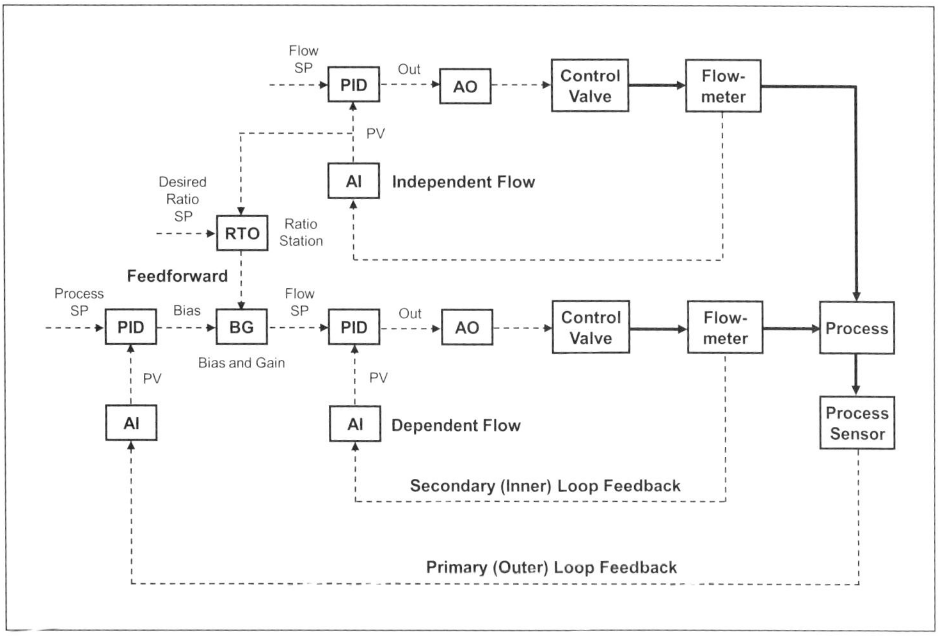

Figure 4-2b. Block Diagram for Ratio (Flow Feedforward) Control

The most frequent disturbance is a change in the flows of the other streams that affect the primary loop. These other flows could be wild or set by other control loops. Since the primary loop has no control over these flows, these flows are termed "independent." The flows manipulated by the primary loop are termed "dependent" flows [2]. Ideally, the independent flows are accurately measured so they are available for flow feedforward control. Here the independent flow is multiplied by a feedforward gain that is the ratio of the dependent flow to the independent flow. Figure 4-2b is a block diagram for the use of flow feedforward in cascade control. Figures 4-2c and 4-2d show examples of flow feedforward for pH control and temperature control. Some examples of flow ratios for feedforward control in the process industry are:

- Coolant/feed flow ratio for crystallizer, cooler, extruder, or exothermic reactor temperature control
- Steam/feed flow ratio for distillation column, evaporator, heater, dryer, or endothermic reactor temperature control

- Distillate/feed or reflux/feed flow ratio for column temperature control
- Reagent/feed flow ratio for pH control
- Additive/feed flow ratio for blend control (e.g., percent solids)
- Air/fuel flow ratio for boiler or furnace combustion control (oxygen control)
- Feedwater/steam flow ratio for boiler drum level control (three-element control)
- Blowdown/feedwater flow ratio for boiler drum total dissolved solids (conductivity control)
- Supply/demand flow ratio for header pressure control

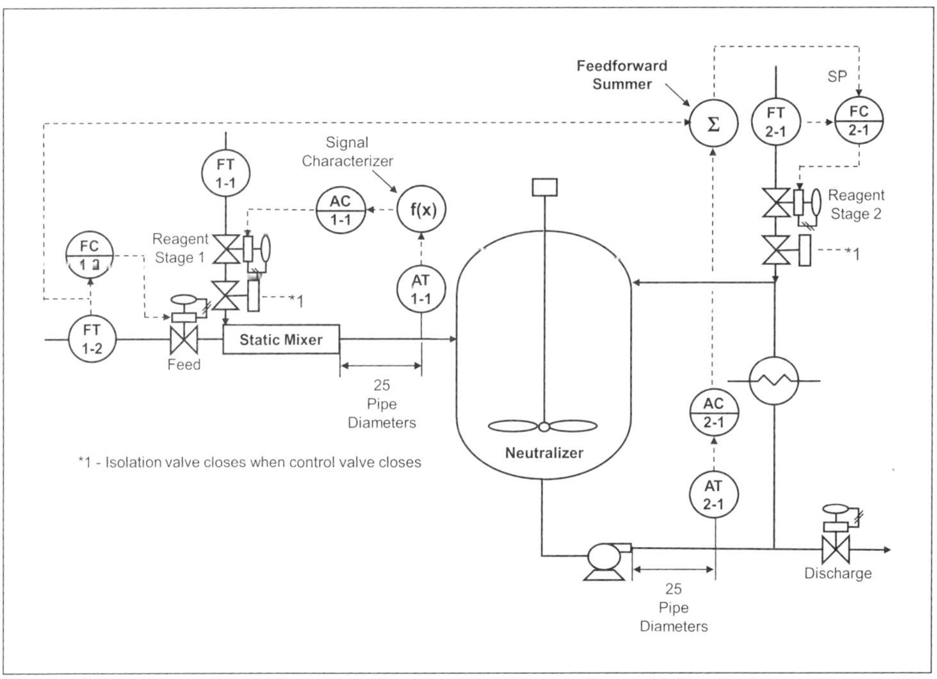

Figure 4-2c. Example of Flow Feedforward for Neutralizer pH Control

If the primary variable is sufficiently indicative of the desired process control condition, random errors (e.g., repeatability and noise) in the flow measurement are more important than the systematic errors. The primary loop will gradually correct for any offset in the flow measurements. However, better flow measurement accuracy translates to better feedforward control and a better response to disturbances. Also, in reactors there may

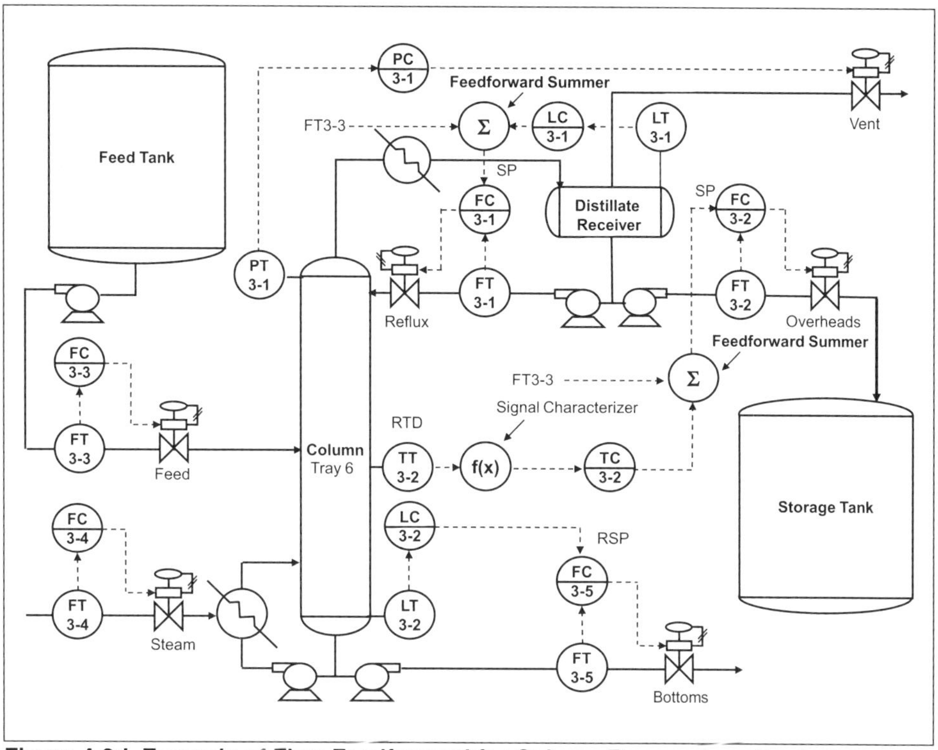

Figure 4-2d. Example of Flow Feedforward for Column Temperature Control

be undesirable side reactions or quality issues undetected by temperature when reactant ratios are not exactly right. Accurate reactant flows are also essential for accurate yield analysis.

Ratio Control

Ratio control is embedded in flow feedforward but is sometimes done without the feedback correction of the primary process loop. This is particularly the case for the many varieties of blend control where an online or at-line analyzer is not available. Ratio control accuracy, including systematic errors in the flow measurements, is important when there is no primary process loop.

Process Analysis

For billing and yield and cost-of-good calculations, accurate flow measurements are needed. As the cost-related importance of a stream increases, the need for flow measurement accuracy increases. The importance can be related to direct cost or to indirect cost from process impact. In other words, some additives have a low cost because the mass flow is low, but the impact on product quality may be high. Since saleable product flow is the ultimate goal, carries the highest value, and is used for accounting and billing, product flow measurements require the utmost accuracy.

Data analytics, such as principal component analysis (PCA), require flow measurements that are repeatable and linear. The use of valve positions instead of flow is generally unproductive because of all the reasons mentioned under installed valve characteristics. The use of split ranged valves is even more problematic for PCA because of the discontinuous and nonlinear behavior from the additional stick-slip and flattening of the valve characteristic at the split range point. Consider a typical split range valve setup for a reverse-acting reactor temperature controller. As the controller output changes from 0 to 75% (the split range point), a fail open (reverse-acting) coolant valve would go from fully open to completely closed. As the controller output changes from 75% to 100%, the steam valve would go from completely closed to fully open. In this case, the cooling valve has three times the effect of the steam valve, so the portion of the controller output for throttling the cooling valve is three times the portion for throttling the steam valve. This selection of the split range point helps compensate for the difference in the magnitude of the relative effects of coolant and steam; however, the valve characteristic is extremely flat as seen in Figure 4-1 and the stiction is greatest when the valve is near its closed position (split range point). The valve may have trouble breaking free of the seat or seal.

Process Modeling

Each of the following types of models benefits from accurate and repeatable flow measurements:

- Projection to Latent Structure or Partial Least Squares (PLS)
- Model Predictive Control (MPC)
- Adaptive Controller Tuning
- Neural Network
- First Principle

None of the models listed will provide reliable predictions if valve positions rather than flow measurements are used. Even first principle models with pressure-flow solvers are notoriously bad at predicting flows through control valves due to uncertainties. First principle models are sensitive to systematic errors in the flow measurements because process parameter and efficiency calculations and the closure of material and energy balances depend on actual flow rather than a deviation in flow.

Offline process simulations use snapshots of flows and primary process variables to verify the model. Online dynamic process simulations use the actual flows as inputs to the model and adapt process parameters in the model to reduce any mismatch between model predictions and plant measurements of primary process variables. For example, in one test case, the influent acid concentration of a waste treatment system was adapted based on matching the predicted and measured reagent/influent flow ratio [3]. Measurement of the reagent flow was particularly important here because there were four reagent valves in parallel, with various degrees of coating and flashing.

Coatings reduce the area for flow and in extreme cases can plug the valve or cause the valve to be stuck. Flashing occurs when the pressure at the smallest area for flow (vena contracta) in the valve drops below the vapor pressure of the fluid. Bubbles form that effectively choke the valve. The bubbles reduce the area for liquid flow and convey an order of magnitude less mass flow than the liquid. If these bubbles implode (collapse), the energy released can do considerable damage to piping and equipment. The noise from flashing and cavitation can violate regulations. A flowmeter upstream of the control valve will normally be at a pressure well above the vapor pressure and be able to measure the true flow. If the flowmeter was installed downstream of the valve in this service, it would be subjected to flashing and cavitation. For these and other reasons, such as a uniform velocity profile, a flowmeter should always be installed upstream of a valve.

Mass Flow Measurement

Because volume varies with temperature and pressure, but mass does not, mass flow is the most accurate, easiest-to-use flow variable in process control. Coriolis flowmeters provide the only true direct measurement of mass flow independent of fluid composition. Coriolis mass flowmeters also provide a density measurement. Measurement accuracy and rangeability for both mass flow and density are the best in industry. The Coriolis density measurement can be also used for an inferential measurement of composition when the concentrations of one or two components have a dominant effect on the fluid density.

Differential pressure, also known as differential head (e.g., orifice, venturi, flow nozzle, and pitot tube) and vortex and magnetic flowmeters, provide volumetric measurements (e.g., cubic meters per minute and cubic feet per minute). Volumetric measurements can be converted to mass flow measurements when the effect of composition, pressure, and temperature on density is quantified. This converted measurement is termed a "compensated" measurement. The inferred mass flow measurement is affected by the time lag and cumulative error of the composition, pressure, and temperature measurements, and data. A pressure and temperature compensated measurement assumes the composition of components in the stream with significantly different specific gravities is relatively constant. Multivariable differential pressure flowmeters use pressure and temperature to provide an inferential measurement of mass flow when the effect of changes in composition on the calculations is negligible.

To measure mass flow in gases, thermal mass flowmeters are an option, particularly for tubing and ductwork. In pipelines, the accuracy of thermal mass flowmeters is generally not as good as the accuracy of multivariable flowmeters with pressure and temperature compensation and discharge coefficient correction.

Thermal Mass Flowmeters

The niche market for thermal mass flowmeters is presently about 2% of total flowmeter sales but the market is expected to increase dramatically for measuring greenhouse gas emissions [4]. The mandatory reduction in greenhouse gas emissions will require flow measurements on the vents from combustion, fermentation, and chemical processes that produce NO_X, CO_X, and SO_X. The growth in applications that measure gas feeds in biopharmaceutical process research and development will also accelerate. Most new complex proteins will be produced by biopharmaceutical processes that require significant amounts of gas flows for dissolved oxygen and pH control. These gas flows must be measured and controlled.

While the details of the product offerings for thermal mass flowmeters is beyond the scope of this book, the technology and equations is presented to provide an understanding of the assumptions and limitations since expectations can exceed actual performance.

A thermal mass flowmeter adds heat to a gas and measures the temperature upstream and downstream of the heater. The sensors may protrude into the flow or be mounted near or on the meter wall. The upstream measurement is representative of the process temperature of the incoming stream. The downstream measurement is designed to be representative of either the heater wall or the process stream temperature after heat addition. In some cases, the heater is integral with the temperature sensor.

The mass flow can be computed from a simple energy balance. Equation 4-2a shows the heat addition is the mass flow multiplied by the specific heat capacity of the fluid and the temperature rise. Equation 4-2b is the solution for mass flow. Equation 4-2c shows that the actual heat addition into the process from the heater is the energy input from the heater multiplied by a correction factor to account for the heat transfer coefficient at the heater surface. The correction factor is a function of meter geometry and fluid heat capacity, thermal conductivity, and viscosity. Whether the flow is laminar or turbulent has a huge effect on the factor calculation. If the meter's surface is not fouled, the factor can be expressed as shown in Equation 4-2c as a function of fluid properties.

One of the major distinctions in meter design is whether the heater output is constant or variable. For a constant heater output, the temperature downstream varies with mass flow. The mass flow is inversely proportional to the temperature rise. For a variable heater output, the heater current is automatically manipulated to keep the downstream temperature a few degrees higher than the upstream temperature to maintain a constant temperature rise. The mass flow is proportional to the heater current.

$$Q_p = W * C_p * (T_2 - T_1) \tag{4-2a}$$

$$W = \frac{Q_p}{C_p * (T_2 - T_1)} \tag{4-2b}$$

$$W = \frac{\kappa(C_p, \mu_p, k_p) * Q_h}{C_p * (T_2 - T_1)} \tag{4-2c}$$

Where:

C_p = specific heat capacity of process fluid (kJoules/°C*kg)

k_p = thermal conductivity of process fluid (kJoules/sec*°C*m)

κ = correction factor for heat input as a function of process fluid heat capacity, viscosity, and thermal conductivity (0 to 1)

μ_p = viscosity of process fluid (centistokes)

Q_p = heat input to the process fluid (kJoules/sec)

Q_h = heat output from the electric heater (kJoules/sec)

T_1 = upstream temperature (°C)

T_2 = downstream temperature (°C)

W = mass flow (kg/sec)

The accuracy of the device is about 1% to 2% for known fluid properties, clean surfaces, and a computable heat transfer coefficient. The rangeability varies from 10:1 to 30:1 [5]. The thermal flowmeter is predominantly used for air and gas flows, where the fluid properties are well known and constant. For gas mixtures where changes in composition cause significant variation in heat capacity and thermal conductivity, the gas composition from "in-line" analyzers with probe sensors and "at-line" analyzers with sample systems may be used to correct the mass flow indication.

Thermal flowmeters are suitable for very small lines (e.g., 1/16 inch) to extremely large pipes and ductwork (e.g., 60 inch). Some meters create a bypass laminar flow to reduce the amount of heat that needs to be added. The heater and temperature sensors are in the bypass element. For ducts, multipoint thermal flowmeters are used to get more a representative measurement. As many as 16 measurement points may be used for combustion emission monitoring in large pipes or smokestacks [4].

In small line sizes, a control valve and proportional-integral (PI) control algorithm is used with the thermal flowmeter to create a mass flow controller (MFC). The integral control valve is a thermal expansion valve, which doesn't have friction or backlash but has a slow response. Consequently, the flow response for large changes in set point can be slow. An MFC is often a secondary flow loop in a cascade control system. The flow controller needs to be tuned for fast action to ensure the secondary flow loop response is five times faster than the primary loop (e.g., DO or pH). However, if the flow controller output changes faster than the thermal expansion valve can move, unstable oscillations will develop. This outrunning of the valve response is likely to occur for large upsets or set-point changes. The PI algorithm should have the "dynamic reset limit" option enabled for the primary loop with the actual flow used as the BKCAL_IN of the primary PID block [6, 7]. This configuration should be extended to the secondary flow loop with the actual valve position used as the BKCAL_IN of the flow PID block. However, the dynamic reset limit option and feedback of actual thermal expansion valve position may not be available in the MFC. The use of dynamic reset limiting is one of the most important overlooked opportunities in process control.

An MFC is widely used on bioreactors for air, oxygen, nitrogen, argon, and carbon dioxide flow control. The set point for the MFC usually comes from a dissolved oxygen (DO) or pH controller. If the mass flow is not being used for first principle modeling, the 1% to 2% accuracy of an MFC is usually sufficient. A multivariable differential pressure flowmeter offers moderately better accuracy (e.g., 0.75%). A Coriolis meter offers the best accuracy (e.g., 0.35% for gas) line size permitting.

Other Flow Measurement Devices

There are more types of flow measurement devices than all of the other common types of pressure, level, and temperature measurement devices combined. The user has quite a task in selecting a flow instrument. This chapter narrows the search to the most reliable and maintainable flow measurement devices for gases and liquids. Consequently, variable area meters (rotameters), positive displacement meters, target meters, and turbine meters (although they have their place) are not discussed, because of their mechanical nature and susceptibility to malfunction from solids, sticky fluids, and debris.

4-3. Differential Pressure Flowmeters

DP Principles

Differential pressure (DP) flowmeters, also called differential producers, are the most common type of flowmeter. DP flowmeters measure the differential pressure (P) caused by an obstruction, called a *primary element*, in the flow stream. The differential pressure is the difference in pressure before the obstruction (high side pressure) and after the obstruction (low side pressure). This pressure change is measured by differential pressure transmitters or multivariable transmitters, which are also known as secondary elements.

The diagram in Figure 4-3a shows an orifice plate primary element and a secondary differential pressure transmitter measuring the differential between the high and low pressure taps. The lines above the pipe are a graphical representation of the pressure in the pipe. The initial working pressure increases as the fluid is restricted in the pipe. As the fluid passes the primary element, the pressure decreases until it reaches the recovery zone. The recovery zone is where the pressure actually increases until it reaches its final state. However, the final pressure never recovers to the levels measured prior to the disturbance caused by the primary element. This difference in initial pressure and final pressure, illustrated in Figure 4-3a, is called *permanent pressure loss*.

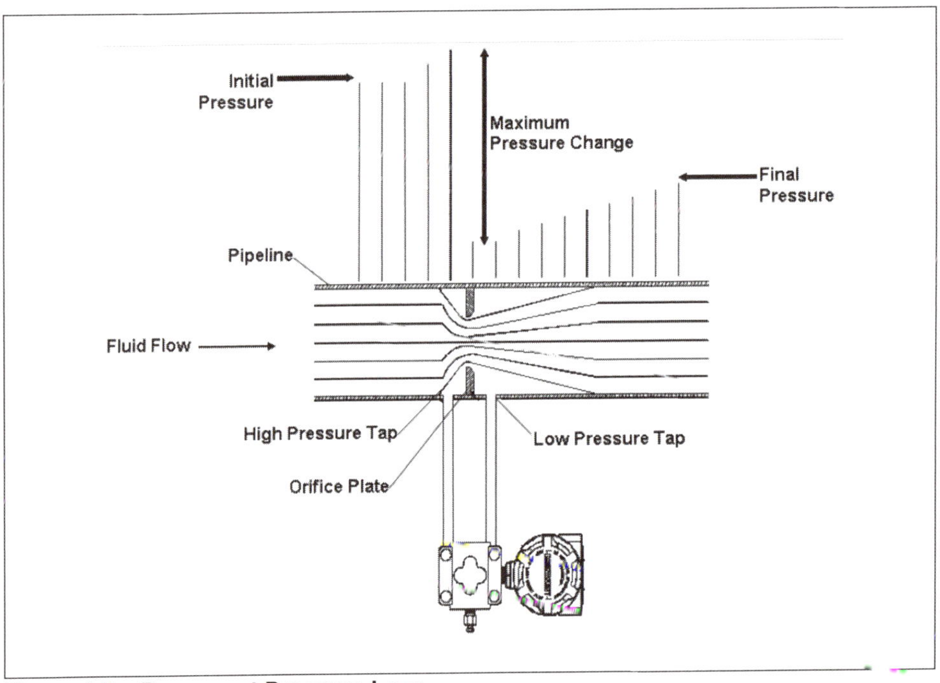

Figure 4-3a. Permanent Pressure Loss

The flow equations used for DP flowmeters are based on Bernoulli's equation (conservation of energy) and the continuity equation (conservation of mass). This is just a way of saying that flow rate (Q) is proportional to the square root of differential pressure (P) (Equation 4-3a):

$$Q \propto \sqrt{\Delta P} \tag{4-3a}$$

Another way to think about this is that flow (Q) is equal to a constant (K) multiplied by the square root of differential pressure (ΔP) (Equation 4-3b):

$$Q = K * \sqrt{\Delta P} \tag{4-3b}$$

Because of this relationship, differential pressure transmitters are used as the secondary element for most DP flowmeters. In actuality, this simplified version of the flow equation can lead to flow error. For more information, refer to the multivariable transmitter section.

DP Selection

There are a number of different types of DP primary elements including orifice plates, averaging pitot tubes, venturi tubes, and flow nozzles, as shown in Figure 4-3b. When choosing the right primary element for the application, it is best to look at the strengths and weaknesses of each.

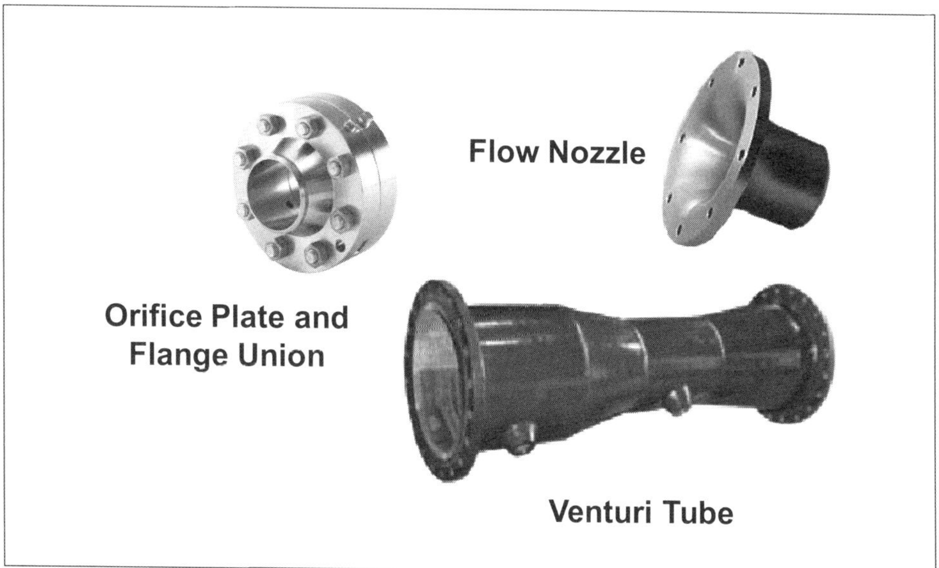

Figure 4-3b. Orifice Plate, Flange Union, Flow Nozzle, and Venturi Tube

In some applications (custody transfer, billing, etc.), compliance with standards is the most important aspect of the primary element selection process. Most of the standards (ISO 5167, AGA Report No. 3, ASME MFC-3M, and ASME MFC-14M) are written for orifice plates, venturi tubes and flow nozzles. Orifice plates can accommodate virtually all clean fluids, but they are not recommended for use with abrasive or sticky fluids. Traditionally, the orifice plate itself is inexpensive but when the required flange union, impulse lines (sensing lines), isolation valves, and straight pipe lengths are added to the system, the installation cost increases significantly.

For optimal measurement performance, orifice plates require a sharp orifice edge at the point where the process fluid passes through the plate.

Integral orifice plates designed to ASME MFC-14M incorporate the basic characteristics of a standard orifice plate, but are specially designed for use with small line sizes. A honed pipe wall reduces the friction between the process fluid and the pipe, enabling a more accurate flow measurement in small line sizes. A conditioning orifice plate provides significant installation advantages as it requires less straight pipe to maintain superior measurement performance.

Venturi tubes are usually used in applications that require minimal pressure drop with reasonable measurement accuracy. Venturi tubes provide very low permanent pressure loss when compared to other DP flowmeters, but they are also larger and significantly more expensive. They are suitable for measuring dirty fluids and slurries that would build up on or clog orifice plates.

Flow nozzles remain precisely calibrated over long periods of time even under severe working conditions. The exact contour of the flow nozzle is not particularly critical for accurate measurement, and for this reason, flow nozzles are often used for the measurement of steam flow and other high-temperature or high-velocity fluid flows where erosion may be a concern.

In other applications, the most important selection criterion is the ability of the primary element to minimize permanent pressure loss. For these applications, an averaging pitot tube is the recommend solution. Averaging pitot tubes are insertion-type DP Flow primary elements that are convenient and relatively inexpensive (Figure 4-3c). Some averaging pitot tubes can be used to measure the flow of fluids in both directions (bidirectional capability) while others can be installed in the process pipe without requiring system shutdown (referred to as a hot tap). Averaging pitot tubes are not recommended for use with dirty fluids, as these can clog the measurement ports and reduce the accuracy.

To achieve the accuracy specifications of the primary element, it is important to comply with its straight pipe requirements. Different primary elements have different straight pipe requirements. For example, orifice plates require a large amount of straight pipe, typically 44 pipe diameters; some configurations require up to 145 pipe diameters. Many times there are space restrictions that prevent the installation of an adequate straight run. In these cases, it would be best to choose a technology that will provide an accurate flow measurement with straight pipe requirements. Con-

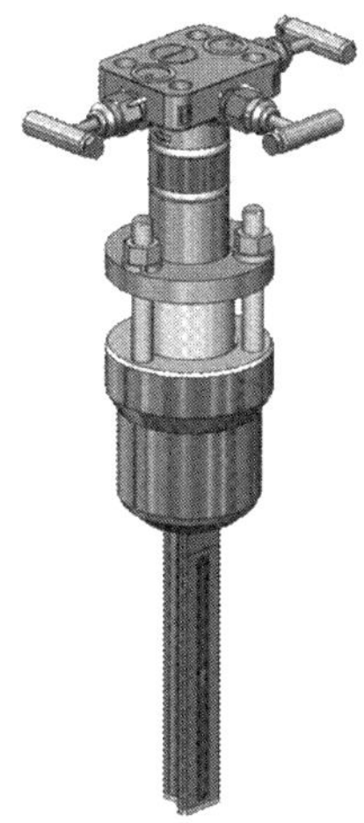

Figure 4-3c. Averaging Pitot Tube

ditioning orifices plate shown in Figure 4-3d can provide a 2:1 reduction in the straight run needed. A magnetic flowmeter offers a 10:1 reduction and a Coriolis flowmeter eliminates the straight run requirement.

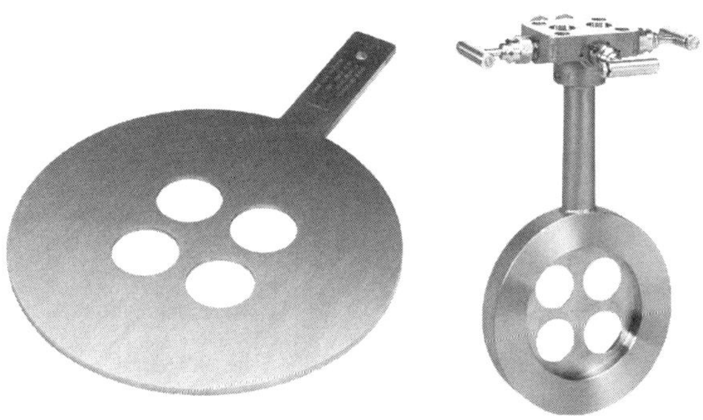

Figure 4-3d. Conditioning Orifice Plate

Installation costs may be the most important aspect of a DP flowmeter. The standard orifice plate meter run is shown in Figure 4-3e.

As you can see in Figure 4-3f, there are a number of components to a DP flowmeter. All of these components increase the installed cost of the flow measurement point. There are proprietary DP flowmeters (integral orifice plate, compact orifice plate, and pitot tube shown in Figure 4-3f) that can

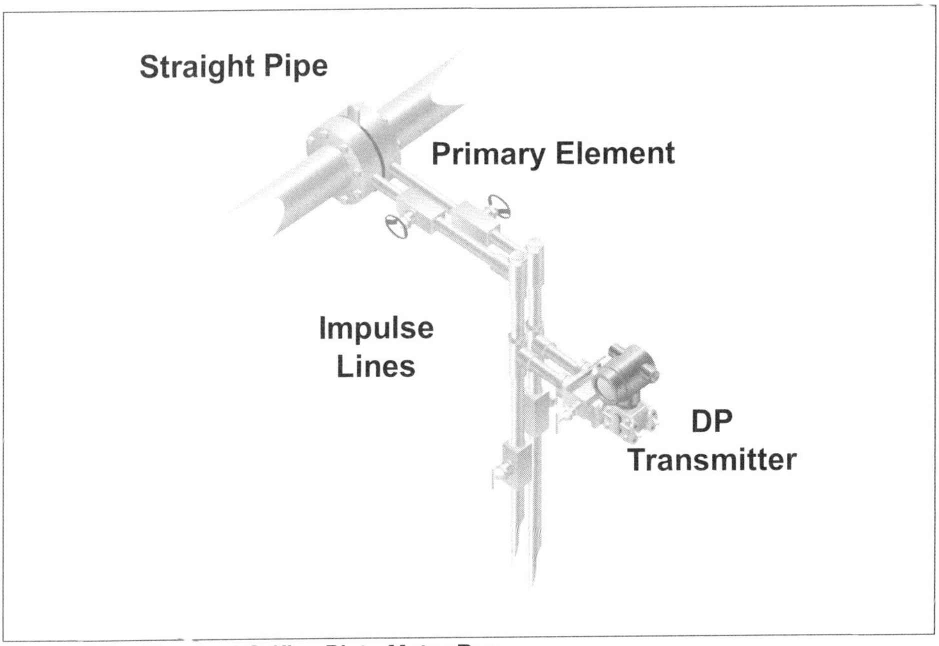

Figure 4-3e. Standard Orifice Plate Meter Run

reduce the cost of the installation with the direct mounting of the transmitter to the primary element (discussed in section on DP Maintenance).

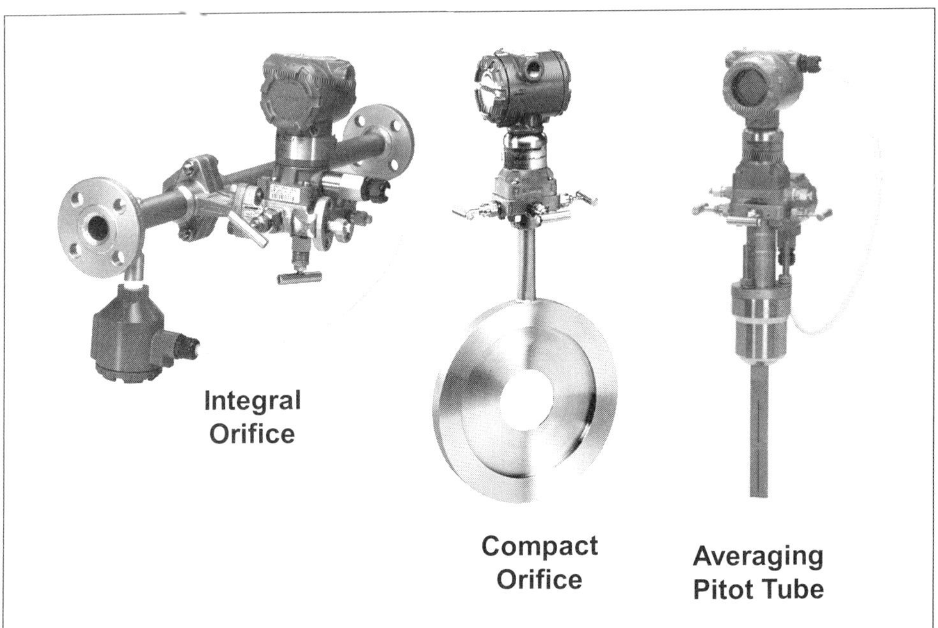

Figure 4-3f. DP Flowmeters with Direct Mounted Transmitter

In general, DP technology is a widely used and globally accepted means of measuring flow. There are advantages and disadvantages relating to life-cycle cost and performance that should be considered when using differential pressure primaries.

DP Specification

There are several different primary elements, and each has its own set of specifications. Here we take a closer look at some of the design details of each major type of element that affect the specifications.

Types

Orifice Plates

Orifice plates come in many different styles from square edge to quadrant and conical edges, to concentric, eccentric and segmental bore types (Figure 4-3g). Each of these serves a particular purpose or is used in specific applications (including lower Reynolds number flows to dirty liquids or wet/damp gas flows). Orifice plates are installed between orifice flanges to hold the primary element in place. There are several different orifice flange styles (weld neck, slip on, raised face and ring type joint), and these have different types of pressure taps (corner taps, flange taps, radius taps, pipe taps) for taking the DP measurement.

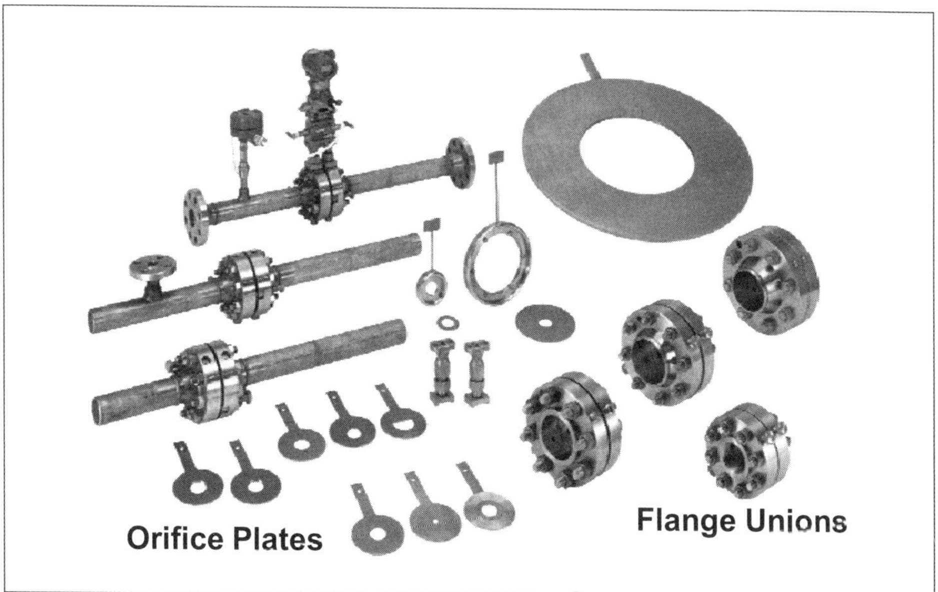

Figure 4-3g. Orifice Plates and Flange Unions

Averaging Pitot Tubes

Averaging pitot tubes come in many different styles including welded, flanged, and "flo-tap" (hot tap) designs (Figure 4-3h). Each has different mounting hardware and different application requirements. For instance, a welded design requires drilling a hole in the pipe and then welding the mounting coupling to the outer wall of the pipe. A flanged design requires a specific flange rating dependent on the process conditions for the application. Flo-tap (hot-tap) designs are unique in the sense that they allow the customer to install or retract the averaging pitot tube while under flowing process conditions and do not require system shutdown for installation.

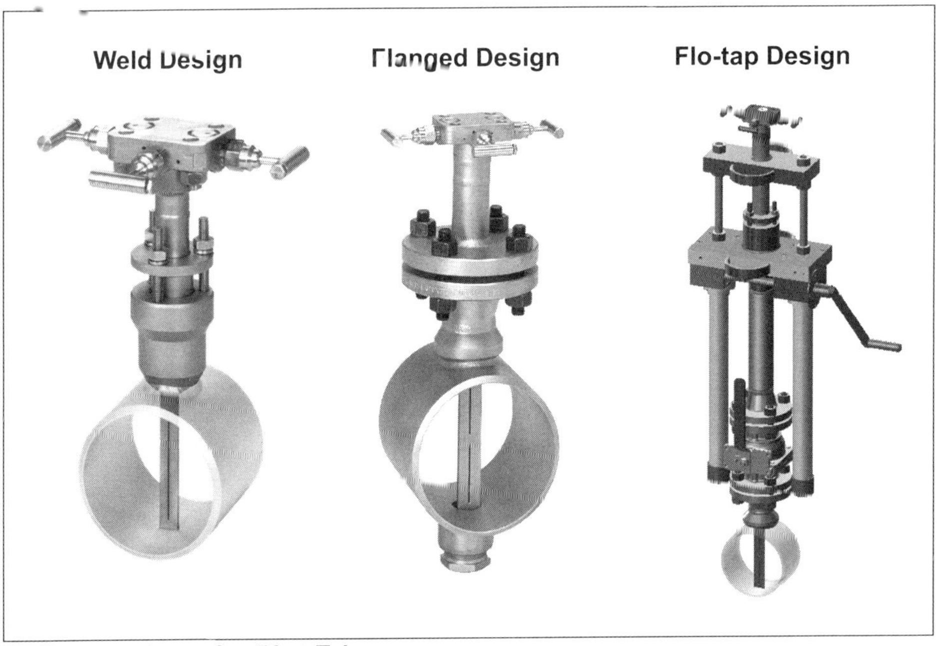

Figure 4-3h. Averaging Pitot Tubes

Venturi Tubes

Venturi tubes are available in flanged, welded and threaded designs. Each design encompasses a converging inlet cone, a throat section and a conical exit (Figure 4-3i). The converging inlet cone gradually decreases the pipe diameter and creates a pressure drop. A high pressure tap is located at the upstream end of the inlet cone. The low pressure tap is located in the throat section. Fluid velocity is neither increasing nor decreasing in the throat section. The conical exit is the point where the pressure increases as the process fluid passes through an increasing cross-sectional area of pipe.

Figure 4-3i. Flanged Venturi Tubes

Flow Nozzles

Flow nozzles are available in a welded design or a flanged design. Nozzles consist of two main sections, an elliptical inlet and a throat section (Figure 4-3j). The elliptical inlet is mounted at the upstream end of the fluid flow. The high pressure tap is located upstream of the elliptical entrance. Fluid velocity increases as fluid enters the inlet, and the pressure decreases. The throat section represents the point of the nozzle where the inlet tapers into a cylindrical shape. This is where the low pressure tap is located.

Figure 4-3j. Flow Nozzles

Performance

The discharge coefficient (C_D) uncertainty (interval within the true value is expected to lie with a stated probability) is based on test data, industry standards, the primary element selected, and manufacturer specifications (Equation 4-3d).

Below is a performance specification range for a few common primary elements:

C_D Uncertainty (Uncalibrated)

Orifice Plates: ±0.50 to ±1.25% C_D uncertainty
(±2.0 to 4.0% full scale)

Averaging Pitot Tubes: ±0.75 to 2.00% of actual flow rate

Venturi Tubes: ±1.0 to 2.0% of actual flow rate
(±0.25% – calibrated)

Flow Nozzles: ±1.0 to 2.0% of actual flow rate
(±0.25% – calibrated)

Rangeability

Rangeability is defined as the ratio of maximum flow to minimum flow. There are two ways to define maximum flow when specifying the rangeability of a flowmeter:

1. The maximum flow rate that the flowmeter can measure.
2. The maximum flow rate that occurs in a particular application.

DP rangeability is generally defined using application maximum flow. It is necessary to understand the difference because the meter maximum is often two to three times the application maximum. Therefore, it is important to determine the correct sizing for your primary element and the application. The following are some typical ranges for a few popular primary elements:

Orifice Plates: typical applications cover a range of 3:1 to 5:1

Averaging Pitot Tubes: typical applications cover a range of 10:1

Venturi Tubes: up to 8:1, but typical applications cover a range of 3:1 to 5:1

Flow Nozzles: up to 8:1, but typical applications cover a range of 3:1 to 5:1

Repeatability

For primary elements, repeatability is generally ±0.1%.

Process Operating Ranges

When defining the operating ranges for primary elements, you need to understand the application requirements as well as the effects of the fluid velocity (or flow profile) and the Reynolds number. As we have seen, there are three types of flow profiles: laminar, transition, and turbulent. Laminar flow occurs when the velocity is low and the fluid flows smoothly with little mixing. The velocity profile (fluid velocity variation with distance across the pipe) is parabolic. The fluid velocity declines rapidly with distance from the center of the pipe and maximum velocity (Figure 4-3k).

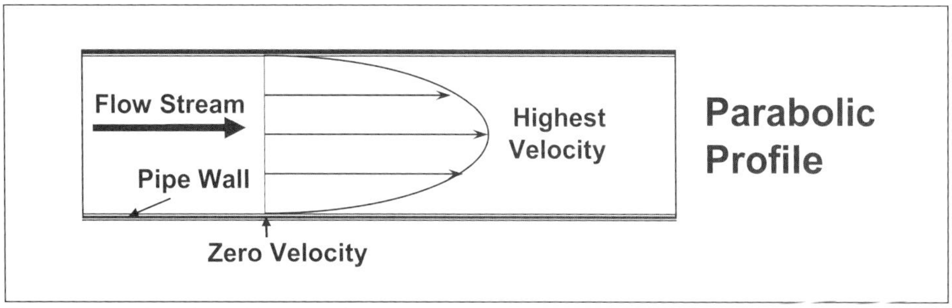

Figure 4-3k. Laminar Flow Profile

Transition flow occurs at a slightly higher velocity than laminar flow but has characteristics of both laminar and turbulent flow in differing proportions with velocity.

Turbulent flow occurs at higher velocities. The fluid velocity profile is flat (fluid velocity is constant) across the pipe (except at the pipe walls), because turbulent eddies form, causing mixing (Figure 4-3l). Turbulent flow is the most common profile encountered in pipes and measured by DP flowmeters.

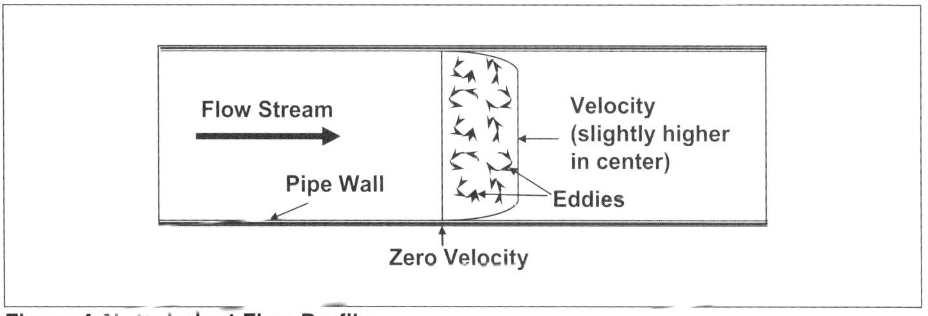

Figure 4-3l. Turbulent Flow Profile

The nature of flow in a pipe depends on the pipe diameter in feet (D), density in lbm/cu-ft (ρ), viscosity in lbm/ft-sec (μ), and the velocity in ft/sec (v) of the fluid. This is referred to as Reynolds number (R_D).

$$R_D = \frac{D * v * \rho}{\mu} \tag{4-3c}$$

The Reynolds number has no dimensions and it may be considered as the ratio of dynamic forces to viscous forces. For the three types of flow, there exists a range for R_D. Velocity profile is determined by the Reynolds number.

1. Laminar – $R_D < 2,000$
2. Transition – R_D 2,000 – 4,000
3. Turbulent – $R_D > 4,000$

Performance specifications for flowmeters are based on a fully developed turbulent flow profile. The following shows some typical operating ranges for a few popular primary elements:

- Orifice Plates, Venturi Tubes, Flow Nozzles:
  Liquid, gas or vapor turbulent flow, for pipe Reynolds numbers greater than the following:
  - AGA-3: 4,000
  - ASME MFC-3M(2): 5,000 and 170 2 D
  - ISO-5167(2): 5,000 and 170 2 D

- Averaging Pitot Tubes:
  Liquid, gas or vapor turbulent flow for Reynolds number 6500 to 25,000.

Maximum Pressure

- Orifice Plates, Venturi Tubes, Flow Nozzles:
  Flange rating per ANSI B16.5 and DIN EN 1092-1. Rated for ASME B16.5 (ANSI) up to Class 2500, DIN up to PN 100, JIS up to 40K

- Averaging Pitot Tubes:
  Rated for ASME B16.5 (ANSI) up to Class 2500, DIN up to PN 100

Minimum and Maximum Temperature

- Orifice Plates: Temperature limit range: –320 to 1200°F (–196 to 649°C)
- Averaging Pitot Tubes: Temperature limit range: –320 to 1200°F (–196 to 649°C)
- *Maximum Velocity*: Depends on the maximum allowable permanent pressure loss and erosion for abrasive suspended solids
- *Maximum Viscosity*: Less than 10 cP (centipoise), but typically in the 1-5 cP range

Wetted Materials of Construction

Wetted materials are defined as those locations on a primary element where the material comes in contact with the process fluid. The flow application often drives the use of a particular material. Most manufacturers provide a wide selection of materials, which should support most process conditions. Some common wetted materials of construction include Stainless Steel (Standard), Alloy C-276, and Alloy 400. Refer to the primary element manufacturer for recommended guidelines on materials for DP applications.

Differential Pressure Flowmeter Primary Element Sizes

1. Orifice Plates: 1/2 to 24 inch and higher (15 to 600 mm)
2. Venturi Tubes: 2 to 48 inch (50 to 1200 mm)
3. Flow Nozzles: 2 to 48 inch (50 to 1200 mm)
4. Averaging Pitot Tubes: 2 to 96 inch (50 to 2400 mm)

DP Considerations

Some advantages include:

- Simple technology proven in a broadest spectrum of applications
- A wide variety of primary elements to serve nearly all types of applications
- May be the only reasonable choice for all types of fluids in large pipelines
- Highly repeatable measurements with excellent response time

- Lower instrumentation hardware cost

Some disadvantages include:

- Troublesome long impulse lines and the potential for more leak points
- Higher permanent pressure loss with some primary elements
- Long straight pipe runs particularly for non-planar elbows upstream
- Low rangeability except for some pitot tubes
- Uncertainty in discharge coefficient and beta ratio

Beta Ratio Uncertainty

ASTM standards permit a 12.5% decrease in wall thickness and a 1% variation in the outside pipe diameter from the nominal value. For a 6-inch pipe this tolerance translates to a 4.5% uncertainty in the cross-sectional area of the pipe [8]. Since beta ratio is the ratio of the orifice diameter to inside pipe diameter, there is a corresponding uncertainty in the beta ratio for an orifice inserted into a pipe. Integral orifice runs and venturi flow tubes can largely eliminate this uncertainty.

Sizing

Sizing DP flowmeters is generally straightforward and easy. Most manufacturers provide sizing software programs to assist in the sizing of these devices. These programs typically provide key information such as minimum and maximum DP, and minimum and maximum measureable flow rates, along with pressure loss requirements.

DP Setup and Calibration

Factory calibration of DP primary elements is an option available with some manufacturers. A complete calibration by a traceable lab will provide performance information about a primary element, but is no guarantee that the same performance and accuracies will be obtained or maintained in the field because of the many process conditions that may be present.

A laboratory calibration provides the user with a performance coefficient for the DP primary element. The coefficient will vary with different primary elements, but will always be between 0 and 1. This coefficient is called by several names, depending on the primary element: discharge coefficient, flow coefficient, or k-factor. The coefficient is defined as the ratio of the actual flow to the theoretical flow (Equation 4-3d).

$$C_D = \frac{ActualFlow}{TheoreticalFlow} \tag{4-3d}$$

The coefficients are well understood and documented for a wide variety of primary elements, such as single-hole orifice plates and averaging pitot tubes. For these elements, a lab calibration is not necessary to reach published accuracies. Other primary elements do require lab calibration for each element since the coefficients are not documented, or cannot be controlled to acceptable levels. Erosion and corrosion can change the edge of an orifice and change the discharge coefficient. Venturi tubes and flow nozzles are less affected in general by erosive and corrosive streams.

DP Installation

The recommended guidelines should be followed as outlined by the primary element manufacturer for straight run requirements and proper flowmeter orientation for each specific application. Proper installation will reduce flow disturbances and increase the accuracy of the flow measurement.

In general, all primary elements are installed and oriented similarly depending on the application and process fluid. Refer to Figures 4-3m and 4-3n for recommended mounting styles.

Horizontal mounting:

- For liquid and steam applications, the primary element should be mounted on the bottom half of the pipe below the horizontal line within the recommended orientation zone.
- For air and gas applications the primary element should be located on the upper half of the pipe above the horizontal line within the recommended orientation zone.

Vertical Mounting: The primary element can be installed in any position around the circumference of the pipe, provided the primary element is positioned properly for bleeding or venting. For liquid or steam, the element should be located for flow up.

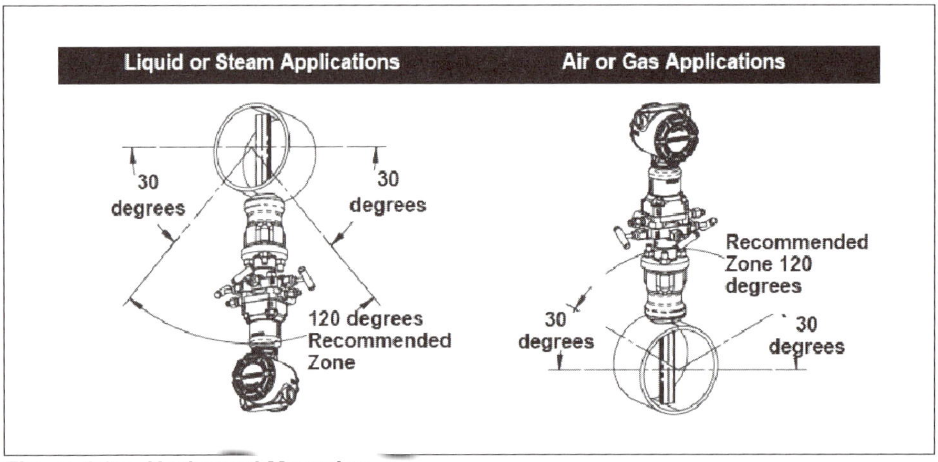

Figure 4-3m. Horizontal Mounting

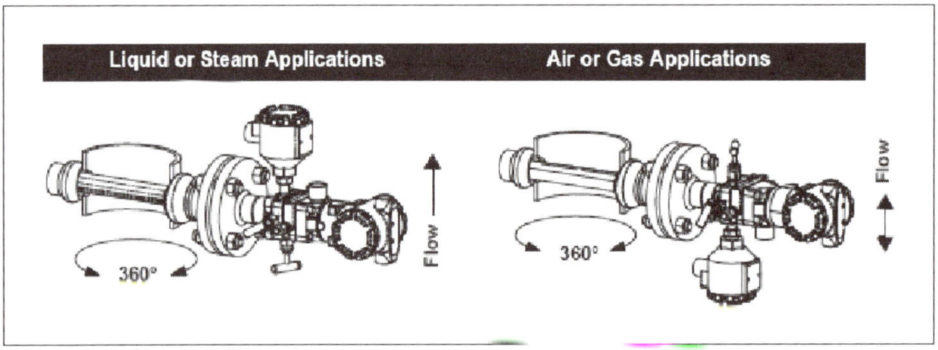

Figure 4-3n.Vertical Mounting

DP Maintenance

When it comes to troubleshooting DP flowmeters, the reliability of a flow measurement point is correlated to items such as impulse tubing and incorrectly sized primary elements. The problem of plugged lines and an increased number of leak points arises when impulse tubing is used on an application.

Plugging may occur in some steam applications using impulse tubing as a result of flashing, the phenomenon in which a fluid quickly changes from a liquid to a vapor state. Precipitate composed of suspended solids found in steam can adhere to the walls of an enclosure when flashing occurs. Consistent flashing causes precipitation accumulation, resulting in plugged impulse tubing. A plugged impulse tube cannot communicate the DP generated by a primary element, resulting in inaccurate measurement. How the primary element connects to the secondary element influences the potential for flashing in a given application.

Additionally, process leaks in impulse tubing result in erroneous readings and affect flow measurement accuracy. The potential for a leak is directly related to the number of connections in the DP flow measurement system. Utilizing direct mounting alternatives is one way to minimize the number of leak points and the troublesome impulse tubing. Direct mounting the transmitter to the DP primary element minimizes the potential for error while eliminating additional hardware and reducing the overall total installed costs.

There are a variety of other maintenance and installation issues, depending on the application. Generic instances are included in the troubleshooting section of most product manuals. See examples in the DP Flowmeter Troubleshooting Guide.

DP Flowmeter Troubleshooting Guide

Improper Installation	Is the flow arrow pointed in the direction of the flow? Verify that the cross reservoirs are perfectly level with one another. Is there sufficient straight run upstream and downstream of the flowmeter?
System Leaks	Check for leaks in instrument piping. Repair and seal all leaks.
Contamination/plugging	Remove the flowmeter and check for contamination.
Closed Valve	Verify that flowmeter high and low pressure connections and meter isolation valves are open. Verify that vent, equalizer, and line valves are properly positioned per the start-up procedure.
Calibration	Is the calibration too high or low for the flow rate?
Entrapped air (liquid applications)	Are there uneven water legs caused by air entrapment in the instrument connections? If so, bleed air.
Misalignment	Misalignment of the flowmeter beyond three degrees will cause an erroneous signal.
Operating conditions	Are the operating conditions in compliance with those given at the time the flowmeter was purchased? Check the flow calculation and the fluid parameters for accuracy. Double-check pipe inside diameter for proper sizing.

4-4. Multivariable Differential Pressure Flowmeters

Multivariable DP Principles

As has been mentioned, differential pressure (DP) flowmeters measure the differential pressure caused by an obstruction, called the *primary element*, in the flow stream. The differential pressure is the difference in pressure before the obstruction and the pressure after the obstruction. This pressure change is measured by differential pressure transmitters or multivariable transmitters, which are also known as secondary elements.

As discussed in the DP flow – Primary Element section, the flow equation used for DP flowmeters is based on Bernoulli's equation. Flow rate (Q) is proportional to the square root of differential pressure (ΔP) as shown in Equations 4-3a and 4-3b.

Although flow is proportional to differential pressure, flow is not equal to differential pressure. The full flow equation contains several factors that are dependent on static pressure, temperature, and differential pressure. Therefore, a pressure and temperature compensated DP flowmeter consists of a primary element, a differential pressure transmitter, a pressure transmitter, a temperature transmitter, a temperature sensor, and a flow computer (Figures 4-4a, 4-4b, and 4-4c).

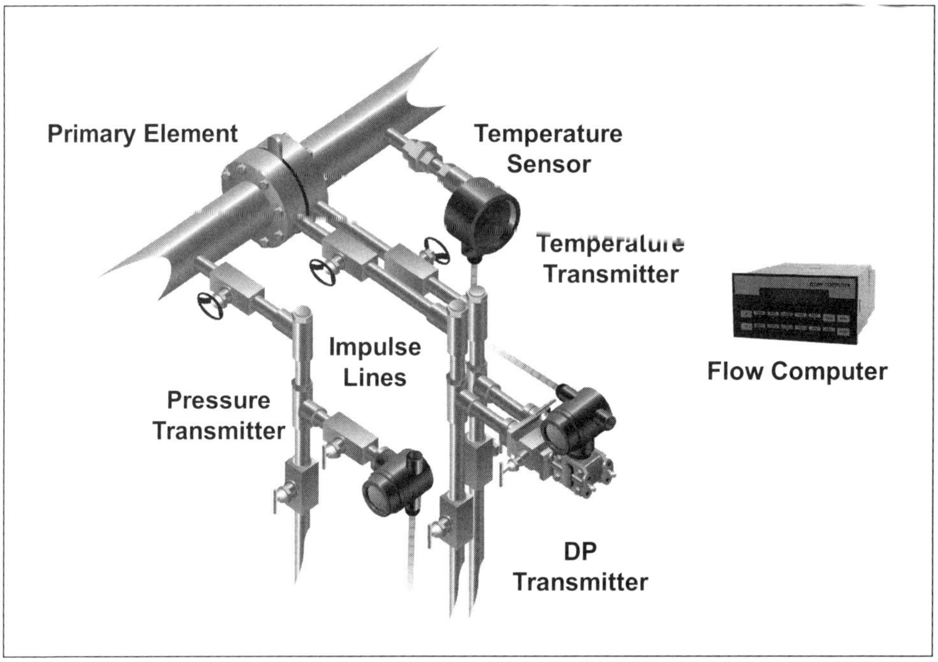

Figure 4-4a. Pressure and Temperature Compensated DP Flowmeter

Multivariable transmitters are designed to simplify the configuration of these separate devices by performing two major functions. The first is that three measurements (differential pressure, static pressure, and temperature) are consolidated into one device. The second function is that the measurements are then used to calculate flow rate, similar to a flow computer. A multivariable DP flowmeter is a much simpler system consisting of a primary element, impulse lines, a multivariable transmitter, and a temperature sensor.

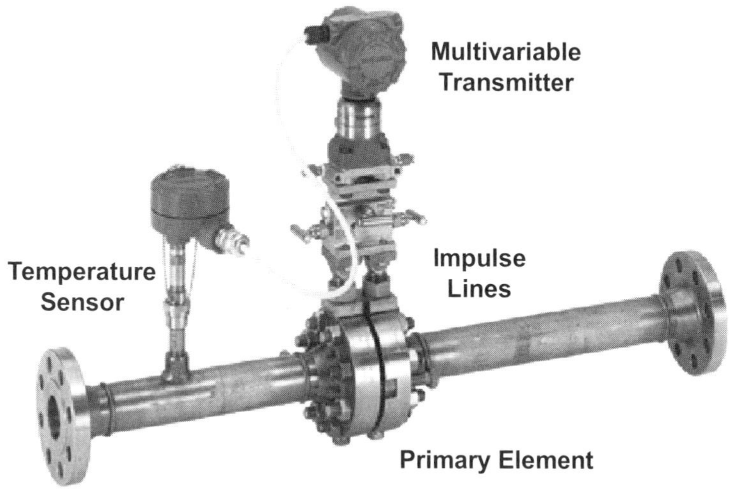

Figure 4-4b. Multivariable DP Flowmeter

Multivariable DP Selection

The specifications, features, sizing methods, and capabilities of transmitters for the three variables are very similar to those of single variable pressure and temperature transmitters. Please refer to Chapters 2 and 3 for more details.

As mentioned earlier, another key feature of multivariable transmitters is that dynamic calculation of fully-compensated mass flow is performed within the device. The flow equations are based on the fluid (steam, water, air, natural gas, etc.), primary element (orifice plate, averaging pitot tube, venturi tube, flow nozzle, etc.) and the relevant standard (ISO, AGA, ASME, DIPPR, etc.). Flow equation parameters include density, viscosity, velocity, Reynolds number, discharge coefficient, velocity of approach, and the gas expansion factor. Figure 4-4d illustrates the flow equation for an orifice plate with the variables and dependencies outlined. The black font indicates the variables that are measured, while the bold gray font indicates the calculated variables.

Although the calculation is complex, multivariable transmitters simplify the process through a user interface. The user interface allows the user to

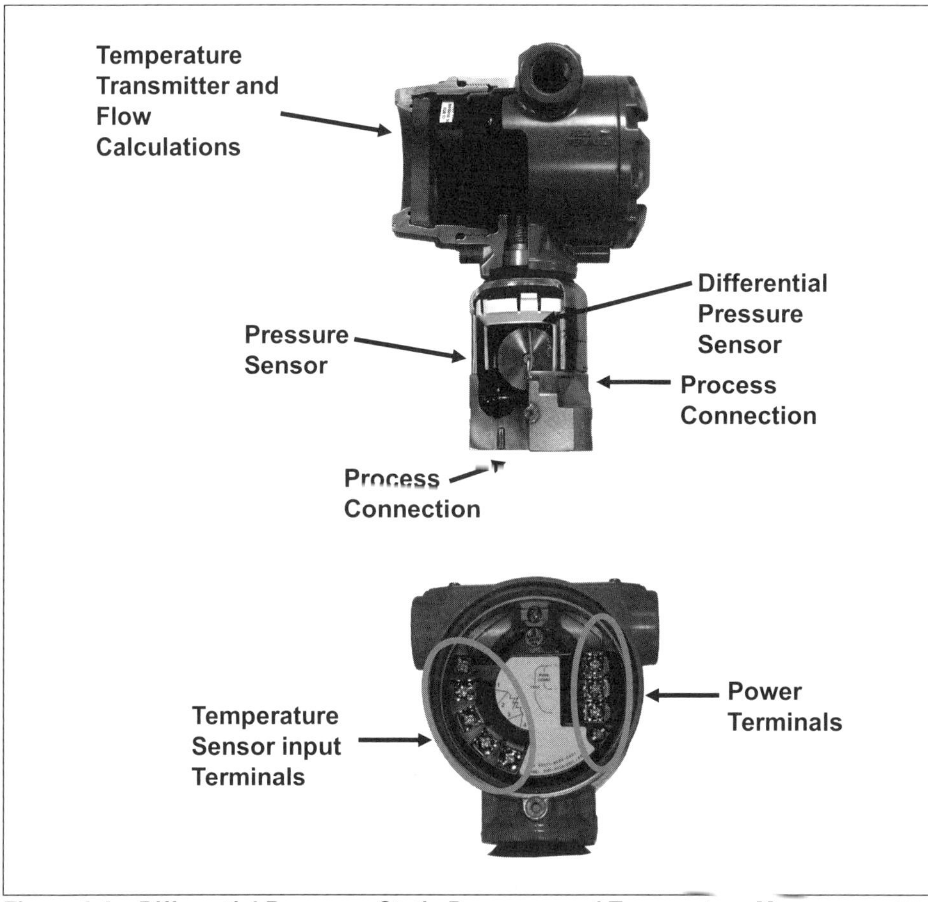

Figure 4-4c. Differential Pressure, Static Pressure, and Temperature Measurements

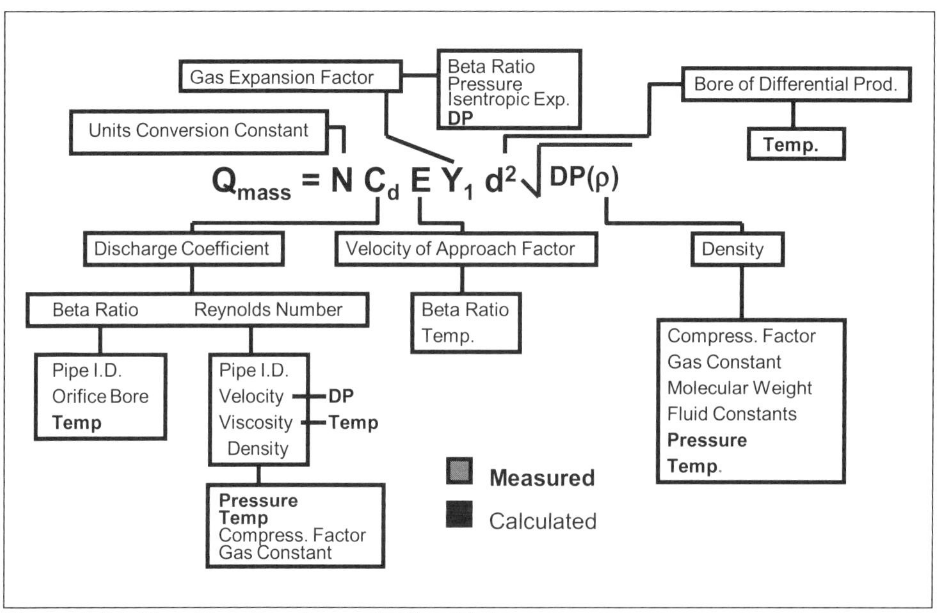

Figure 4-4d. Flow Equation for an Orifice Plate

configure the transmitter for the specific flow application without having to know all the values listed in Figure 4-4d. Additionally, because multivariable transmitters compensate for factors and conditions affecting mass flow, many traditional sources of DP flow uncertainty are eliminated. This is especially important in gas and steam applications because compensation for density, discharge coefficient (C_d) and gas expansion (Y_1) plays a large part in the flow measurement's accuracy.

Multivariable DP Specification

The specifications for multivariable transmitters are similar to those for single variable transmitters. Please see Chapter 3 for pressure transmitter and Chapter 2 for temperature transmitter for specifications, process connections sizes and types, wetted materials of construction, performance, and process operating ranges. Where multivariable transmitters differ is in their calculation of flow performance. A typical performance specification is ±1% of mass flow rate with a 14:1 turndown.

Multivariable DP Considerations

As we have seen, there are four basic methods for achieving DP flow measurement.

- The first method is to take the square root of differential pressure (single variable differential pressure transmitter). This leads to a flow measurement error due to the lack of compensation for changes in pressure, temperature, Reynolds number, etc.
- The second method is to take differential pressure, static pressure, and temperature measurements and perform a flow calculation in the control system. The problem with this process is that it takes a large amount of engineering resources to implement and maintain. Because of the complexity and the drain on engineering resources, the flow equations programmed usually do not account for all factors of the flow equation. Additionally, most control systems poll the transmitters once per second; thus, flow rate information is updated at a slower rate and response time is affected.
- The third method is to measure differential pressure, static pressure, and temperature and wire the individual devices to a flow

computer. This method requires additional components and complicated wiring.

- The fourth method is to use a multivariable DP flow transmitter. This method eliminates the need for separate differential pressure, static pressure and temperature transmitters a well as a flow computer. It also frees up control system loading by moving flow calculation functionality to the device. Because the measurements and flow calculations are performed in the device, better accuracy and faster response times can be achieved.

Sizing

Choosing the right multivariable transmitter is determined by the application. The primary element sizing will determine the differential pressure range of the multivariable transmitter. Please see the DP Flow – Primary Element section for more information. The pressure and temperature ranges of the device will also depend on the application. A wide variety of configurations is available to meet most applications.

When choosing a multivariable transmitter, a major consideration is flow measurement accuracy. Be sure to choose one that meets the accuracy requirements of the application.

Multivariable DP Setup and Calibration

The factory calibration of transmitters is generally a closely guarded secret of each manufacturer as a potential competitive advantage. How transmitters are calibrated at the factory determines their performance specifications in the field.

Calibrating an instrument in the field is needed for commissioning the device and for ongoing maintenance. Calibrating multivariable devices is as easy as calibrating single variable transmitters. However, calibration of all three variables must be performed separately. The configuration and calibration of multivariable transmitters includes the following tasks as provided by the manufacturer's operating manual:

- Basic Setup (tag information, etc.)
- Set Process Variable Units – flow rate, totalized flow, differential pressure, static pressure, and temperature

- Set Primary Variable
- Re-Range
- Set Damping
- Calibrate Each Sensor (differential pressure, pressure and temperature)
    - Perform Sensor Trim
    - Perform Zero or Lower Sensor Trim
    - Calibrate the 4–20 mA output
    - Perform 4–20 mA Analog Trim
    - Perform 4–20 mA Scaled Output Trim

Also see the pressure transmitter and temperature transmitter sections for additional information.

Multivariable DP Installation

Measurement performance depends on proper installation of the transmitter, impulse piping, and process temperature sensor. Mount the transmitter close to the process and use minimum piping to achieve the best performance. Also, consider the need for easy access, personnel safety, practical field calibration, and a suitable transmitter environment. Install the transmitter to minimize vibration, shock, and temperature fluctuation. In general, all primary elements should be installed and oriented similarly depending on the application and process fluid.

- Liquid Flow Measurement (Figure 4-4e)
    - Place taps to the side of the line to prevent sediment deposits on the process isolators.
    - Mount the transmitter beside or below the taps so gases/vapors vent into the process line.
    - Mount drain and vent valves to allow DP to be vented and flushed.

- Gas Flow Measurement (Figure 4-4f)
    - Place taps in the top or side of the line.
    - Mount the transmitter beside or above the taps to drain liquid into the process line.

- Steam Flow Measurement (Figure 4-4g)
    - Place taps to the side of the line.

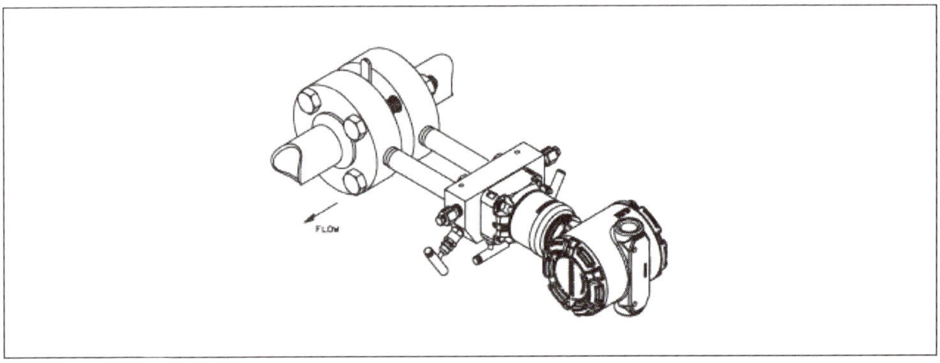

Figure 4-4e. Liquid Flow Measurement

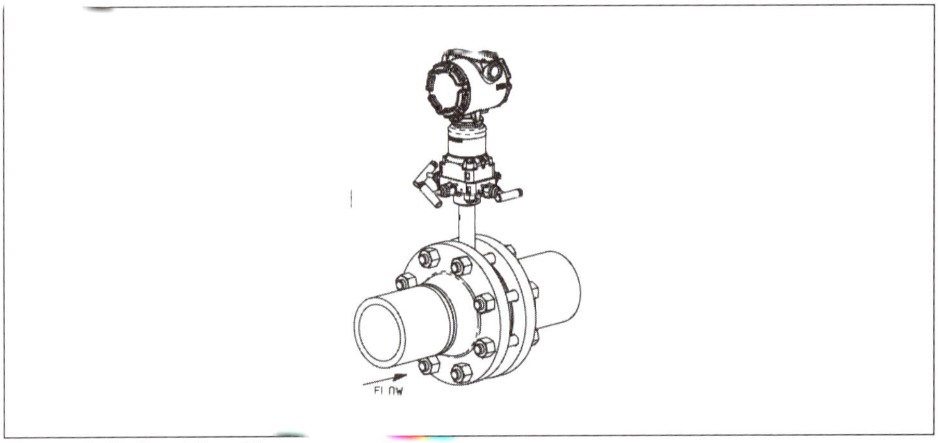

Figure 4-4f. Gas Flow Measurement

- Mount the transmitter below the taps to ensure that impulse piping will remain filled with condensate.
- In steam service above 250°F (121°C), fill impulse lines with water to prevent steam from contacting the transmitter directly and to ensure accurate measurement at start-up.

Multivariable DP Maintenance

Multivariable transmitters have many of the same maintenance concerns discussed in Section 4-3 for differential pressure flowmeters but with some additional requirements for the pressure and temperature measurements. The use of smart (intelligent) transmitters can simplify the process of diagnosing problems. Multivariable transmitter diagnostics typically include user entered sensor limits, accuracy verification, test calculation, measurement quality and limit status, and sensor temperature.

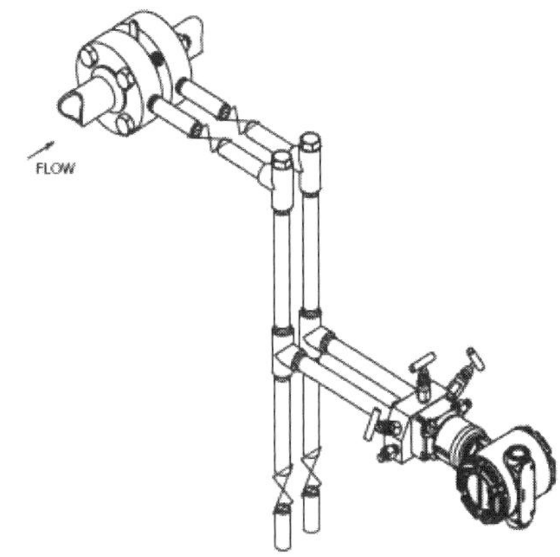

Figure 4-4g. Steam Flow Measurement

User Entered Sensor Limits

User entered sensor limits provide early detection that the resistance temperature detector (RTD) sensor may be failing. Figure 4-4h illustrates this early detection. Fixed sensor limits are the full range of an RTD (-185°C to 815°C or -300°F to 1500°F). If the temperature reading goes outside the sensor limit range, the transmitter will either go into alarm mode or into back-up temperature mode depending on how it is configured.

If it is configured to go into back-up temperature mode, the multivariable transmitter will continue to operate and flow will continue to be calculated using the user entered fixed temperature value. Otherwise, if configured to do so, the multivariable transmitter will go into alarm.

With the user entered sensor limits diagnostic, the sensor limits range band can be narrowed based on the application requirements. If the temperature reading is outside the set range, the transmitter will go into back-up temperature mode or alarm, based on configuration. Because the user entered values narrow the band, earlier detection of a failing RTD is achieved. Figure 4-4h demonstrates this capability.

Accuracy Verification

Checks are made within the flow computer to ensure the application is within the stated accuracy and is in compliance with the standards. There are many checks performed, including verification that natural gas components are configured to meet applicable standards such as ISO 5167, AGA Report No. 3, ASME MFC-3M and ASME MFC-14M. There are many other checks including equation of state for the fluid and velocity.

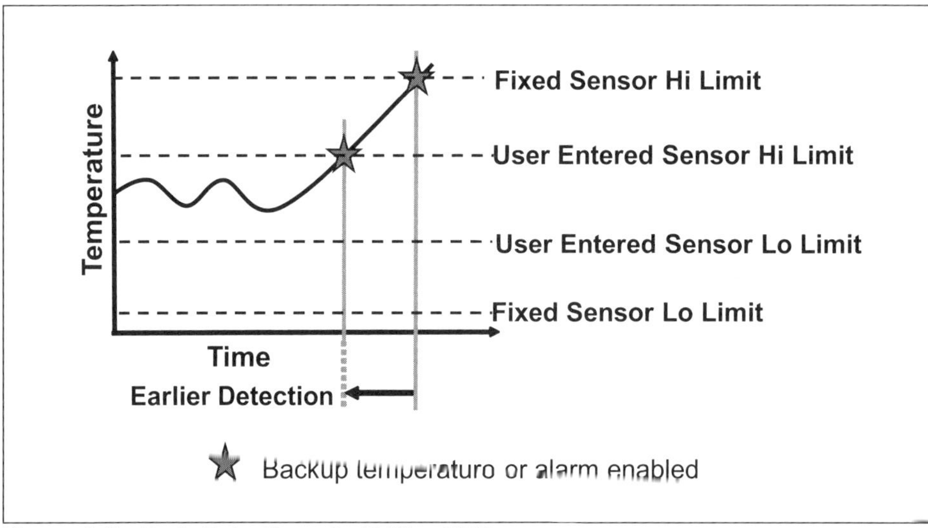

Figure 4-4h. Early Detection of a Failing RTD

Test Calculation

Test calculation allows the user to verify the transmitter is correctly calculating flow. This diagnostic is performed by fixing differential pressure, static pressure, and temperature to values that match up to a flow calculation. The transmitter will then output the flow rate given the fixed differential pressure, static pressure, and temperature. The flow value can be compared to a sizing sheet to verify accuracy.

Measurement Quality and Limit Status

Multivariable transmitters can provide measurement quality and limit status for each variable.

Possible measurement quality readings:

- Good – Displayed during normal device operation.
- Poor Accuracy – Indicates the accuracy of the variable measurement has been compromised. Example: The module temperature sensor failed and is no longer compensating the differential pressure and static pressure measurements.
- Bad – Indicates a process variable measurement has failed. Example: A differential pressure, static pressure, or process temperature sensor failure.

Possible Limit Status Readings

- Not Limited – Displayed during normal device operation.

- High Limited – Indicates the current variable reading has gone above the transmitter's maximum possible reading and is no longer representative of the actual variable measurement.
- Low Limited – Indicates the current variable reading has gone below the transmitter's minimum possible reading and is no longer representative of the actual variable measurement.
- Constant – Indicates the variable reading is set to a fixed value. Example: The totalizer has been stopped.

Sensor Temperature

Sensor temperature as an output can be used to control heat tracing (e.g., steam or electrical heating of insulation blankets or enclosures) or diagnose device overheating.

Multivariable DP Flowmeter Troubleshooting Guide

Transmitter milliamp output is zero	Verify power is applied to signal terminals Check power wires for reversed polarity Verify terminal voltage is 12 to 42.4 VDC Check for open diode across test terminal on terminal block
Transmitter not communicating	Verify the output is between 4 to 20 mA and the indication is between process saturation levels Verify clean DC power to transmitter (Max AC noise 0.2 volts peak to peak) Check loop resistance, 250–1321 Loop Resistance = (Power supply voltage - transmitter voltage) / loop current Check if unit is at an alternate HART address
Transmitter milliamp output is low or high	Verify applied process variables Verify 4- to 20-mA range points and flow configuration Verify output is not in alarm or saturation condition An analog output trim or sensor trim may be required
Transmitter will not respond to changes in measured process variables	Check to ensure that the equalization valve is closed Check test equipment Check impulse piping or manifold for blockage Verify primary variable measurement is between the 4- to 20-mA set points Verify output is not in alarm or saturation condition Verify transmitter is not in Loop Test, Multidrop, Test Calculation, or Fixed Variable mode
Digital Variable output is low or high	Check test equipment (verify accuracy) Check impulse piping for blockage or low fill in wet leg Verify transmitter sensor trim Verify measured variables are within transmitter limits
Digital Variable output is erratic	Check application for faulty equipment in process line Verify transmitter is not reacting directly to equipment turning on/off Verify damping is set properly for application
Milliamp output is erratic	Verify power source to transmitter has adequate voltage and current Check for external electrical interference Verify transmitter is properly grounded Verify shield for twisted pair is only grounded at one end
Transmitter output is normal, but LCD is off and diagnostics indicate an LCD problem	Verify LCD is installed correctly Replace LCD
Transmitter indicating a flow value and/or DP value during no flow condition	Zero DP sensor Verify DP Low Flow Cutoff setting

4-5. Vortex Flowmeters

Vortex Meter Principles

A vortex flowmeter (Figure 4-5a) uses a bluff body (body that is not streamlined), or *shedder*, placed in the fluid flow stream that causes vortices to form based on the von Karman Effect (Figure 4-5b). This same phenomenon causes a flag to flap in the wind. The shedder acts as the primary device. As the fluid flows around the shedder, velocity increases and pressure decreases on one side, while velocity decreases and pressure increases on the other side. The alternating forces cause vortices to form that are picked up by the sensing mechanism. The fluid flow rate is obtained from the frequency detected by the vortex sensor (e.g., pressure or piezoelectric sensor) of the alternating forces, which is directly proportional to the velocity of the fluid.

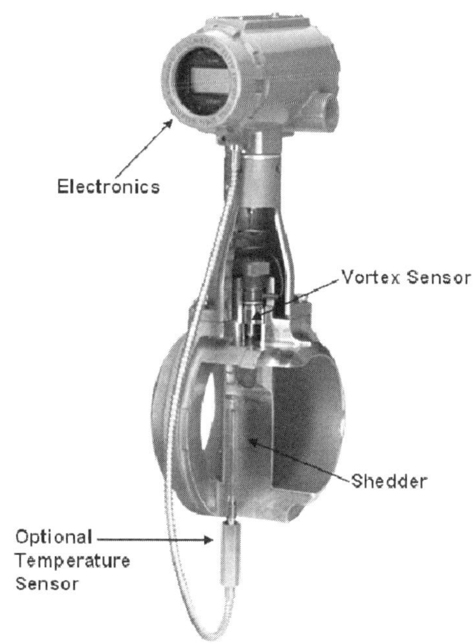

Figure 4-5a. Vortex Flowmeter

The electronics convert the frequency into a flow rate with the desired engineering units. Some vortex flowmeters are available with an optional temperature sensor that can be used for measuring the process temperature and providing temperature compensated mass flow rates.

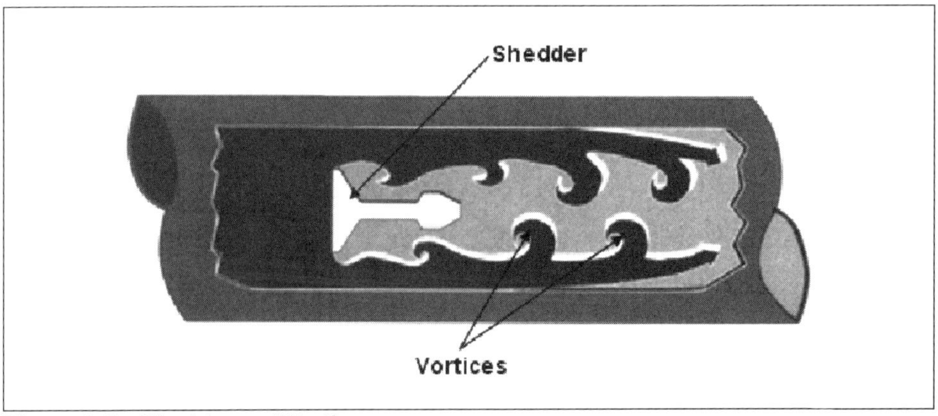

Figure 4-5b. Shedder and Vortices

Vortex Meter Selection

In general, vortex flowmeters have good accuracy and rangeability. Vortex meters produce only a small pressure loss in the pipeline. Their versatility can help simplify spare parts inventory and maintenance training. Vortex flowmeters are fairly simple and have no moving parts and few connections where leaks may occur. Their installation cost is low, and there is no need for the meter body to be winterized. Vortex flowmeters are often thought of as good all-purpose flowmeters because they can be used with liquid, gas, or steam without recalibration. Vortex meters have straight piping runs similar to differential pressure flowmeters. The accuracy is affected by swirling, distortions of the velocity profile, and changes the kinematic viscosity of the fluid.

Vortex Meter Specifications

Performance

Accuracy:

- Liquid: ±0.50 to 0.75% of volumetric flow rate for Reynolds number greater than 20,000
- Gas/Steam: ±1.0 to 1.5% of volumetric flow rate for Reynolds number greater than 20,000
- Repeatability: ±0.10 to 0.25% of actual flow rate
- Resolution: 0.02% of actual flow rate
- Rangeability: Up to 25:1, but in typical applications only up to 15:1
- Drift: ±0.1% of volumetric flow rate over one year

Process Operating Ranges

- Minimum and Maximum Flow Velocity (cross-sectional area divided by flow):
    - Liquid: Typically from about 0.2 to 7.6 meters/sec (0.7 to 25 fps) and Reynolds number must be greater than 10,000
    - Gas/Steam: Typically from about 2 to 76 meters/sec (7 to 250 fps) and Reynolds number must be greater than 10,000
- Maximum Pressure: Rated for ASME B16.5 (ANSI) up to Class 1500, DIN up to PN 160, JIS up to 40K
- Minimum and Maximum Temperature: -200 to 427°C (-330 to 800°F)
- Maximum Viscosity: Less than 10 cP, but typically in the 1-5 cP range

Wetted Materials of Construction
Stainless Steel (Standard), Nickel Alloy, Carbon Steel

Meter sizes
15 to 300 mm (0.5 to 12 inch)

Process Connections

Figure 4-5c. Flanged Vortex Flowmeter

Vortex Meter Considerations

Vortex flowmeters have minimum requirements for velocity that vary with the viscosity and density of the measured fluid. In general, they are recommended for measurement of fluids with Reynolds numbers over 10,000 (well-developed turbulent flow). They are not useful at very low

Figure 4-5d. Wafer Vortex Flowmeter

Figure 4-5e. Welded Vortex Flowmeter

flow rates or with fluids with very low Reynolds numbers because of the absence of vortex formation. If the meter is properly sized, it may be necessary to install a reducer to increase velocity at the point where the vortex meter is measuring flow. Some vortex flowmeters even come with reducing flanges built into the meter body. In addition, vortex meters are not recommended for the measurement of abrasive fluids or fibrous slurries because of the lack of vortex formation and the need to keep the shedder bar free of erosion.

Sizing

Sizing vortex flowmeter applications is critical because of minimum requirements for velocity. Most vortex flowmeter manufacturers provide sizing software programs to accommodate the need for sizing every vortex flowmeter application. These sizing programs will provide key information such as minimum and maximum measureable flow rates, minimum and maximum accurate flow rates, pressure loss, and back pressure requirements. As mentioned above, it may be necessary to select a vortex flowmeter smaller than the process piping to increase velocity.

Vortex Meter Setup and Calibration

Vortex flowmeters are typically factory calibrated with water flows to determine the K-Factor. The K-Factor is the proportionality constant that relates the measured frequency to a volumetric flow rate and will have engineering units of pulses per volume (liter, gallon, etc). The K-Factor is dependent on the meter body geometry, not the process fluid, so the same K-Factor applies to liquids, gases, and steam. In general, field calibration is not required since there are no moving parts to wear out over time. If field calibration is necessary, it typically requires a meter proving skid (portable platform with equipment and measurements to perform a reference flow calculation) or other reference flowmeter in the same line. Figures 4-5f and 4-5g show asset management system (AMS) windows for configuration and signal processing.

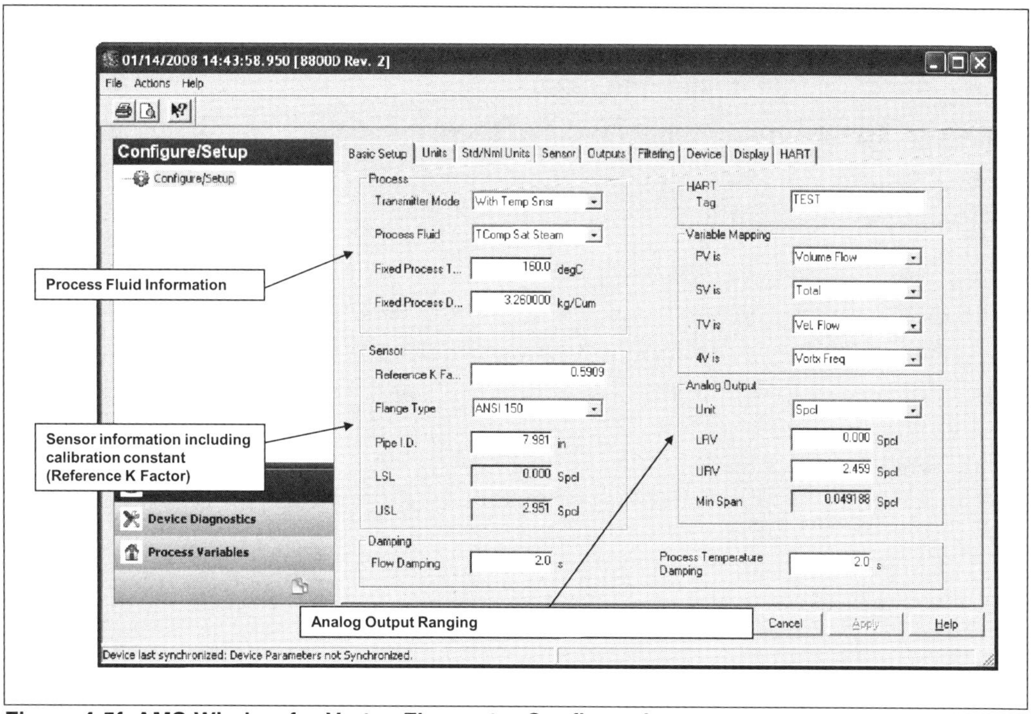

Figure 4-5f. AMS Window for Vortex Flowmeter Configuration

Navigation through the AMS windows is straightforward by selection of the tabs on the main window. Figure 4-5h is a flowchart showing the integration of the configuration decisions with the installation activities to provide an overall view of the process.

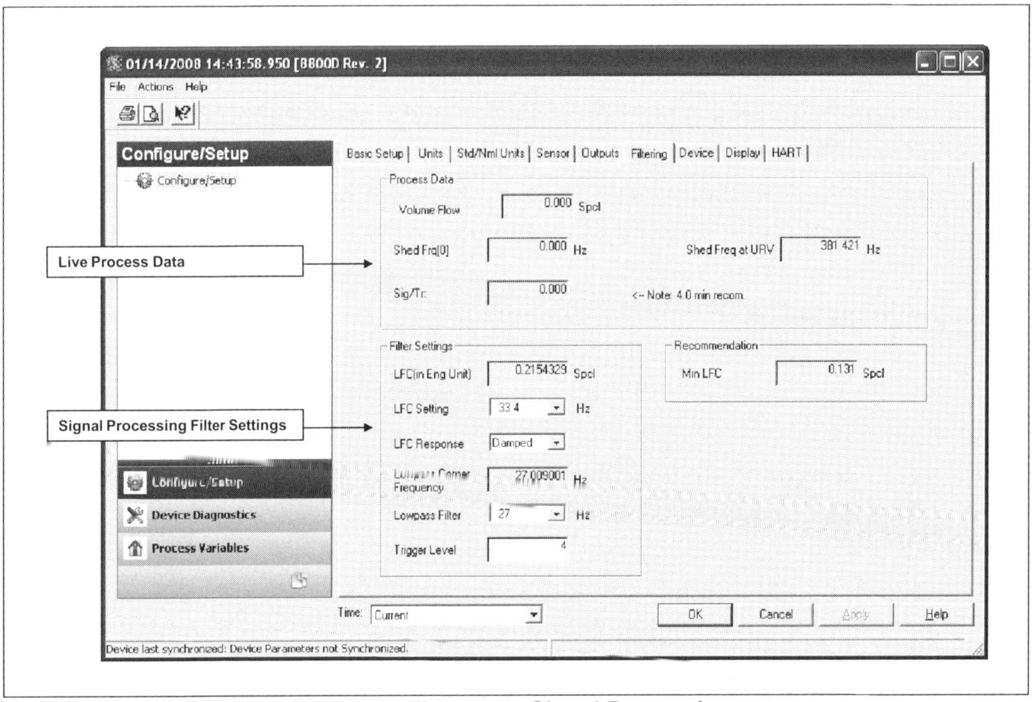

Figure 4-5g. AMS Window for Vortex Flowmeter Signal Processing

Vortex Meter Installation

It is important to design process piping so the meter body will remain full, with no entrapped air. The vortex flowmeter can be installed in any orientation without affecting accuracy. However, the following are guidelines for certain installations.

Vertical Mounting

If the vortex flowmeter will be installed in a vertical orientation:

- Install upward or downward flow for gas or steam.
- Install upward flow for liquids.

Horizontal Mounting

For steam and liquids with low solids content, it is recommended to have the flowmeter installed with the electronics to the side of the pipe. This will minimize potential measurement errors by allowing the condensate or solids to flow under the shedder bar without interrupting the vortex shedding.

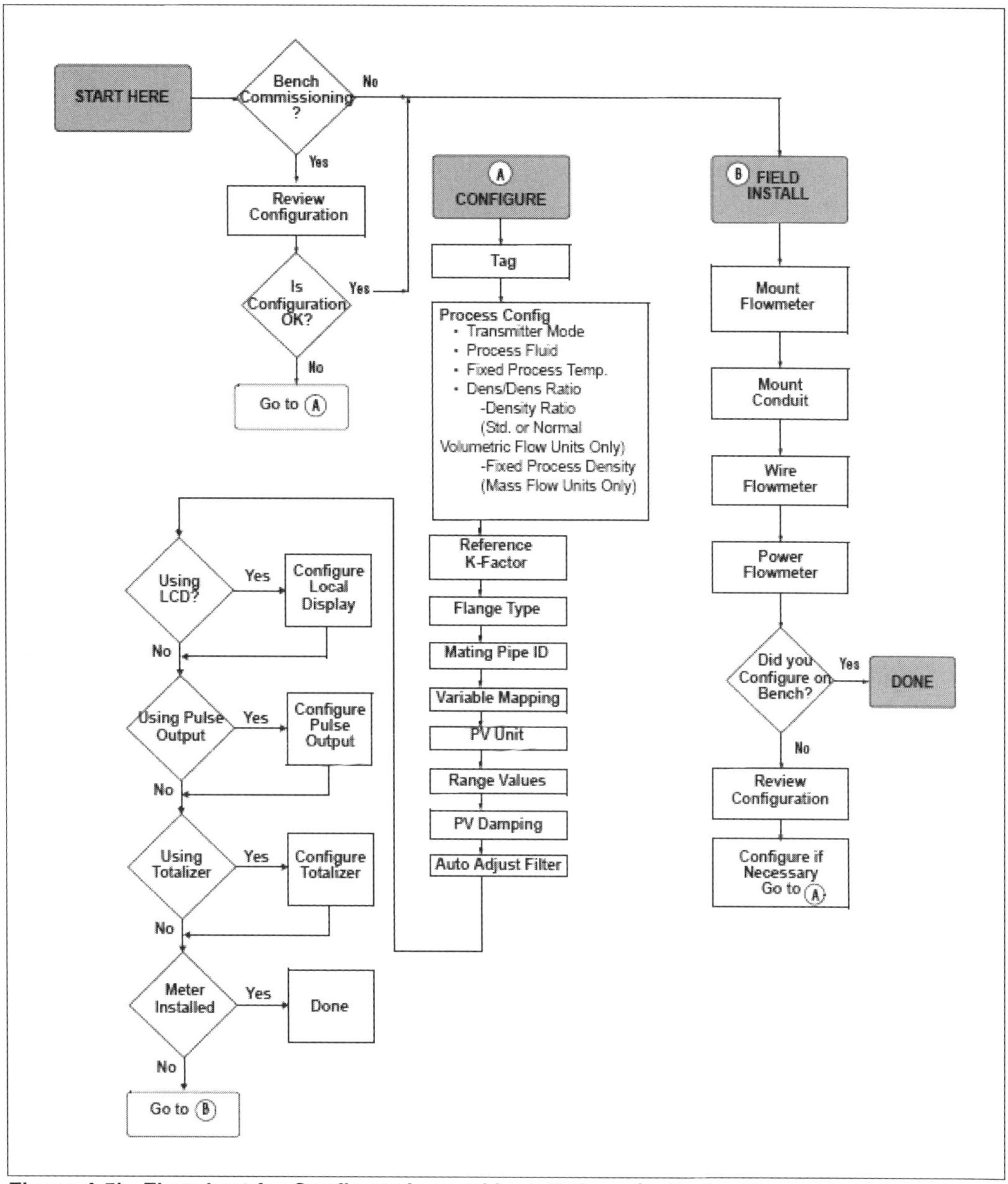

Figure 4-5h. Flowchart for Configuration and Installation of a Vortex Flowmeter

High Temperature Mounting

The maximum temperature for integral electronics is dependent on the ambient temperature where the flowmeter is installed. The electronics must not be subjected to temperatures outside of their temperature limits.

The following orientations are recommended for applications with high process temperatures:

- Install with meter body beside or below the process pipe.

- Insulation around the pipe may be necessary to maintain ambient temperature below electronics temperature limits.

Steam Installations

Avoid installation in piping low points. Such conditions may cause a water-hammer condition at start-up due to trapped condensation.

Vortex Meter Maintenance

The troubleshooting of a vortex flowmeter should be done in the order of easiest to hardest components to verify. Start by verifying that the application sizing was done with the correct process conditions and the results are acceptable for the vortex flowmeter in use. Next, verify the electronics by reviewing the configuration and performing output tests. Some vortex flowmeters have flow simulation capabilities to simplify the electronics verification process.

If the application sizing and electronics are verified, then any problem must exist in the sensor or in a process abnormality. Sensors can usually be verified using signal strength measurements or physical tests. Common process abnormalities include two-phase flow, cavitation, blockages, or low flow rates. Figures 4-5i through 4-5k show the asset management system (AMS) windows for diagnostics.

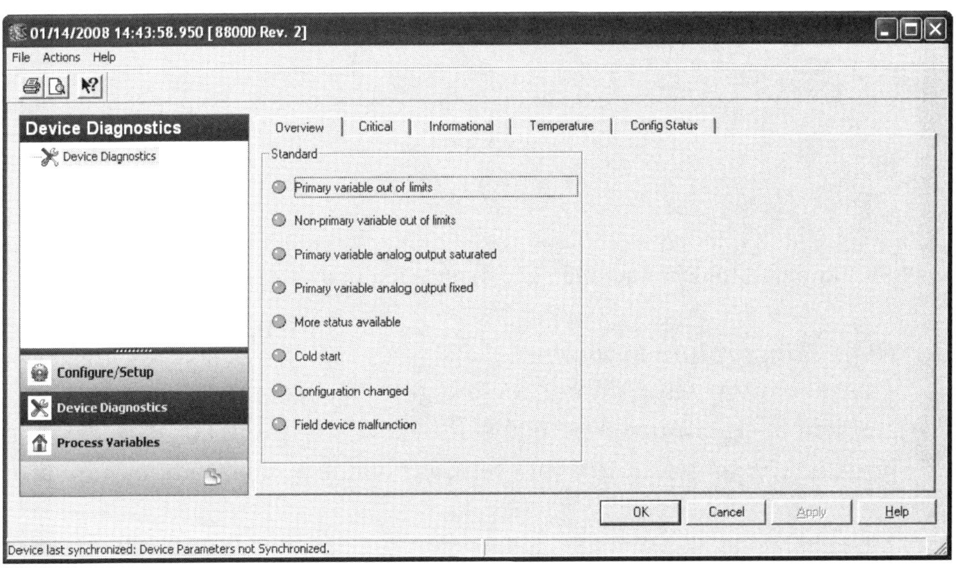

Figures 4-5i. AMS Window for Vortex Flowmeter Diagnostic Messages (Primary Variable Out of Limits)

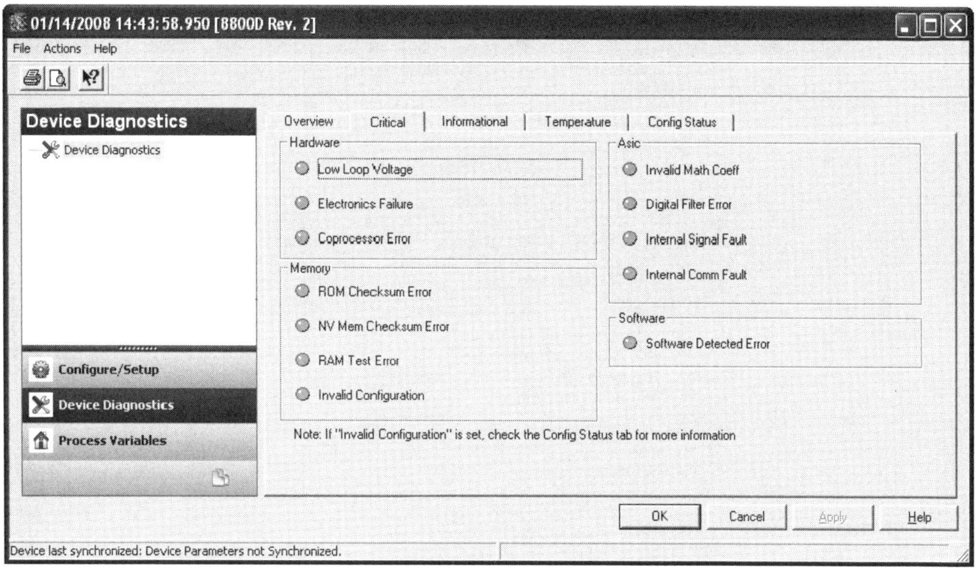

Figures 4-5j. AMS Window for Vortex Flowmeter Diagnostic Messages (Low Loop Voltage)

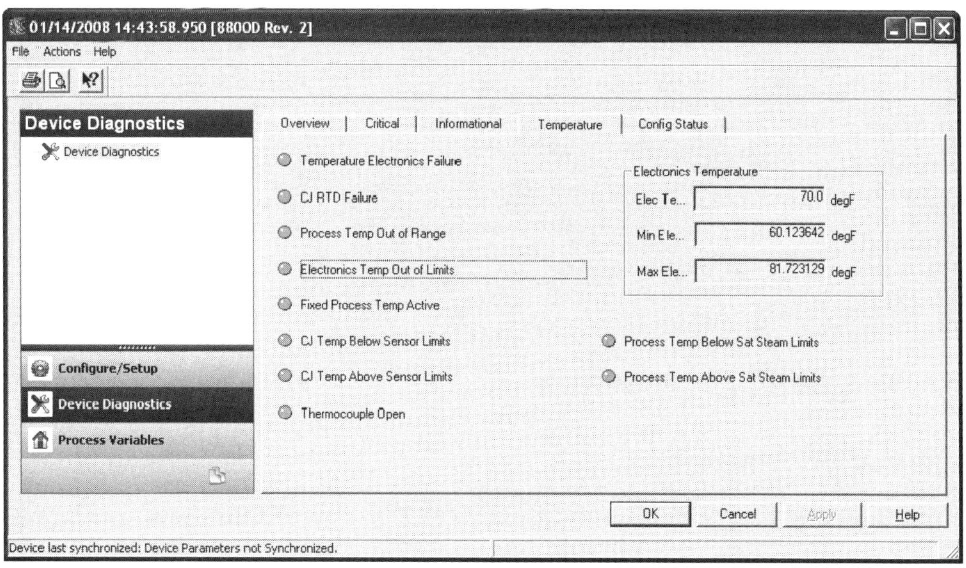

Figures 4-5k. AMS Window for Vortex Flowmeter Diagnostic Messages (Electronics Temp Out of Limits)

Vortex Meter Troubleshooting Guide

Symptom	Corrective Action	
Communication problems with HART-based Communicator	• Check for a minimum of 16.8 V dc at transmitter terminals. • Check communications loop with HART-based communicator. • Check for loop resistor (250 to 1000 ohms). • Check for transmitter in multidrop mode.	• Check for transmitter in burst mode. • Remove pulse connection if you have a three wire pulse installation. • Replace electronics.
Incorrect 4–20 mA Output	• Check for minimum 10.8 Vdc at transmitter terminals. • Check URV, LRV, Density, Special Units, LFC– compare these inputs with the sizing program results. Correct configuration. • Perform 4–20 mA loop test.	• Check for corrosion on terminal block. • Replace electronics if necessary. • Refer to "Advanced Troubleshooting" on page 5-3. • See Appendix C: Electronics Verification for electronics verification procedure.
Incorrect Pulse Output	• Check that 4–20 mA output is correct. • Check pulse counter specifications. • Check pulse mode and scaling factor. (Make sure scaling factor is not inverted).	• Perform pulse test. • Select pulse scaling so that pulse output is less than 10,000Hz at URV.
Error Messages on HART-based Communicator	See alphabetical listing in the Error Messages Table for the communicator starting on page [illegible] · Diagnostic Messages	
Flow in Pipe, No Output	Basics • Check to make the sure that the meter is installed with the arrow in the direction of process flow. • Perform basic checks for Incorrect 4–20 mA Output Problem (see Incorrect 4–20 mA Output). • Check and correct configuration parameters in this order: Process Config - transmitter mode, process fluid, fixed process temperature, density/density ratio (if required), reference K-factor, flange type, mating pipe ID, variable mapping, PV unit, range values - (URV, LRV), PV damping, auto filter adjust, pulse mode and scaling (if used). • Check sizing. Make sure flow is within measurable flow limits. Use Instrument Toolkit for best sizing results. • Refer to "Advanced Troubleshooting" on page 5-3 • See Appendix C: Electronics Verification for electronics verification procedure. Electronics • Run a self test with a HART-based interface tool. • Using sensor simulator, apply test signal. • Check configuration, LFC, trigger level, STD vs. actual flow units. • Replace electronics.	Application Problems • Calculate expected frequency (see Appendix C: Electronics Verification). If actual frequency is the same, check configuration. • Check that application meets viscosity and specific gravity requirements for the line size. • Recalculate back pressure requirement. If necessary and possible, increase back pressure, flow rate, or operating pressure. Sensor • Check torque on sensor nut (32 ft-lb). For 1-8 inch meter body with ANSI 1500 flanges torque on sensor nut should be 50 ft-lbs. • Inspect coaxial sensor cable for cracks. Replace if necessary. • Check that sensor impedance at process temperature is > 1 Mega-Ohm (will function down to 0.5 Mega-Ohms). Replace sensor if necessary ("Replacing the Sensor" on page 5-16). • Measure sensor capacitance at SMA connector (115-700pF).

5-2

4-6. Magnetic Flowmeters

Magmeter Principles

Magnetic flowmeters, also called *magmeters,* provide obstruction-less flow measurement that is ideal for metering any electrically conductive process fluid.

The operating principle of magmeters is based on Faraday's law of induction (Equation 4-6a and Figure 4-6a), which states that a voltage will be induced in a conductor moving through a magnetic field.

$$E = k * B * D * V \tag{4-6a}$$

The magnitude of the induced voltage (E) is directly proportional to the conductor velocity (V). Magnetic field coils placed on opposite sides of the pipe wall generate a magnetic field (B). As the conductive process fluid moves through the magnetic field generated by the coils, electrodes sense the induced voltage. The distance between electrodes represents the conductor width (D). An insulating liner prevents the signal from shorting to the pipe wall. The output voltage from the magnetic flowmeter sensor is amplified and sent to a magnetic flowmeter transmitter where the signal is conditioned and sent to the control system. Figure 4-6b shows the functional construction.

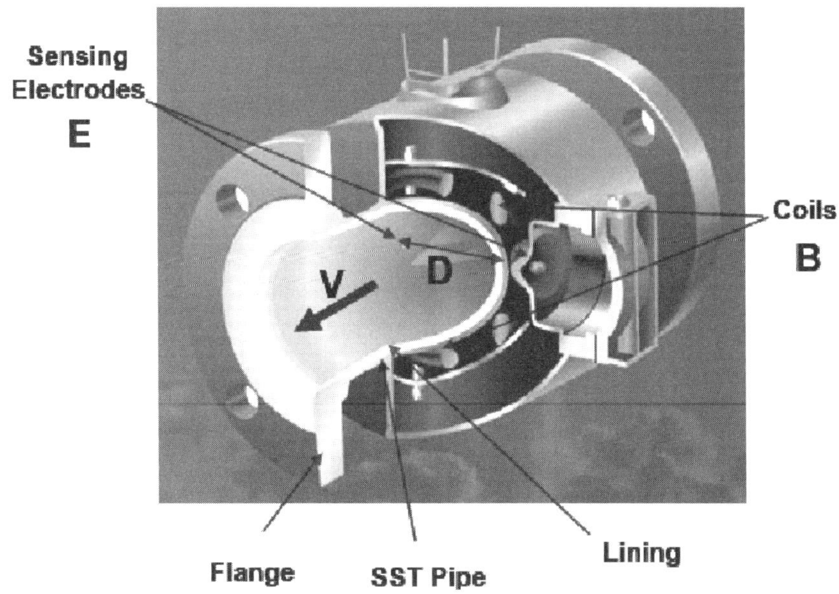

Figure 4-6a. Faraday's Law of Induction

Magmeter Selection

In general, magnetic flowmeters have excellent accuracy and rangeability. As was mentioned, magnetic flowmeters produce no pressure loss in the pipeline. The wide range of liner and electrode materials makes them suitable for any conductive liquid application. Any stream that has some process or utility water usually has an electrical conductivity above the low limit. Deionized water and many pure monomer streams are not conductive enough. Magnetic flowmeters are fairly simple and have no moving parts, making them ideal replacements for mechanical meters. Their installation cost is low, and there is no need for the meter body to be win-

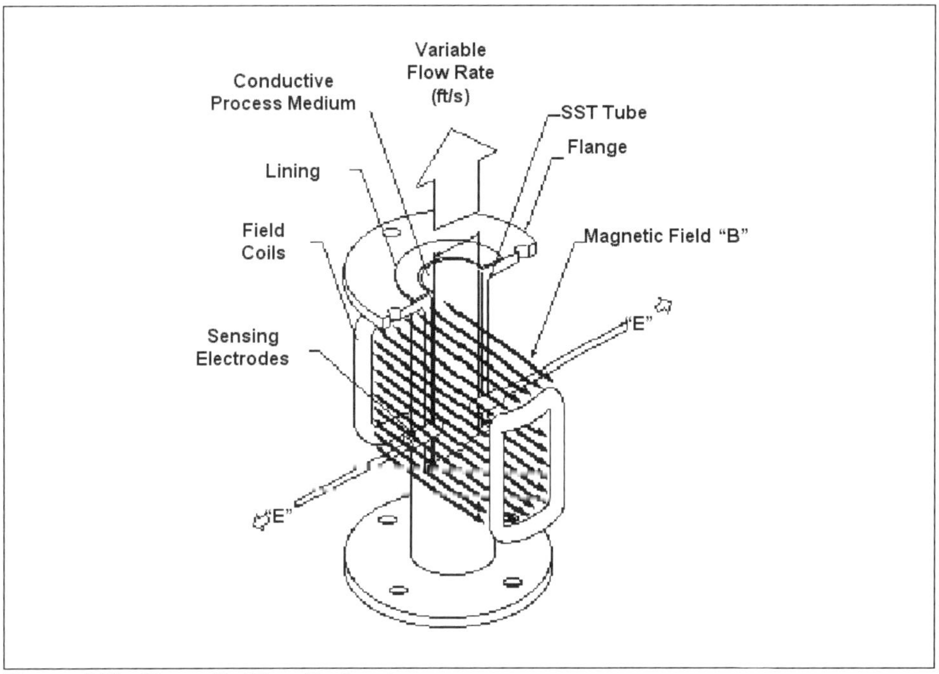

Figure 4-6b. Magnetic Flow Technology

terized. Magnetic flowmeters are often the first choice for flow measurement technology when the fluid is electrically conductive.

Magmeter Specification

Performance

- Accuracy: ±0.15 to 0.50% of volumetric flow rate for velocities from 0.3 to 12 meters/sec (1 to 40 fps), below 0.3 meters/sec (1 fps) accuracy degrades but often exceeds other volumetric flow measurement technologies.

- Repeatability: ±0.10 to 0.25% of volumetric flow rate.

Resolution: Magnetic flowmeter sensitivity is generally around 0.01 ft/sec. The resolution percentage is dependent on the configured range of the meter. The typical maximum measurable velocity for any magmeter is around 40 ft/sec, making the resolution % 0.025%.

- Rangeability: Up to 1000:1, but in typical applications only up to 30:1

- Drift: ±0.1% of rate over six months

Process Operating Ranges

- Minimum and Maximum Flow Velocities:
    - Typically from about 0.1 to 4 meters/sec (0.3 to 12 fps)
    - Absolute range from 0.01 to 12 meters/sec (0.04 to 40 fps)
- Maximum Pressure: Rated for ASME B16.5 (ANSI) up to Class 2500, DIN up to PN 160, JIS up to 40K, various sanitary fittings including Tri-Clamp, IDF, SMS, and DIN fittings.
- Minimum and Maximum Temperature: Liner dependent, but typically ranges from -29°C to 180°C (-20°F to 350°F). Ceramic liners can be installed for higher process temperatures.
- Maximum Viscosity: No viscosity limits.
- Minimum Conductivity: 20 µSiemens/cm for water; 5 µSiemens/cm for other processes are typical. Special meters are available that can measure lower conductivity fluids.

Wetted Materials of Construction

Liners

- PTFE (Teflon)
- ETFE (Tefzel)
- PFA
- Polyurethanes
- Linatex or Natural Rubber
- Neoprene
- Hard Rubber
- Ceramic

Electrodes

- 316L SST
- Nickel Alloy 276
- Titanium
- Tantalum
- Platinum / Iridium

Meter Sizes

0.5 inch (13 mm) up to 48 inch (1.2 m) is common, larger sizes are available.

Process Connections

Magnetic flowmeter bodies have flanged, wafer, or sanitary connections for mounting inline in a process pipe (Figures 4-6c, 4-6d, and 4-6e).

Figure 4-6c. Flanged Magnetic Flowmeter

Figure 4-6d. Wafer Magnetic Flowmeter

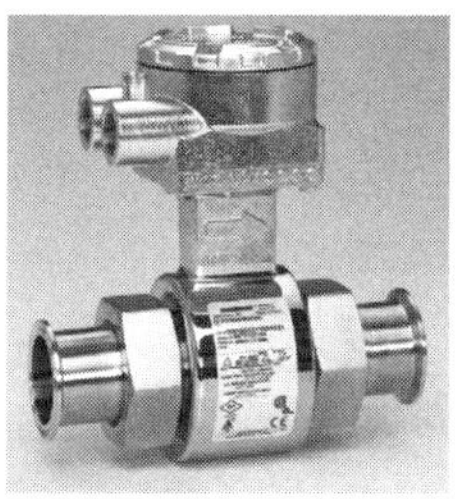

Figure 4-6e. Sanitary Magnetic Flowmeter

Magmeter Considerations

Magnetic flowmeters require less straight pipe run upstream of the meter than inline flowmeter technologies except Coriolis, and also tend to be the most forgiving when minimum straight runs are not met. They provide excellent performance and rangeability in conductive fluid applications.

As mentioned, they do not generate any pressure loss, and can handle laminar as well as turbulent and transitional flow. Additionally, they are an excellent technology for slurry flows such as pulp stock or mining slurries. Things to watch for include low fluid conductivity, potential for coating of the electrodes, and material compatibility with the process.

Sizing

Because of their wide rangeability, magnetic flowmeters will often be the same size as the process piping. Typically this practice is acceptable, unless the process has the potential to coat the electrodes. In these applications, the magnetic flowmeter should be sized so that normal operating fluid velocity is between 2 and 3.3 meters/sec (6 and 10 fps) to create a self cleaning effect on the head of the electrodes. This may result in the magmeter being one or two line sizes smaller than the piping.

Magmeter Setup and Calibration

Magnetic flowmeters are typically factory calibrated with water flows to determine the calibration number or K-Factor. The calibration number is the proportionality constant that relates the measured voltage to the velocity of the fluid flowing through the meter. The calibration number is dependent on the magnetic field strength and the distance between the electrodes, not the process fluid, so the same calibration number applies to all conductive liquids. In general, field calibration is not required since there are no moving parts to wear out over time. If field calibration is necessary, it typically requires a proving skid or other reference flowmeter in the same line. New diagnostic capabilities are available with some magnetic flowmeters that can extend the interval between calibrations or even eliminate the need for field calibrations. Figures 4-6f through 4-6i show asset management system (AMS) windows for configuration of the magnetic flowmeter for an application.

Magmeter Installation

It is important to design process piping so the meter body will remain full, with no entrapped air. A magnetic flowmeter can be installed in any orientation without affecting accuracy. However, the following are guidelines for certain installations.

Vertical Mounting

If the magnetic flowmeter will be installed in a vertical orientation, install so that fluid flow is upward through the sensor.

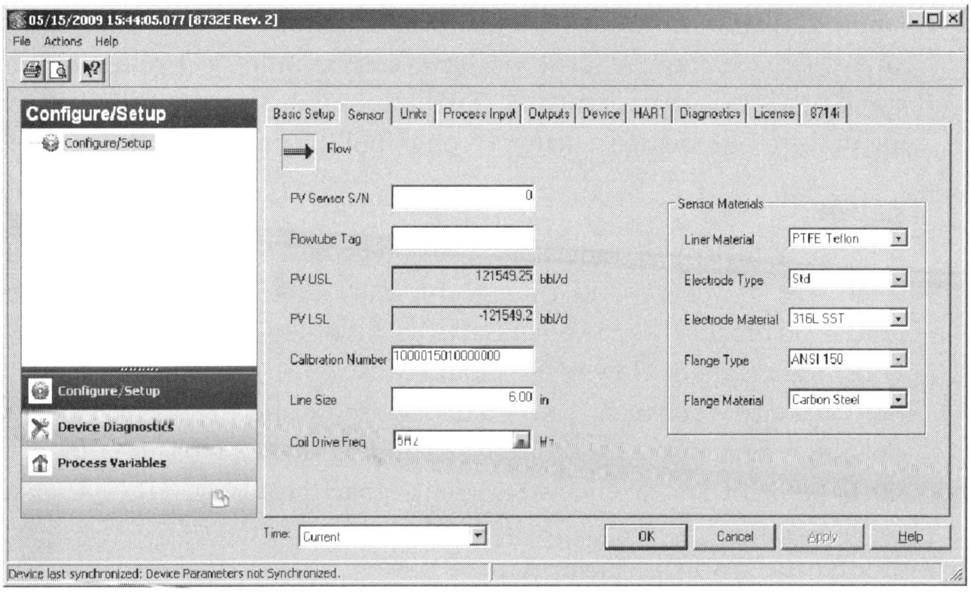

Figure 4-6f. AMS Window for Magnetic Flowmeter Sensor Configuration

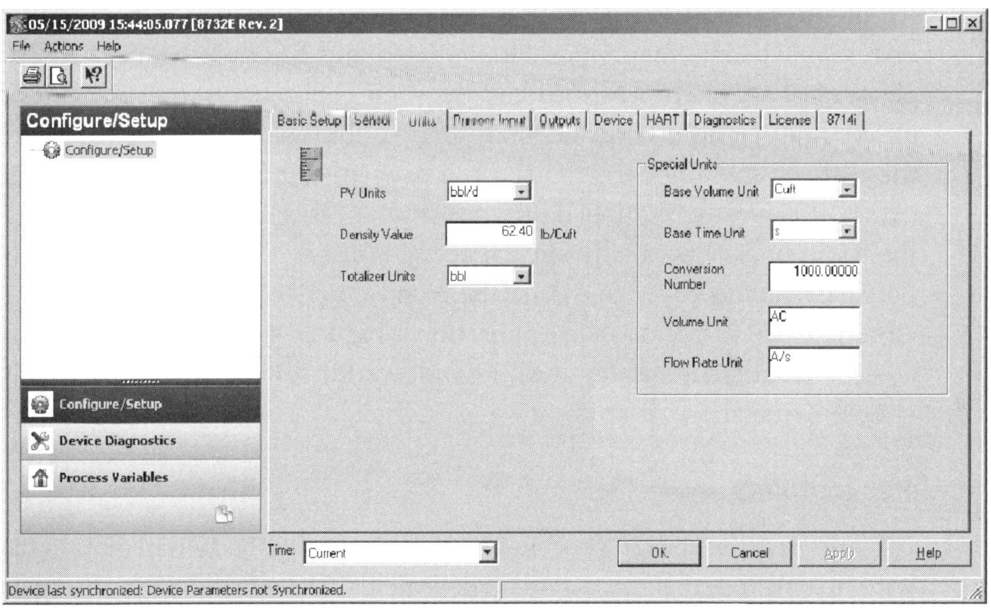

Figure 4-6g. AMS Window for Magnetic Flowmeter Units Configuration

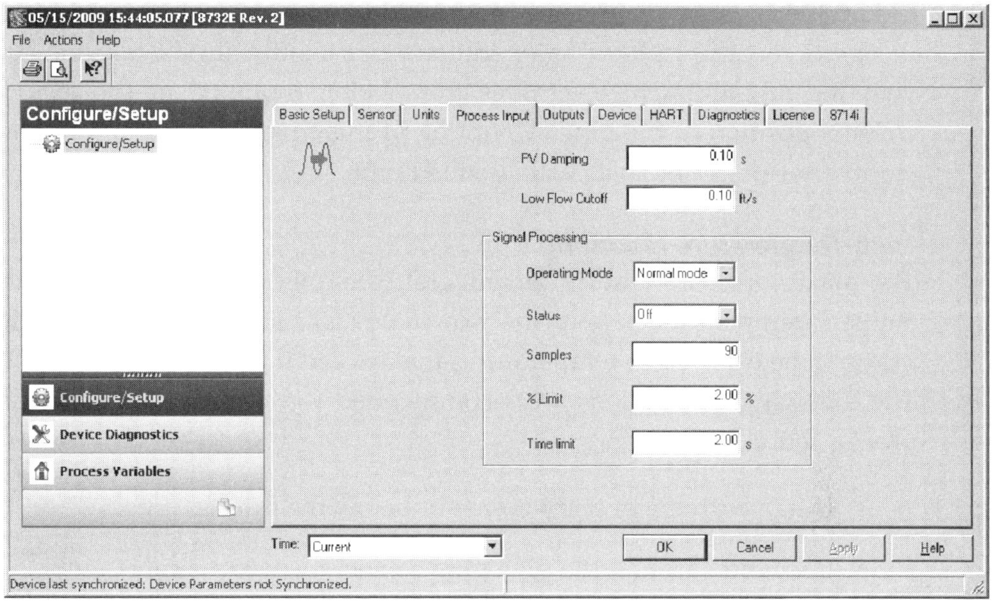

Figure 4-6h. AMS Window for Magnetic Flowmeter Process Input Configuration

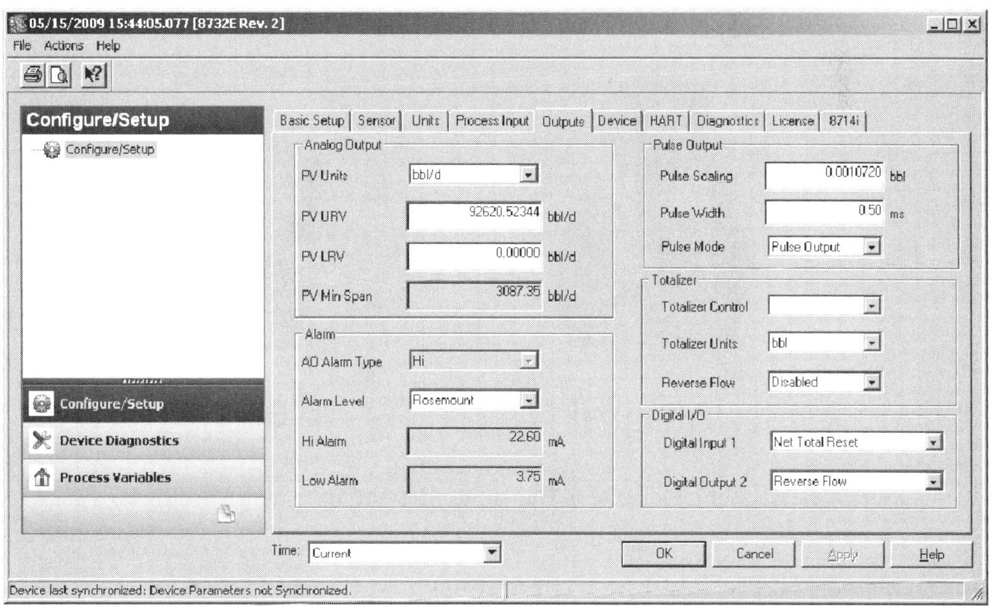

Figure 4-6i. AMS Window for Magnetic Flowmeter Outputs Configuration

Horizontal Mounting

Mount the sensor body in a low point in the process piping to ensure that the meter remains full during operation. Care should also be taken to prevent the measurement electrodes from being mounted at the top (12 o'clock) position of the pipe. Mounting in this manner can result in false empty pipe detection in a partially filled pipe.

High Temperature Mounting

The maximum temperature for integral electronics is dependent on the ambient temperature where the flowmeter is installed. The electronics must not be exposed to temperatures that exceed their temperature limits. The following orientations are recommended for applications with high process temperatures.

- Install with meter body beside or below the process pipe.
- Insulation around the pipe may be necessary to maintain ambient temperature below electronics temperature limits.

The meter can have an integral (direct) mounted transmitter that eliminates the need to install and wire a remote transmitter (Figure 4-6j).

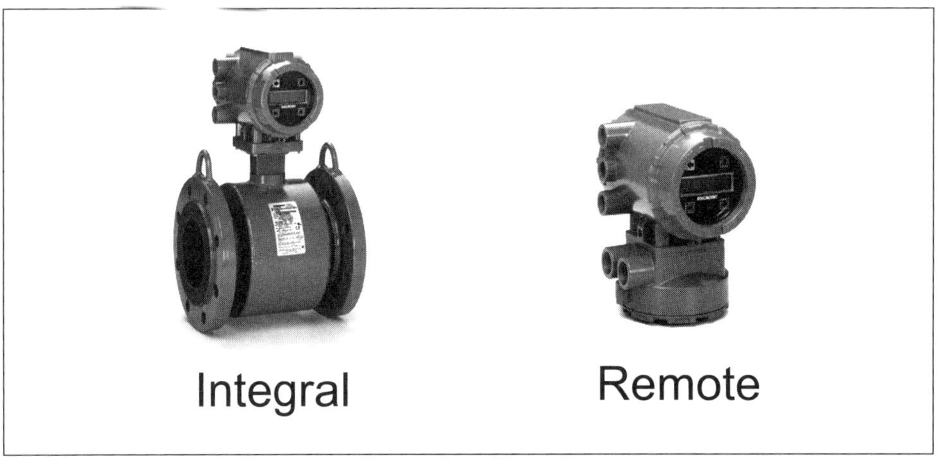

Figure 4-6j. Magnetic Flowmeter Transmitter Mounting

The meter body and process fluid must be connected to a reference ground both upstream and downstream. Metal unlined piping can provide a ground path from the meter to the fluid via the piping flanges. Ground straps connect the flanges upstream and downstream to the meter body (Figure 4-6k). If pipe supports do not provide a path from the pipe to a good ground then additional ground straps should be used to connect the pipe to a good ground. For lined or plastic piping where the process is not in contact with a metal pipe wall, grounding rings that are in contact with the process fluid must be inserted between the meter and the pipe flange upstream and downstream. These rings are then connected by straps to the meter and flanges (Figure 4-6l). Alternatively, grounding electrodes can be used (Figure 4-6m).

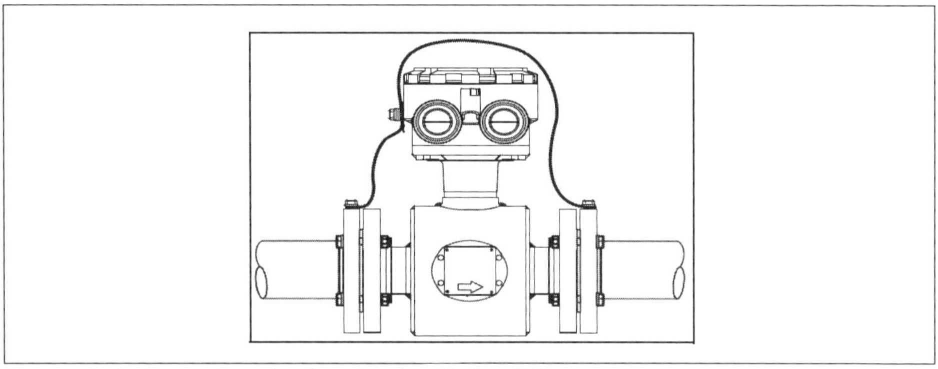

Figure 4-6k. Grounding for Conductive Unlined Pipe

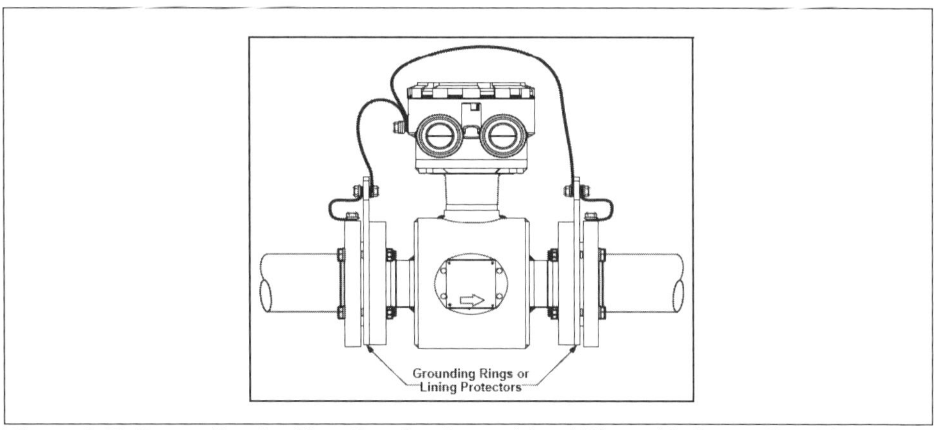

Figure 4-6l. Grounding for Non-Conductive or Lined Pipe with Grounding Rings

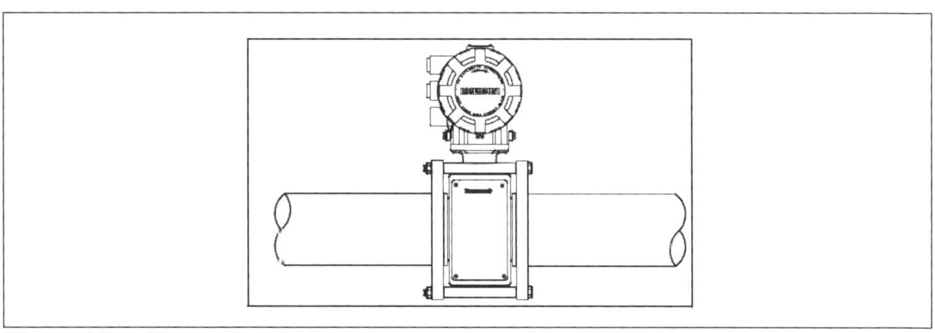

Figure 4-6m. Grounding for Non-Conductive or Lined Pipe with Grounding Electrodes

Magmeter Maintenance

Troubleshooting a magnetic flowmeter should be done in the order of easiest to hardest components to verify. Start by verifying that the transmitter is configured correctly for the application by reviewing the configuration and performing output tests. After verifying the electronics, verify the wiring between the transmitter and sensor (in remote mount applications), and the process grounding connections of the sensor to the process piping. Wire terminations should be tight and clean. Moisture should not be present in the junction box at the transmitter or the sensor. Process ground references are required for the transmitter to "zero" out any stray voltages that may be present in the process stream.

Sensors can usually be verified using resistance measurements of the coil and electrode circuitry. After verification of the transmitter, wiring connections, and the sensor are complete, if a problem was not identified, the source is likely process related. Common process abnormalities include low fluid conductivity, process fluids with EMI noise, partially filled pipes, or chemical reactions. Many magnetic flowmeter transmitters have diagnostic capabilities that are capable of diagnosing many of these common process issues, and in some cases even performing meter verifications without requiring external equipment. Figures 4-6n through 4-6q show asset management system (AMS) windows for diagnostics.

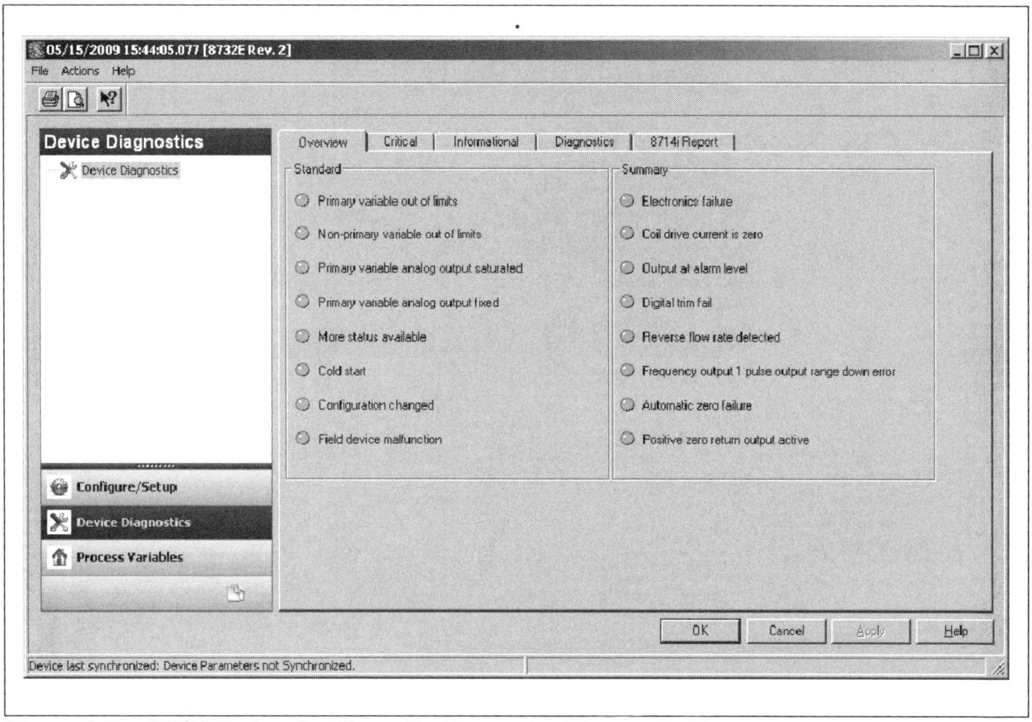

Figure 4-6n. AMS Window for Magnetic Flowmeter Diagnostic Messages

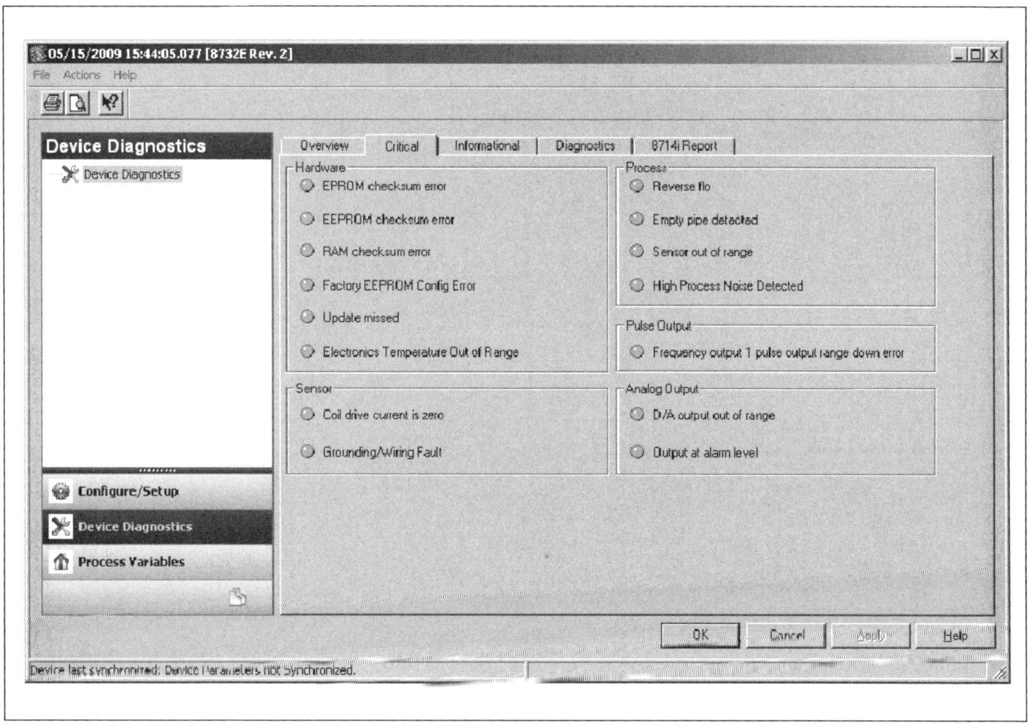

Figure 4-6o. AMS Window for Magnetic Flowmeter Diagnostic Messages

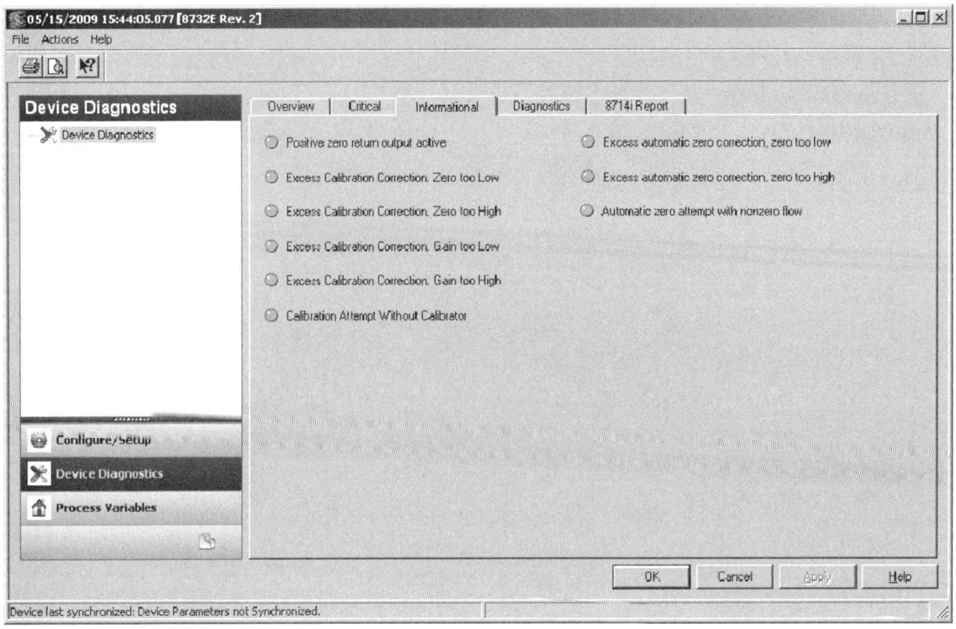

Figure 4-6p. AMS Window for Magnetic Flowmeter Diagnostic Messages

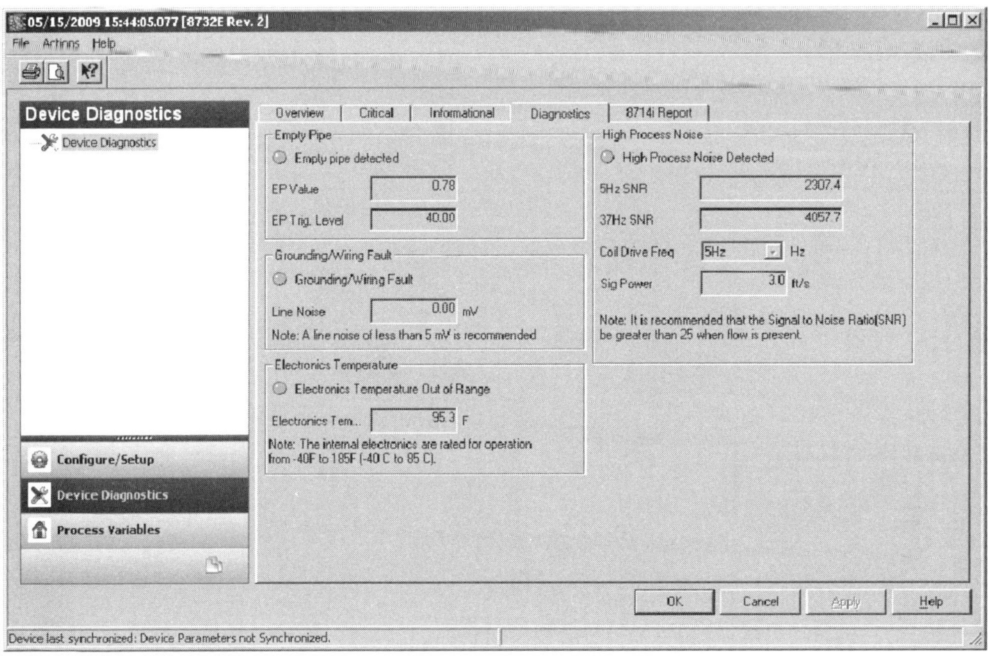

Figure 4-6q. AMS Window for Magnetic Flowmeter Diagnostic Messages

4-7. Coriolis Mass Flowmeters

Coriolis Meter Principles

Mass Flow

A dual-tube Coriolis meter (Figure 4-7a) consists of a sensor with two flow tubes that are vibrated in opposition to each other and a transmitter. The frequency of oscillation of the sensor tubes is detected by coil assemblies called pick-offs—one on the inlet side of the tubes and one on the outlet side of the tubes.

When the tubes are filled with fluid, but there is no flow, the inlet and outlet sides of the tubes are subject to the same force operating in the same direction. The inlet and outlet pick-offs register oscillations at the same time. However, when there is flow through the tubes, it is accelerating (changing direction) on the inlet side and decelerating on the outlet side. This produces a slight twist in the flow tubes, which causes the inlet pick-off to register before the outlet pick-off. The magnitude of the time delay between the inlet and outlet sides of the tubes is directly proportional to the mass flow of the fluid.

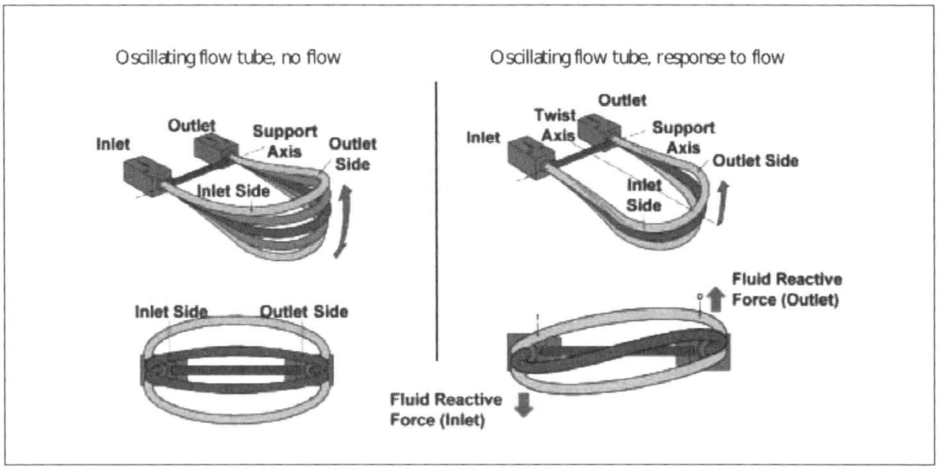

Figure 4-7a. Dual Tube Coriolis Sensor

Density

The density of the fluid can be directly determined by changes in the natural frequency of the tubes' oscillations. The natural frequency is based on the mass of the flow tubes themselves, plus the mass of the fluid. When the total mass increases, the natural frequency decreases.

Because the volume of fluid contained within the flow tubes is constant, and because the mass of the flow tubes is constant, the only cause of a change in total mass (and by extension the natural frequency) is a change in fluid density.

Temperature

Most Coriolis meters include an RTD for process temperature measurement. The temperature value can be used to compensate measurement for the effect of temperature on flow tube stiffness. This effect is typically measured at the factory and included in the factory calibration.

Volume Flow

Coriolis meters can calculate a highly-accurate volume flow measurement based on the direct mass flow, density, and temperature measurements.

Concentration

Coriolis meters can also calculate a highly accurate concentration measurement when the concentration of one or two components has a dominant effect on the fluid density.

Coriolis Meter Selection

As mentioned, Coriolis sensors provide direct measurement of both mass flow and density, with an included direct temperature measurement, and can derive volume flow and concentration. Coriolis meters can provide a bidirectional flow measurement.

Fluid viscosity, Reynolds number, and velocity profile do not affect Coriolis sensor measurements. Coriolis meters can measure the actual density and mass flow of a stream that results from a mixture of fluid states (gas, liquid, and solid). For example, Coriolis meters can measure the true density and mass flow over a specified concentration range of bubbles and solids. Coriolis sensors offer the best accuracy and turndown of any flowmeter. Coriolis meters can provide a more accurate totalization of charges to operations than loss in weight calculations form load cells.

Coriolis sensors have no moving parts and require no periodic maintenance except for cleaning in applications with fouling or coating problems. There are no flow conditioning or straight-run requirements for Coriolis meters. Coriolis meters can provide in-situ verification of meter performance without the need for special tools or process downtime.

The life-cycle cost of a Coriolis meter comes down to mostly the cost of the meter and wiring. If you expand the concept of a Coriolis meter to include process performance benefits, the life-cycle cost is often negative in that the reduction in cost of goods from an improvement in process efficiency and/or the increase in revenue from greater process capacity outweighs the initial cost of the Coriolis meter.

For process streams, a Coriolis meter is the right choice if the materials of construction match the process requirements and a large size (e.g., > 4 inch) does not cause the meter cost to exceed the savings in life-cycle cost from reduced maintenance and better process performance. For utility streams, accurate mass flow and density measurements may not be important.

Coriolis meters come in various tube designs. Besides the dual U-tube and straight tube designs shown in Figure 4-7a there are other geometric configurations for the tubes as shown in Figure 4-7b. The straight tube design is particularly used for streams containing particles to prevent impact and erosion at the bends and accumulation in the tube.

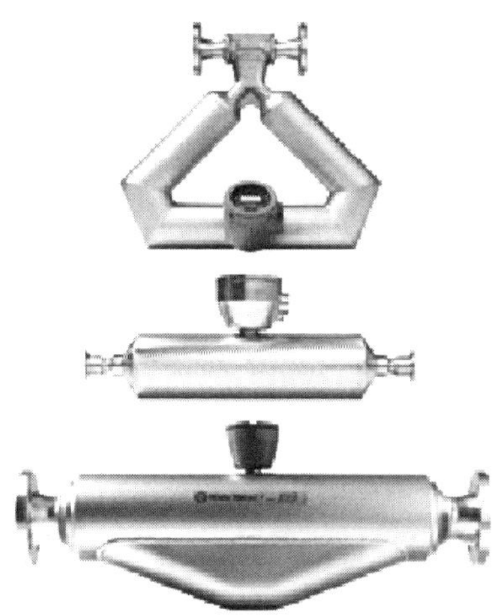

Figure 4-7b. Various Coriolis Sensors

While the Coriolis sensor is inherently multivariable (mass, density, volume, and typically temperature), it can be paired with either a single variable or multivariable transmitter, depending on application requirements. A typical installation does not require a temperature sensor or a flow computer; pressure data can be input to the transmitter if pressure compensation is required. For applications that require highly accurate temperature correction, an external temperature measurement may also be input to the transmitter.

Coriolis Meter Specification

Performance

1. Accuracy:
   - Mass and Volume Flow:
     - ±0.05% for liquid
     - ±0.35% for gas
   - Density: ±0.0002 g/cm^3
   - Temperature: ±1°C ± 0.5%
2. Repeatability: ±0.025% of actual flow rate
3. Resolution: 0.017%
4. Rangeability: 200:1
5. Drift: None

Process Operating Ranges

1. Minimum and Maximum Mass Flow Rates: 0.01 kg/h to 2,550,000 kg/h (0.022 lb/hr to 5,621,792 lb/hr)
2. Maximum Pressure:
   - Sensor rating up to 119 bar (1730 psi) for typical stainless steel, 197 bar (2855 psi) for nickel alloy, and 413 bar (6000 psi) for high-pressure models
   - Secondary containment rated for ASME B31.3 up to 58 bar (850 psi)
3. Minimum and Maximum Temperature: 240 to +350°C (400 to +662°F)
4. Minimum and Maximum Velocity: See Min/Max Flow Rates (velocity is mass flow rate divided by the product of the cross-sectional meter volume and density)
5. Minimum and Maximum Viscosity: None

Wetted Materials of Construction

304L, 316L, C-22 Ni Alloy, Titanium, Super Duplex

Meter sizes
2mm to 300mm (0.1 to 12 inch)

Coriolis Meter Considerations

As has been mentioned, Coriolis meters have no special pipe-run requirements, are immune to variations in flow rate and fluid composition and physical properties, and are straightforward to install. The maintenance requirements are negligible, if cleaning is not needed, calibration for regulatory compliance purposes is not necessary, and pressure and temperature is either within the calibration range or used for correction of the calibration. Coriolis meters are suitable for nearly all process control applications where they meet the process requirements for meter size and materials of construction. They are available in high-temperature and high-pressure models, and with a variety of wetted materials for process compatibility. Coriolis meters maintain their high measurement accuracy over a wide turn-down.

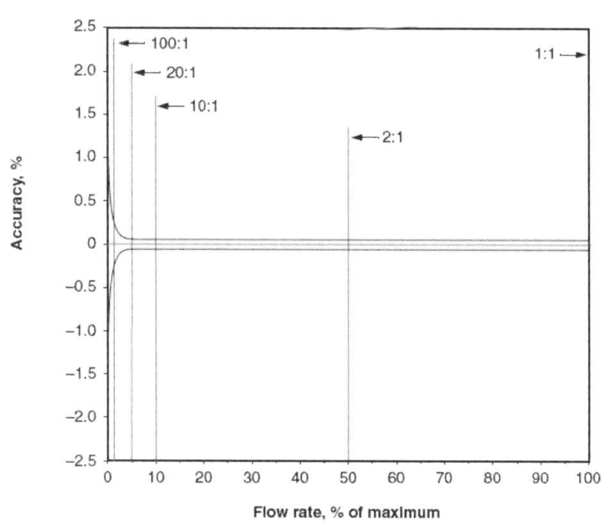

Figure 4-7c. Coriolis Meter Measurement Accuracy

The direct density measurement provided by Coriolis meters can provide precise, in-line concentration measurement without the need for laboratory analysis or additional devices. This feature is often of great value in the food and-beverage and oil-and-gas industry segments.

It is possible to measure abrasive fluids and slurries (e.g., sand) with straight tube Coriolis meters with no impact on reliability. Some Coriolis designs can provide improved measurement performance in applications involving entrained gas in liquids.

Environmental and installation effects occasionally experienced when using Coriolis meters can be divided into two general categories: zero effects, which are flow rate dependent; and span effects, which are flow rate independent.

Zero effects

Zero effects are caused by zero instability or inaccuracies in the zero value resulting from any of several factors. Zero effects are flow rate-dependent because the zero value is a component of the total mass flow calculation; therefore, as the flow rate decreases, the zero value becomes a larger portion of the total result, and zero effects are magnified. Consequently, performance at lower flow rates will magnify effects involving zero stability or errors in the zero value.

Temperature, for example, can cause a zero error. An operating temperature that is significantly different than the calibration temperature causes a random error in the zero value. This effect can be zeroed out at operating conditions.

Span effects

Because span is a constant ratio, span errors will be manifested as a direct error in mass flow calculation regardless of flow rate (i.e., they are flow rate-independent).

Pressure, for example, can cause a span error. This is due to the fact that pressure stiffens the flow tubes. Flow tube stiffness directly affects a Coriolis meter's ability to produce a flow signal (see Principles). As process pressure increases, tube stiffness increases. In a field installation, elevated pressure will cause a Coriolis meter to under-register when compared to a known reference (negative bias). Although this effect cannot be zeroed out, it can be compensated for by a pressure measurement input.

Sizing

Coriolis meter sizing is highly important because of the need to balance flow rate, accuracy, and pressure loss. In general, Coriolis meter manufacturers provide sizing software programs to accommodate the need for sizing Coriolis meter applications. These sizing programs consider key information such as minimum and maximum flow rates, desired accuracy, and maximum pressure loss.

Coriolis Meter Applications

Applications and fluids measured include (among others) batching, bioreactor feeds, bleaching agents, carbonation and bottling, catalyst feeds, compressed natural gas, concentration measurement, coatings, combustion control, continuous blending, custody transfer, drilling mud quality and returns, endpoint detection, evaporator control, fermentation, hygienic applications, interface detection, leak detection, loss control, mass balance, net oil measurement, pill coating, mixing, dilution, fueling, sterile gas feeds, and vial filling.

Soft Drink Blending in the Food and Beverage Industry

Soft-drink blending is one type of continuous blending application (Figure 4-7d). Soft drink process lines can include few or many ingredients. More sophisticated process lines include CO_2 injection and the ability to handle multiple recipes. Almost all producers require concentration measurement in °Brix (for High Fructose Corn Syrup, or HFCS), hygienic conditions, a packaging solution, and tight control of temperature, from very hot (for pasteurization) to very cold (for storage).

Typical measurement and control challenges associated with soft drink blending include:

- High fluid viscosity
- Wide viscosity range
- Wide turndown requirements to accommodate the pressure differences between recipes
- Limited space and opportunities for instrumentation

In addition, producers have the following business challenges to remain competitive:

- Increase throughput
- Improve quality and consistency
- Reduce waste
- Minimize raw material inventories

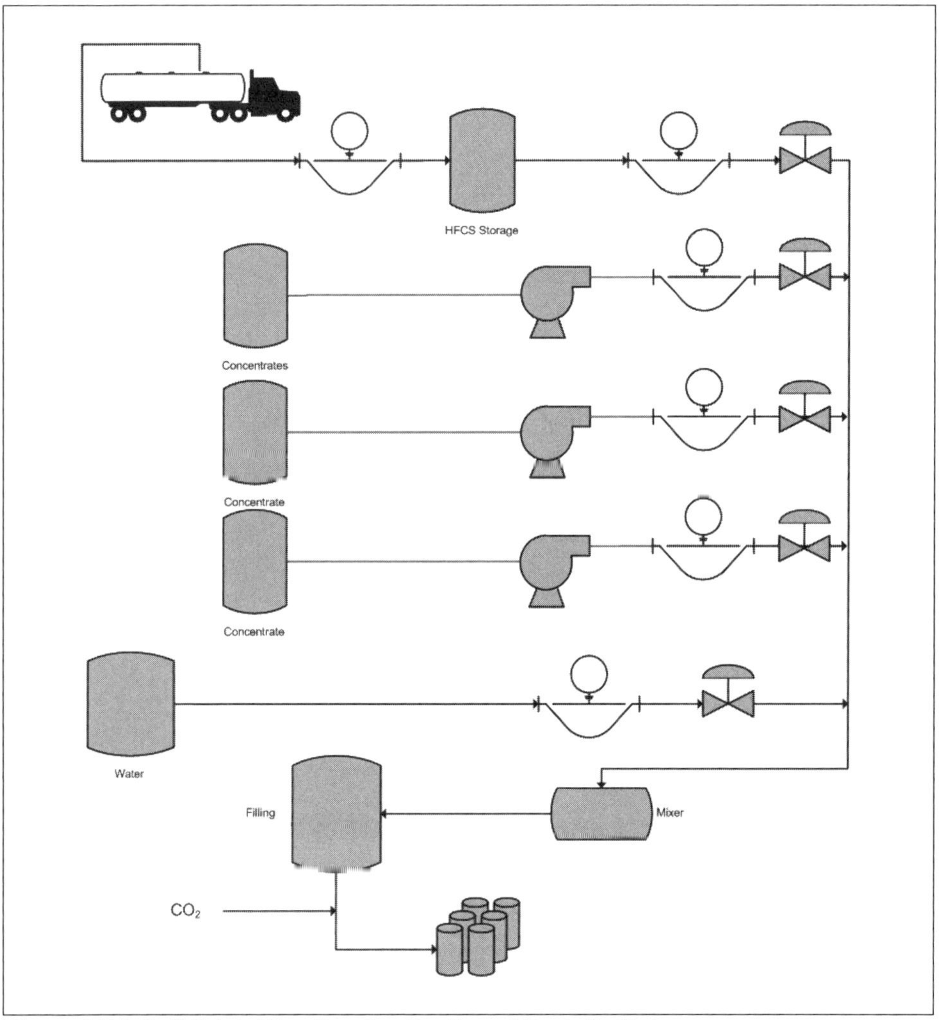

Figure 4-7d. Soft Drink Blending Application of Coriolis Meters

Coriolis flowmeters have successfully addressed these challenges:

- Coriolis meters can calculate °Brix (a form of concentration measurement) from inline process data. This reduces or eliminates the delays associated with manual sampling and laboratory analysis.
- Coriolis meters are immune to changes in temperature or viscosity. One meter can measure any fluid in any process condition.
- A wide turndown is inherent in Coriolis technology.

- Coriolis solutions save space. A Coriolis multivariable device can replace several other instruments while providing equal or better measurement, and Coriolis solutions are available in a variety of footprints and mounting options.
- Coriolis solutions are easy to clean, require little maintenance, and have a low incidence of failure. The result is increased process uptime.
- Coriolis solutions provide accurate real-time measurement across a wide variety of process conditions, allowing manufacturers to shift set points closer to the target, reduce off-spec product, and reduce ingredient waste.
- Coriolis solutions enable tighter process control, allowing manufacturers to predict material use and manage inventories more accurately and efficiently.
- Coriolis solutions can be implemented with batching and filling enhancements to support packaging applications.

LDPE (Low Density Polyethylene) Production in the Chemical Industry
Low Density Polyethylene (LDPE) is a very common plastic used in packaging films, wire and cable coating, injection molding, housewares, bottles, toys, insulation, and plastic bags. It is characterized by high flexibility, high impact strength, and relatively low heat resistance.

LDPE is produced from ethylene and a catalyst in a high-pressure (20,000 to 35,000 psi) high-temperature (400 to 500°F) continuous process (Figure 4-7e). The reaction is fast and exothermic, requiring careful temperature control for both product quality and plant safety. Temperature control mechanisms include feeding the ethylene into the reactor at a controlled rate, and constant drawing off the polymerized result (polyethylene) to maintain a chemical balance. Both require accurate measurement.

Because ethylene is expensive and only 15–20% of the ethylene is converted to polyethylene, significant effort is devoted to ethylene recovery.

Manufacturer challenges include:

- Maintaining plant and personnel safety.
- Maximizing output and optimizing materials utilization. This requires precise reaction chemistry and precise control of the

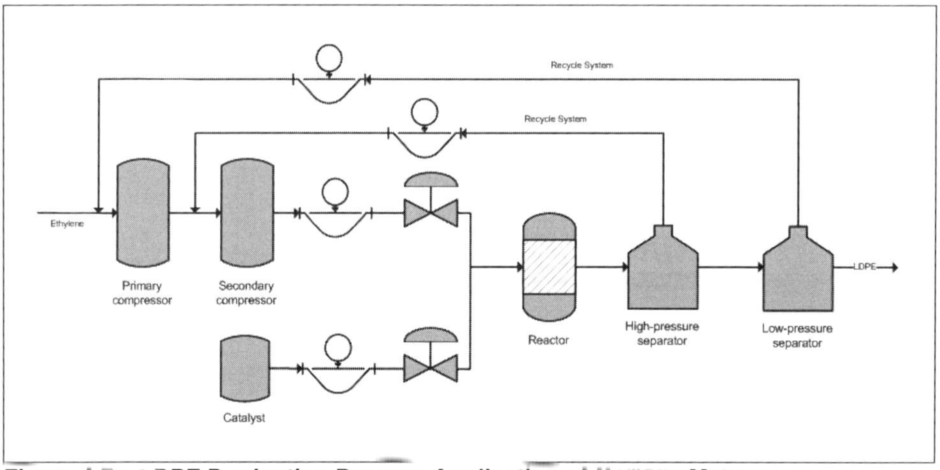

Figure 4-7e. LDPE Production Process Application of Coriolis Meters

ethylene feed and the catalyst feed into the reactor. Different grades of LDPE and different solvents (used for catalyst delivery) require different balances and feed rates.

- Measuring processes that are both high-pressure and low flow.
- Tracking costs accurately. Ethylene and other feeds are very expensive. Slight errors in vendor measurement can result in very large errors in billing. Accurate measurement enables better internal accounting and better vendor relationships.
- Optimizing the ethylene recovery system.

Coriolis solutions:

- Measure accurately across wide variations in temperature, pressure, and flow rate.
- Do not require process shutdown for maintenance or meter verification.
- Minimize instrumentation and therefore minimize potential leak points.
- Meet custody transfer accuracy requirements for most process fluids in most countries in the world.
- Are available with custody transfer security enhancements and other specializations.

Alcohol Recovery in the Food and Beverage Industry

The quintessential distilling application in the United States is beer. At one plant, approximately 100,000 gallons are processed each day, in the production of several different brands.

Alcohol is lost:

- During product changeovers
- During CIP (Clean In Place) cycles
- During filling operations
- In tank bottoms
- In waste yeast

Lost alcohol is costly in several respects:

- It incurs waste disposal costs.
- It has a negative impact on the environment.
- It doesn't take advantage of the waste alcohol market. Waste alcohol can be used as a fuel additive if it meets the proof requirements (190 proof, or 95.5% alcohol).

To sell reclaimed alcohol, the producer must extract it, treat it, distill it to the required proof, and then ensure that the alcohol content is measured accurately for invoicing and taxation purposes. Steam is required for distillation. Manual sampling for proof is time-consuming and, although accurate for the sample, is not accurate enough for continuous process control.

Coriolis flowmeters have been used successfully in the alcohol reclamation and sales process. An eight-week test period compared Coriolis proof readings (a form of concentration measurement) with lab readings and found that variation was within acceptable limits (Figure 4-7f). As a result, manual sampling is performed only during process audits. The realtime proof measurement enables more consistent alcohol content, narrower setpoints, reduced steam and energy usage, and higher volume output.

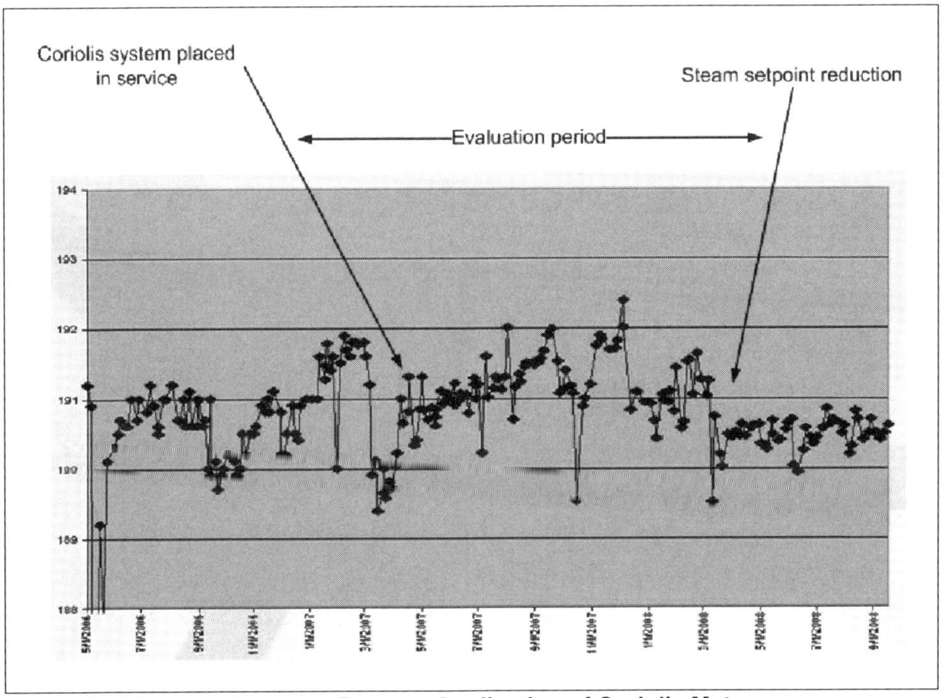

Figure 4-7f. Alcohol Recovery Process Application of Coriolis Meters

Coriolis Meter Setup and Calibration

Coriolis meters are zeroed at the factory, and in many cases, the factory zero can be used in the field.

Coriolis meters are also calibrated at the factory for mass flow and density measurement. Mass flow calibration is achieved by direct comparison to either a gravimetric reference or a master meter reference. Volume flow calibration is achieved as the result of the combination of the mass flow and density calibrations.

The factory calibration typically includes both on air and water before shipment to set calibration factors (Figure 4-7g). A temperature coefficient is used to automatically compensate for any effect of temperature on the stiffness of the flow tube material. The result of the factory calibration is the linear relationship defined by the two points.

Water calibration is transferable to all other process liquids, so recalibration in the field is unnecessary. Because there are no moving parts, no period recalibration is required to maintain accuracy, although calibration verification may be required to comply with regulations. For regulatory proving purposes, a proving skid or other meter in the same line may be used.

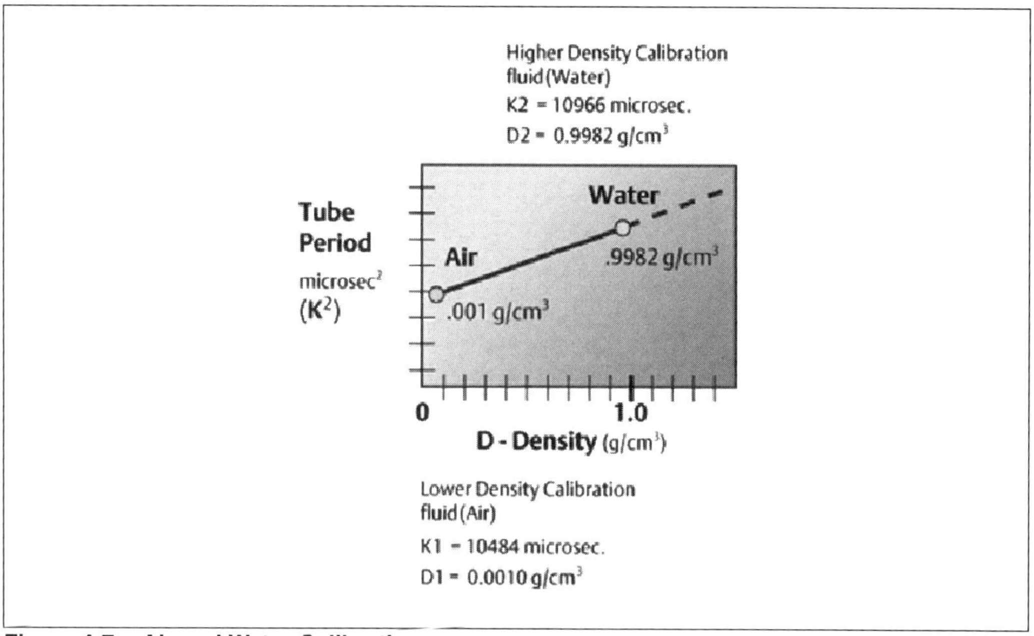

Figure 4-7g. Air and Water Calibration

Figures 4-7h, 4-7i and 4-7j show the AMS screens of a Coriolis meter for the configuration of flow, density, and analog out, respectively.

Coriolis Meter Installation

For measurement accuracy, the tubes of a Coriolis sensor must be full at all times. The sensors can be installed in any orientation as long as this requirement is met. For some applications, it may be beneficial to position the flow tubes upward or downward. For applications in which the sensor must be self-draining, installation in a vertical pipeline is recommended.

Installations must not exceed the ambient and process temperature limits of the sensor.

Coriolis transmitters may be direct-mounted to the sensor or may be remote, with no performance impact and little maintenance impact because process data is communicated electronically rather than mechanically.

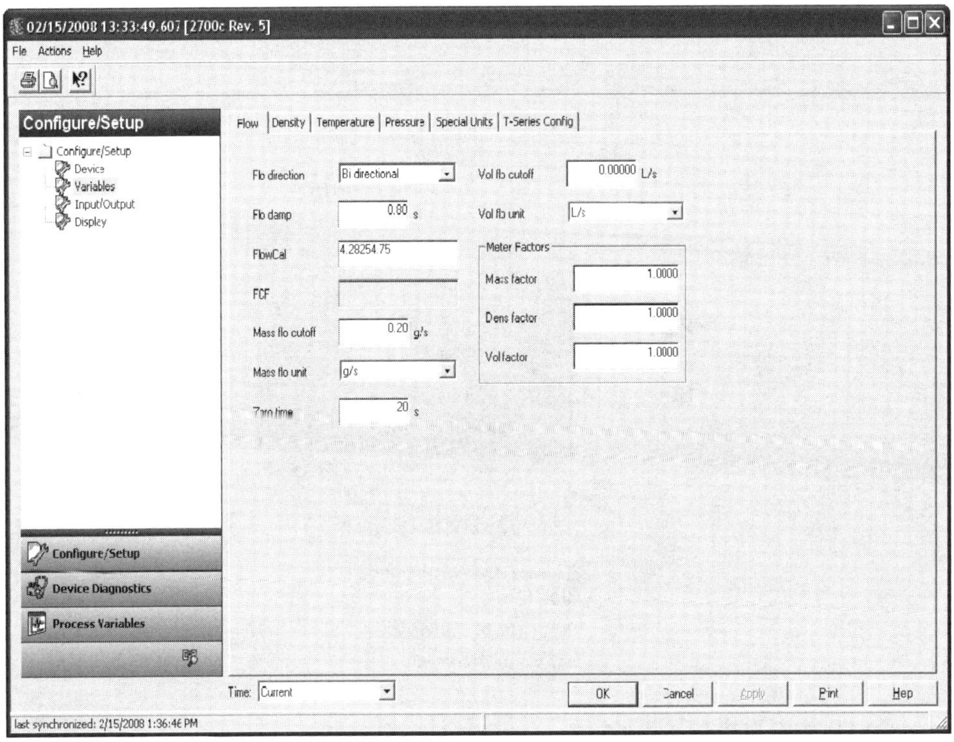

Figure 4-7h. AMS Window for Coriolis Flowmeter Flow Configuration

Coriolis Meter Maintenance

Because a Coriolis measurement has no moving parts and no components inside the process fluid, maintenance typically includes only periodic cleaning, calibration, or proving as required by the application. Any required cleaning is simplified by the design, which is essentially a tube.

Sensor tubes can be damaged by corrosive process fluids or particles that erode the sensor lining. In-situ meter verification provided by some manufacturers can check the structural integrity of a flowing (not flow-simulated) Coriolis meter against original factory settings while the sensor is in-line. Meter verification can store, track, and trend results over time, and can predict meter maintenance issues before operations are interrupted.

Figures 4-7k and 4-7l show the AMS screens of a Coriolis meter for advisory messages and process variables, respectively.

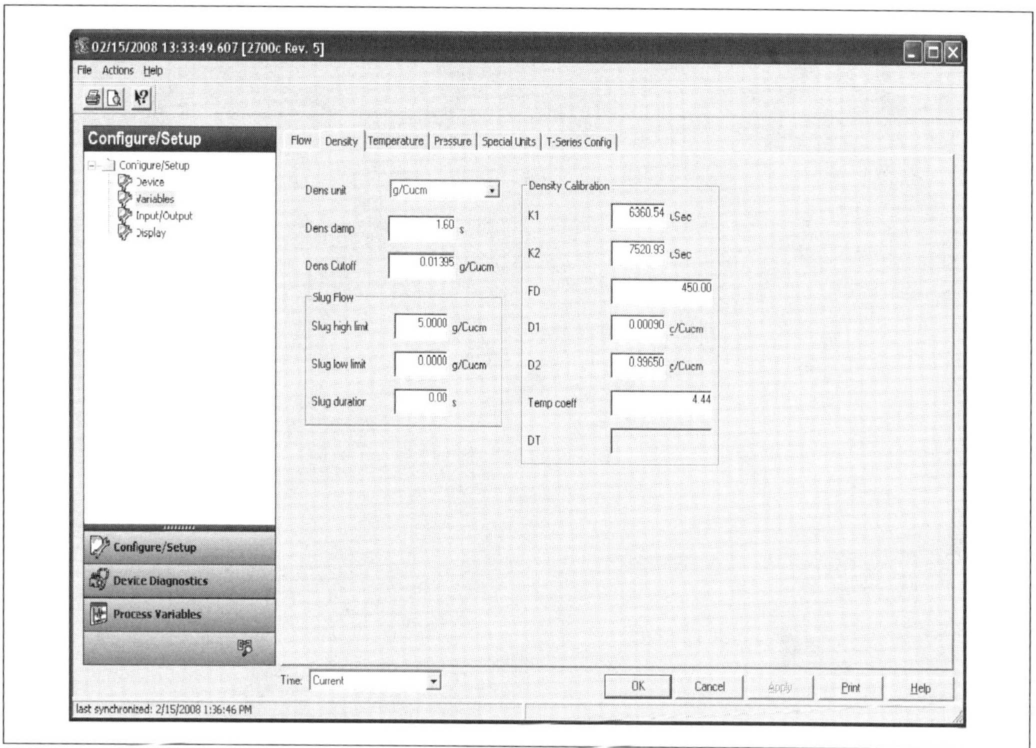

Figure 4-7i. AMS Window for Coriolis Flowmeter Density Configuration

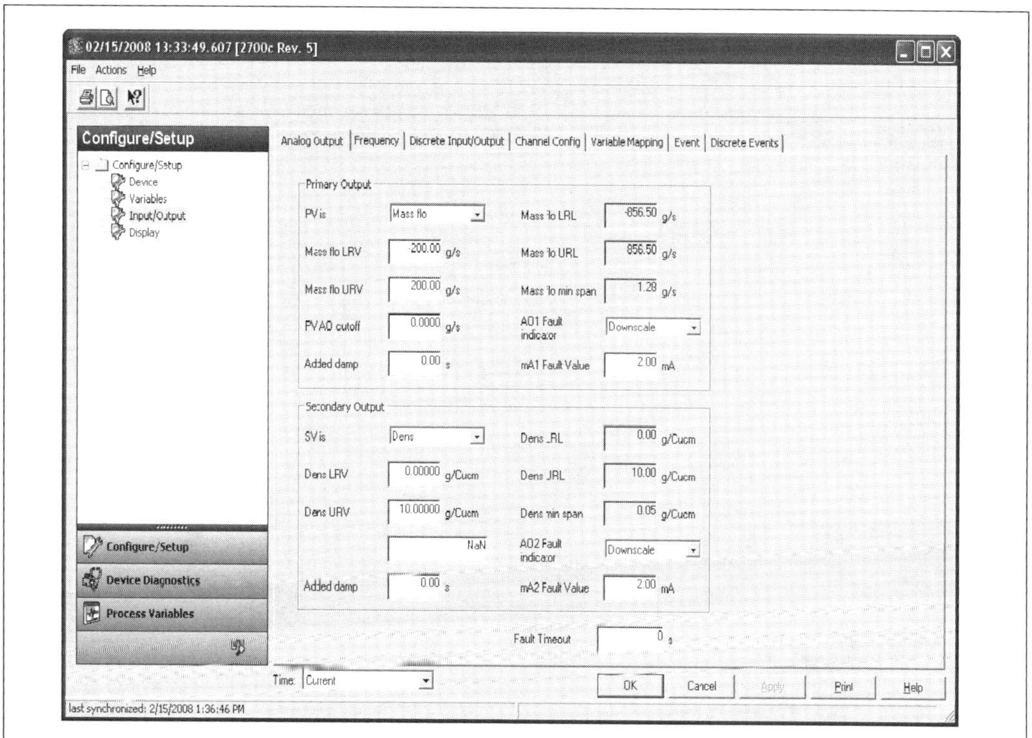

Figure 4-7j. AMS Window for Coriolis Flowmeter Analog Output Configuration

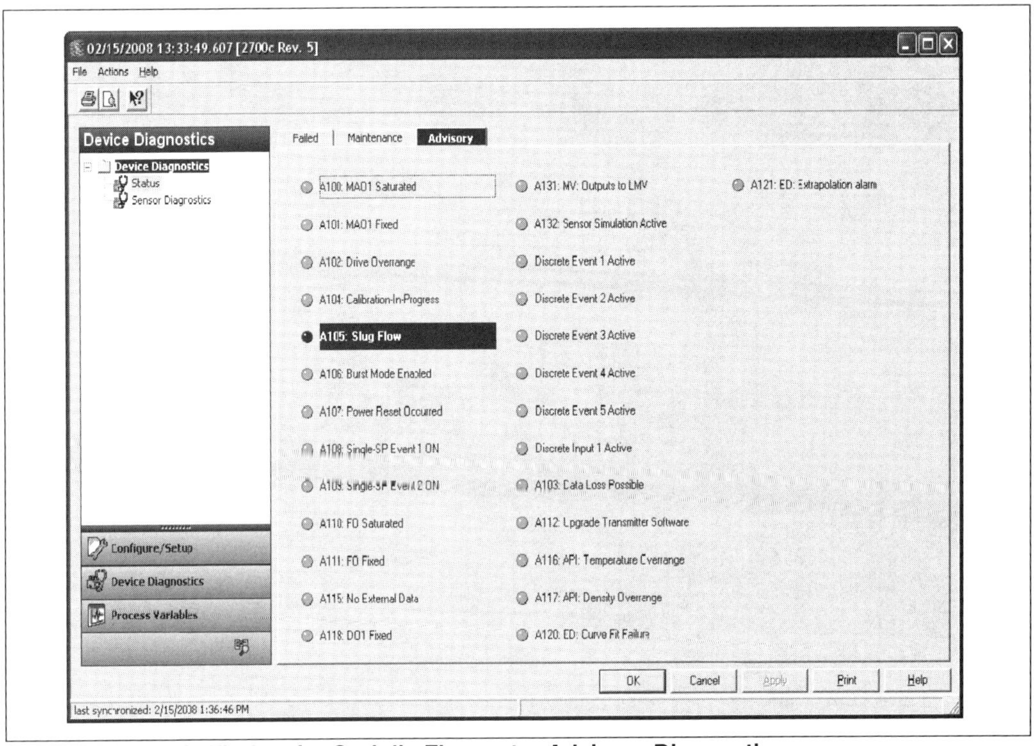

Figure 4-7k. AMS Window for Coriolis Flowmeter Advisory Diagnostics

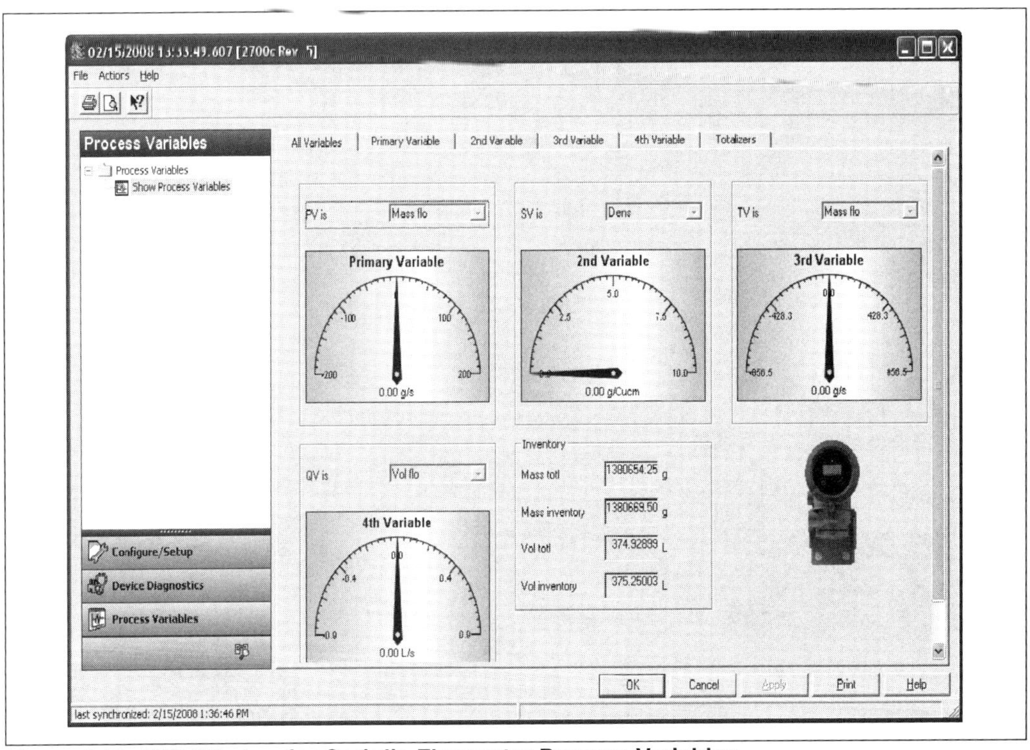

Figure 4-7l. AMS Window for Coriolis Flowmeter Process Variables

4-8. Questions

1. Why are flow measurements important in the process industry?
2. What physical properties of a process fluid affect the accuracy of a thermal mass flowmeter?
3. What type of flowmeter has the longest straight upstream pipe run requirements?
4. For a 60-inch flue gas duct, and an accuracy and rangeability requirement of 2% and 20:1, what type of flowmeter can be used?
5. For a 60-inch pipe with a variable liquid composition and an accuracy and rangeability requirement of 1% and 10:1, what type of flowmeter can be used?
6. For a 36-inch pipe with an abrasive slurry and an accuracy and rangeability requirement of 1% and 4:1, what type of flowmeter can be used?
7. For a 10-inch pipe with a pure monomer and an accuracy and rangeability requirement of 0.25% and 15:1, what type of flowmeter can be used?
8. For a 24-inch pipe with process water and various chemicals and an accuracy and rangeability requirement of 0.25% and 30:1, what type of flowmeter can be used?
9. For a 4-inch pipe with a process liquid and an accuracy and rangeability requirement of 0.1% and 100:1 or a drift requirement of less than 0.1% per year, what type of flowmeter can be used?

4-9. References

1. Blevins, T. L., McMillan, G. K., Wojsznis, W. K., and Brown, M. W. *Advanced Control Unleashed*. Research Triangle Park: ISA, 2004.
2. McMillan, G. K., "What Have I Learned? – Ratio Control." at www.modelingandcontrol.com, Mar. 30, 2009 (Part 1) and Apr. 7, 2009 (Part 2).
3. McMillan, G. K. and Sowell, M. S. "Virtual Control of Real pH." *Control*, Nov. 2007 at http://www.controlglobal.com/articles/2007/385.html.
4. Yoder, J. "A Market Prime for Boom - Thermal Flowmeters to Gain from Environmental Push." *Flow Control Network*, Mar. 2009. http://www.flowcontrolnetwork.com/issuearticle.asp?ArticleID=327

5. Cascetta, Furio and Vigo, Paola, *Flowmeters - A Comprehensive Survey and Guide to Selection*. ISA, 1988.
6. McMillan, G. K. "What Have I Learned? – Cost and Source of Oscillations (Part 3)." at www.modelingandcontrol.com, May 4, 2009.
7. McMillan, G. K. "What Have I Learned? – Cascade Control" at www.modelingandcontrol.com, Mar. 4, 2009 (Part 3), Feb. 23, 2009 (Part 2), Feb. 16, 2009 (Part 1).
8. Kopp, J. G., "How to Get the Accuracy You Expect from Flowmeters," *Instruments and Control Systems (I&CS)*, Sept., 1989

For further reading on flow measurement the following references are recommended:

9. Spitzer, D. W. *Industrial Flow Measurement*. Research Triangle Park: ISA, 1990.
10. Miller, R. W. *Flow Measurement Engineering Handbook, 2nd Edition*, McGraw-Hill, 1989.
11. Benedict, R. P., *Fundamentals of Temperature, Pressure, and Flow Measurements*, 3rd Edition. Wiley-Interscience, 1984.
12. International Organization for Standards *ISO 5167 Measurement Of Fluid Flow By Means Of Pressure Differential Devices Inserted In Circular Cross-Section Conduits Running Full*, Part 1: General principles and requirements, 2nd edition, ISO 5167-1:2003(E), Part 2: Orifice Plates, 1st Edition, ISO 5167-2:2003(E), Part 3: Nozzles and Venturi nozzles, 1st edition ISO 5167-3:2003(E), Part 4: Venturi tubes, 1st edition ISO 5167-4:2003(E).
13. The American Society of Mechanical Engineers, *ASME MFC-3M-2004 Measurement of Fluid Flow in Pipes Using Orifice, Nozzle and Venturi*. 2004.
14. The American Society of Mechanical Engineers, *ASME MFC-14M Measurement of Fluid Flow Using Small Bore Precision Orifice Meters*. 2003.
15. American Gas Association, *AGA Report No. 3, Orifice Metering of Natural Gas and Other Related Hydrocarbon Fluids*. Part 1: General Equations and Uncertainty Guidelines, 3rd edition, second printing, Jun. 2003; Part 2: Specification and Installation Requirements, 4th edition, second printing, 2003.

5

Level Measurement

5-1. Learning Objectives

A. **Appreciate the importance of level measurement for key applications.**

B. **Gain an overview of accuracy requirements.**

C. **Understand the relative capabilities of level measurement devices.**

D. **Learn about the latest technological advances in level measurement.**

E. **Find out how to select specific level measurement devices.**

F. **Understand important aspects of process applications.**

G. **Gain some insight into how to improve measurement performance.**

H. **Learn the basic installation requirements.**

I. **Learn terminology and issues to intelligently discuss industrial applications.**

5-2. Introduction and Overview

Level measurements are important for accounting, inventory, residence time, and material balance control. The accuracy requirement varies more for level than for any other measurement. In storage tanks and the overhead receivers for distillation columns, changes in level as small as 0.1 inches (2.5 mm) may be significant for accounting and/or material balance control. In surge tanks, the level may just need to be kept between the low and high alarm points. However, a surge tank or any vessel operated near the high or low limits due to production rate or residence time requirements demands precise level measurement.

The term "tank" is commonly used. However, process engineers usually reserve this term for storage or surge applications. For equipment with an operating level, the more general term is "vessel," which includes column receivers, crystallizers, evaporators, neutralizers, and reactors. Except for the specific terms "hydrostatic tank gauging," "storage tank," "surge tank," and "weigh tank," this chapter will use the more general term "vessel" to help promote an understanding of the wider scope of level measurement opportunities.

Storage Tanks

Raw material and product storage tanks require the best level measurement accuracy, particularly resolution, sensitivity, and repeatability when used in the calculations for:

- Inventory accounting
- Custody transfer
- Batch charges
- Continuous feed rates
- Material balances

Load cells were long considered to be the best means of inventory measurement for these tanks. Since load cells incur a large engineering and installation cost for the special piping, support, and calibration system, there has been an effort to seek alternatives that offer as good or better performance on the plant floor.

The accuracy of a load cell installation is about 0.25% if just an electronic calibrator is used. Weigh scale readings of tank trucks, as they enter and exit a plant, that are used to fill or empty storage tanks in production units can achieve an accuracy of 0.1% if no material is left in the lines and the fuel consumption before and after readings is negligible. If dead weights can be loaded on top of the tank that total up to 25% of the span, the accuracy can be improved to 0.05% but a crane and uniform weight distribution are required. These accuracy statements assume there are no issues with the tank installation such as vibration, deflection, or restriction of tank movement. Vibration can originate from nearby equipment, floors, agitation, boiling, or gaseous reactions. Deflection can result from wind, solar heating, filling, thermal expansion of piping, and tilt of the structure. Restrictions in movement can occur from piping and support connections that are not flexible enough [1].

A hydrostatic tank gauging (HTG) system is capable of 0.05% level measurement accuracy by the use of multiple differential pressure (DP) transmitters, temperature measurements, and strapping tables (multiple circumference measurements) for the tank. One DP measures total liquid head and a second DP measures density when the level is above its upper process connection (see Section 5-3). The stated accuracy assumes a homogeneous fluid and that no errors are introduced by sensing lines or capillary systems [2]. The strapping tables provide the cross-sectional area versus level and may include a temperature correction for expansion or contraction of the walls. The results can be expressed in level, volume, and mass [3].

Radar gauges for inventory measurements are capable of an exceptional accuracy of 0.04 inches (1.02 mm). Since radar is a surface level measurement, the accuracy in percent of span gets better as the span increases. For a 15 foot (3.8 m) tall tank, the 0.04 inch accuracy corresponds to an incredible accuracy of 0.01% of span. When used in combination with a strapping table and a Coriolis meter for density measurement in a recirculation line, a radar device is capable of an unexcelled calculated mass or volume measurement in addition to level measurement.

Columns

One of the biggest and least recognized opportunities for sensitive level measurement and tight level control is overhead receivers on distillation columns. The most widely used column strategy control is the direct material balance scheme shown in Figure 5-1 where temperature manipulates the distillate flow from the overhead receiver. The level in the receiver is controlled by the manipulation of reflux flow to the top of the column [4].

Vapor flow from the reboiler bubbles up through the column, condenses in the overhead exchanger, and accumulates in the overhead receiver. Reflux flows down the column and accumulates in the sump. A decrease in vapor flow from a decrease in steam to the reboiler or a sudden shift in wall temperature from a cold wind or cold rain will cause a decrease in overhead receiver level. For tight level measurement and control, a small change in level will quickly translate to a change in reflux flow that will balance the change in vapor flow. This inherent self-regulation provides some internal reflux control and helps decouple the energy balance from the material balance. Since reflux flow is typically higher than distillate flow (reflux/

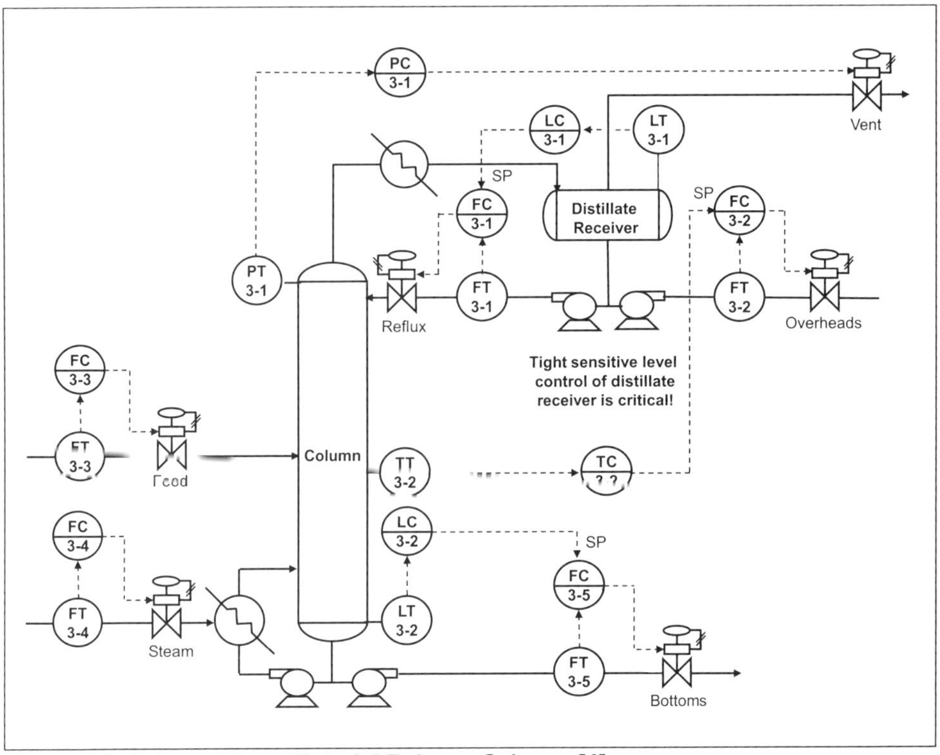

Figure 5-1. Column Direct Material Balance Scheme [4]

distillate ratio > 1), the level controller has plenty of muscle to keep the level exactly at set point.

The performance of the temperature loop depends upon a sensitive level measurement and tight control. When the temperature loop makes a change in the distillate flow, the change in controller output has no effect on column temperature until the overhead receiver controller makes a change in the reflux flow. If the level control is tight, the correction to the column is fast. In fact, to bring the level back to set point there is some lead-lag action because the reflux must be driven momentarily past the balance point where reflux flow equals the distillate flow to keep the level constant. What is important here is not total accuracy but the ability of the level device to consistently distinguish as small a change in the level as possible. The level measurement must have excellent resolution, sensitivity, and repeatability (precision) and negligible noise or sufficiently filterable (e.g., high frequency) noise.

Many overhead receivers are horizontal vessels that have incredibly large and variable cross sectional areas. What may seem like a negligible change in level (0.1 inch; 2.5 mm) can correspond to a significant change in inventory. If users recognize the allowable controller gain is so much higher than their comfort range, there is no tuning issue despite the variable area of the horizontal vessel. The user just needs to set the highest controller gain that does cause fluctuations in the controller output from noise to exceed the deadband of the final element. Set-point rate of change limits can be added to avoid banging the reflux valve and upsetting the operator and column.

For an example of the unrecognized opportunity, consider the following case. When an analog control system on a column was replaced with a distributed control system (DCS), the proportional band of 100% was mistakenly entered as a gain of 100 for the overhead receiver level controller. The performance of the column improved dramatically and the improvement was thought to be due to the use of the DCS even though the control strategy had not changed. The same effect could have been gained with a proportional band of 1% [5].

Crystallizers and Reactors

From the 1960s to the 1980s, load cells in weigh tanks were the way to ensure the charges to reactors were accurate. Accuracy was typically 0.1% to 0.25%. Loss in weight was used to determine the charge and to provide a mass flow rate. Today, DP level and radar level transmitters with Coriolis flowmeters provide a lower cost alternative for accurately feeding critical unit operations.

In batch and continuous crystallizers and reactors pushed to capacity, the level is operated close to its high limit. Operation above this limit is undesirable because it may be above the heat transfer area and may allow some loss of material into the overhead vent system due to foaming and swelling from bubbles, sloshing from agitation, and liquid entrainment in the vent gas.

For batch operations, a higher level offers a higher product mass per batch cycle. For continuous operations, a higher level enables a higher feed rate for the same residence time requirement. Conversion to product in continuous reactor vessels depends on the reactants or super-saturated components having enough time to complete the product or crystal formation. Most reaction and crystallization rates are not instantaneous. The conversion depends on residence time (inventory divided by throughput flow). Level should be kept proportional to production rate to keep the conversion constant. Excessive residence time can increase byproducts, product degradation, and crystal agglomeration.

Surge Tanks

The normal purpose of a surge tank is to provide surge capacity. The level should be allowed to move up and down to absorb mismatches in process equipment operating rates upstream and downstream of the surge tank. However, when production rates are high or low, it may be desirable to operate at the tank's level limits. There may be residence time limits similar to those stated above for reactors and crystallizers to keep the product concentration and quality constant or enable the product formation rate to continue to meet production rate requirements. Operation at level or residence time limits put greater demands on level measurement accuracy.

Admittance Measurements

Some level measurements are relegated to a small part (e.g., < 5%) of the market. A prime example is the admittance level measurement. An overview of the technology is offered here to provide an understanding of how this type of measurement is used in industrial processes. The user should contact the supplier for more information on this type of level measurement.

Admittance measurements have their origin in capacitance probes that measure a change in capacitance as being proportional to the level of the dielectric between the capacitance probe and the vessel wall. If the vessel wall is non-metallic, a concentric probe is used or a reference conductor is inserted to form the other side of the capacitor. Capacitance level measurements are sensitive to coatings and changes in the dielectric constant.

Radio Frequency (RF) admittance probes measure the resistance in addition to the capacitance to compensate for coatings. If the coating is electrically long enough, the resistive and capacitive components of the coating have equal impedances. The impedance of the coating is subtracted from the total impedance to get the capacitance of just the level [6]. The effective electrical length of the coating depends upon the thickness and length of the coating relative to the diameter of the sensing element. Thin and short coatings still introduce errors. Special probe designs and smarter transmitters can do better at minimizing these errors. Admittance probes that can ignore process coatings offer an alternative for sticky fluids and solids that would coat the diaphragms or plug the sensing lines of DP transmitters. If the process fluid's dielectric constant is known and constant and if the fluid's conductivity is always greater than 20 µS/cm, the standard admittance level transmitter can be calibrated at the factory based on the installation geometry and desired level measurement range.

For the broad spectrum of chemicals, the electrical properties of process liquids are not well enough defined, particularly for multi-component mixtures. There are enough unknowns that a factory calibration is insufficient. An actual process calibration, where the process fluid has to be varied by known amounts, is time consuming and possibly not feasible. A manual level measurement made by inserting a measuring stick through a manway or by observing a level gauge reading is no longer permitted because these methods and devices are not safe enough for most chemicals.

Admittance probes can be enhanced to measure and cancel out the effect of the dielectric constant online. The online measurement of dielectric constant can be done as illustrated in Figure 5-2 by the lower portion of the level probe that extends below the minimum level or by a separate side entry probe mounted below the level probe [6]. In both cases, the online measurement of dielectric depends upon a level of process fluid existing at the dielectric sensor with the same composition as in the level range. The accuracy is about 1% of span. Errors in the compensation develop for non-homogeneous fluids, which can be caused by non-ideal mixing in the heel of the vessel, sparges (agitation) or bubbles, layering fluids from differences in density, and the settling of solids. Rigs for insertion and removal and supports to prevent whipping are needed for long probes.

Perspective

The right level measurement device for the application is the most maintainable and reliable device that meets the objectives of accuracy, sensitivity, repeatability, and what actually needs to be measured (e.g., liquid level, bubble level, head, mass, or volume).

Older technologies that involve mechanical devices, such as displacers and floats, are not discussed here because of maintenance issues. Nuclear devices (e.g., gamma ray sources and detectors) provide a viable alternative for the external measurement of the level of extremely difficult or hazardous materials. Changes in density and wall thickness, and the variation in path lengths from point sources to the strip detectors, introduce errors.

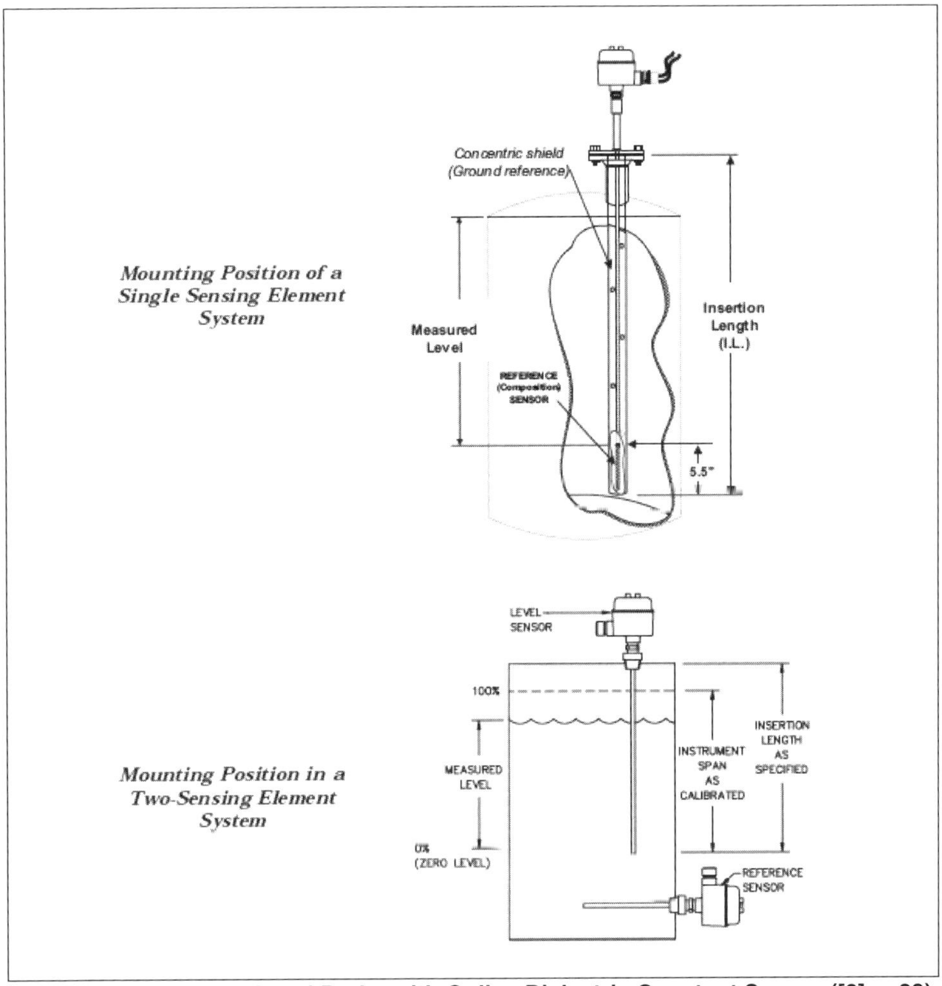

Figure 5-2 Admittance Level Probe with Online Dielectric Constant Sensor ([6] p. 29).

Probably the biggest reason they are not often used is the perceived hassle with regulations on radiation source permits, handling, and operation.

If an accurate measurement of liquid surface level including swell is required, a radar level device is the best bet if the dielectric constant is above a minimum and any foam is relatively dry. If liquid head rather than surface level is important to prevent cavitation in pumps and to maintain dissolved gas concentrations, then a DP level measurement is most likely the right choice. Also for applications where the level just needs to be kept within limits, a DP level transmitter is often sufficient. If the fluid tends to coat or plug, remote seals and a filled capillary system for the DP can eliminate sensing line and sensing element problems. A bubbler with the DP can be used provided the bubbler gas rate is regulated to prevent spikes in the reading and introducing a variable pressure (see discussion of bubblers in Chapter 1) [7]. Unfortunately, common purge regulators only control supply pressure to within 10% [8].

An ultrasonic device is a good choice if there are no line of sight problems [8]. If an accurate measurement of inventory or change in inventory is the goal, as mentioned a radar-Coriolis system should be considered. If the application has a low dielectric constant, involves extremely hazardous materials (e.g., hydrogen cyanide), or operates at an extremely high temperature, then completely external measurements, such as nuclear devices, may be the only viable approach.

5-3. Differential Pressure Level

DP Level Principles

Pressure transmitters are the most commonly used technology to measure liquid level. Because there is a linear relationship between hydrostatic head pressure and liquid level, the level in a tank can be calculated by taking a simple pressure measurement. As an example, a submarine can calculate what ocean depth it is at by measuring the amount of water pressure.

Level is simply the pressure (liquid head) divided by the specific gravity.

$$\text{Level} = (\text{Pressure}) \div (\text{Specific Gravity of Process Fluid})$$

Open Vessel Applications

For applications where the top of the vessel is open or vented to atmosphere, a single gauge pressure transmitter mounted near the bottom of the tank can be used to measure level (Figure 5-3a).

Example:

- Tank is filled with Cotton Seed Oil (Specific Gravity = 0.929)
- Pressure reading = 1300 mm (51.2 inches) H_20
- Level = (1300 mm) / (0.929) = 1400 mm (55.1 inches) of Cotton Seed Oil

The transmitter is only able to measure the pressure from the liquid that is above its location at the high-pressure process flange. For the previous example, assume the process flange is located 500 mm (19.7 inches) from

Figure 5-3a. Open-Vessel Sample Installation

the bottom of the tank. In this case, the total level in the vessel would be the amount measured by the transmitter plus the fixed amount below the transmitter (1400 mm + 500 mm = 1900 mm (75 inches) of Cotton Seed Oil).

Closed Vessel Applications

If the vessel is pressurized, a gauge transmitter will not be able to distinguish between how much of the measured pressure is from the liquid level and how much is from the internal pressure of the vessel. A differential pressure transmitter solves this problem, as it compensates for the vessel pressure. Impulse piping or remote seals with capillary may be required to transfer the low-side pressure (vessel pressure) reading to the transmitter (Figure 5-3b).

DP Level Selection

A pressure-based level system commonly consists of three components: the pressure transmitter, one or two remote seals, and an oil-filled connection between the remote seal(s) and the transmitter (Figure 5-3c).

Remote Seals

A remote seal is a flexible diaphragm that comes in direct contact with the process fluid in the vessel. Pressure is applied to the remote seal, causing it

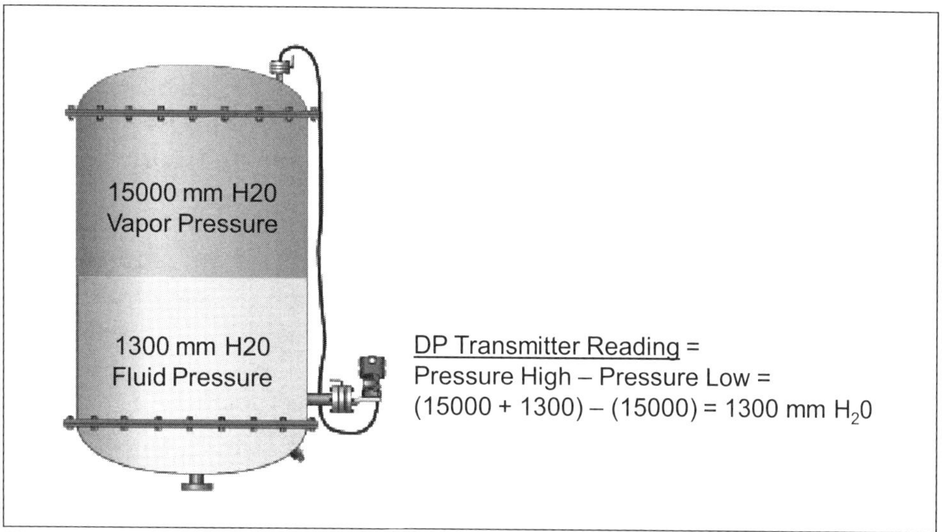

Figure 5-3b. Pressurized-Vessel Sample Installation

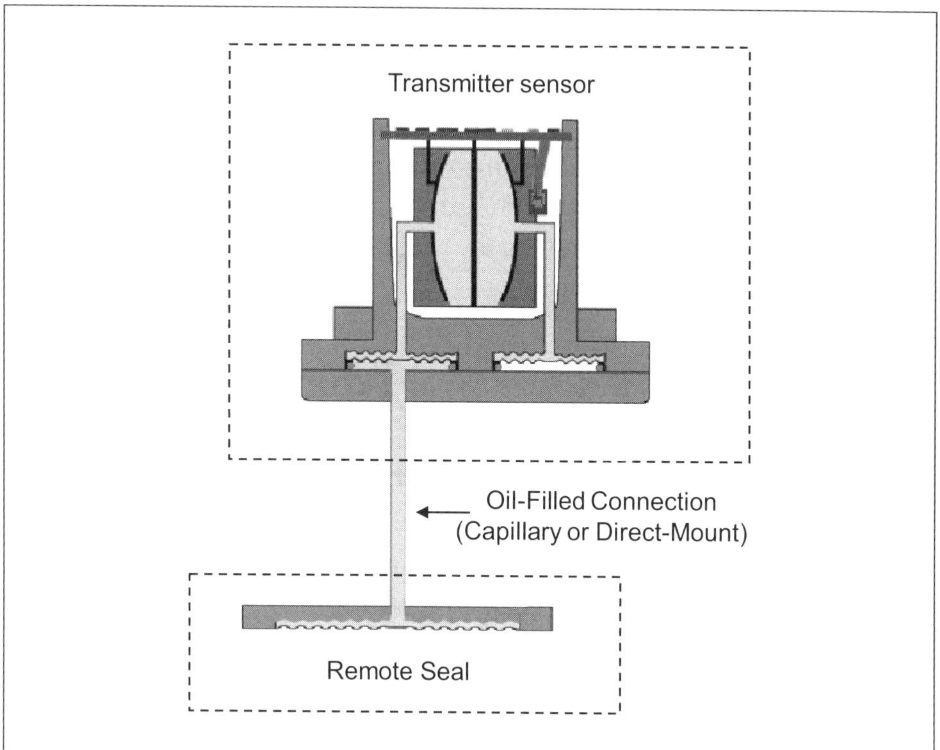

Figure 5-3c. Cross-sectional Diagram of a Basic Transmitter/Remote Seal System

to deflect. The pressure signal is then transferred through the oil-filled connection to the transmitter sensor. Thus, remote seals can be thought of as "extensions" of the pressure transmitter. Common reasons to use remote seals include:

- There is a need for a differential pressure measurement across a long physical distance (such as a pressurized vessel, distillation column, etc.)
- The process temperature is outside of the normal operating range of the transmitter.
- The process fluid is corrosive and would require frequent transmitter replacement or specific materials of construction not available on the transmitter.
- The process fluid contains suspended solids or is viscous and may plug impulse piping.
- The application requires the use of hygienic process connections.

Remote Seal Connection

As mentioned, an oil-filled connection is needed between the transmitter and the remote seal. Two common connection types include direct-mount and capillary. Direct-mounting a remote seal offers many benefits over capillary including better time response, better accuracy, and easier installation. A capillary may be required when making a DP measurement to bring the low-side pressure to the transmitter (Figure 5-3d).

Figure 5-3d. Example of a DP Transmitter Used for Measuring Level in a Closed Vessel.

This configuration contains one direct-mount connection and one capillary connection.

Transmitter

The transmitter measures the pressure signal that is transferred from the remote seal through the capillary or direct-mount connection. A gauge style transmitter can be used for open vessel applications, while DP transmitters should be used for closed vessel and other applications that require the use of two remote seals.

DP Level Specifications

Performance

The overall accuracy of a DP level system depends on the process and ambient conditions where it is installed. Changes in temperature from day to night and season to season cause the fill fluid in a DP level system to expand and contract in volume and change in density. Below are the common parameters that impact the performance of a DP level system:

- Seal Temperature Effect: Any change in fill fluid volume due to expansion or contraction of the fluid with temperature is either absorbed by the diaphragm of the remote seal or is exerted as back pressure on the transmitter sensor. The back pressure exerted on the transmitter sensor will cause measurement error.
- Head Temperature Effect: As temperatures change, the density of the fill fluid in the remote seal system will also change. This change in density (liquid head of fill seen by the DP) will affect the installation zero offset reference pressure that is applied on the low-side of the DP transmitter, and consequently the transmitter output will change.
- Time Response: The length and diameter of the capillary and the type of fill fluid used all impact how responsive a remote seal system will be to changing process conditions.
- Transmitter Total Probable Error: The pressure transmitter itself has a certain degree of measurement uncertainty depending on the type, sensor range, and installed conditions.

Ultimately, all of the above parameters acting together determine how a DP level system as a whole is going to perform under installed conditions. Calculating installed performance by hand is difficult and complex. Electronic sizing software is available to help with the seal specification process and to quantify installed performance.

Fill Fluids

The fill fluid that is used in the construction of the remote seal system impacts the overall temperature limits and time response of the entire assembly. Each fill fluid has a specific temperature range in which it can operate. If a fill fluid goes below its lower temperature limit, it will become too thick to transmit changes in pressure. If the fill fluid goes above its upper limit, it could break down. Below are examples of temperature limits for some common fill fluids in a capillary system:

- DC 200 Silicone: -49 to 401°F (-45 to 205°C)
- DC 704 Silicone: 32 to 599°F (0 to 315°C)
- Inert (Halocarbon): -45 to 320°F (-49 to 320°C)
- Syltherm XLT: -102 to 293°F (-75 to 145°C)
- Glycerin and Water: 5 to 203°F (-15 to 94°C)
- Neobee M-20: 5 to 437°F (-15 to 225°C)

Process Connection Types

Remote seal systems are available in three major process connection types:

- Flanged: 1- to 4-in. (DN 25 – DN 100)
- Threaded: ¼-in. NPT to 1.5-in. NPT
- Hygienic:
  - Tri-Clamp: 1.5- to 4-in. (38 to 102 mm)
  - Vessel Spud connections

Wetted Materials of Construction

Remote seals are available in a wide variety of wetted materials. Common materials include:

- 316L SST
- Alloy C-276
- Tantalum
- Alloy 400
- Duplex 2507
- Nickel 201

Remote seals can also have special coatings for unique process applications. Examples include:

- Gold plating: prevents hydrogen permeation

- PTFE coating: prevents the process fluid from sticking to the seal surface

DP Level Considerations

Application Strong Points

DP level systems have a main advantage in that they can be used in almost any liquid level application. Because a DP level system infers level by measuring pressure, it is not affected by agitators, foam, bubbles, and other common vessel disturbances. Additionally, the variety of remote seal types, sizes, material, and fill fluids allow DP level systems to be installed in the harshest of conditions. In extreme hot and cold temperatures, deep vacuums, and viscous processes, a properly specified DP level system can meet the needs of problematic applications. However, it is possible to misapply a DP level system.

Watch-Out #1: Understanding Your Process

When specifying a DP level system, it is absolutely critical to understand the operating conditions of the process. For example, as mentioned each fill fluid is designed to operate in a specific pressure and temperature range, and specifying the wrong fill fluid may cause sub-optimal performance or even system failure. Sizing calculations should always be run on new applications to ensure the DP level system will meet measurement requirements.

Watch-Out #2: "One-Size-Fits-All"

Rarely does a "one-size-fits-all" DP level system work throughout a plant or facility. Understanding the characteristics of each application and specifying a unique DP level system for each installation will result in better plant performance.

Watch-Out #3: Myths of a "Balanced System"

A "balanced system" is a DP level configuration where identical remote seals and lengths of capillary are used on the high and low-side process connections (Figure 5-3e). The theory behind the balanced system is that any expansion or contraction of fill fluid volume will happen on both the high and low sides of the transmitter, resulting in no seal temperature effect. However, a balanced system does not compensate for head temperature effect, and the time response of the system may suffer due to excessive lengths of capillary on the high-pressure side connection.

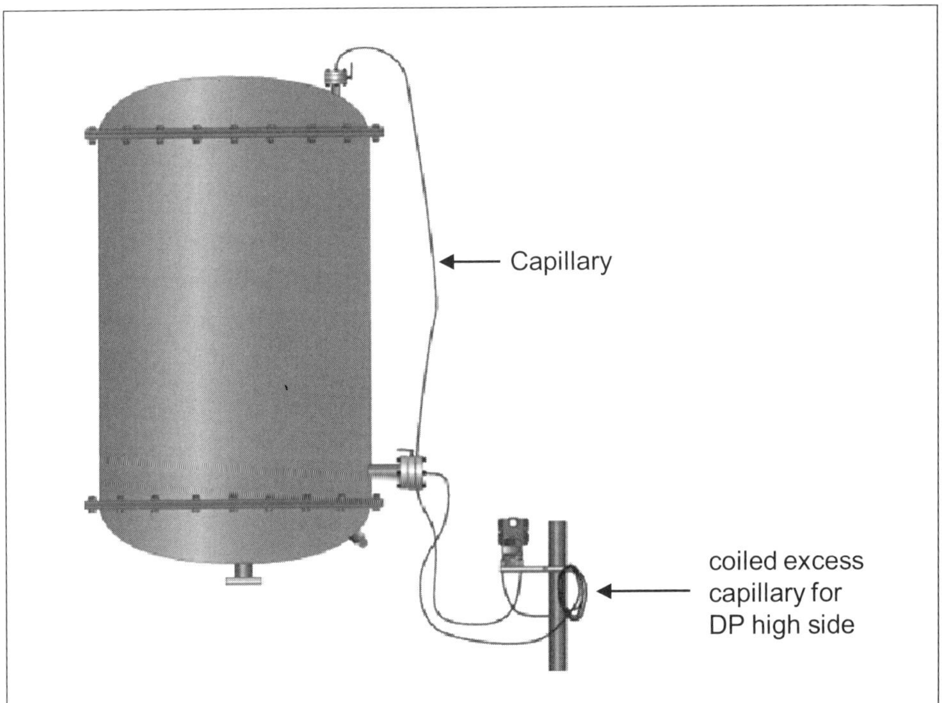

Figure 5-3e. Balanced Remote Seal DP Systems (do not compensate for head temperature effect and often result in poor time response due to excess capillary)

A better alternative is to go with an asymmetrical design (Figure 5-3f). A common approach is to direct mount the transmitter to the high-pressure side connection while using capillary on the low-pressure side connection. A direct-mount system is designed so that the seal temperature effect and head temperature effect work against each other, resulting in less total net temperature effect. Additionally, time response tends to be quicker due to the elimination of the capillary on the high-pressure side connection.

DP Level Setup and Calibration

The most commonly used signal output for pressure transmitters is 4–20 mA with a digital HART signal. DP level is no different since level measurements are inferred from pressure measurements. A DP level system is calibrated and installed much like a standard pressure transmitter. The instructions below provide simple steps for installing and calibrating a DP level pressure transmitter.

1. Perform a bench configuration of the transmitter, including range points.

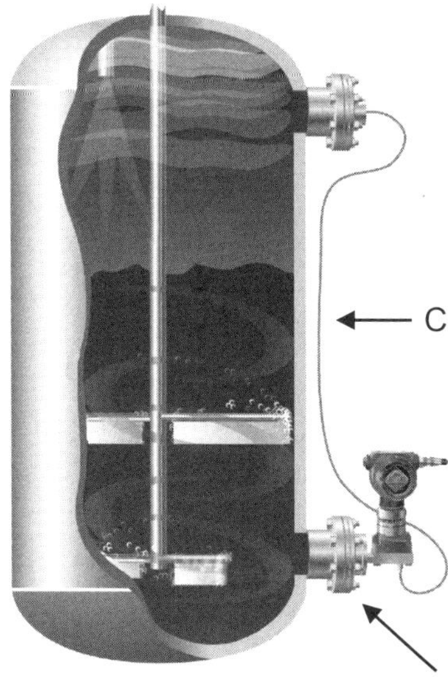

Figure 5-3f. An Asymmetrical DP Level System (offers better overall performance, faster time response, and eliminates the need for transmitter mounting hardware)

2. Zero the transmitter using a HART communicator. Make sure both seal diaphragms are on the same plane and facing each other.
3. Install the high-side seal/transmitter on the bottom fitting and the low-side seal on the top fitting, making sure the gaskets are properly installed.
4. Turn transmitter power back on.
5. Re-range the transmitter.

DP Level Installation

Pressure transmitters are usually mounted near the bottom of the vessel, on a side wall, or on the underside of a suspended vessel. Mounting transmitters on outlets or drainage pipes, should be avoided. Mounting transmitters on a suction of a pump is particularly troublesome due to bubbles and the pressure drop when the pump starts. As long as the seals are slightly recessed and protected from the major impacts of agitation, bottom-mounted pressure transmitters work well in vessels that are subject to turbulence from agitation.

Pressure transmitters should be mounted where sediment will not build up on the measurement surface. Pressure transmitters are immune to foam layers, which generally do not have enough weight to contribute to the pressure measurement.

Any vapors above the fluid are compensated for in a DP system and do not interfere with the measurement.

The installation and maintenance of DP transmitters with sensing lines (impulse lines) may be impractical or impossible for vessels whose process connections are not accessible (e.g., underground tanks). Remote seals and capillaries allow the transmitter to be located for accessibility.

The best installation practice for DP level consists of:

- Maintaining the integrity of the seal and eliminating accidental seal damage by keeping the seal cover on prior to installation.
- Utilizing direct mounting of transmitters to the vessel when applicable.
- Specifying asymmetrical assemblies instead of balanced systems and filled leg systems (wet and dry legs)
- Using wireless level transmitters (Figure 5-3g) for remote storage tanks and hard to access areas to eliminate the costly and difficult installation and maintenance.

DP Level Maintenance

Many pressure transmitters come equipped with various alerts and diagnostics that can be used in DP level applications. One such example is a configurable pressure alert. A user-defined "window" can be set that defines the normal operating conditions of the vessel. If the measured pressure goes above or below the normal operating window, the pressure transmitter will send an alert back to the control room. As an example, this sort of alert could be used to prevent a vessel from being overfilled and causing spill-over.

As another example, some pressure transmitters are equipped with an internal temperature sensor that monitors the ambient temperature. A temperature alert could be set up to make sure that the ambient temperature is not too hot or cold for the fill fluid used in the DP level system. For

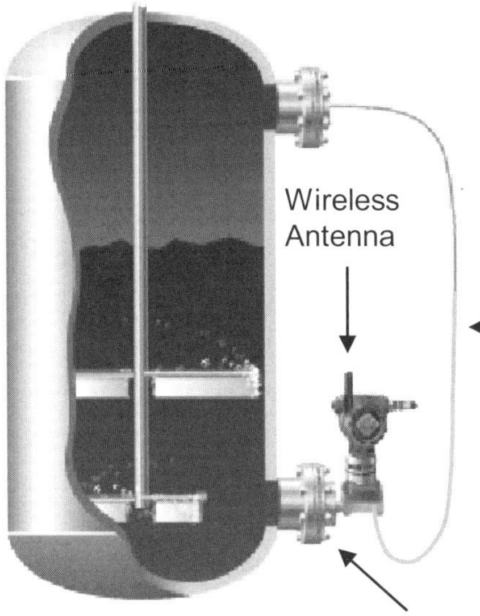

Figure 5-3g. A Wireless Asymmetrical DP Level System (offers lowest life-cycle cost)

example, a low alert could be set up to notify the control room that electric heat tracing needs to be turned on.

Installing a DP level system that is not properly specified for the given application can result in a number of issues including:

- Slow time response
- Output drift
- Remote seal diaphragm damage
- Process clogging and coating

Nearly all problems common to DP level measurement can be avoided by specifying the right assembly from the beginning. This includes:

- Gathering the process information, including operating pressures and temperatures, for each application
- Selecting a remote seal diaphragm material that is compatible with the process fluid

- Selecting a fill fluid that will work over the entire operating range of the application
- Specifying a DP level system and validating performance using electronic sizing software

Once a DP level system has been specified and installed, routine maintenance should be exercised to protect the remote seal diaphragms. Keep the diaphragms clean of process build-up by using flushing connections or performing routine service. Use the protective diaphragm covers when removing the DP level system from service, and never let anything sharp or abrasive come in contact with the diaphragm surfaces. Following these basic guidelines will maximize your chances of obtaining long-term success with DP level technology.

5-4. Radar Level

Radar Level Principles

Radar (radio detection and ranging) technologies transmit a microwave signal from a radar device mounted on top of a vessel to the surface of the material held inside. The transmitted signal is reflected back to the device and the gauge measures the distance (and determines the level) by differentiating the transmitted and returned signals. The level measurement is determined by using the reference height of the gauge (distance from zero level to the radar sensor as shown in Figures 5-4a and 5-4b) minus the distance to the surface.

Radar level devices are available in two basic versions: non-contacting (free radiating) (Figure 5-4a) and guided wave (Figure 5-4b). Each type has distinct characteristics that result in their use in different kinds of applications.

Radar Level Selection

Both non-contacting and guided wave radar devices provide a top-down direct measurement of level. Both can be used with liquids, sludges, slurries, and some solids. Radar level devices are ideal for applications where a direct measurement is needed. A key advantage of radar is that no compensation is necessary for changes in density, dielectric, or conductivity of the vessel contents. Changes in pressure, temperature, and vapor space

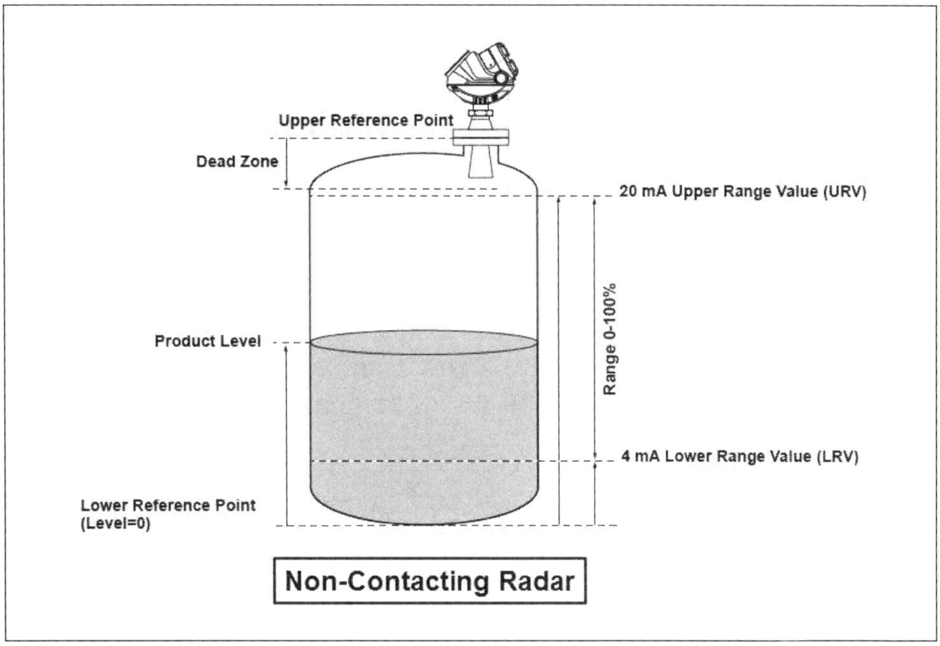

Figure 5-4a. Non-Contacting Radar

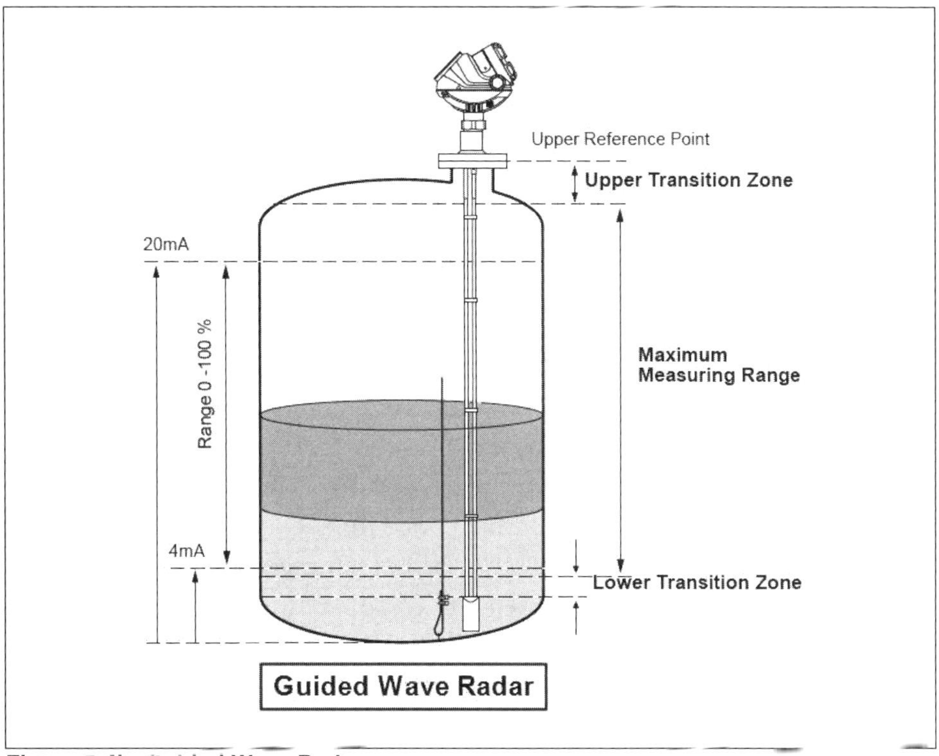

Figure 5-4b. Guided Wave Radar

conditions have no impact on the accuracy of radar measurements. In addition, radar devices have no moving parts so maintenance is minimal.

Non-Contacting (Free Radiating) Radar

Non-contacting radar sends a signal through the vapor space that bounces off the surface and returns to the gauge. Because it is non-contacting, its susceptibility to corrosion is limited and it is an ideal choice for viscous, sticky, or abrasive fluids. Non-contacting radar can frequently be used in vessels with agitators. It can be completely isolated from the process and used with isolation valves. Most vendors offer non-contacting versions that can be used in applications ranging from 1 to 50 meters (3.3 to 164 feet).

Non-contacting radar is available using two basic techniques: pulse (time of flight) radar and FMCW, or frequency modulated continuous wave. Pulse based radar offers a repeatable measurement and has lower power requirements to make the basic measurements. The power requirement for FMCW is higher than for pulse, but it is also a more robust measurement (less susceptible to interferences) and is more accurate.

The frequency ranges of non-contacting radar can impact its performance more than the technique used. A lower frequency reduces sensitivity to vapor, foam, and contamination of the antenna, whereas a higher frequency keeps the radar beam narrow in order to minimize influence from nozzles, walls, and disturbing objects (Figure 5-4c). Beam angle/width is inversely proportional to antenna size. The beam width at a given frequency will decrease as the antenna size increases.

Non-contacting radar comes with a wide selection of antenna types including cone antenna, rod antenna, PTFE process seal, and parabolic antenna (Figure 5-4d). The antenna choice, along with the frequency, influence the signal strength of the radar pulse. The antenna and frequency are selected to get the best performance for the application and the available process connection. The cone (or horn) antenna is most commonly used since they handle a broad range of applications. The rod antenna is only available in low frequency and allows vessel access through small process connections. The PTFE process seal is an inexpensive solution for corrosive environments or heavy vapors. Last, the parabolic antenna has the best antenna gain, and the most focused radar beam. This makes it suitable for low dielectric applications, like solids measurements and/or long measuring ranges (up to 50 m [164 ft]).

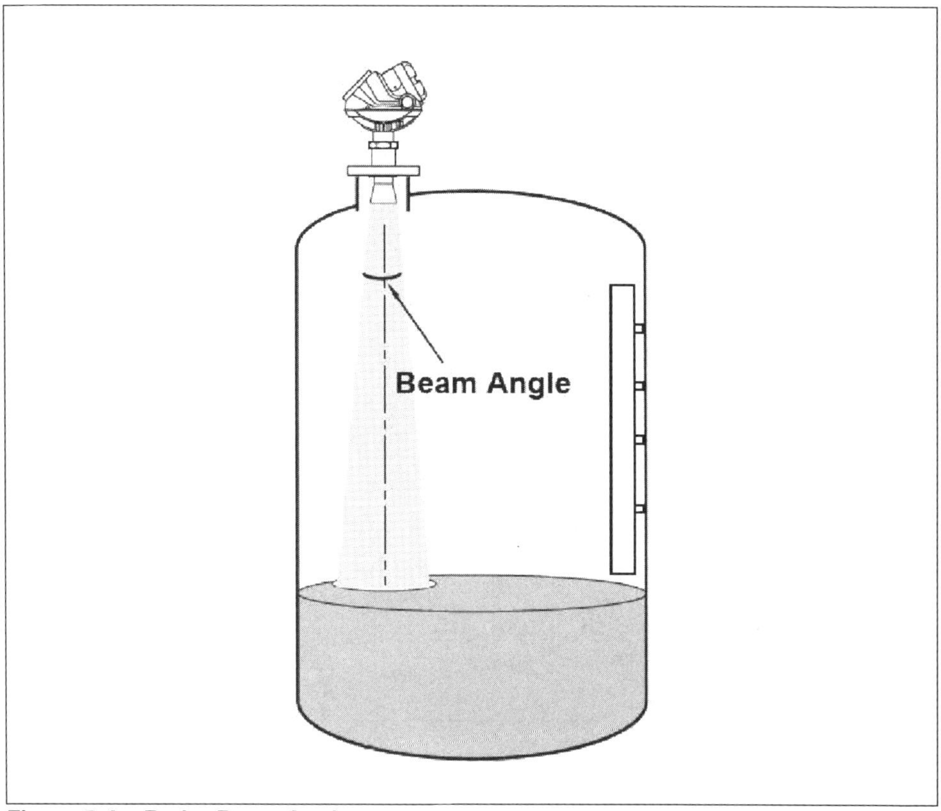

Figure 5-4c. Radar Beam Angle

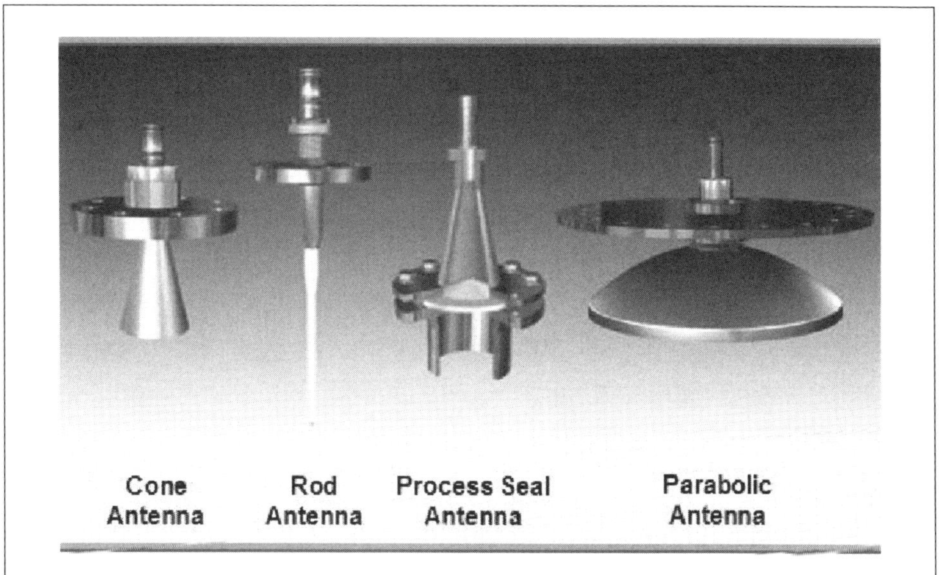

Figure 5-4d. Antenna Types

Guided Wave Radar

Guided wave radar (GWR) is also called *time domain reflectometry* (TDR) or *micro-impulse radar* (MIR). GWR sends a low energy pulse down a probe or cable that bounces off the surface and back to the device (Figure 5-4e). Since the probe guides the signal in each direction, a lower energy pulse can be used and there is a high signal-to-noise ratio. Because a portion of the energy continues to travel through low dielectric fluids, a second pulse can be detected from a second surface that has a higher dielectric. This characteristic makes guided wave radar a good technique for measuring liquid/liquid interfaces such as oil and water and measuring through some foams. Guided wave radar can be used in vessels with tight geometry, smaller tanks, and long access nozzles. It also works well in low dielectric, turbulent applications. Because it is not dependent on reflecting off a "flat" surface, it works well with many powders and grains as well as with liquids with slanted surfaces caused by vortices.

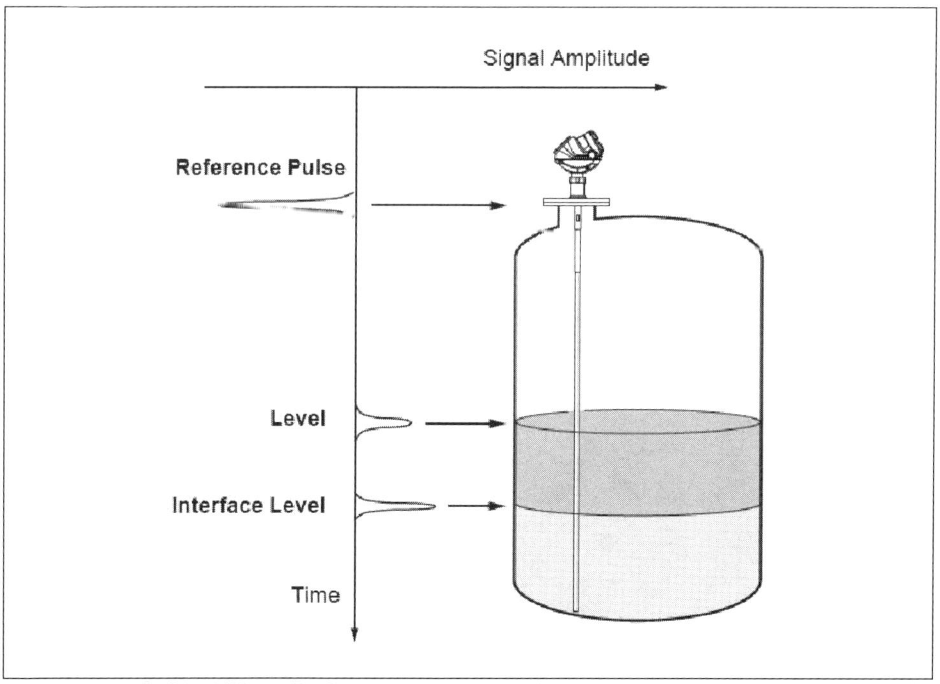

Figure 5-4e. Guided Wave Radar: Level and Interface

Guided wave radar is available with a wide selection of probe types including single lead (rigid and flexible), coaxial, and twin lead (rigid and flexible). Just as the antenna selection is important for non-contacting radar, probe choice is important for optimizing guided wave radar performance in various application and vessel conditions.

Single lead probes are the most tolerant of coating and deposits. However, compared to other probe types, they are more affected by obstructing objects close to the probe. The rigid single lead with a centering disc is a good fit for most level or interface chamber installations. Generally, single lead probes are sufficient for most applications.

Coaxial probes are the optimal solution when there are obstructing objects close to the probe, high turbulence, and liquid or vapor spray. Wave propagation is between an inner rod and an outer pipe. This means the entire signal is contained within the coaxial making the surface pulse stronger, which is important for very low dielectric fluids. Coaxial probes can be used to avoid foam since the probe acts similar to a stilling well. This probe type is for liquids that are not sticky and comes with the option of being perforated for level and interface measurements or easier cleaning.

Last, the twin lead probe uses two parallel rods. The wave propagates between the rods. The highest concentration of the signal is close to the rods, in the middle, meaning the wave propagation is less dispersed than for a single lead probe. The twin probe handles viscous media better than the coaxial probe; however, sticky media still present a risk of bridging between the probes, leading to a false measurement. It handles disturbing objects better than single probes but cannot be in direct contact with metallic objects. Both the twin- and single-lead probe can be obtained in a rigid and in a flexible design (Figure 5-4f). A flexible design allows a longer measuring range and requires less headroom above the vessel for installation.

Radar Level Specifications

Non-Contacting (Free Radiating) Radar

Non-contacting cone antennas are available with process connection sizes ranging from 2- to 8-in. flanged, 3- to 8-in. tri-clamp and bracket mount. A cone antenna may be 316/316 L SST, Alloy C-276, or Alloy 400, with a vessel seal of either PTFE or quartz. A rod antenna may use flanged 2- to 8-in., bracket mounted, 1.5- to 2-in. threaded connections, or 2- to 3-in. tri-clamp connections. Materials of construction include PTFE, PVDF, all-PFA, or partially PFA and partially 316/316 L SST. Process seal antennas are available in connection sizes ranging from 2- to 6-in. flanged, in PTFE with a variety of O-ring seal materials. The parabolic antenna is available in 10- to 18-in. (25.4 to 45.7 cm) sizes with PTFE for the vessel seal. The specific

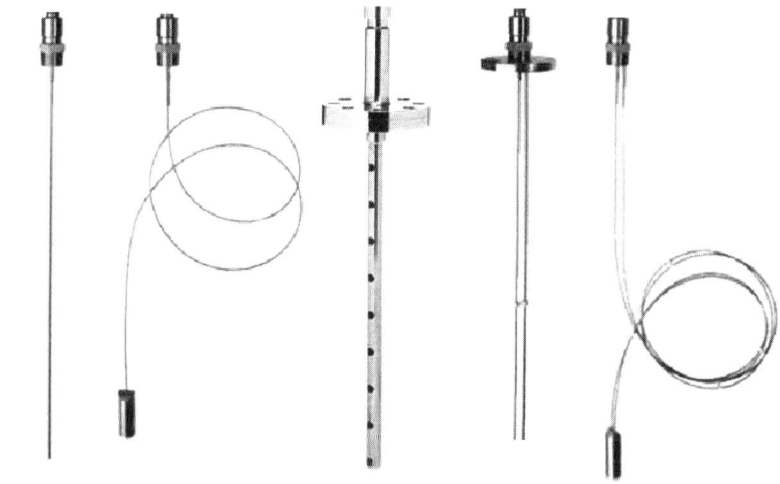

Figure 5-4f. Probe Types

availability of each antenna type will depend on the vendor and whether the radar is high or low frequency.

Reference accuracy ranges anywhere from ±0.04 in (1 mm) to ±0.6 in (15 mm) and repeatability is typically around ± 0.04 in (1 mm). Exact performance will depend on vendor, radar frequency, antenna size, product dielectric constant, and measuring range. Process operating limits range from full vacuum to 2320 psi (160 bar) and temperatures of -76°F (-60°C) to 752°F (400°C). The operating limits are dependent on vendor, antenna type, and materials.

Guided Wave Radar

Guided wave radar with rigid single probes is available in process connection sizes of 1-, 1.5-, or 2.0-in. threaded and 2- to 6-in. flanged. Probe material options include 316/316 L SST, Alloy C- 276, Alloy 400, and PTFE covered probe. Probe length can range anywhere from 1.3 ft to 10 ft (39.6 to 254 cm). A flexible single probe can be used for anything over 3 ft. 1 in. (1 m) and up to 164 ft. (50 m) in length. The flexible single probe comes with the same process connection options, but with material options limited to 316/316 L SST and PTFE covered. Rigid and flexible twin probes are available for process connections of 1.5-in. and 2- to 6-in. flanged. Probe material options include 316/316 L SST. Rigid twin probes can range anywhere from 1 ft. to 10 ft. (30.5 to 305 cm), while a flexible twin probe can be used for anything over 3 ft. 4 in. (1 m) and up to 164 ft (50 m) in length. A coaxial probe (including perforated option) can be purchased in process connection sizes of 1- to 1.5-in. threaded and 2- to 5-in. flanged. Probe material options include 316/316 L SST, Alloy C- 276, and Alloy 400. Coaxial probe length ranges from 1 ft. 4-in. (0.4 m) to 19 ft. 8-in. (6 m).

Reference accuracy ranges anywhere from ±0.12-in. (3 mm) to ±0.4-in. (10 mm) and repeatability is typically around ± 0.04-in. (1 mm). Exact performance will depend on vendor, probe type, product dielectric constant, and measuring range. Process operating limits offer pressure limits as high as 6250 psi (130 bar) @ 70°F (20°C) and temperature limits as high as 800°F (427°C) @ 2000 psi (133 bar) and as low as -320°F (-196°C) @ 5000 psi (345 bar). The operating limits are dependent on vendor, probe type, and probe material.

Radar Level Considerations

The dielectric properties of the medium and the surface conditions can impact the measurement. The user should review application details with the supplier to determine the degree of the effect and possible solutions.

Non-Contacting (Free Radiating) Radar

As has been mentioned, non-contacting radar provides a top-down, direct measurement to the surface with no calibration required. It has no moving parts and is unaffected by changing density, conductivity, temperature, pressure, viscosity, or pH. Since it does not contact the process, it is more resistant to corrosive processes and can be installed in vessels with agitators and other obstructions.

Non-contacting radar needs a clear view of the surface with a smooth, unobstructed, unrestricted mounting nozzle. The measured surface needs to be relatively flat, not slanted. Non-contacting radar gauges can handle agitation, but their success will depend on a combination of the fluid properties and the amount of turbulence, the overall range, the antenna size, and the frequency.

Guided Wave Radar

Similar to non-contacting radar, guided wave radar provides a top-down, direct measurement to the surface with no calibration required. It also has no moving parts and is unaffected by changing density, conductivity, temperature, pressure, viscosity, or pH. Guided wave radar is more resistant to the effects of foam, vapors, and turbulence than is non-contacting. It can also be used to measure interfaces and solids, powders, and granules. Since the radar pulses are guided by a probe, guided wave radar is a technology with few installation rules and an easy swap into existing and small openings. Therefore it works well in displacer and bypass cages for both interface and level measurements.

Guided wave radar is a contacting measurement and therefore the process fluid must be compatible with the probe material. Additionally, the probe must be mounted with sufficient clearance from agitators and other obstructions in the vessel. The specific clearance will depend on probe type.

While each type of radar has unique advantages, both are subject to application restraints. Some vendors offer assistance in choosing the best radar type, model, and options for a specific application. Understanding the application conditions and matching them to the correct type of radar is the first step to a successful measurement. Beyond that, proper installation and configuration play key roles.

Radar Level Setup and Calibration

Neither non-contacting nor guided wave radar units need field calibration. However, configuration of vessel height, vessel geometry, product dielectric, and probe length (for guided wave only) are all important to obtaining accurate level measurements. Factory configuration of radar units is an option available with some manufacturers and requires that the application information be obtained upfront.

Radar units often come with configuration software or local operating interfaces for configuration. Others can be configured with handheld configuration devices, and some units can interact with control systems for remote configuration. While field configuration is not always necessary, it can improve results and may be required for some applications. The Asset Management System (AMS) windows shown in Figures 5-4g through 5-4j guide the user through the configuration and setup for vessels (commonly referred to as tanks).

Radar Level Installation

Non-Contacting (Free Radiating) Radar

For radar measurements, as has been stated, good installation is a key to success. The following mounting recommendations should be observed in

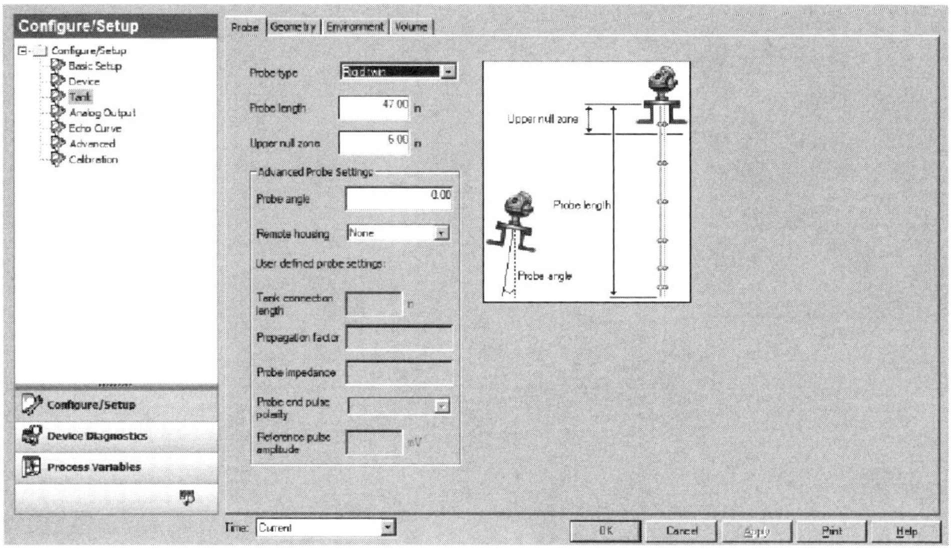

Figure 5-4g. AMS Configuration Window

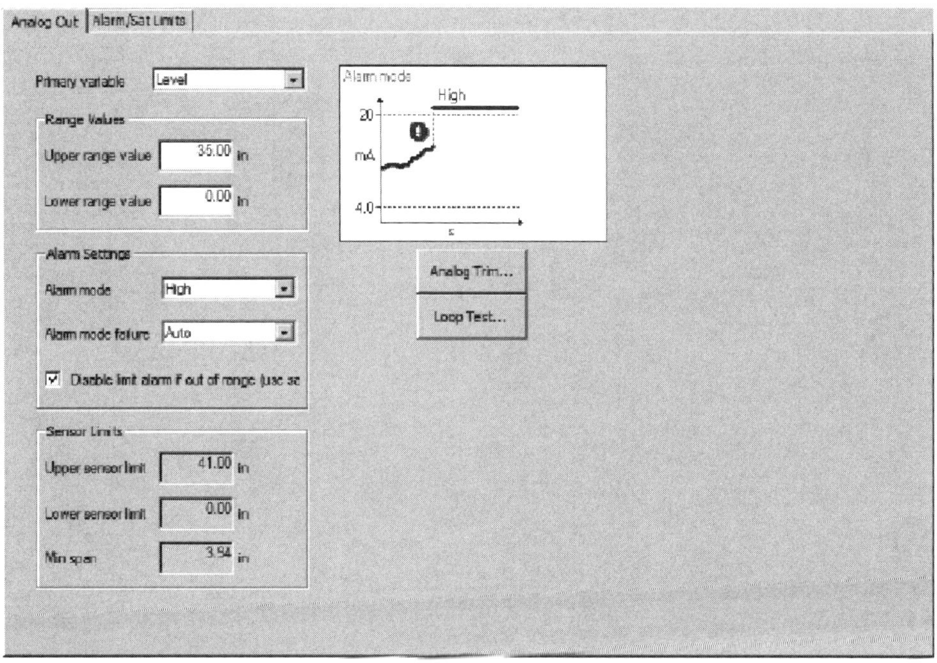

Figure 5-4h. AMS Configuration Window

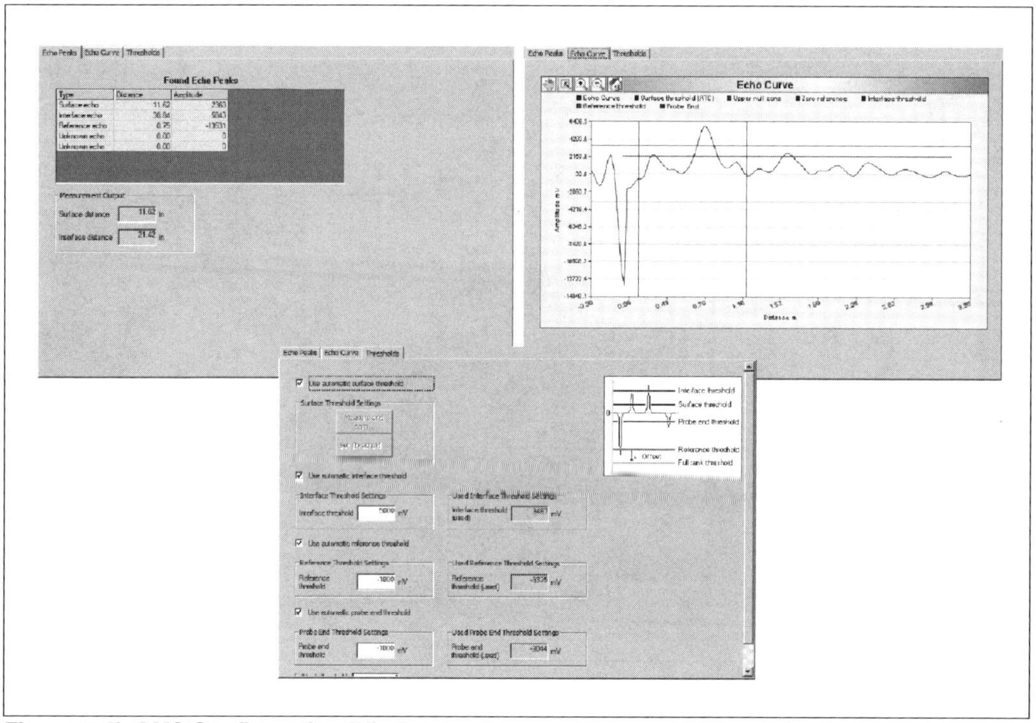

Figure 5-4i. AMS Configuration Window

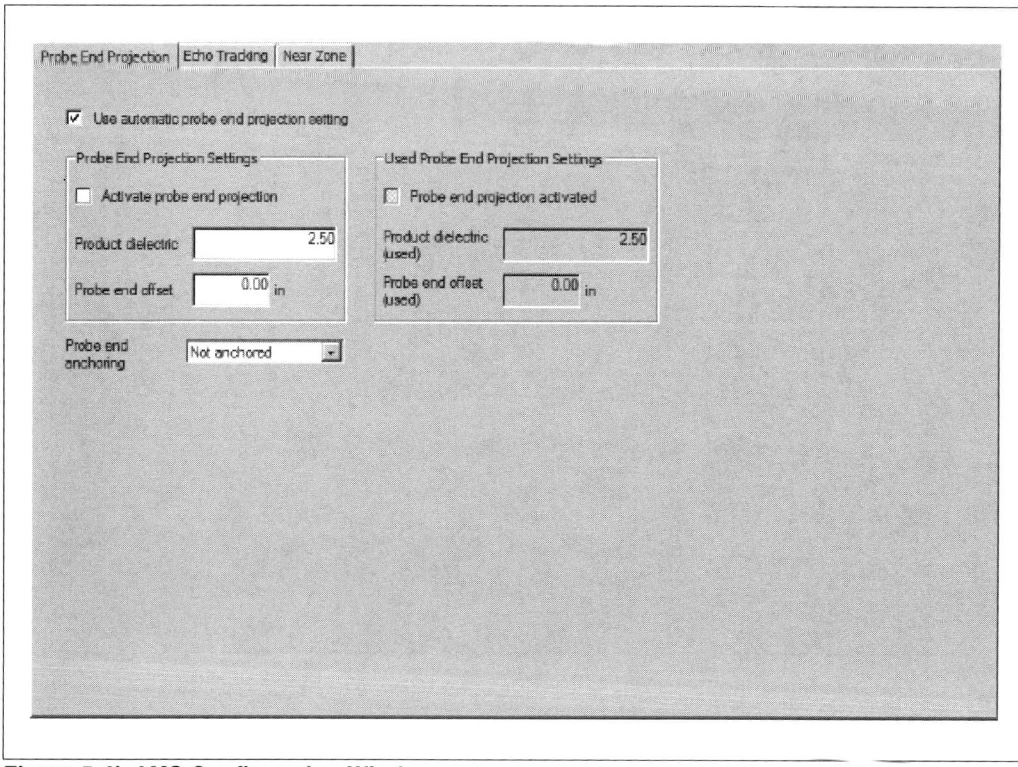

Figure 5-4j. AMS Configuration Window

order to get optimized performance for a given vessel environment (Figures 5-4k and 5-4l):

- Nozzles should be located where view to the surface is unobstructed.
- Avoid mounting on nozzles at the center of the vessel as this can cause double bounces leading to incorrect measurements.
- Disturbing objects and filling inlets creating turbulence should be kept at a distance, outside the signal beam.
- It is important that the end of the antenna projects inside the vessel. If this is not possible, then use a radar with high frequency so that the beam angle is narrower.
- Match the antenna size to the nozzle pipe diameter.
- A bypass chamber or stilling well can be used to avoid interference from disturbing objects, turbulence, or foam (Figure 5-4n).
- If valves are present, use high frequency radar only.

Valves are sometimes required on vessels both for safety and so that instrumentation can be isolated for service. If a valve is present on a nozzle where a non-contact radar transmitter is used, then a high frequency (> 24 GHz) radar should be used. This is because high frequency radar can be used when the antenna is recessed in a nozzle while lower frequency units cannot tolerate recession of the antenna. If lower frequency units are recessed, then excessive disturbances to the signal will occur. The valve must be a full port valve such as a ball valve or fully recessed gate valve. This minimizes disturbances to the radar signal.

As vessel height increases, the nozzle diameter and antenna size should increase. A larger antenna provides a more concentrated radar beam and a larger receiving area for the signal, making the signal stronger and more suitable for longer distances (Figure 5-4m).

As has been mentioned, low frequency radar waves cut through condensation, foam, and vapor better than high frequency, but high frequency provides a more concentrated beam and is recommended for use in narrow vessels, tall and narrow nozzles, and vessels with disturbing obstacles (Figure 5-4m).

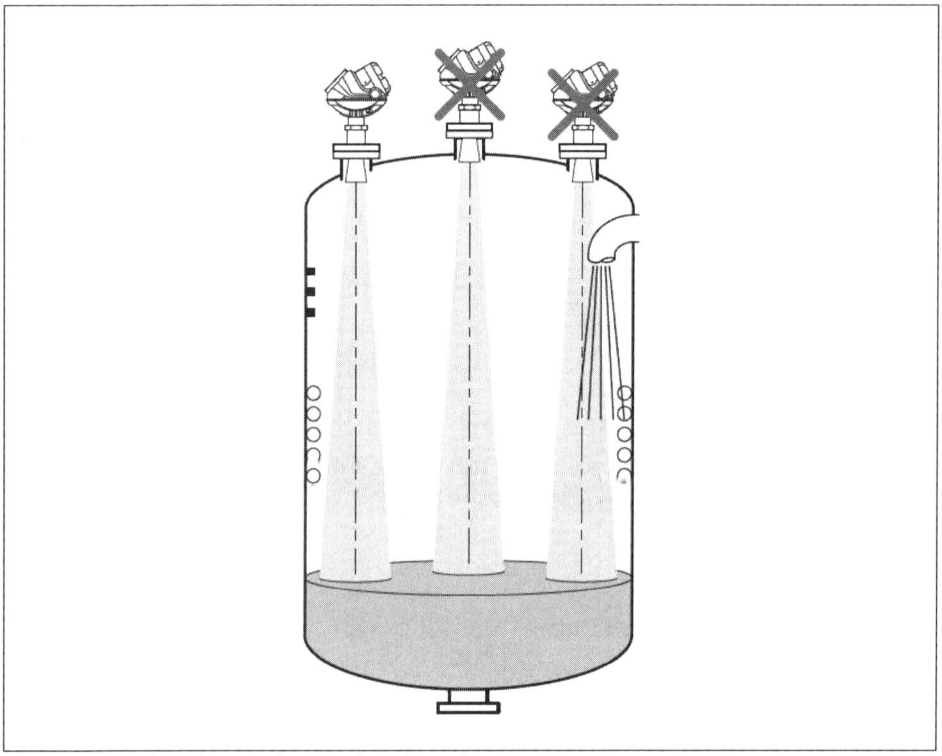

Figure 5-4k. Non-Contacting Radar Mounting Positions

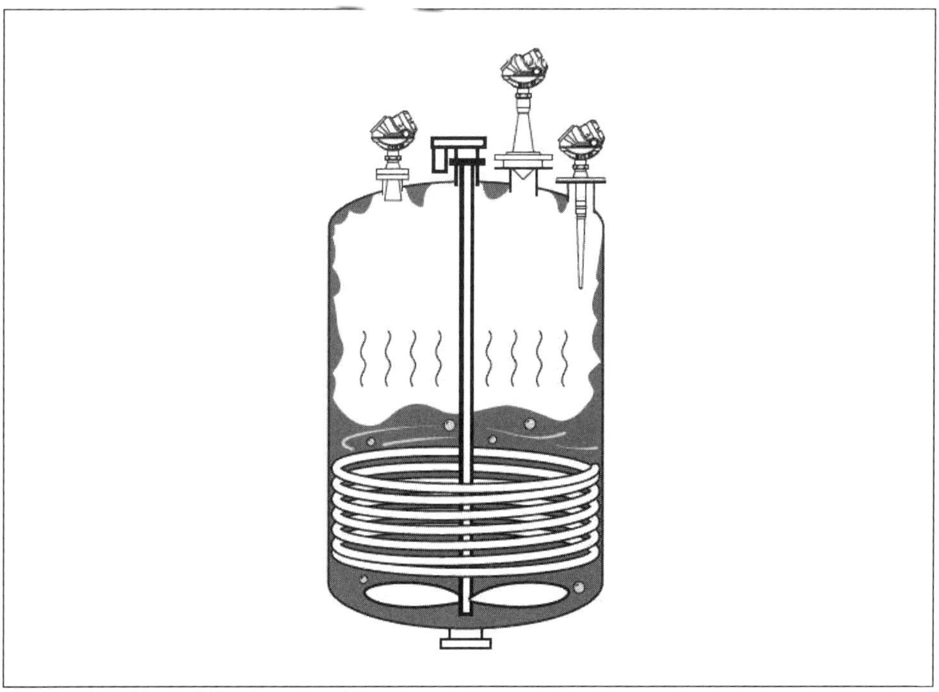

Figure 5-4l. Vessel Environment

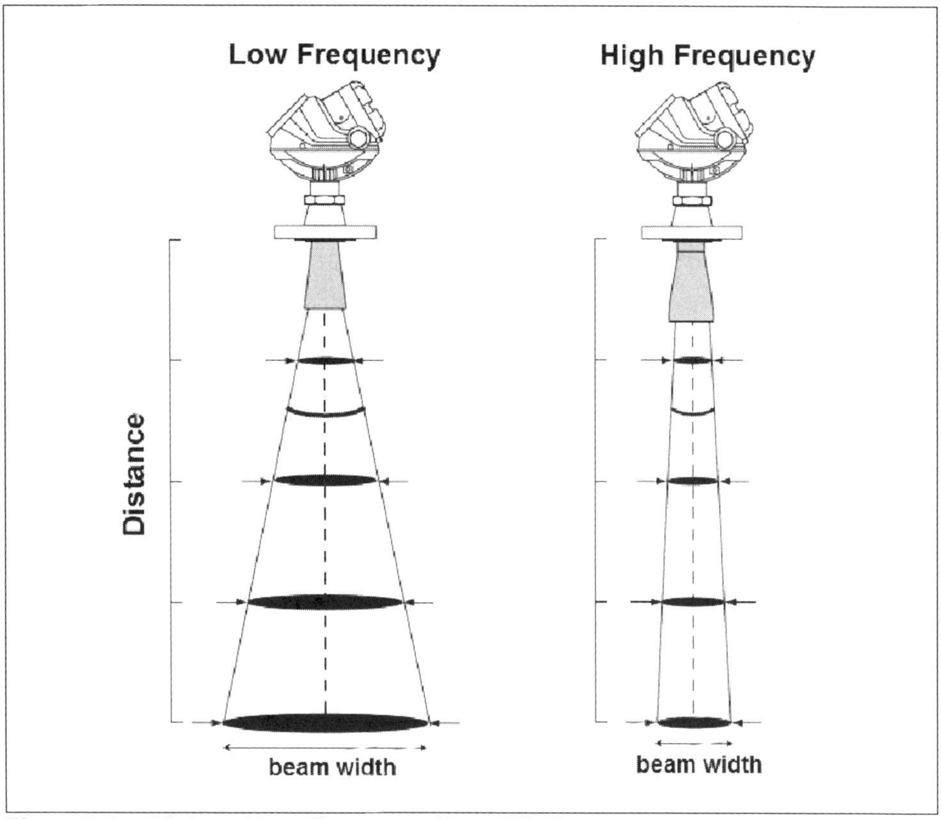

Figure 5-4m. High and Low Frequency Beam Widths

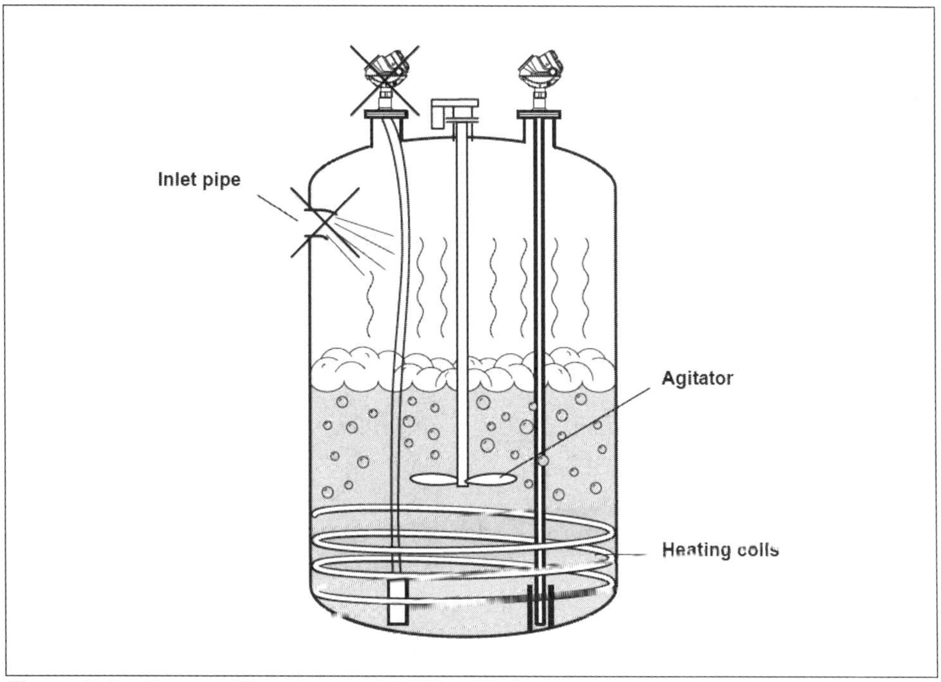

Figure 5-4n. Bypass Chamber and Stilling Well

Guided Wave Radar

Proper installation is important for guided wave radar measurements as well. The following mounting considerations should be observed in order to get optimized performance (Figure 5-4o):

- Do not mount close to inlet pipes.
- Do not mount close to agitators. If the probe can move to within 30 cm (11.8 inches) of an agitator, a probe tie-down is recommended.
- If the probe tends to sway due to turbulent conditions in the vessel, the probe should be anchored to the vessel bottom.
- Avoid mounting close to heating coils.
- Make sure that the nozzle does not extend into the vessel.
- Make sure that the probe does not come into contact with the nozzle or other objects in the vessel.
- Position the probe such that it is subject to a minimum of lateral force.
- Allow a minimum clearance to wall or disturbing objects of about 4 in. (100 mm). No clearance is necessary for coaxial probes.

Also it is important to avoid long nozzles and/or reduced nozzles unless using coaxial (Figure 5-4p).

For optimal single lead probe performance in non-metallic (plastic) vessels, the probe must be mounted with a metal flange, or screwed into a metal sheet (d>8 in./200 mm) if the threaded version is used (Figure 5-4q). It is also important to consider the Electromagnetic Compatibility (EMC) performance of the transmitter since single leads are more susceptible to interference in non-metallic vessels.

A rigid single probe is the best option for mounting in bypass chambers or stilling wells (Figure 5-4r).

For solids measurements, additional mounting considerations should be observed (Figure 5-4s):

- Do not mount close to inlet pipes in order to avoid product filling wear on the probe.

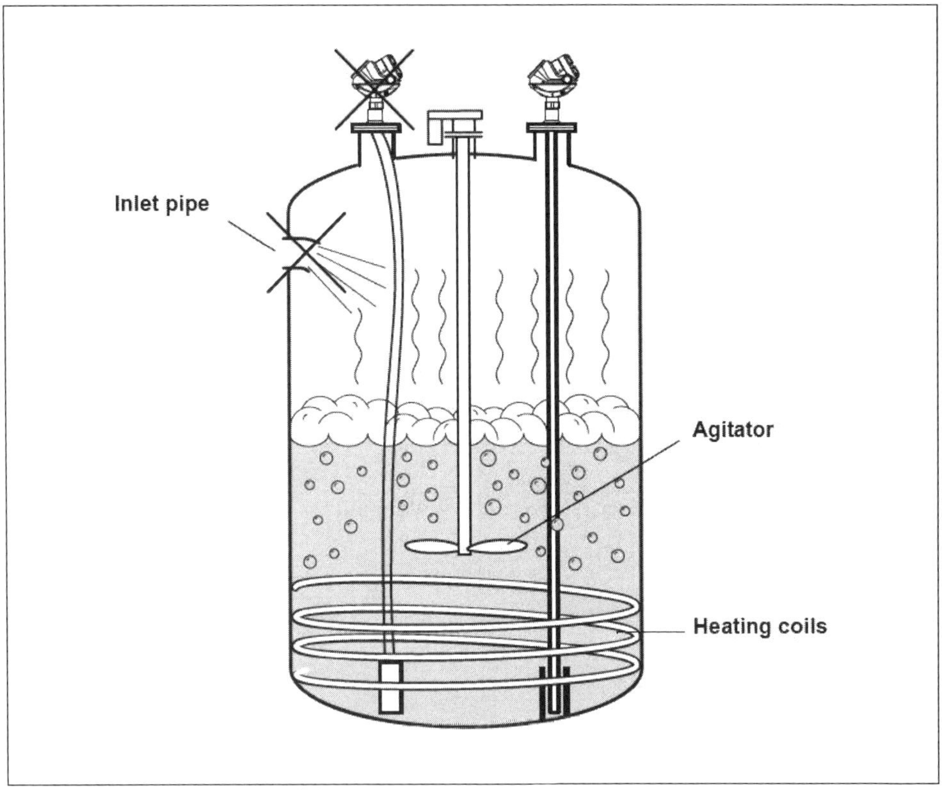

Figure 5-4o. Mounting Position

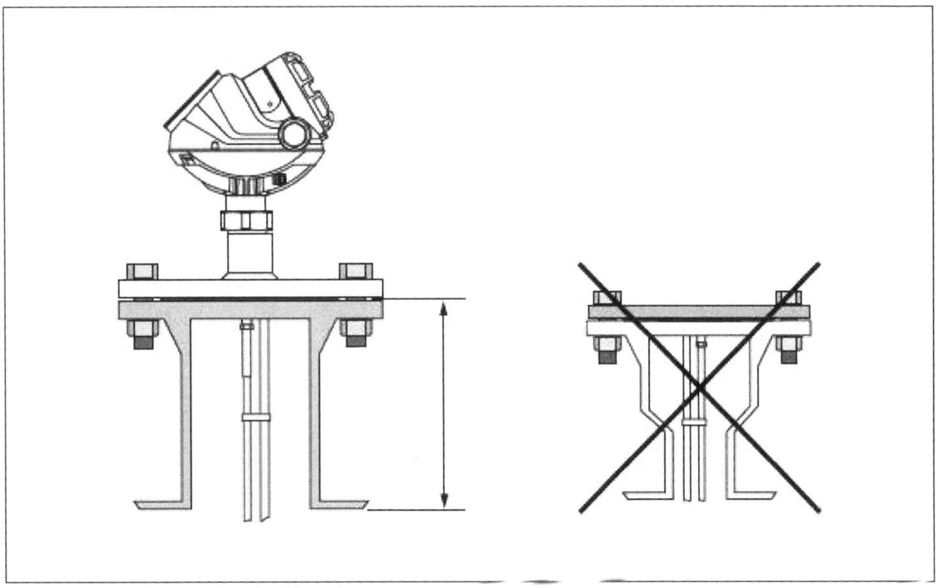

Figure 5-4p. Mounting in Nozzles

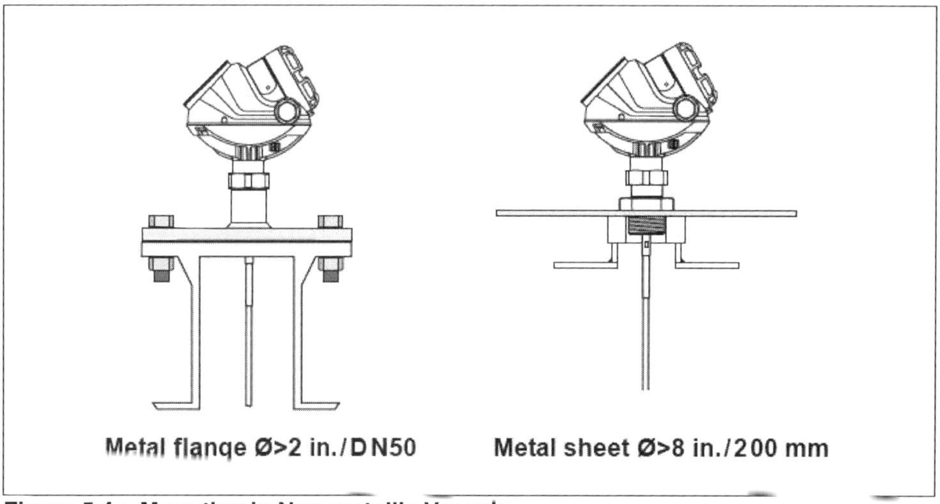

Figure 5-4q. Mounting in Non-metallic Vessels

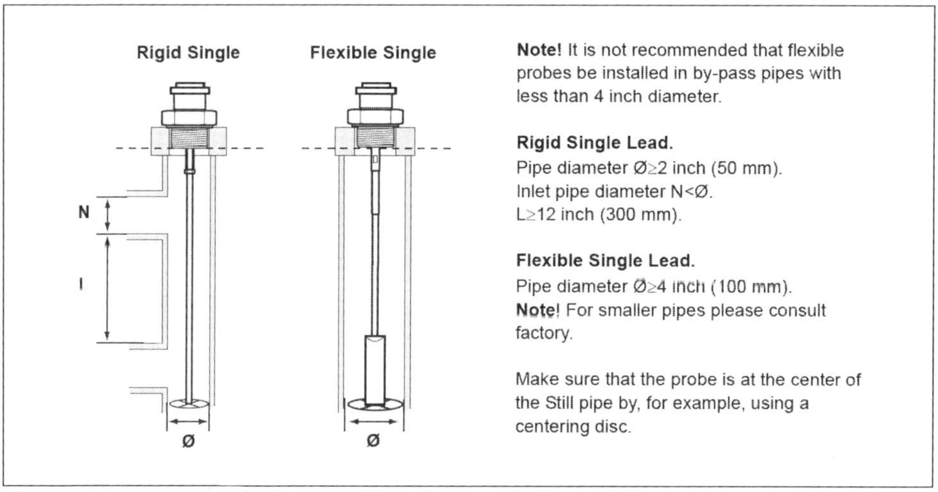

Figure 5-4r. Mounting in Bypass Chambers or Stilling Wells

- Regularly check to ensure the probe has no defects from damage or wear.
- It is recommended that installation is carried out when the vessel is empty.
- In concrete vessels the distance L between the probe and the wall should be at least 500 mm (20 in.).
- In order to stabilize the probe against side forces, it is recommended to fix the probe to the vessel bottom.

- For solids, consider using the thicker probe option (availability varies with vendor) since it has higher tensile strength. The probe should have a sag of 1 in./100 in. (1 cm/m) to prevent probe damage.
- Avoid anchoring in solids vessel over 98 ft. (30 m) in height since tensile loads are much stronger for anchored probes.
- Product build-up on the vessel walls close to the probe may interfere with measurements. Choose a mounting position which makes sure that the probe is not in contact with, or very close to, the product build-up.

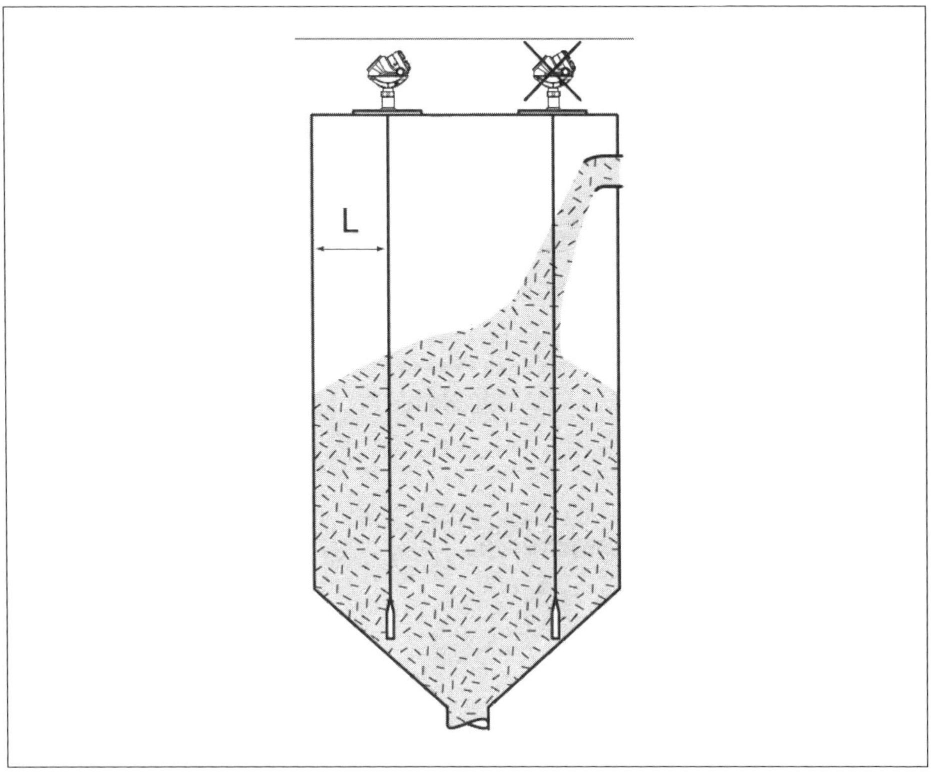

Figure 5-4s. Mounting in Solids Applications

Radar Level Maintenance

If non-contacting or guided wave radar units are not performing as expected, the application characteristics (process media, mounting options, and process conditions) may not be compatible with the chosen radar device or the configuration is incorrect. Other issues could indicate that the transmitter needs cleaning or that process conditions have changed since installation. It is recommended to save a baseline plot and initial configuration file to use as a comparison for troubleshooting applications. General instances are included in the troubleshooting section of most product manuals. The Asset Management System (AMS) windows in Figures 5-4t through 5-4w are useful for monitoring secondary process variables and system health.

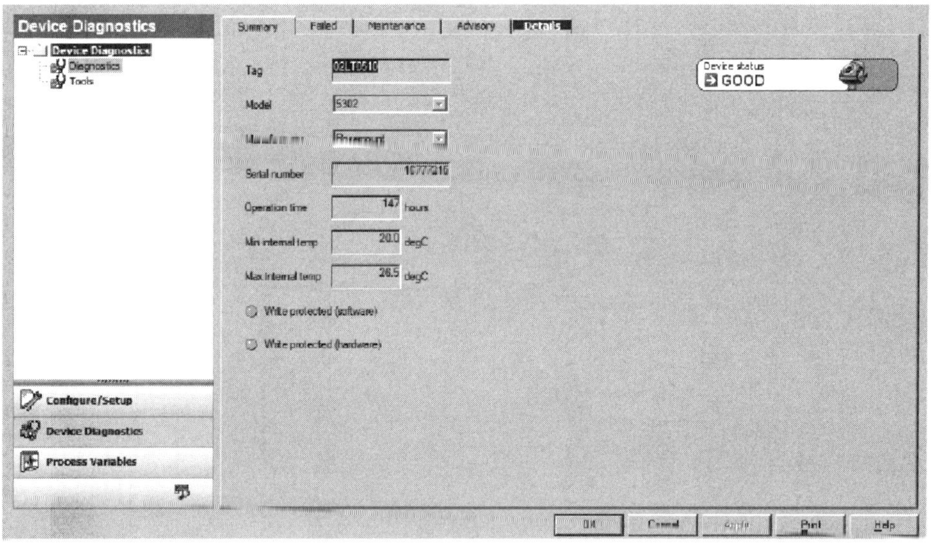

Figure 5-4t. AMS Windows for Monitoring System Performance and Health

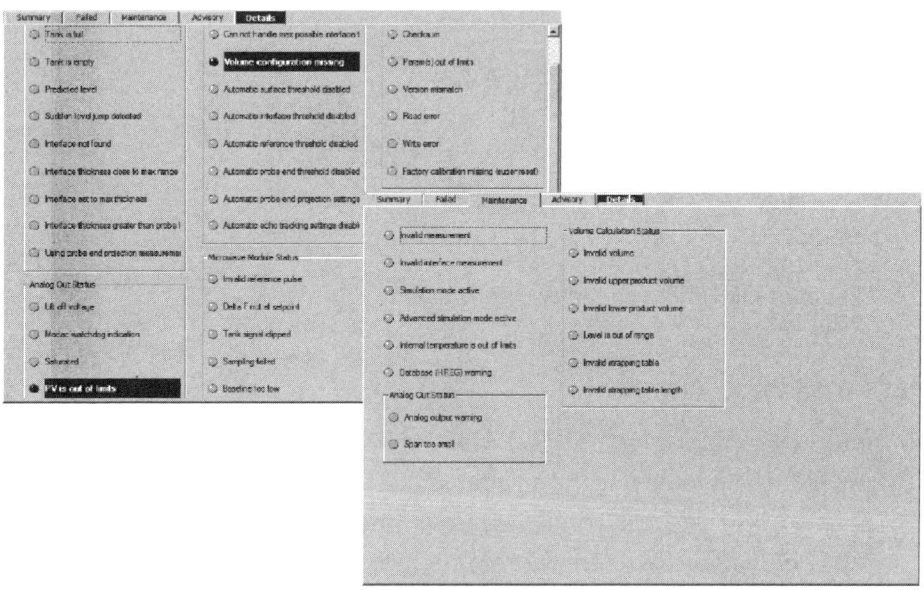

Figure 5-4u. AMS Windows for Monitoring System Performance and Health

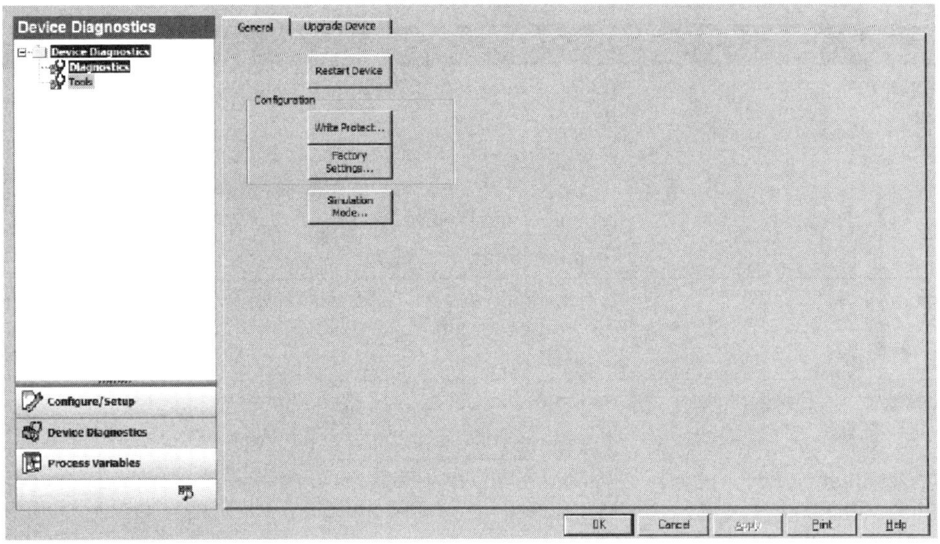

Figure 5-4v. AMS Windows for Monitoring System Performance and Health

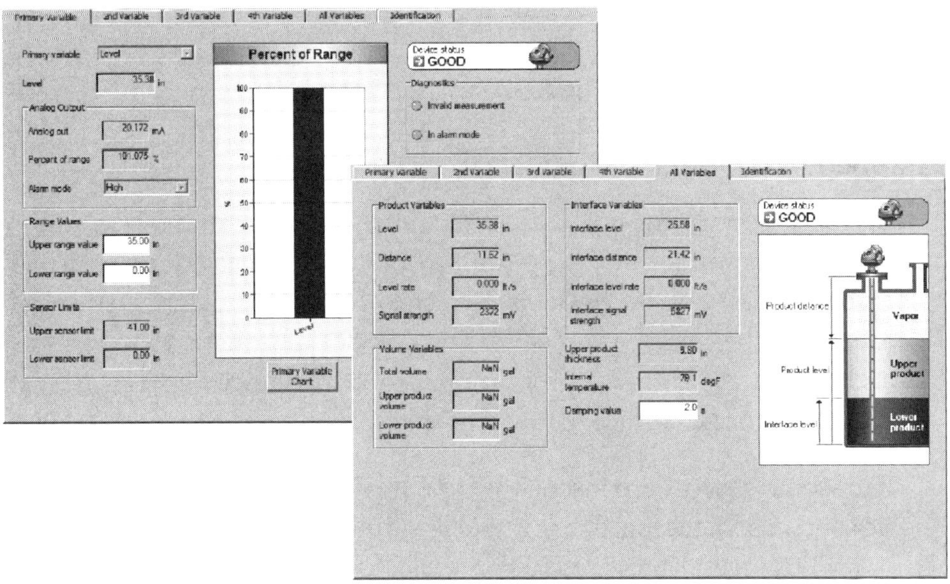

Figure 5-4w. AMS Windows for Monitoring System Performance and Health

Radar Level Troubleshooting Guide

Improper Installation	Is there sufficient clearance through nozzle and along vessel walls? Are there inlet pipes near radar unit? For non-metallic vessels, is there a metal plate or flange being used? (guided wave) Is the antenna inside the vessel? (non-contacting)
Configuration	Check that vessel height, vessel geometry, product dielectric, and probe length (guided wave only) have been properly configured.
Probe/Antenna Contamination	Is there excessive build-up on the antenna/probe? Was the process properly sealed (e.g., O ring, etc.)?
Obstacles/Interference	Are there any false echoes in the vessel that have not been accounted for? (non-contacting) Could the probe be touching the wall or any other metallic obstacle? (guided wave)
Operating Conditions	Are the operating conditions in compliance with those given at the time the radar was purchased? Has the process changed?

5-5. Ultrasonic Level Measurement

Ultrasonic Level Principles

In continuous ultrasonic measurements, a sound pulse is transmitted about once per second from the sensor to the surface of the measured material. The sound pulse is reflected from the material surface and returns to the sensor as an echo. The amount of time that it takes to send and receive the signal is proportional to the distance between the sensor and the material surface. Ultrasonic devices operate at around 20 kHz.

Ultrasonic devices are insensitive to many problematic liquid characteristics such as changing density and conductivity, viscosity, corrosion, and optical changes. The speed of sound is, however, affected by the presence of dust, steam, process vapors, and higher temperatures and pressures. Therefore, ultrasonic devices work best in environments where there is a clear or consistent path between the transmitter and the surface. Continuous ultrasonic devices are limited to pressures from slightly negative to about 45 psi and temperatures of less than 200°F.

Ultrasonic Level Selection

Ultrasonic transmitters have no moving parts and they are easy to install by simple mounting on the top of a vessel. Because they are noncontact, ultrasonic measurements are usually not affected by product buildup or coating and the sensing element is not likely to be contaminated. To further protect the device from vapors which could corrode the surface, some ultrasonic devices are capable of operating through a PTFE protective covering. The sensor is often is made of an inert material such as PTFE or PVDF.

More recent devices with microprocessor capabilities provide sophisticated echo processing algorithms and higher energy sensors. This not only makes the devices very easy to set up and use, but it also results in greater accuracy.

Ultrasonic level can be used in many applications. Storage vessels where extreme accuracy is not required, corrosive chemicals, and viscous or sticky liquids are some examples. Other applications of interest include:

- All water applications, such as water and wastewater (W&WW) plants, but also utilities of industrial plants
- Open air applications (sumps, reservoirs, filter beds, flow channels, etc.)
- Water based (aqueous) liquids
- Chemicals (non-vaporous for closed vessels)
- Liquids soluble in water

Ultrasonic Level Specifications

Ultrasonic level sensors typically have a range capability of up to 50 ft. (15 m) and can be used in applications with an operating pressure of -4 psi (-0.25 bar) to about 45–50 psi (3 bar) and with operating temperature of about -22 to 158°F (-30 to 70°C).

Typical measurement accuracy is +/- 5 mm (0.2-in.) for ranges up to 1 m (39.4-in.) and 0.25% of measured distance for ranges > 1 m. Loop powered devices require power of 12 to 30 VDC.

Ultrasonic devices can be used in process flange connection of 2-in. or larger. There should be an air gap between the pipe wall and sensor.

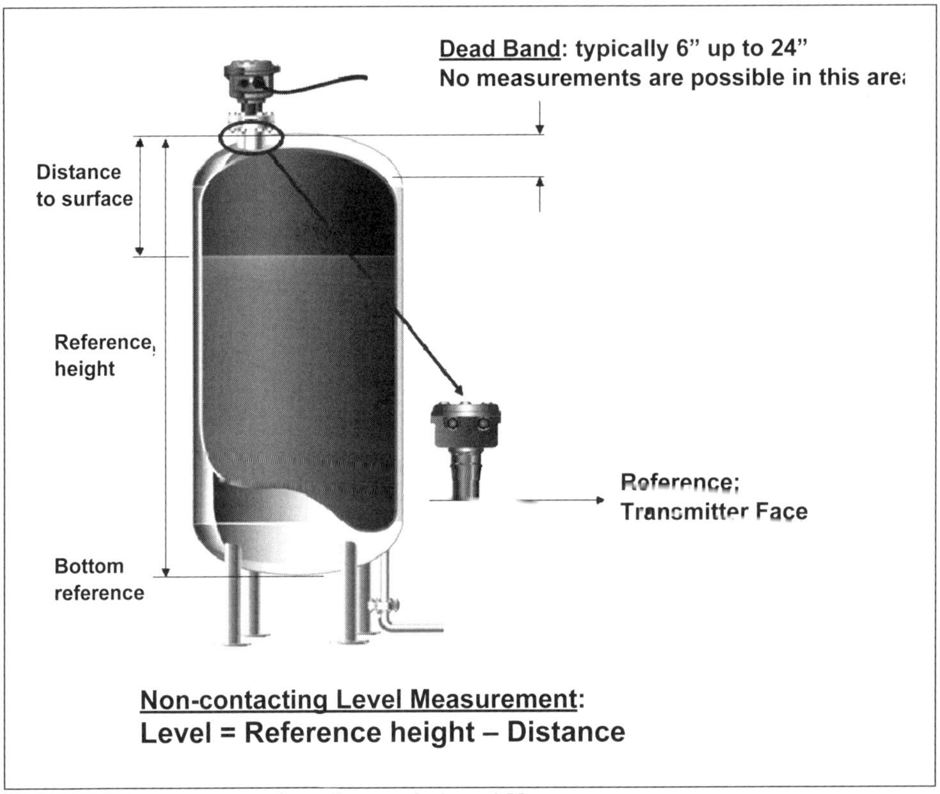

Figure 5-5a. Non-Contacting Ultrasonic Level Measurement

Ultrasonic Level Considerations

As a proven technology, ultrasonic is one of the most cost effective means of level measurement. The transmitter is mounted on the top of the vessel for direct measurement. It has no moving parts, which means virtually no maintenance is required. As mentioned, the ultrasonic signal is not affected by the density, dielectric or conductivity of the measured liquid.

Since ultrasonic devices are not in contact with the liquid, they can easily be used for dirty, viscous or coating applications. With non-metallic wetted parts, they can be used with many corrosive materials.

There are several application factors to consider in order to optimize the use of non-contact continuous ultrasonic devices. Any evenly distributed material in the vapor space will result in a proportional change in the speed of sound through it. A common solution is to apply a gas blanket over the vapor space to try to maintain a constant environment. This will allow the speed of sound to remain constant and the instrument can be calibrated to provide an accurate output.

Another process condition which affects ultrasonic measurements is the ability of the surface of the material to reflect the sound waves. Foam on the surface is a condition commonly encountered by ultrasonic devices. The sound waves can reflect off the foam instead of the actual liquid surface. Alternatively, the foam can sometimes absorb the sound waves and no signal is returned. Ultrasonic level measurement is generally not recommended for use when foam is expected to be present.

If the surface is not horizontal, then the sound waves may not be reflected back directly to the instrument. Solids often have a slanted surface and the angle of repose can be sufficient to cause only part or none of the echo to be seen by the instrument. Turbulent surfaces can have a similar affect and can even absorb the sound waves. To overcome this problem, a stilling well can be installed to provide a more stable surface. Vortexes caused by agitation can also be severe enough to cause the echo to be misdirected. Dust and other particles in the vapor space can scatter the sound waves.

Ultrasonic Level Setup and Calibration

Factory calibration of ultrasonic transmitters is an option available with some manufacturers. A complete calibration by a traceable lab will provide performance information, but is no guarantee that the same performance and accuracy will be obtained or maintained in the field because of the many process conditions that may be present.

When installing ultrasonic devices, the unit itself can be used as an electronic tape measure to measure the distance from its mounting position to a target, such as a strike plate at a fixed distance or the empty vessel bottom. This information can be used to verify the validity of measurement or to input vessel information for the level calculation (Level = vessel reference height – distance).

Ultrasonic Level Installation

When installing an ultrasonic transmitter, care must be taken to ensure that there is a clear vertical path between the instrument and the surface. The sound waves may reflect off any devices or obstructions present and that point will be interpreted as the level measurement. The sensor should be mounted away from the wall of the vessel with a 2 ft (0.6 m) clearance commonly suggested. Center of vessel mounting locations should be avoided. One final consideration in installing ultrasonic devices on open vessels in outdoor applications is the exposure of the device to wind. Wind can literally blow the sound wave off course at either the initial time of transmission or the echo before it is received by the transmitter. Baffles or stilling wells can be installed to limit this problem.

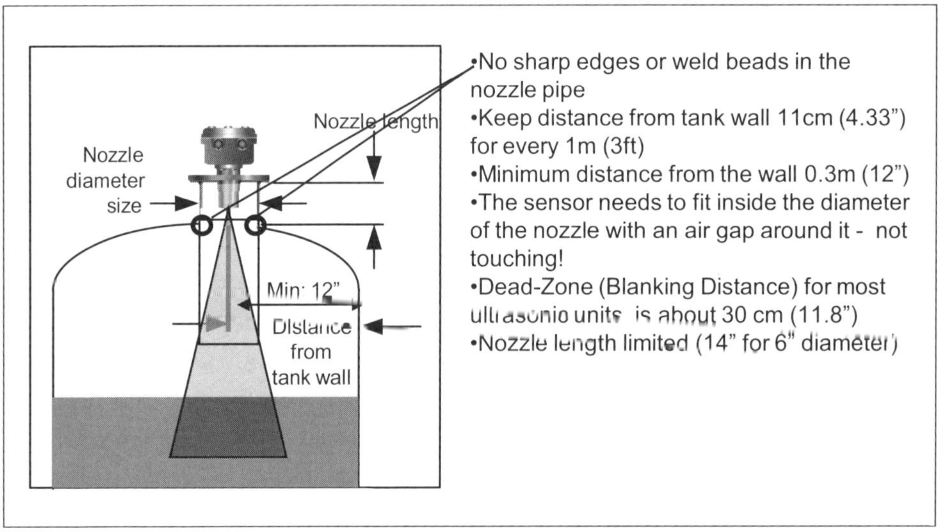

Figure 5-5b. Mounting Considerations for Optimum Performance

Ultrasonic Level Maintenance

Ultrasonic Level Troubleshooting Guide

Improper Installation	Verify that the sensor is perpendicular to the level surface. Verify that the sensor is at least 2 ft. from the side wall of the vessel and has a clear, unobstructed view of the surface. Verify that the sensor is not in contact with the metal nozzle.
Operating conditions	Are the operating conditions in compliance with the specifications of the unit?
No Display	Check the power supply. Ensure there is a minimum of 12 VDC at the instrument terminals. Check to be sure that the cable insulation is not preventing contact at the terminal block
No Level Reading	Check to ensure the instrument is ticking about once per second. If there is no ticking, the instrument should be replaced.
Calibration	Is the calibration too high or low for the vessel height?

5-6. Questions

1. Why are level measurements important?
2. What can be done to eliminate sensing line problems in DP level measurements?
3. In what way does the dielectric constant (DC) of the media affect the radar measurement?
4. Are radar gauges safe?
5. Does radar work on foam?
6. Will two or more radar units in one vessel conflict with each other?
7. Given application requirements, what is the best type and size of ultrasonic meter and given a maintenance problem, what is the fix?

5-7. References

1. BLH Electronics *Electronic Weigh Systems Handbook.* BLH Electronics HDBK 002-1, 1986.
2. Fuhrman, J. "Demystifying DP level Control." *Control,* May 1997.
3. Schumacher, M. "Hydrostatic Tank Gauging Theory and Practice." *Control,* Jan. 1990.
4. McMillan, G. K. and Toarmina, C. M. *Advanced Temperature Measurement and Control*. Research Triangle Park: ISA, 1995.
5. McMillan, G. K. *Good Tuning - A Pocket Guide, 2nd Edition*. Research Triangle Park: ISA, 1995.
6. Schuler, M. "*A Practical Guide to Radio Frequency Level Controls.*" Horsham: Drexelbrook Engineering Company, Publication 420-1-531, 1989.
7. Stichweh, L. "Gas Purged DP Transmitters for Liquid Level and Flow." *InTech,* Research Triangle Park: ISA, Nov. 1992.
8. Parker, S. "Measuring Liquid Level – A Guide for CPI Users." *Chemical Engineering,* May 1995.

6

pH Measurement

6-1. Learning Objectives

A. Appreciate the extraordinary rangeability and sensitivity of pH measurements.

B. Be introduced to the effect of ion concentrations and temperature on solution pH.

C. Understand the role resistances play in the accuracy of pH measurements.

D. Learn about the latest technological advances in transmitters and electrodes.

E. See how wireless transmitters can eliminate spikes in indicated pH values.

F. Gain some insight into the lifecycle cost savings of wireless transmitters.

G. Be introduced to the importance of wireless flexibility and portability for R&D.

H. Find out how to select electrodes for a variety of difficult process conditions.

I. Learn terminology and issues to intelligently discuss industrial applications.

J. Visualize how asset management systems are used for setup and calibration.

K. Understand calibration methods and effectiveness.

L. Learn the basic types and advantages of process connections.

6-2. Introduction and Overview

The concept of the pH measurement as the negative logarithm of the hydrogen ion concentration in a solution was first published in 1909 by Soren Sorensen of the Carlsberg Laboratory of Copenhagen. 100 years later, the glass pH electrode is the most prevalent online composition measurement in the process industry. The logarithmic relationship offers an incredible rangeability and sensitivity, far beyond the capability of any other common field measurement. For the 0–14 pH scale, the glass electrode can measure a 14 orders of magnitude change in hydrogen ion concentration. At 7 pH, the glass electrode can detect changes in the hydrogen ion concentration to the eighth decimal place.

While the concept of pH measurement that was first published 100 years ago remains essentially the same, significant technological improvements in glass formulation and construction and reference electrode design have increased pH electrode life and performance. Considerable experience has been gained in terms of electrode installation methods for difficult process conditions as the scope of pH applications has expanded. Yet much of this practical expertise has not been documented and the ability of wireless devices and embedded diagnostics and calibration records to eliminate common problems and provide smarter and more effective maintenance has not been explored.

Nearly every industrial plant has pH measurements to meet environmental regulations on wastewater. For many processes, accurate pH measurement and control is important for water and product quality. This chapter will focus on demanding industrial applications that require 0.01 to 0.02 pH accuracy despite harsh conditions from steam-in-place sterilization in biopharmaceutical processes and high reaction temperatures and high salt, polymer, and crystal concentrations in chemical processes.

6-3. Principles

This section has considerable detail on what affects the performance and life-cycle cost of a pH measurement since the pH electrode requires more calibration and maintenance than any other common type of sensor in the process industry. A simple electrical circuit analogy is used to explain many of the non-ideal effects in an industrial application.

While we are accustomed to thinking of pH as a measurement of hydrogen ion concentration, in reality pH is a measurement of hydrogen ion activity as defined by Equations 6-1 and 6-2. Most pH measurements are in extremely dilute water solutions where the hydrogen ion concentration is the same as the hydrogen ion activity. In Equation 6-3, this one to one relationship corresponds to a unity activity coefficient. The ion concentration in most water and waste treatment applications is so low that no distinction between hydrogen activity and concentration is necessary. If the solution gets crowded with ions, the activity of the hydrogen ions decreases, and the activity coefficient drops below one. For example, in chemical reactors and evaporators, the concentration of ions can be high enough to make the activity coefficient significantly less than one. In these cases, the distinction is important because the pH may change for a change in total ion concentration even though the hydrogen ion concentration is constant.

The product of the hydrogen $[H^+]$ and hydroxyl ion $[OH^-]$ concentration in a water solution depends upon the water dissociation constant (pK_w) per Equation 6-4. The water dissociation constant changes with temperature about -0.03 per °C which for a given hydroxyl ion concentration causes the pH to change -0.03 per °C for strong basic solutions. This change in solution pH with temperature is not corrected by the standard temperature compensator. For example, an 8.3 pH hot waste stream at 50°C (122°F) that cools down to 25°C (77°F) by the time it reaches the effluent discharge will be above 9 pH, possibly in violation of an environmental constraint. The change in the dissociation constants of weak acids and weak bases with temperature also affects pH. The dissociation constants of strong acids and bases are beyond the pH scale so changes in these constants have a negligible effect on pH.

$$a_H = 10^{-pH} \tag{6-1}$$

$$pH = -\log(a_H) \tag{6-2}$$

$$a_H = \gamma * c_H \tag{6-3}$$

$$c_H * c_{OH} = 10^{-pK_w} \tag{6-4}$$

Where:

a_H = hydrogen ion activity (gm-moles per liter)

c_H = hydrogen ion concentration (gm-moles per liter)

c_{OH} = hydroxyl ion concentration (gm-moles per liter)

γ = activity coefficient (1 for dilute solutions)

pH = negative base 10 power of hydrogen ion activity

pK_w = negative base 10 power of the water dissociation constant (14.0 at 25°C)

Glass Electrode

The most popular pH sensor in the process industry is the combination electrode that consists of a glass measurement electrode surrounded by a reference electrode. A spherical or hemispherical glass bulb has a pH sensitive surface that develops a millivolt potential (E_1) that is proportional to the pH of the liquid in contact with it. The inside surface of the glass electrode also develops a potential that corresponds to a 7 pH buffer. The sensing of pH depends upon a thin hydrated layer on each glass surface. If the surface in contact with the process has the same composition and hydration as the surface on the interior of the bulb, the difference in the millivolt potentials between the interior and exterior glass surfaces can be described by Equation 6-5, which is a simplification of the Nernst equation. The standard temperature compensator corrects for the change in the millivolts generated per pH unit by the temperature factor in Equation 6-5. Typically an RTD (resistance temperature detector) sensor embedded in the electrode is used to provide automatic compensation that effectively keeps the temperature factor at 298°K (25°C; 77°F) in Equation 6-5. At 25°C we have Equation 6-6 with the often stated potential for the glass measurement electrode of –59.16 millivolts/pH. If you plot pH versus millivolts per Equation 6-6, the intercept is at 7 pH and 0 millivolts. The slope is –59.16 millivolts/pH, which corresponds to 100% efficiency. As the pH electrode ages, the slope and efficiency decrease.

$$E_1 - E_2 = 0.1984*(T + 273)*(7 - pH_1) \tag{6-5}$$

at T = 25°C:

$$E_1 - E_2 = -59.16 * (pH_1 - 7) \tag{6-6}$$

Where

E_1 = potential developed at exterior glass in contact with process solution (millivolts)

E_2 = potential developed at interior glass in contact with 7 pH buffer (millivolts)

pH_1= process solution pH (pH)

T = temperature of glass electrode (°C)

Reference Electrode

There must also be a reference electrode whose internal electrolyte is in contact with the process to complete the circuit through the process liquid. For the combination electrode shown in Figure 6-1 there is a concentric porous ring called a "liquid junction" around the glass bulb for the potassium chloride reference electrolyte to move through to reach the process. Since potassium and chloride ions migrate at the same rate (equitransferent), the ions arrive at the process at the same time so that the potential (E_5) at the liquid junction is zero. Since the reference junction is porous by design, process ions can migrate into the reference electrode. Process ions rarely migrate at the same speed. An excess of process ions builds up in the liquid junction until the potential becomes large enough to stop the further migration of ions. The liquid junction potential (E_5) changes until this equilibrium is reached. For dilute low ionic strength aqueous solutions between 4 and 10 pH, this potential is quite small (e.g., 15 millivolts). As process ions move into the liquid junction, they can plug the junction. Plugging reduces the contact of the reference electrolyte with the process, which increases the resistance (R_6) of the junction. If process ions can continue to move freely inside the reference electrode, additional extraneous potentials are formed. Chemical attack of the internal silver-sliver chloride reference electrode can occur, which changes the internal reference potential (E_4).

This contamination of the internals of the reference electrode is commonly termed "poisoning." Reference electrodes in chemical and waste treatment processes typically have multiple junctions to protect the inner sanctum of the internal electrode and use gels or porous solids to retard the migration of process ions. When acids or bases participate in chemical reactions, the acid and/or base concentration is large and consequently the ionic strength is high enough to cause a relatively large liquid junction potential. These chemical reactions often require tight control (e.g., 0.02 pH). If the liquid junction potential is larger than the allowable pH measurement error, a flowing reference junction is used where a continual flow of electrolyte into the process is maintained to sustain a small constant liquid junction potential. Another application for a flowing junction is water so pure there is concern about insufficient ions in the process resulting in a high fluid electrical resistance. In this case, the flow of electrolyte into the process is used to sustain electrical continuity between the reference and measurement electrodes.

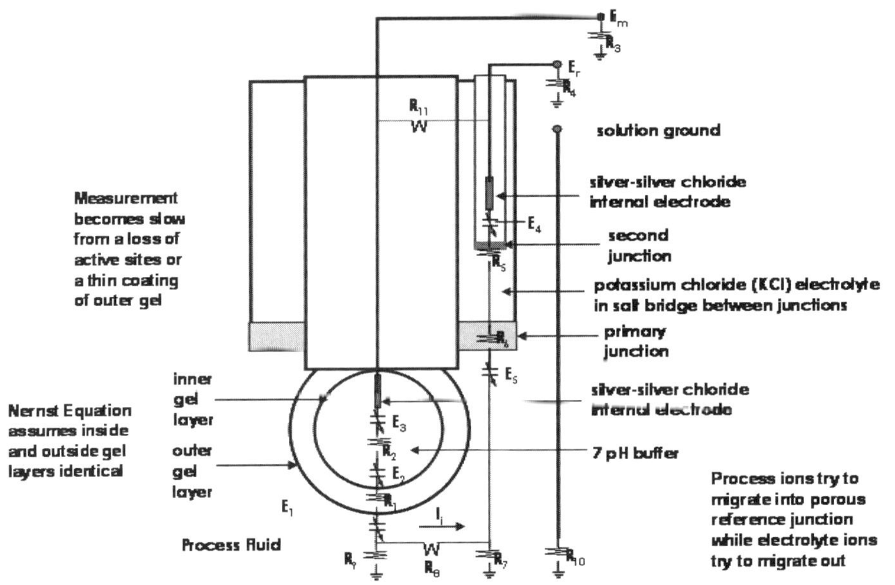

High acid or base concentrations can affect glass gel layer and reference junction potential.
Increase in noise or decrease in span or efficiency is indicative of glass electrode problem.
Shift or drift in pH measurement is normally associated with reference electrode problem.

Figure 6-1. Combination pH Electrode

In Figure 6-1, the potential (E_1) developed at the glass electrode surface is indicative of the process pH. Changes in the hydrated glass surface layer activity and shunt resistance (R_{11}) are corrected by a slope calibration adjustment. Changes in the liquid junction potential (E_5) are corrected by a zero (offset) adjustment.

Resistances

The resistances in the path between the internal electrodes in the reference and measurement electrodes shown in Figure 6-1 are normally all small compared to the glass measurement electrode resistance of 20 to 200 megohms. The current flow is extremely small, making the potential drop across these resistances negligible. The process fluid resistance (R_8) is negligible except for non-aqueous solvents or ultra pure water. The resistances to earth ground should all be large enough to suppress noise from electromagnetic interference and ground loops.

A decrease in the high impedance shunt resistance (R_{11}) can cause a loss of the electrode efficiency (percent of the theoretical slope). Impedance is the complex ratio of voltage to current. The real part of the complex number is resistance and the imaginary part is reactance. The degree to which the measured potential is decreased will depend on the relative magnitudes of the glass electrode impedance and the impedance of the shunt. In one case of bad sensors with low shunt resistances, at room temperature the sensors had low slopes (short-spanned), but at a higher temperature the sensors had adequate slopes for that temperature. This was due to the decrease in glass impedance at higher temperature, which decreased the effect of the high impedance shunt.

How bad can it be? If you look at the resistance (R_{11}) of the high impedance shunt as a multiple of the resistance (R_1) of the high impedance glass electrode ($R_{11} = n * R_1$), you can get an idea of how much the measured electrode efficiency would be lowered:

n	Efficiency
100	99%
50	98%
20	95%
10	91%
5	67% (bad electrode)

The decrease in the shunt resistance (R_{11}) can be accelerated by aging due to the exposure of O-rings and seals to a harsh process, which could attack the materials of construction and introduce some leakage, as well as the effects of prolonged heating on the internal potting materials, which could break down over time. The decrease has little to do with the glass electrode itself, but with the integrity of the sensor construction over prolonged periods. It is also likely that at least some high impedance shunting may be built into the sensor.

6-4. Selection

Transmitters

pH transmitters are commonly called *analyzers*. While technically a pH transmitter and electrode is an online analyzer with a probe, the term "transmitter" will be used as defined in Chapter 1 for industrial measurements. The term "analyzer" will be reserved for laboratory and portable devices, which generally provide only an indication and no transmission of pH (no output signal).

In the selection of a transmitter, key features should be considered. The transmitter should provide solution pH temperature compensation in addition to the standard Nernst temperature compensation. Additionally, the transmitter should provide assistance during buffer calibrations and diagnostics for the transmitter and its electrodes.

- A configurable solution temperature compensation should correct the pH measurement for changes in solution pH, which prevents the error of making a pH adjustment to a hot solution to correct a pH offset due to temperature alone, and eliminates the seemingly inexplicable difference between the lab and online measurements in hot processes.
    - All solutions above 7 pH will exhibit at least some decrease in pH with increasing temperature, due to changes in the dissociation constant of water, which can be compounded by changes in the dissociation constants of weak bases.
    - Solutions below 7 pH can also exhibit temperature dependence due to changes in weak acid dissociation constants with temperature.

- The sensor millivolt signal and temperature, as well as the calibration constants (sensor slope and zero offset) should be accessible as an aid in troubleshooting.

- The calibration routines provided by the transmitter should allow time for sensor stabilization, and include temperature compensation for buffer pH changes with temperature to avoid calibration errors.

- The transmitter should include diagnostic fault alarming for the sensor and transmitter itself, as well as for calibration errors.

Advanced Diagnostics

The main reason for using transmitters with advanced diagnostics (Read/Write mode) is to have pre-calibrated pH sensors in the field and the ability to track and evaluate tendencies in diagnostics data. This information can be used for preventive maintenance.

Usually the sensor calibration data set is recorded on the sensor side and displayed on the instrument. Therefore, in order to use the advantages of the Read/Write mode a pH sensor with embedded memory chip is required. This sensor feature is implemented through an advanced pre-amplifier design. A pre-amplifier has traditionally been used with pH sensors to increase the low voltage (high impedance) signal as close to the point of measurement as possible. In the selection of a smart sensor for advanced diagnostics, the user should look for these features:

- The communication should be established once the sensor with embedded memory chip is connected to a transmitter capable of Read/Write. The transmitter should automatically set the pre-amplifier option to the sensor.
- The auto-detection screen should notify the user that the sensor with embedded memory chip is detected.
- Calibration history should be accessible by the user. At least five most recent calibration data sets should be stored in the memory.
- Every calibration should be recorded (Write mode). Every data set includes, but is not limited to time stamp, method of calibration, slope, offset, temperature, glass impedance, and reference impedance.
- Factory calibration of the sensor should be accessible in the Read mode only. The factory calibration data set includes, but is not limited to, serial number of the sensor, time stamp, calibration method, slope, pH offset, temperature offset, glass impedance, and reference impedance.

- Accessing calibration data should allow live pH and temperature readings.

Advanced diagnostics loop capability is most useful from a maintenance point of view. It allows the user to reduce overall maintenance expenses by a more optimum scheduling of sensor cleaning, rebuilding, or replacement. The Read/Write mode is a standard feature of most modern generation wireless transmitters.

Wireless

The benefits from the use of wireless transmitters are greater for pH than for other types of measurements because of propensity for spikes from ground loops and electromagnetic interference (EMI). In some cases the spikes disappear when the electrode is removed and inserted in a glass lab container, indicating that one side of the ground loop is through the process. In most cases, the ground noise is gone when the electrode is left in place and a lab meter is connected to the electrode cable instead of the transmitter wiring. This indicates that the other side of the ground loop is somehow established in the transmitter wiring. Whenever there is a problem, the wiring is always a suspect.

The cost savings in the elimination of wiring design and installation are obvious for new projects. Less obvious are the savings in maintenance and relocation costs in existing installations. Cable does not need to be relocated to different or new cable trays and conduits and re-terminated at different or new panels. WirelessHART devices automatically report alerts without interrogation and can reconfigured and monitored remotely. Often WirelessHART devices take advantage of the latest advances in electrode technology and diagnostics. For example, the smart WirelessHART pH transmitter in Figure 6-2 has a much finer resolution (e.g., 0.001 pH) and a more accurate monitoring of glass impedance than previous models of transmitters.

Wireless transmitters offer a significant opportunity to develop pH applications in the lab and readily move them to the field. The wireless transmitter is first used in the lab with electrodes in a container with the process solution whose composition and temperature are changed to match the range of expected operation. The accuracy and durability of the electrode can be tested for normal and extreme process conditions. Different electrode designs can be tried for optimum life and performance. The compensation of solution pH for changes in temperature and ionic

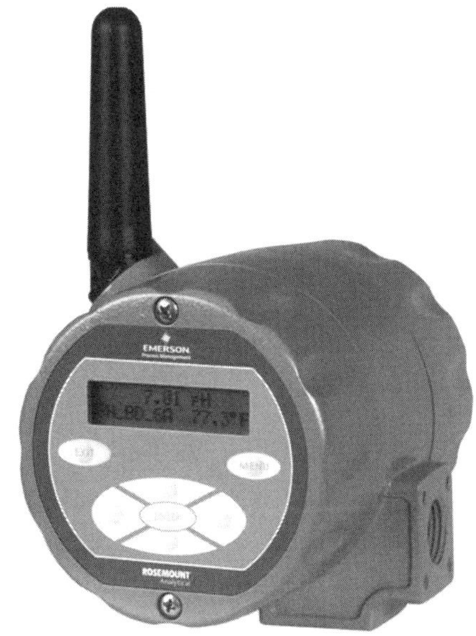

Figure 6-2. Wireless pH Transmitter with Advances in Technology and Diagnostics

strength can be identified. Inferential measurements based on pH can be developed. The wireless transmitter communicates with the DCS so all of the lab test results can be historized and utilized by tools in the DCS, such as neural networks and data analytics. When the testing is done, the transmitter with its solution pH compensation can be moved to the field. The pH measurement location can be experimentally determined that provides the most representative pH measurement with the optimum velocity and fastest response. If the lab route is bypassed and a pH measurement needs to be tried out in the field to see if the electrode will work in a tough application, a wireless transmitter eliminates the need to install wiring and have an intelligent input card for the trial installation.

WirelessHART pH devices conduct a measurement every second but are set up to transmit an updated value only if the change in measurement from the last transmitted value is greater than the "resolution" setting or the elapsed time since the last transmitted value is greater than the "refresh time" setting. The "refresh time" provides periodic reporting, usually every 15 seconds for pH. The "resolution" setting provides exception reporting on an as-needed basis, which can reduce the number of transmissions by an order of magnitude. Battery life can be extended to five or more years.

Some chemical processes need a 0.01 to 0.05 pH resolution setting, whereas for chemical waste treatment a 0.1 pH resolution setting should be good enough. In any case, the resolution setting should be larger than the noise amplitude and smaller than the allowable deviation from set point in order to deal with process disturbances. Where they are suitable, coarser resolution settings correspond to a greater reduction in communications and commensurate increases in battery life.

The untethered portability of wireless transmitters is particularly useful in the dynamic environments of process research and development and flexible manufacturing. For example, a pilot plant at the University of Texas for research into new technologies for CO_2 capture and demonstration of customer designs for column packing has documented significant savings in relocation and setup costs to meet changing equipment configurations and control objectives. WirelessHART devices are being used for the same reasons on single-use-bioreactors (SUB) mounted on carts that can be moved for connection to analyzers and other process equipment.

Wireless transmission can eliminate wiring termination mistakes and ground loops. While only about 10% of wired installations suffer from ground potentials and noise, these problems are difficult to track down. As has been mentioned, when a lab meter is connected to the electrode, the noise or shift in pH often goes away, which indicates that one could get rid of the problem by eliminating the wiring. A recent comparative test of conventional wired versus new WirelessHART pH transmitters on a single-use-bioreactor had some interesting test results. The SUB contained a mammalian cell culture that was very sensitive to pH. A resolution setting of 0.01 pH was requested but resolution was mistakenly set at 0.01%, which translated to about 0.001 pH. The pH loop was able to control to within 0.002 pH of set point while reducing the number of communications by 60% by the use of exception reporting with the incredibly fine 0.001 pH resolution setting in the WirelessHART transmitter, which is beyond the stringent 0.02 pH control requirement of bioprocesses. Confirming suspicions that noise from grounding problems could be eliminated with wireless transmitters, spikes in the conventional wired transmitter trend did not appear in the WirelessHART transmitter pH trend chart. Figure 6-3 shows the SUB test results.

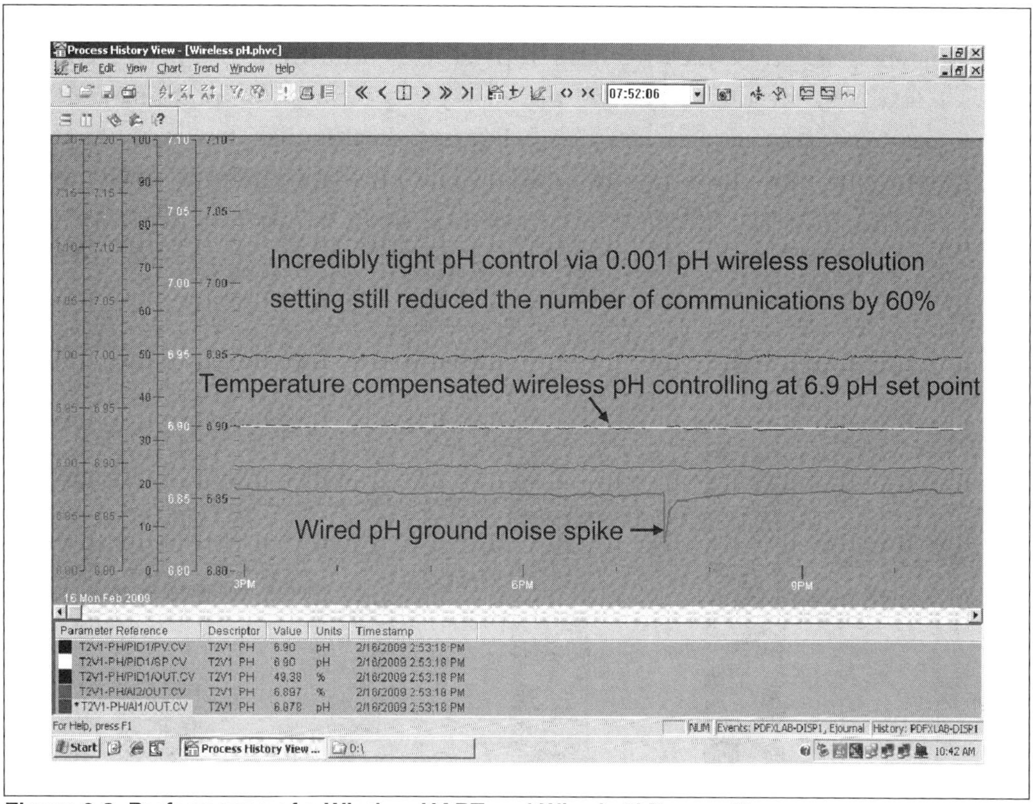

Figure 6-3. Performance of a WirelessHART and Wired pH Transmitter

Electrodes

There are many different types of measurement and glass electrodes due to the wide range of operating conditions and accuracy requirements in the process industry. New designs have increased the life, accuracy, and speed of response for electrodes subjected to high process temperatures, high salt concentrations, low conductivity samples (e.g., pure water), sticky fluids, particles, and sterilization. We first look at the major developments in electrode design and then summarize the electrode selection considerations for difficult application conditions.

The selection of the right glass largely depends on the pH range and temperature. Some suppliers offer low pH glass that minimizes the acid error below 1 pH, high pH glass that minimizes the alkalinity (sodium ion) error above 10 pH, low temperature glass that prevents an excessively high glass resistance below 10°C (50°F), a high temperature glass that slows down the aging process that accelerates for temperatures above 25°C (77°F), and a glass that withstands repeated steam-in-place (SIP) sterilizations. Many suppliers offer a single general-purpose glass that addresses many of these issues well enough to avoid the complication of users selecting, stocking, and replacing a multitude of special glasses. However, the life of general-purpose glass in high temperature and SIP applications can be unacceptable. Figure 6-4a shows that using a high temperature glass electrode can double electrode life expectancy for temperatures greater than 50°C (122°F). Figure 6-4b shows that the response time of a high-temperature glass electrode is 10 seconds versus 100 minutes for a general-purpose glass electrode subjected to high process temperatures.

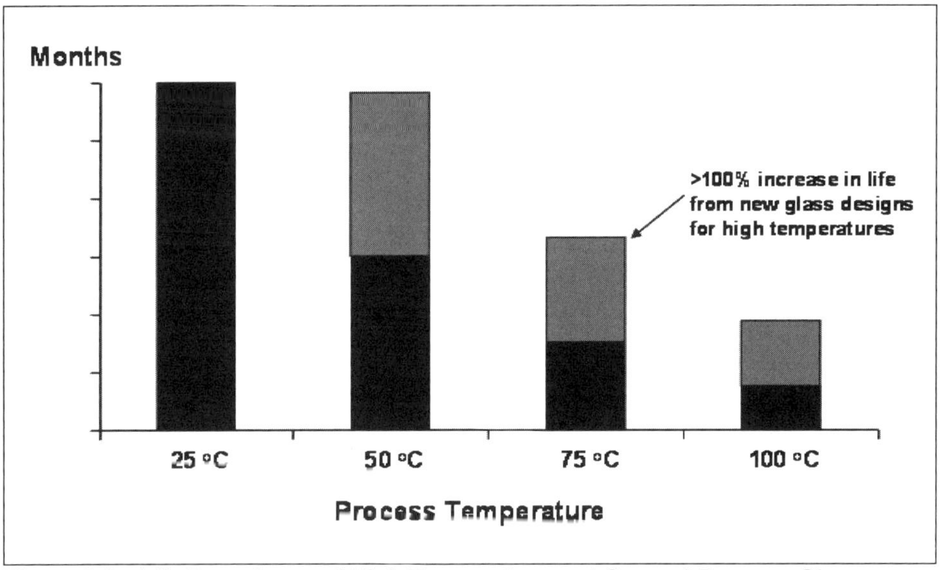

Figure 6-4a. Life Expectancy of High Temperature and General-Purpose Glass Electrodes

Flat glass electrodes are promoted by some suppliers to minimize coatings and abrasion. The fluid velocity, electrode shroud, location, and probe holder geometry play a bigger role than glass surface design in preventing coatings and abrasion. Most of the success in reducing coatings has been achieved by a narrowing of the probe holder, thereby increasing the fluid velocity at the surface. Flat glass electrodes have a more limited pH range and thinner glass with stress points. Spherical or hemispherical bulbs are more accurate and rugged but user perception still creates a sizeable, although declining, market for flat glass.

The selection of the reference electrode has the ultimate goal of keeping the reference potential as constant and small as possible despite extremes in process composition and the presence of solids or sticky fluids. High and variable concentrations of acids, bases, ions, and non-aqueous sol-

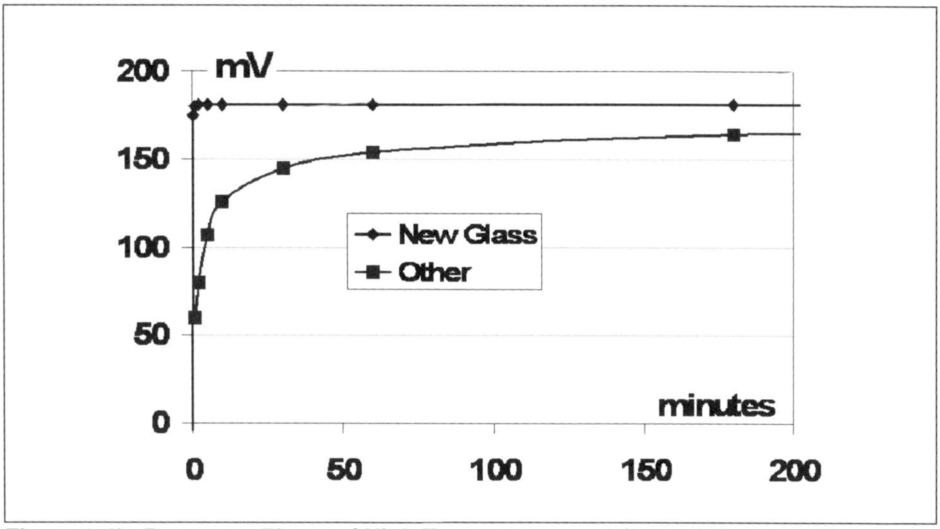

Figure 6-4b. Response Times of High Temperature and General-Purpose Glass Electrodes

vents can cause large and fluctuating liquid junction potentials. Ultra low conductivity solutions increase the resistance of the path from the reference to the glass and are prone to EMI, fluid velocity induced errors, and variable liquid junction potentials. Coatings and particles can plug the reference and create a large reference resistance and liquid junction potential. Finally, contamination of the reference with process chemicals is extremely disruptive to the reference electrode potential and can seriously degrade the reference internals to the point of failure.

Most raw water and waste treatment applications can use general purpose glass and standard multiple junction reference electrodes. Most modern electrodes, such as shown in Figure 6-5, have wide range pH glass and a triple reference junction. For waste streams that contain cyanides, bromides, iodides, sulfides, or nitrates that can cause precipitation of the silver, a special electrolyte and internal electrode are used.

Alternatively, a solid reference electrode can be used. Figure 6-6 shows a reference electrode with solid internals, an embedded electrolyte, and a helical tortuous pathway to drastically slow down the contamination process. These reference electrodes typically have a large reference junction area that helps prevent coatings from greatly affecting the reference junction resistance and potential. The increase in life expectancy and reliability and reduction in maintenance costs have made solid references popular in difficult chemical production processes where pH is important. Some processes with high concentrations of ions (high ionic strength), sticky poly-

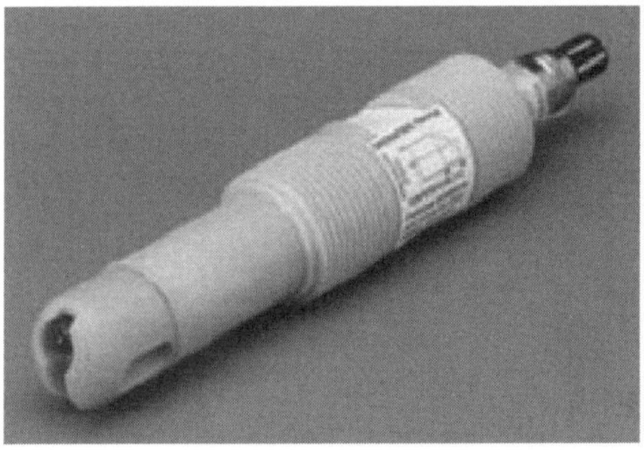

Figure 6-5. General-Purpose Glass Electrode with Triple Junction Reference

mers or monomers, and particles, have necessitated the replacement of gel and liquid reference electrodes within weeks or even days after being put in service. The high propensity for coating, plugging, and contamination of the reference in such processes is a major point of pain for the user. Some plants with these processes and problems have standardized on the use of solid references in all pH applications even though a general-purpose electrode would have worked fine in the easier applications.

As detailed in Figures 6-4a and 6-4b, the life expectancy and response time of pH electrodes at high temperature can be greatly improved by going to high temperature glass formulations and electrode construction. The addition of a readily removable reference junction enables the life of the reference electrode to be extended to match the increased life expectance of the glass. The reference junction can be popped out, the reference refilled, and a new reference junction inserted as shown in Figure 6-7. This rebuildable technology has advantages over disposable sensors from the perspective of total cost of ownership.

In food, drug, and beverage applications, the pH electrode may need to be steamed in place (SIP), cleaned in place (CIP) or autoclaved. As mentioned, the sustained high temperature from sterilization causes premature aging and failures of pH electrodes. Conventional electrodes going through only a single sterilization show a significant increase in response time, offset, and drift. Progress has been made in glass formulations and electrode structure for these applications. The sterilizable electrode shows significantly less drift (Figure 6-8) and a faster response than older sterilizable electrode designs.

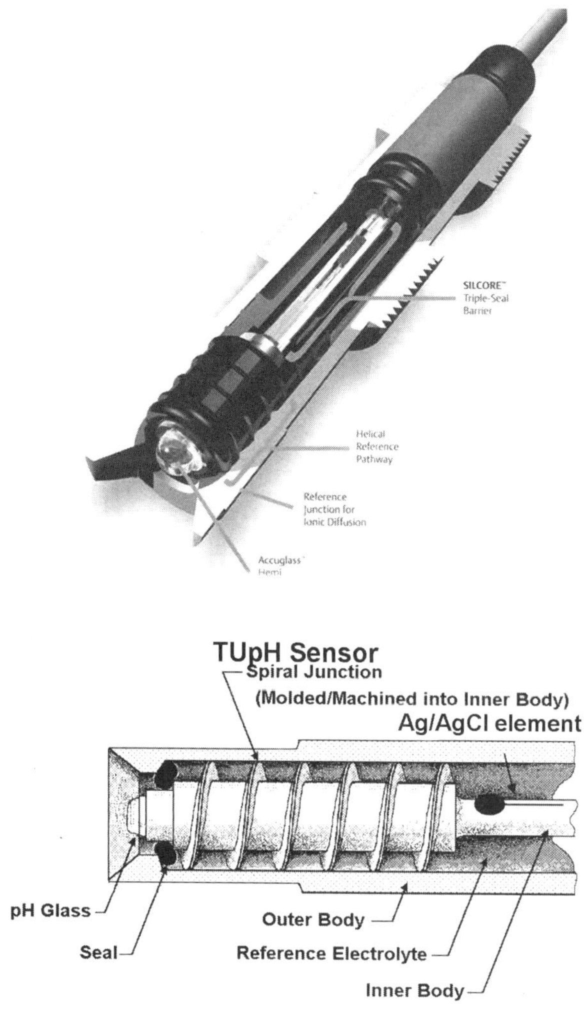

Figure 6-6. Solid Reference with Large Reference Junction Area

Process pH

pH measurements in water solutions below 2 pH or above 12 pH are strong acid or base solutions with concentrations approaching the percentage range. These concentrations are usually better measured using conductivity measurements.

- Very low pH measurement can suffer errors from acid effects on the pH electrode and large potentials on the liquid junction of the reference electrode.

Figure 6-7. High Temperature Glass Electrode with Removable Reference Junction

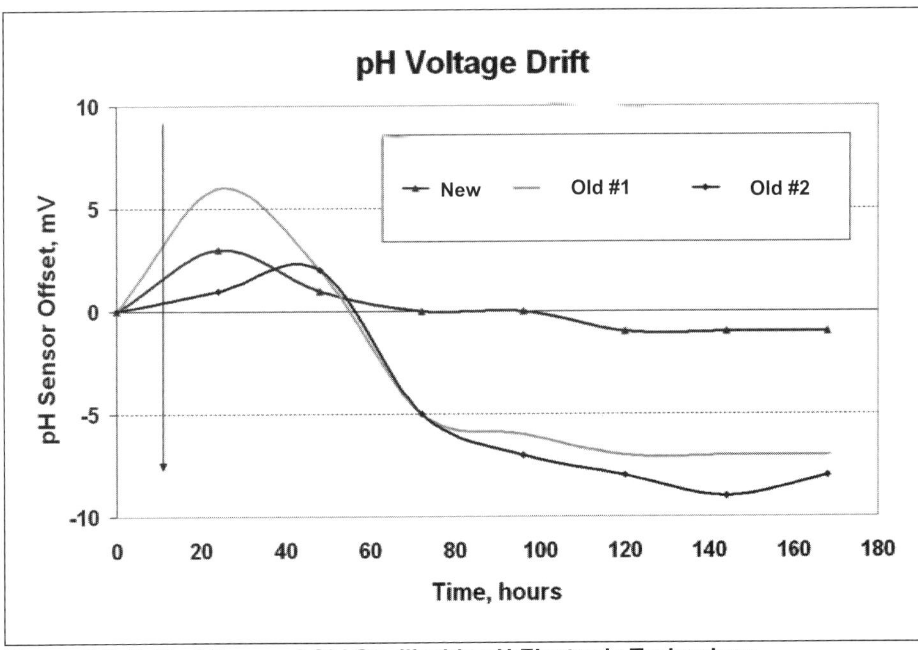

Figure 6-8. Drift of New and Old Sterilizable pH Electrode Technology

- In addition to liquid junction potential, high pH solutions can severely shorten the life of the pH electrode, especially at high temperatures, and the measurement itself is subject to alkalinity errors (also termed sodium ion errors), which offset the indicated pH below the actual pH value.

Process Temperature

As has been mentioned, exposure to high temperatures shortens the life of pH electrodes and lengthens the response time, leading to a sluggish response to pH changes. The traditional approach to high temperature situations has been to use a sample stream with sample cooling, which increases installation and maintenance costs and creates measurement time delays. Today there are alternatives available:

- pH electrodes have been designed for continuous high temperature exposure to provide a reasonable life under these circumstances (Figure 6-7).
- A second case is SIP (Steam in Place) pH sensors, which, as described earlier, are designed to withstand a relatively short exposure to high temperature, primarily for sterilization in the biotech industry (Figure 6-8).

Process Concentrations

To properly function, a reference electrode should maintain a standard potential, which is a function of the silver ion dissolved in its fill solution, and maintain electrolytic contact with the process by diffusion of potassium chloride ions through its liquid junction. The potassium and chloride ions in a reference fill solution diffuse through a reference at an equal rate, so there is a net charge balance across the liquid junction. Potassium chloride is said to be equitransferent.

- Ions from the process diffusing into the liquid junction are likely not equitransferent, and so a potential, called the liquid junction potential, develops in opposition to the charge imbalance created by unequal diffusion rates of positive and negative ions.
- Certain ions in a process, the most common being sulfide, bromide, iodide, and cyanide, can precipitate or complex the silver ions upon diffusing into the reference electrode, which can effect the functioning of the reference electrode in two different ways:
    - Poisoning: Reduction of the silver ion concentration in the innermost portion of the reference electrode causes

silver ions on the silver-silver chloride wire to dissolve, and once all of the silver chloride has dissolved a large offset in the potential (and the pH measurement) is produced, and the reference electrode must be replaced.

- Plugging: A second effect can be precipitation of silver within the liquid junction, blocking the diffusion of ions into the process and thus opening the pH measuring circuit, and causing the pH measurement to drift.

- Reference electrodes have been designed to retard the migration of process ions into the reference electrode by using multiple liquid junctions, an outer reference fill of potassium nitrate to lower silver ion concentration, and gelled fill solutions to prevent ion transport by thermal convection.

- In extreme cases, where there are high concentrations of harmful ions in the process, a flowing reference is required, which uses a flow of fill solution into the process to make electrolytic contact to prevent incursion of process components into the reference. Flowing reference electrodes are maintenance intensive due to the need to keep an adequate reservoir of fill solution, and it is necessary to maintain sufficient pressure on the fill solution to overcome process pressure.

Process Conductivity

Low conductivity processes (conductivity less than 50 μS/cm) can create problems for pH measurements, largely due to effects on the liquid junction of the reference.

- Large and unstable liquid junction potentials can develop in low conductivity water samples due to spurious concentration changes within the liquid junction.

- A pH sensor designed for low conductivity processes must be used for these applications, and the current approach for dealing with liquid junction instability is to use a flowing reference or a pressurized reference to ensure that the liquid junction is always filled with reference fill solution.

- pH sensors for low conductivity processes are almost always mounted in a flow-through configuration (described in the "Installation" section) to better control sample flow rates and prevent contamination of very pure water with electrolyte from the reference electrode Figure (6-9).

- Applications involving the detection of acidic or basic contaminants in ultra-pure water (conductivity near 0.055 μS/cm) are better served using conductivity measurements, which are more sensitive to low levels of contaminants and much easier to apply than pH.

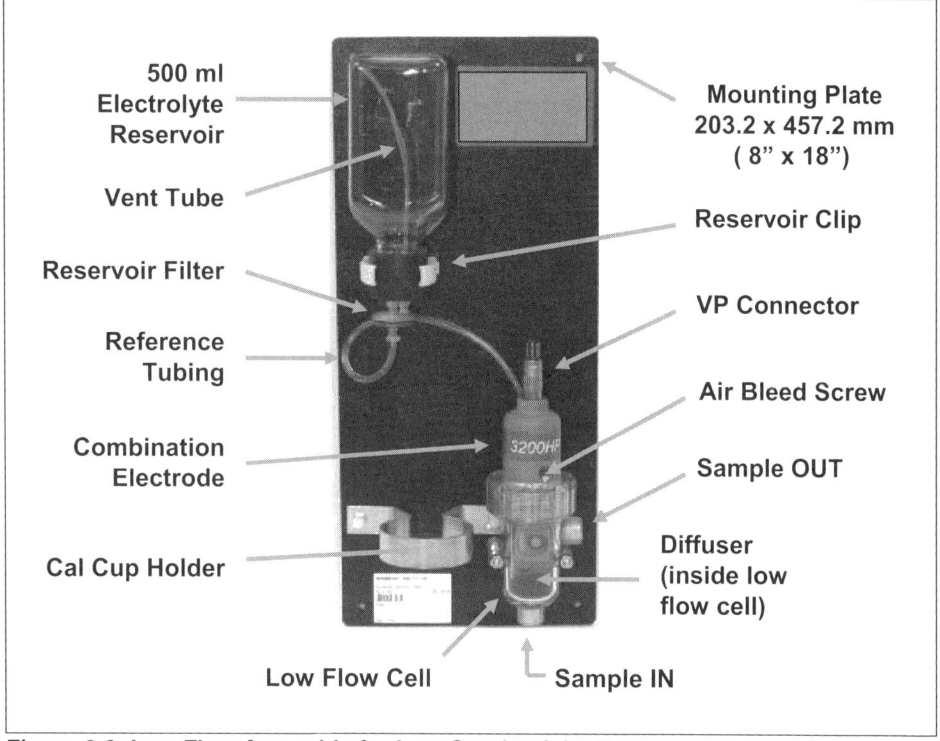

Figure 6-9. Low Flow Assembly for Low Conductivity Processes

Non-Aqueous and Mixed Solvents

Generally, pH sensor design and the interpretation of a pH measurement are based on the assumption that measurements are being made in water solutions.

- When the water content is below 40%, the pH reading can be shifted due to solvent effects on the hydrogen ion activity, and the pH electrode can lose sensitivity due to drying, which make it necessary to periodically rewet the electrode with water.

- Fully non-aqueous solvents can greatly shift the pH scale, acidic solvents shifting it downward, basic solvents shifting it upward, while solvents that are neither acidic nor basic can extend the pH range (e.g. -4 to 18 pH).
- The dissociation constants of acid and bases will change with changes in the concentration of the solvent and water.
- The dissociation constant and hence the neutral point for a non-aqueous solvent is different than that for water. For example, the dissociation constant (pK_W) for pure methanol is 16.84 versus a dissociation constant of 14 for water at 25°C (77°F). The neutral point of a methanol solution is 8.42 pH at 25°C.
- The interface between the liquid junction of the reference and a non-aqueous or mixed solvent can result in large junction potentials.
- When considering applying a pH sensor to a mixed solvent process, the materials of construction of the sensor should be reviewed to avoid solvent attack on O-rings and seals, which can result in leaks and sensor failure.
- The application itself should be examined in the laboratory before going to an on-line measurement, to prevent surprises on-line from the above and other nonlinear effects that are a function of solvent concentration.

Sensor Coating and Cleaning

Undissolved solids and liquids in a process solution can coat a pH sensor, increasing its response time or even making it unresponsive to changes in solution pH.

- A sample flow rate of 5 ft/sec (30.5 cm/sec) or greater can reduce the accumulation of fouling material, through the turbulence created.
- Alkaline deposits can be removed with dilute acid (5% HCl) or vinegar, while acid deposits can be removed with mild caustic (1% NaOH caustic soda).
- Organic deposits (oil and grease) can often be removed with a detergent solution, but more tenacious coating may require the

use of a solvent, which should be carefully chosen to avoid attack on the sensor's O-rings and seals.

- The sensor's exposure to any cleaning solution should be limited to avoid contamination of the liquid junction by the cleaning solution.

6-5. Specifications

The exact specifications depend upon the manufacturer. The following example is representative of the range of operating conditions, performance, process connections, and wetted materials of construction offered.

Sensor Specifications

- General Purpose pH Sensor (disposable)
    - Glass options: general purpose, high pH, hydrogen fluoride (HF) resistant
    - Process Connections: 1-in. MNPT
    - Materials of Construction: Tefzel®1, glass, ceramic and Viton®1
    - Temperature Compensation: Automatic 0 to 85°C (32 to 185°F)
    - Maximum Pressure: 790 kPa (100 psig) at 65°C (154°F)
    - Pre-amplifier: embedded memory chip
- Triple Junction pH Sensor (disposable)
    - Glass options: general purpose, high pH
    - Process Connections: 1-in. MNPT
    - Materials of Construction: Tefzel®1, titanium, glass, ceramic, Viton, Teflon, EPDM, 316 SS
    - Temperature Compensation: Automatic 0 to 100°C (32 to 212°F)
    - Maximum Pressure: 790 kPa (100 psig) at 100°C (212°F)
    - Pre-amplifier: embedded memory chip
- High Temperature pH Sensor (re-buildable)
    - Wetted Materials: Titanium, Ryton3, Teflon®1, glass, and EPDM. Options: Viton®4 or Kalrez®5
    - Operating Temperature: 5°C to 155°C (41 to 311°F)
    - Maximum Process Pressure and Temperature: 285g kPa (400 psig) at 155°C (311°F)

- Process Connections: Insertion/Submersion, 1-inch compression process connector
- Process Connection: Retraction, can be inserted through a 1¼-in. or 1½-in. ball valve
- Maximum Pressure at Retraction or Insertion 524 kPa (64 psig) 21 inch (53.3 cm) sensor length, 343 kPa (37 psig) 36 inch (91.4 cm) sensor length
- Pre-amplifier: embedded memory chip

- Coating Resistant pH Sensor (disposable)
  - Glass options: general purpose, high temperature glass
  - Process Connections: 1-in. MNPT
  - Materials of Construction: Tefzel®1, titanium, polypropylene, glass, Viton, Teflon, Kalrez, EPDM, 316 SS
  - Temperature Compensation: Automatic 0 to 100°C (32 to 212°F)
  - Maximum Pressure: 790 kPa (100 psig) at 100°C (212°F)
  - Pre-amplifier: embedded memory chip

- Low Conductivity pH Sensor Assembly (disposable)
  - Accuracy: ±0.05 pH
  - Noise: < 0.02 pH; Drift: < 0.05 pH per week
  - Conductivity Range: > 0.4 S/cm (0.16 S/inch)
  - Sample Temperature: 0°C to 70°C (32°F to 158°F)
  - Sample Pressure: 5 to 10 psig (34 to 69 kPa) – drain to atmosphere
  - Sample Flow Rate: 60 to 180 ml/min (1 to 3 gph)
  - Wetted Materials: Silicone, polycarbonate/ polyester, glass, stainless steel, PVDF, Viton
  - Reference Electrode: Flowing, open capillary
  - Process Connections: ¼-in. NFPT

- Steam-in-Place (SIP) pH Sensor (disposable)
  - Temperature Range: up to 105°C (221°F) (steam up to 140°C, 284°F)
  - Pressure Range: max 600 kPa (72 psig) at 105°C (221°F)
  - Wetted Materials: Glass, ceramic and EPDM O-ring USP VI
  - Process Connector: Pg 13.5 connector

- Dimensions: 12 mm (0.47 inch) OD, shaft length 120 mm (0.95 inch), 225 mm (8.9 inch), or 325 mm (12.8 inch)

pH Transmitter Specifications

- pH Range: 0 to 14
- Accuracy: ±1 mV @ 25°C (77°F) ±0.01 pH
- Repeatability: ±1 mV @ 25°C (77°F) ±0.01 pH
- Stability: 0.25%/year @ 25°C (77°F)
- Temperature resolution: 0.1°C (99.9°C) [0.18°F (211.8°F)]; 1°C (100°C) [1.8°F (212°F)]
- Digital Communications:
    - Fieldbus (pH): Four AI blocks assigned to pH, temperature, reference impedance, and glass impedance.
    - HART (pH): PV assigned to pH, SV, TV, and 4V assignable to pH, temperature, mV, glass impedance, reference impedance, or RTD resistance.
- Diagnostics: The internal diagnostics include:
    - Calibration Error
    - Sensor Failure
    - High Temperature Warning
    - CPU Failure
    - Low Temperature Warning
    - Input Warning
    - ROM Failure
    - Glass Impedance Warning
    - Glass Impedance Failure
    - Reference Impedance Warning
    - Reference Failure
- Advanced diagnostics (Read/Write) when connected to a sensor with embedded memory chip. Advanced diagnostics include:
    - Sensor auto-detection
    - Factory calibration (Read only)
    - Calibration history (five calibration data sets, read and write)

- Support for four calibration methods

6-6. Setup and Calibration

pH transmitters are factory calibrated for an accurate millivolt potential measurement, but due to variations in pH sensors (e.g., differences in offset and slope) and aging effects, initial and routine pH sensor calibration is necessary. Transmitters can be set up and calibrated by handheld communicators or an asset management system (AMS). An example of an AMS window for the configuration of the pH measurement range and custom compensation of the solution pH for temperature is shown in Figure 6-10. An AMS window for the configuration of the diagnostics is shown in Figure 6-11, and windows for buffer calibration are shown in Figures 6-12a and 6-12b.

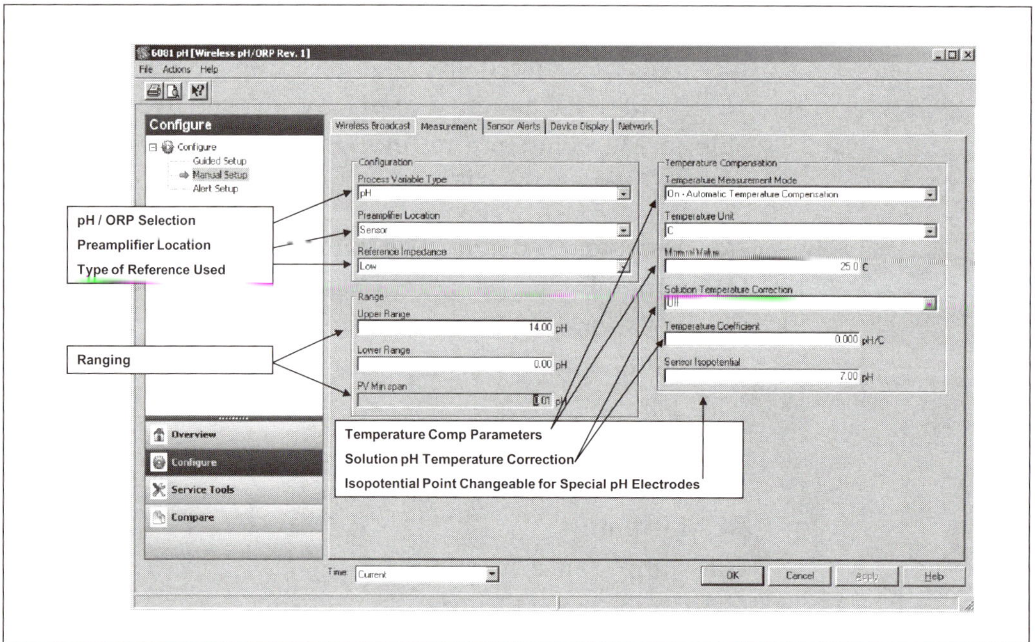

Figure 6-10. AMS Window for Configuration of pH Range and Solution pH Temperature Compensation

Buffer Calibrations use two buffer solutions, usually at least 3 pH units apart, which allow the pH transmitter to calculate a new slope and zero value corresponding to the particular characteristics of the sensor, to more accurately derive pH from the millivolt and temperature signals. The AMS windows illustrating the steps for buffer calibration are shown in Figures

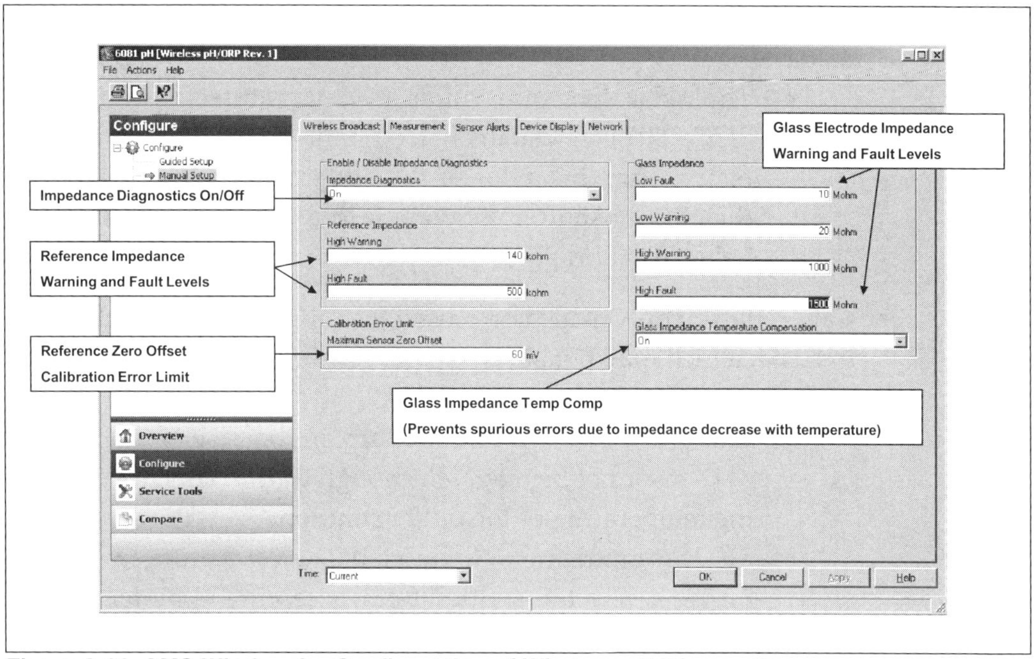

Figure 6-11. AMS Window for Configuration of Wireless pH Diagnostics

6-12a and 6-12b. The AMS windows in this chapter reflect the new capabilities of a wireless transmitter.

- The slope and zero value derived from a buffer calibration provide an indication of the condition of the glass electrode from the magnitude of its slope, while the zero value gives an indication of reference poisoning or asymmetry potential, which is an offset within the pH electrode itself.
    - The slope of pH electrode output tends to decrease from an initial value relatively close to the theoretical value of -59.16 mV/pH, largely due in many cases to the development of a parallel high impedance short (low resistance path) within the sensor, which forms a shunt of the electrode potential.
    - Zero offset values will generally lie within ± 15 mV due to liquid junction potential; larger deviations are indications of poisoning.
- Buffer solutions have a stated pH value at 25°C (77°F), but as mentioned the stated value changes with temperature, especially for stated values that are 7 pH or above. The buffer value

at the calibration temperature should be used or errors will result.

- The pH values of a buffer at temperatures other than 25°C (77°F) are usually listed on the bottle; or better, the temperature behavior of the buffer can be loaded into the pH transmitter allowing it to use the correct buffer value at calibration.

- Calibration errors can also be caused by buffer calibrations done in haste, which may not allow the pH sensor to fully respond to the buffer solution.
    - This will cause errors, especially in the case of a warm pH sensor not being given enough time to cool down to the temperature of the buffer solution.
    - pH transmitters employ a stabilization feature, which prevents the transmitter from accepting a buffer pH reading that has not reached a prescribed level of stabilization, in terms of pH change per time. This feature helps eliminate the common problem of not allowing enough time for the pH to approach the buffer value that leads to a lower observed slope and higher than necessary slope correction during calibration.

pH Standardization: Standardization is a simple zero adjustment of a pH transmitter to match the reading of a sample of the process solution made using a laboratory or portable pH analyzer. Standardization eliminates the removal and handling of electrodes and the upset to the equilibrium of the reference electrode junction. Standardization also takes into account the liquid junction potential from high ionic strength solutions and non-aqueous solvents in chemical reactions that would not be seen in buffer solutions. For greatest accuracy, samples should be immediately measured at the sample point with a portable pH analyzer. If a lab sample measurement value is used, it must be time stamped and the lab value compared to a historical online value for a calibration adjustment. The middle signal selected value from three electrodes of different ages can be used instead of a sample pH. Periodic standardization based on data analytics can be automated, greatly reducing the need for samples and freeing up operations, lab, and maintenance technicians. The AMS windows illustrating the steps for grab sample standardization are shown in Figures 6-13a and 6-13b.

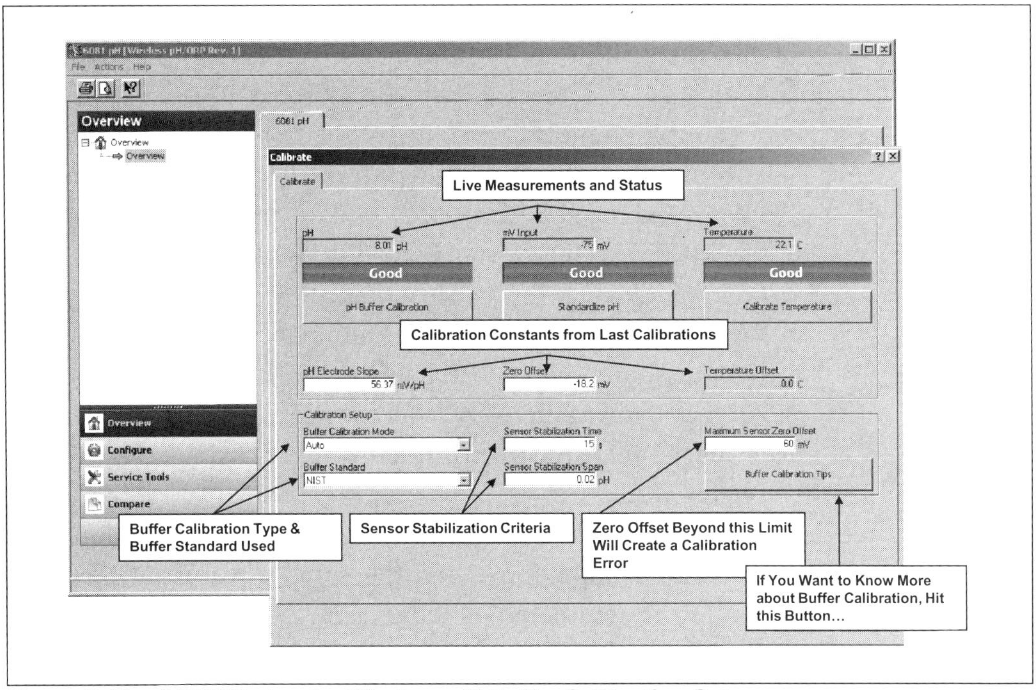

Figure 6-12a. AMS Window for Wireless pH Buffer Calibration Setup

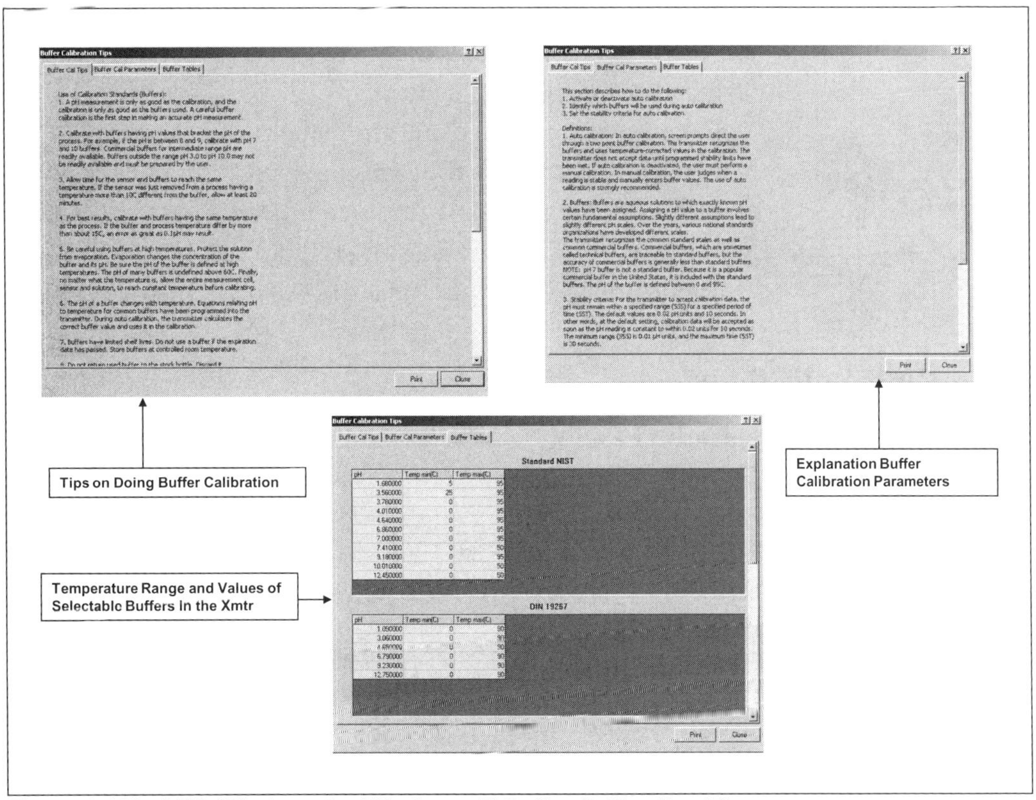

Figure 6-12b. AMS Windows for Wireless pH Buffer Calibration Tips

- Standardization is most useful for zeroing out a liquid junction potential, but some caution should be used when using the zero adjustment.
    - A simple standardization does not demonstrate that the pH sensor is responding to pH, as does a buffer calibration, and in some cases, a broken pH electrode can result in a believable pH reading, which may be standardized to a grab sample value.
    - A sample can be prone to contamination from the sample container or even exposure to air; high purity water is a prime example. A referee measurement must be exposed to a flowing sample using a flowing reference electrode.
    - A reaction occurring in the sample may not have reached completion when the sample was taken, but will have completed by the time it reaches the lab.
    - Discrepancies between the laboratory measurement and an on-line measurement at an elevated temperature may be due to the solution pH being temperature dependent. Adjusting the transmitter's solution temperature compensation (not a simple zero adjustment) is the proper course of action.
    - It must be remembered that the laboratory or portable analyzer used to adjust the on-line measurement is not a primary pH standard, as is a buffer solution, and while it is almost always assumed that the laboratory or portable analyzer is right, this is not always the case.

6-7. Installation

In general, pH sensor location should be chosen to meet the following objectives:

1. Safe and convenient access for calibration and maintenance to enable periodic calibration, cleaning, and replacement of the electrode

2. Process conditions (e.g., temperature and pressure) consistent with electrode capability to prolong electrode life

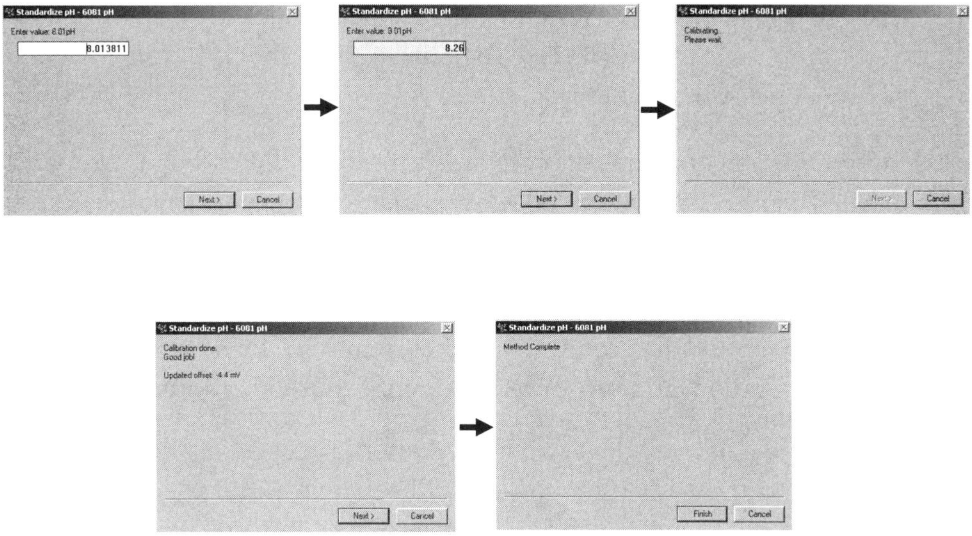

Figure 6-13a. AMS Windows for Wireless pH Standardization Steps

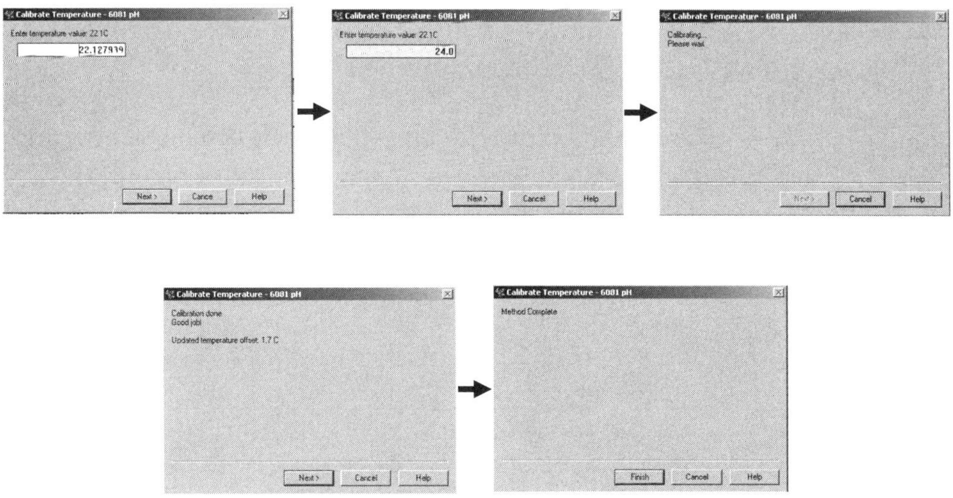

Figure 6-13b. AMS Windows for Wireless pH Temperature Standardization Steps

3. Representative mixture (no stagnation, layering, or short circuiting of reagent) to prevent old or extraneous pH indications

4. Optimal fluid velocity to minimize electrode coating, abrasion, and response time

5. Minimum transportation time delay to reduce loop deadtime

pH sensors should be mounted at least 10° above horizontal to promote good reference electrode function (gravity assistance to electrolyte) and to avoid air bubbles in the pH electrode fill solution that is in contact with the pH-sensitive surface of the electrode.

Submersion Mounting involves an electrode at the end of a conduit that extends below the lowest possible surface level in a process vessel or pond.

- It is the most inconvenient mounting configuration because the sensor and the conduit must be removed for maintenance or calibration, and sensor replacement can be labor intensive.
- Care must be taken to make a watertight seal between the sensor and the conduit, and backfilling the conduit is advisable to avoid leakage of process fluid into the cable end of the sensor, which can cause shorts.

Insertion Mounting has the sensor mounted in a pipe tee as shown in Figure 6-14, or through the side of a pipe or vessel, in a location where electrodes will be fully immersed in a representative process mixture and air or bubble entrapment can be avoided.

- The sensor should be provided with a quick-disconnect mounting accessory or a cable connector to allow it to be withdrawn from the process without having to unscrew the sensor and consequently not having to remove the cable from the transmitter or junction box to avoid twisting and damaging the sensor cable.

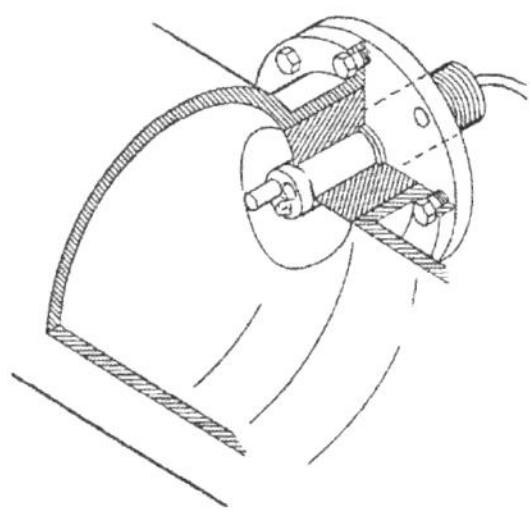

Figure 6-14. Insertion pH Electrode

Retraction Mounting allows a pH sensor as shown in Figure 6-15 to be mounted in an insertion configuration and be withdrawn by pulling it back behind a valve, which can be closed, which eliminates the need to drain the pipe or vessel.

- Withdrawal can be done manually or using a pneumatic or electric actuator.
- The pressure and temperature limits, and the operational procedure of the retraction device should be strictly observed to prevent ejection of the sensor or release of process fluid.

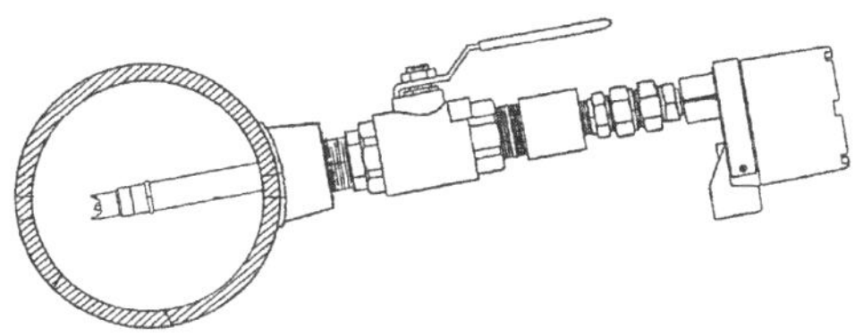

Figure 6-15. Retractable pH Electrode

Flow-through Mounting can consist of a sensor as shown in Figure 6-16 with female threads or flanges to mate with process piping, or can simply be an insertion sensor mounted in a tee or flow chamber.

- The term flow-through is usually taken to mean a pH sensor mounted in a side stream off the main process, and is the only mounting choice when sample cooling, pressure reduction, or filtration and separation is required to make the measurement.
- When fitted with isolation valves, this mounting configuration is the most convenient because it allows the sample flow to be shut off and the line drained for sensor calibration and maintenance.
- The major drawbacks to flow-through mounting are the added expense of providing a sample bypass, the delay introduced by the transport time of getting the sample to the sensor, and the loss of sample to drain, unless provision is made to return the sample to the process.

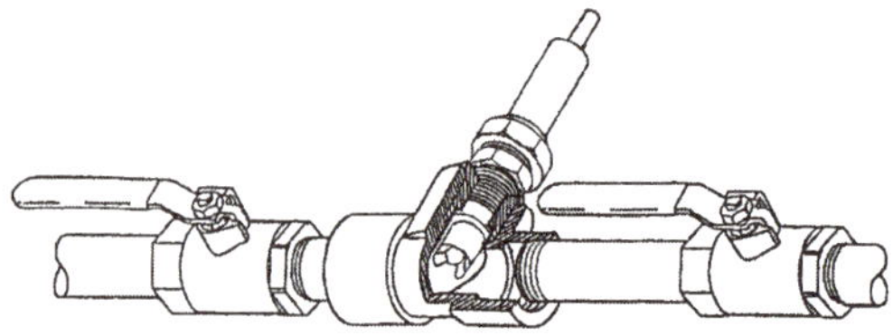

Figure 6-16. Flow-Through pH Electrode

6-8. Maintenance

Middle Signal Selection

The best solution in terms of reliability, maintainability, and accuracy is the use of three pH electrodes with different durations of process service and automatic selection of the middle value for the loop process variable (PV). The return on investment for the additional electrodes from improved process performance and reduced life cycle costs is typically more than enough to justify the additional capital costs in biological and chemical processes if the electrode life expectancy has been proven to be acceptable in lab tests for harsh conditions.

The use of the middle signal inherently ignores a single failure of any type including the most insidious failure, one that gives a pH value equal to the set point. The middle value reduces noise without the introduction of lag from a damping adjustment or signal filter and facilitates monitoring the relative speed of the response and drift, which are indicative of a measurement and reference electrode coating, respectively.

The middle value used as the loop PV for well tuned loops will reside near the set point regardless of drift. A drift in one of the other electrodes is indicative of plugging or poisoning of its reference. If both of the other electrodes are drifting in the same direction, the middle value electrode probably has a reference problem. If the change in pH for a set-point change is slower or smaller for one of the other electrodes, it indicates a coating or loss in efficiency, respectively for the subject glass electrode.

Measurement Noise

High input resistance will make the measurement more susceptible to electromagnetic interference and ground loops, particularly for older transmitter designs. High input resistance can be caused by dehydration of the glass surface or an extremely low electrical conductivity of the sample stream (e.g., pure water).

The agitation required to provide the degree of uniformity of ion concentration commensurate with the extraordinary sensitivity of the pH measurement goes way beyond the norm of what is considered as good mixing. Consequently, measurement noise is common for operation on the steep portion of the titration curve due to concentration gradients and fluctuations. Since the noise is normally relatively fast, it can be readily attenuated. For wireless transmitters, the attenuation should be done by way of transmitter damping adjustment rather than a DCS input or PID block filter so that the exception reporting used for wireless transmitter signals is not triggered by measurement noise. Middle signal selection as previously mentioned can reduce the noise without the lag time introduced by transmitter damping or DCS signal filtering.

Diagnostic Information

The first level of diagnostic information in a pH transmitter is the fundamental measurements made by the transmitter and sensor, which include millivolt input, temperature, and the resistance of the temperature element. Figure 6-17a shows an AMS dashboard for an overview of pH, temperature, and battery life. Figure 6-17b shows an AMS dashboard for a summary of the primary and secondary HART variables. Figures 6-18a and 6-18b are AMS windows for diagnostic messages and explanations.

The source of a questionable pH measurement can be determined as being due to the temperature measurement or the millivolt measurement by examining their values, as in the case of a failed temperature sensor, which indicates a process temperature that is improbably low or high.

Smart pH transmitters also provide internal diagnostic routines, which can provide the user with real time fault and warning information. These can be divided into the following categories:

- Transmitter diagnostics alert the user to electronic or CPU problems with the transmitter itself, which can either indicate that the validity of the measurement has been compromised, or in the extreme situations, that the transmitter electronics should be replaced.

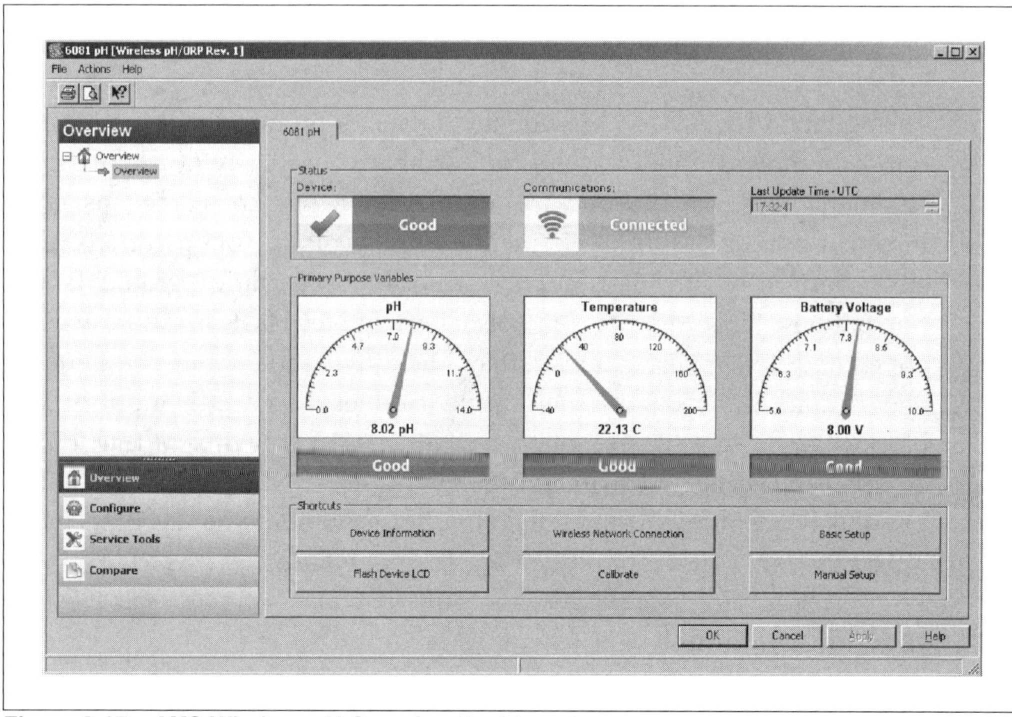

Figure 6-17a. AMS Wireless pH Overview Dashboard

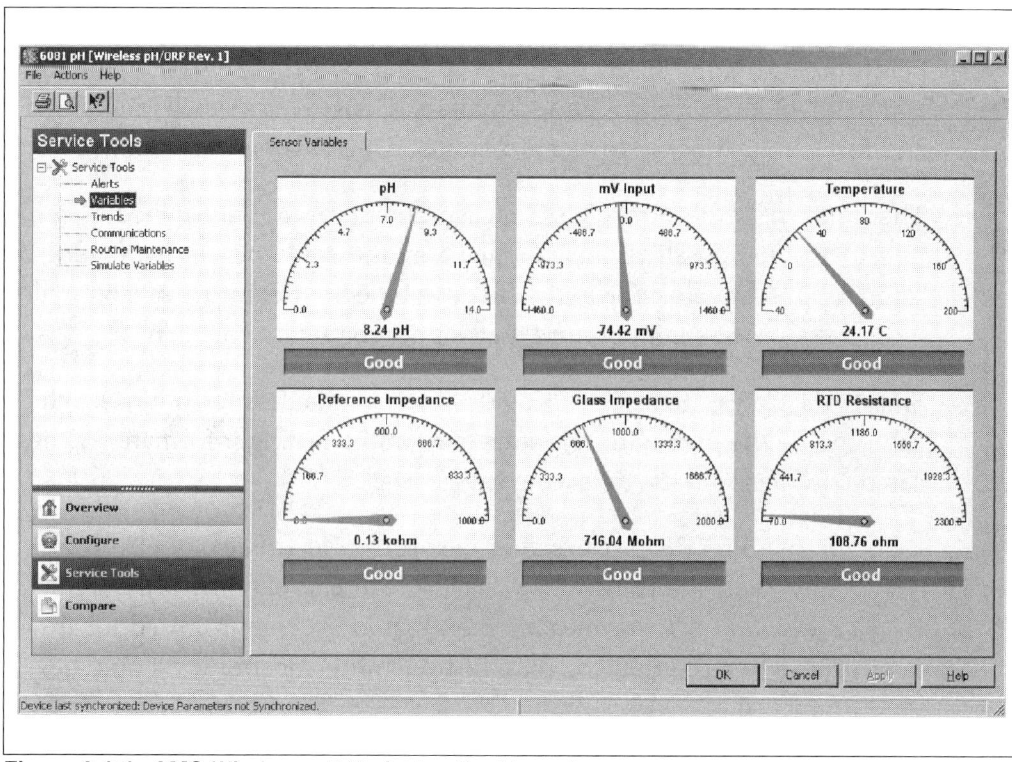

Figure 6-17b. AMS Wireless pH Variables Dashboard

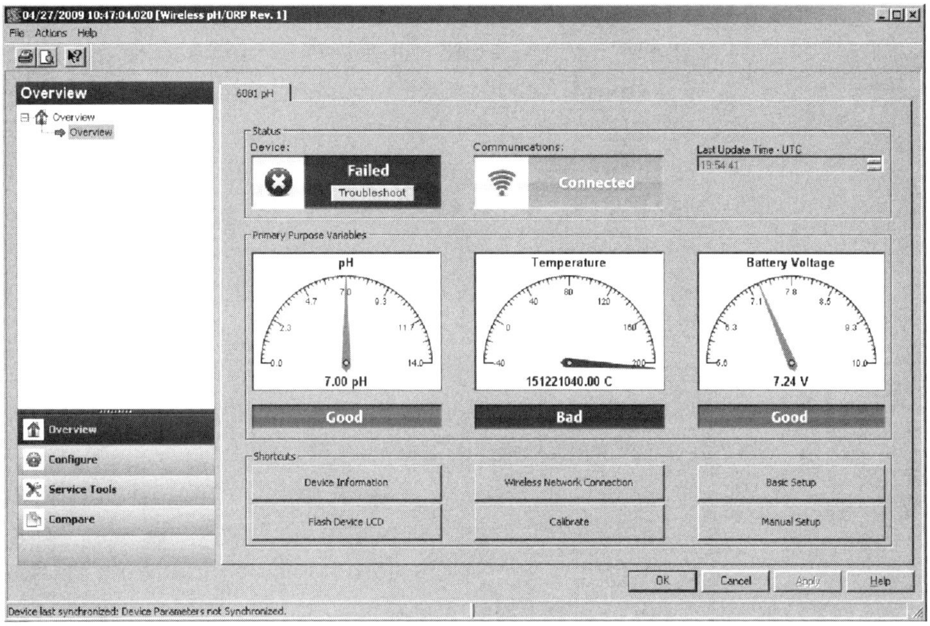

Figures 6-18a. AMS Window for Wireless pH Diagnostics

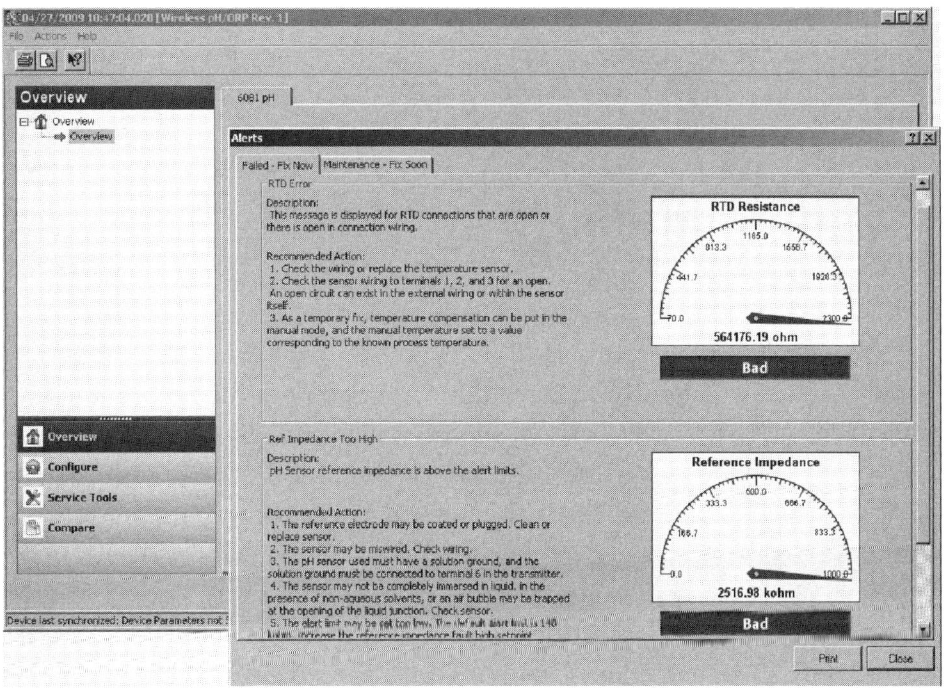

Figures 6-18b. AMS Window for Wireless pH Diagnostics

- Sensor diagnostics can indicate problems with the pH sensor itself, or problems with the process.
    - These include millivolt measurements out of range, which indicate a sensor has been miss-wired, or there is a major sensor problem.
    - Temperature measurements will trigger alerts if they are too low or high and outside the operating range of the measurement, as well as failures in the temperature element such as shorts, opens, or miss-wiring.
    - Calibration errors are triggered by sensor calibration constants (slope and zero) being out of the normal range, which indicates a sensor failure or procedural errors on the part of the user.
    - Advisory messages do not indicate a problem with transmitter or sensor function, but do let the user know that calibrations, simulations, and other events are occurring or have taken place.
- Glass pH electrode impedance diagnostics:
    - Glass pH electrode impedance is in the range of 10s to 100s of megohms and is strongly temperature dependent.
    - A crack or break in the glass electrode will create a short, which will result in an impedance of 1 megohm or less.
    - Detecting broken or cracked pH electrodes requires impedance measurement circuitry with temperature compensation and the simple setting of a low impedance alarm.
- Calibration history:
    - Having factory calibration and periodic calibration data sets along with the associated time stamp is useful for establishing the service trend and scheduling preventive maintenance and/or replacement of the sensor.
    - Table 6-1 shows the calibration history data that can be used for preventive diagnostics. It shows an example of glass electrode aging. The glass impedance increases and the sensitivity of the sensor (slope) gradually decreases with time. This trend indicates that sensor

replacement needs to be scheduled within a couple of months.

Table 6-1. Advanced pH Diagnostics (Calibration Record)

History	Time Stamp	Calibration Method	Slope mV/pH	Offset (mV)	Glass Impedance, (mOhm)	Reference Impedance, (kOhm)	Temperature (°C)
1-Latest	120 days	Auto-cal	56.0	11	820	13	25
2	90 days	Auto-cal	57.5	10	760	10	28
3	60 days	Auto-cal	57.1	9	670	15	26
4	30 days	Manual	58.4	10	550	10	27
5-Oldest	2 days	Manual	59.0	3	480	3	30
Factory Calibration	0.0	Auto-cal	59.3	2	500	2	25

Asset Management

If the transmitter is provided with digital communications such as HART or Foundation Fieldbus, asset management software can provide a remote window into transmitter function and settings. Diagnostic information can accessed remotely, and diagnostic alerts provided to operators and control systems in real time using asset management software.

In addition to simply alerting the user of a problem, the software can provide information on the root cause of the problem by combining the alert with live measurements and configuration information, which allows troubleshooting to be done before making a trip to the transmitter. Remote access to transmitter configuration makes it possible to review measurement and temperature compensation settings, and transfer a complete configuration to a transmitter from another transmitter or from a configuration template composed in the software. Calibration can be performed remotely based on a grab sample analysis or knowledge of the process fluid composition during a particular stage of a process. Asset management software records all of the above information with timestamps, which makes it possible to review events and configurations conveniently without the need to keep paper records.

6-9. Questions

1. Why are pH measurements important in the process industry?
2. What process conditions change the process solution pH for a given concentration of acids and bases?
3. Why are combination non-flowing junction electrodes used in industry?
4. How is the electrical continuity of the reference and measurement electrode achieved?
5. What causes plugging and poisoning of the reference electrode?
6. What are the assumptions in the Nernst equation used in the literature for pH measurement?
7. What is the primary suspected cause of a decrease in electrode efficiency when the glass surface is in good condition?
8. Why is the use of wireless transmitters particularly advantageous for pH?
9. What can be done to improve the performance and life of electrodes subjected to high process temperatures and steam sterilization?
10. What are the options to prevent poisoning of the reference electrode?
11. What can be done to minimize the coating and abrasion of the glass electrode?
12. What must be done for pure water (low conductivity) applications to insure a stable reference junction potential and to minimize noise?

6-10. References

1. Bates, R.G. *Determination of pH Theory and Practice*. Hoboken: John Wiley and Sons, 1964.
2. Galster, H. *pH Measurement Fundamentals, Methods, Applications, Instrumentation*. Weinheim: VCH Verlagsgesellschaft mbH, D-6940, 1991.
3. Westcott, C. C. *pH Measurements*. Academic Press Inc., 1978.

4. McMillan, G. K. and Cameron, R. A. *Advanced pH Measurement and Control, 3rd Ed.* Research Triangle Park: ISA, 2005.

5. Feng, C. D. "Aging Mechanism of pH Sensing Glass in High Temperature Environment." *Proceedings of the ISA 53rd Analysis Division Symposium*, April 2008.

6. McMillan, G. K. "Understand Some Basic Truths of pH Measurement." *Chemical Engineering Progress*, Oct. 1991.

7. Moore, R. L. "Good pH Measurements in Bad Process Streams." *Instruments and Control Systems* (I&CS), Dec. 1990.

8. Martin, M. S. "How to Measure pH in Mixed and Non-aqueous Solutions." *Today's Chemist at Work*, Jun. 1986.

9. McMillan, G. K. "What Have I Learned – Unexpected Wireless Benefits." at http://www.modelingandcontrol.com/ Feb. 9, 2009.

10. Starostin, N. "New Trend in pH-Technology." *Processing Magazine*, Apr. 2008.

11. McMillan, G. K. and Weiner, S. "The Secret Life of pH Electrodes –Part 1." *Control*, Feb. 2009.

12. McMillan, G. K. and Weiner, S. "The Secret Life of pH Electrodes –Part 2." *Control*, Mar. 2009.

13. McMillan, G. K. and Weiner, S. "The Secret Life of pH Electrodes –Part 3." *Control*, Apr. 2009.

14. McMillan, G. K. "Is Wireless Process Control Ready for Prime Time?" *Control*, May 2009.

7

Final Element Fundamentals

7-1. Learning Objectives

A. Appreciate the importance of using the latest final element technology.

B. Be introduced to the relative capability of modern final elements.

C. Understand how final element, system, and process design affect performance.

D. Learn about the latest capabilities of digital positioners and variable speed drives.

E. Understand factors that affect final element energy use and turndown.

F. Gain insight into how the size of the change in an input signal affects response.

G. Find out how to reduce the dead time and limit cycles in final elements.

H. Learn terminology and issues to be able to intelligently discuss industrial applications.

I. Visualize how valve installed flow characteristic changes with process conditions.

7-2. Introduction and Overview

A final element is the last device in a control loop that causes a change in the process, principally by means of changing a flow. Final elements are an essential part of nearly every process control system. Without final elements, there is no way of controlling the process. We could not change operating points or correct for disturbances. There may be several layers of control loops but it is usually a flow that a final element ends up changing in a process. The most notable exceptions are heater or electrode current and mixer speed.

The term "final control element" is often used when referencing control valves. By far, the most common final element is the control valve, with its attendant positioner, actuator, and other components. Variable speed peristaltic pumps are used for the exceptionally small flows of benchtop and pilot plant operations. Other types of variable speed positive displacement pumps are used for small additive and reagent flows in production.

Axial and centrifugal blowers, fans, and pumps are used for the flow ranges normally associated with gas and liquid streams in industrial plants. A variable speed drive (VSD), particularly in large utility flow applications, can save energy by the elimination of a control valve and its pressure drop. However, the energy savings is usually overestimated for process streams [1].

A damper can reduce the cost of the final element or fit in a non-circular duct. Dampers are commonly used in HVAC systems, boilers, furnaces, and scrubbers to manipulate air and vent gas flows. Dampers have a lower pressure drop than a control valve but generally the performance (e.g., rangeability, resolution, sensitivity, speed, and seal) of a damper is not as good as a control valve. The leakage and limited dynamic response and materials/ruggedness of construction of dampers relegate their application to mostly utility and vent systems.

The dynamic response of a final element and its effect on process control will be the dominant theme in this book. The selection and specification process does not generally consider requirements for the dynamic response of the final element, but the ability of the final element to make smooth, accurate small or large changes in flow rate in response to a load change or a set-point change from the control system is important for loop performance.

Additionally, the final element should not create flow oscillations. Oscillations (periodic fluctuations) in a process flow can translate to process variability whose magnitude is proportional to the product of valve resolution or deadband, the slope of a valve's installed flow characteristic, and the process gain [2, 3, 4, 5, 6]. The details of this relationship and its consequences in terms of process oscillations and cost will be examined in section 7-5 on Dynamics.

The following examples provide an overview of the role of final elements for common loops in the process industry in terms of the manipulation of flows.

1. Gas pressure – gas flows into or out of equipment or piping
2. Dissolved oxygen – air and oxygen sparge flows into equipment or ponds
3. Liquid level – liquid flows into or out of a vessel
4. Liquid pressure – forward or recirculation flows in piping
5. pH – reagent flows to equipment or piping
6. Equipment temperature – coolant, hot fluid, or steam flow to coils or jacket
7. Exchanger temperature – coolant, hot fluid, or steam flow to tubes or shell or bypass flow around exchanger
8. Bioreactor substrate control – amino acid, glucose, and nutrient flow
9. Inline composition – additive or blend flows to pipeline mixer (e.g., static mixer)
10. Boiler, furnace, or kiln temperature – fuel flow to burner
11. Boiler, furnace, or kiln oxygen – air flow to burner (air/fuel ratio)
12. Boiler drum total dissolved solids (conductivity) – blowdown flow

Flow can be manipulated by changing the driving force (pressure) or the restriction (in a valve or damper, the cross-sectional area of the flow path). For solids, the flow path should be as straight and streamlined as possible. In an electrical analogy, pressure is voltage, the restriction is resistance, and the flow is current. The area of the restriction in a valve or damper is varied to change resistance to flow. The speed of a motor drive is varied to change the output rate and, hence, pressure from prime movers, such as blowers, compressors, fans, and pumps.

The valve positioner is a critical part of a control valve. In a modern plant, a smart digital positioner is used to provide tight control of valve position and diagnostics. The positioner uses the feedback from the actuator shaft position to match a desired valve position by the manipulation of air flow to the valve actuator. Basic design, diagnostics, performance, and tuning will be discussed in the section on Digital Positioners.

This book will focus on flow control using final elements that get a valve position set point from a distributed control system (DCS) or programmable logic controller (PLC) that resides in a control room. The following overview is offered of final elements that are integrated into field devices that do not depend on a control room signal for the regulation of flow. In these devices, the valve is integrated along with a feedback measurement and control mechanism or algorithm into a self-contained field control system. The primary examples are the fieldbus flow loop, mass flow controller (MFC), temperature regulator, and pressure regulator. These field devices provide a faster final element manipulation because the communication and execution delays are minimal. Also, the sensing and actuation delays and lags are negligible for the regulator and MFC. However, in a regulator, the user cannot see the controller set point, process variable, and output nor choose the controller structure or settings.

Flow control loops can be designed into fieldbus devices, but the final element is still functionally separate and final element implementation considerations are similar.

The mass flow controller, described in Section 4-2 of Chapter 4 on flow measurements, has a thermal expansion type of valve, a thermal mass flow measurement, and a proportional-integral (PI) flow controller. An MFC is frequently used to control air, oxygen, inert gas, and carbon dioxide mass flows to bioreactors.

Temperature and Pressure Regulators

Another type of final element, the temperature regulator, has a thermal expansion valve to manipulate the amount of steam flow for heating. This type of regulator is mostly used on packaged equipment to save on instrumentation costs; however, for performance, reliability, and maintainability, temperature sensors and loops are recommended.

The pressure regulator has a plunger type of valve with a simple proportional control mechanism. This regulator is widely used on non-corrosive streams. A pressure regulator is particularly suitable for remote locations or as a low-cost solution for fast pressure control. There are no signals. The pressure is set by using a wrench to turn a bolt that compresses a spring. Careful regulator and piping design and the installation of upstream and downstream pressure measurements help mitigate the consequences of the regulator's lack of remote adjustment and monitoring.

Self-acting pressure regulators are completely self-contained field devices for controlling pressure by manipulating the amount of flow through the valve body. The regulator performs all the functions of a measurement instrument, control system, and final element. Back pressure regulators and pressure reducing regulators control the upstream and downstream pressure, respectively. For the pressure reducing regulator, the pressure downstream is loaded onto the bottom side of a diaphragm as shown in Figure 7-1 by means of a channel on the outlet port. In this regulator, there is a disc that moves up into a seat to restrict flow. In a control valve, there is normally a plug moving down into a seat to restrict flow. A higher downstream pressure retracts the disc, moving it into the seat and reducing the annular area for flow. A bolt on top that pushes a plate to compress the spring is turned to adjust the pressure set point by changing the spring force acting to extend the disc, moving it away from the seat. A "pressure loaded" regulator is offered where the case with a bias spring is pressurized. The pressure set point for this regulator can be remotely set by a current to pneumatic transducer (I/P).

The self-operated pressure regulator is a proportional-only flow controller with an effective gain of about 4 [1]. *Pilot operated* regulators amplify the change in process pressure and have an effective gain of about 20 [1]. Since there is no integral action, there is a steady-state offset from set point called *droop*. Droop increases with the flow requirement and decreases with the regulator gain. It is less for pilot operated regulators than for self-operated regulators.

Regulators are extremely fast since there are no time delays between sensing element, transmitter, control system, and closure component. Regulators are also quite inexpensive because of their simplicity.

The main problems with pressure regulators are reliability, visibility, and flexibility. If a diaphragm breaks, a pressure-reducing regulator will go wide open over-pressurizing downstream piping and equipment and a back pressure regulator will go completely closed over-pressurizing upstream piping and equipment [1]. There is no measurement, set point, or controller output to monitor or trend. While failures are obvious from the drastic consequences, performance problems, such as the plugging of the pressure sensing tap or a bad spring setting, are extremely difficult to recognize and diagnose because there is usually not a pressure transmitter near the regulator. Additionally, you cannot tune or select the control action. Regulators are also sensitive to the available pressure drop. The materials of construction are quite limited. For these and other practical reasons, pressure regulators are predominantly used on utility flows such as air, nitrogen, water, fuel gas, and steam.

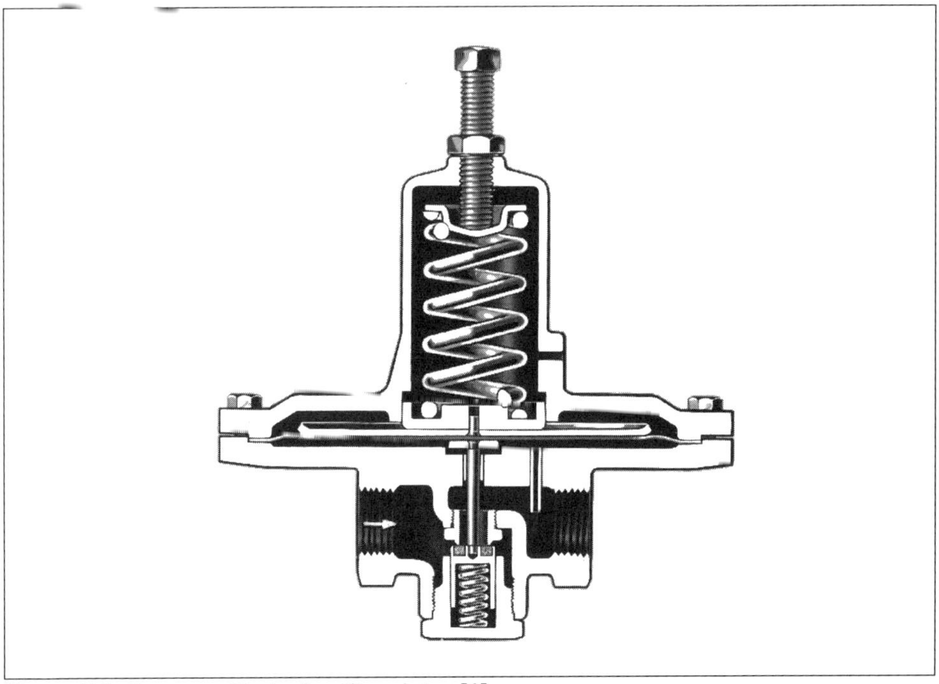

Figure 7-1. Pressure Reducing Regulator [1]

7-3. Valves, Dampers, and Vanes

Control valves, dampers, and vanes manipulate flow by changing the restriction to flow. Reducing the flow by decreasing the flow path cross sectional area is called "throttling." The effect of the change in area for throttling is lumped with geometric effects into a parameter called flow coefficient (C_v). While the flow coefficient is used for valves and dampers with all types of fluids, the basis is the number of gpm of water a valve can pass for a 1 psi pressure drop.

The closure component (also known as closure member and trim) that changes the cross-sectional area may be a plug, diaphragm, boot, ball, or disk in a valve, or a wafer, blade(s), airfoils, or vanes in a damper. Valves are classified as sliding stem or rotary control valves based on the movement of the closure component. Sliding stem valves linearly move a stem that moves a plug, presses a diaphragm, or pinches a boot. Rotary valves rotate a stem that turns a plug, ball, or disk.

A control valve changes its resistance to flow by changing the restriction to flow. Note there is also system resistance from flowmeters, inline equipment (e.g., filters, static mixers, or heat exchangers), piping, and other valves (e.g., manual and on-off) in the flow path. The pressure drop due to system resistance (frictional losses) is proportional to the volumetric flow rate squared. The volumetric flow rate through a control valve is proportional to the flow coefficient (C_v) and the square root of the pressure drop across the control valve as detailed in Equation 7-1.

The pressure drop across a control valve is the difference between the valve inlet pressure (P_1) and outlet pressure (P_2). The pressure drop available to the control valve is the difference between the pump curve and the system curve at a particular flow as illustrated in Figure 7-2. The system curve is the sum of the static head (liquid head from elevation changes and vessel pressure downstream) and the pressure drop from system resistance. The pump curve is the discharge pump head (pressure) at a particular speed with flow. The discharge pressure decreases with flow and falls off sharply at flow rates above the rated maximum for the pump.

A change in the valve stroke (position) changes the flow coefficient. The change in coefficient moves the operating flow rate on the x axis and changes the available pressure drop. A control valve is sized by selecting a maximum flow coefficient that gives the maximum process flow for the pressure drop available at this flow rate per the system and pump curves.

$$q = N_1 * F_p * C_v * \sqrt{\frac{(P_1 - P_2)}{G_f}} \tag{7-1}$$

Where:

C_v = flow coefficient (dimensionless)

F_p = piping geometry factor (dimensionless)

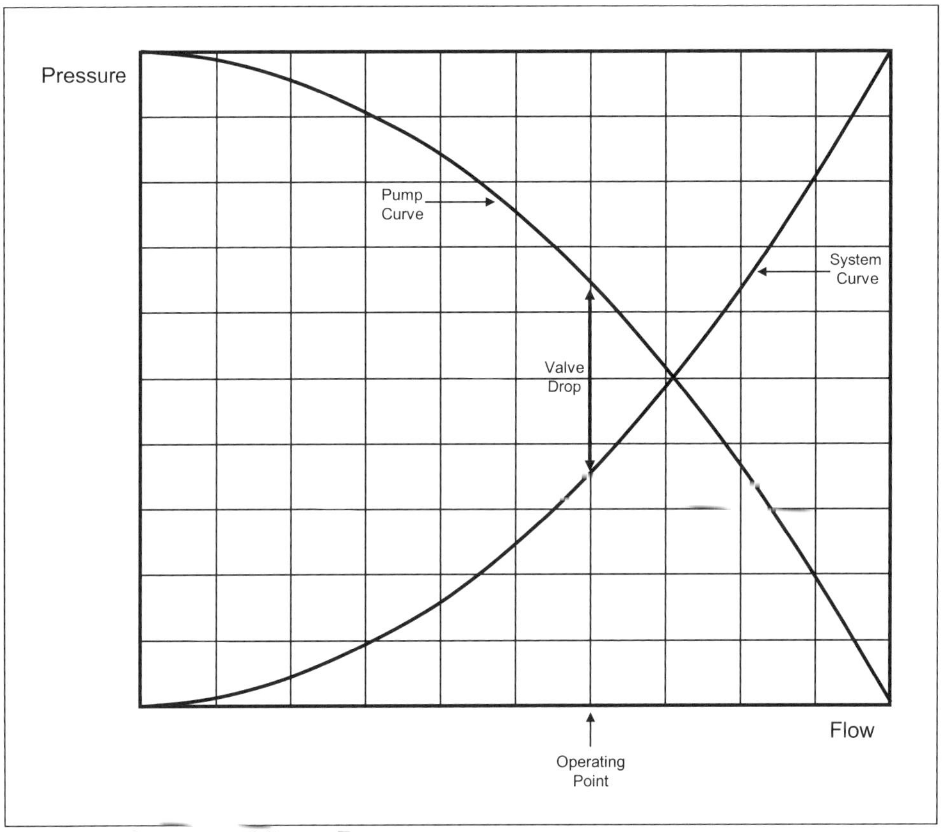

Figure 7-2. System and Pump Curves

G_f = specific gravity of fluid (dimensionless)

N_1 = numerical constant that accounts for dimensions (1.0 gpm/psi$^{0.5}$)

P_1 = valve upstream pressure (psia)

P_2 = valve downstream pressure (psia)

q = volumetric flow (gpm)

Throttling Versus Piping Valves

This book makes a distinction between throttling and piping valves. Throttling valves are designed by a control valve manufacturer for throttling service. Piping valves are designed by piping manufacturers to provide either full flow or no flow (on-off service). If the user is only considering price and leakage and an inherent compatibility with piping specs, the piping valve appears to be an attractive choice. Even for coarse throttling requirements, however, as we will see, it is the wrong choice because of the oscillations inflicted on the loop and the process. To complicate matters, control valve manufacturers have bought piping valve manufacturers. Fortunately, the original company name is often retained so the user can trace the history of the company.

The predominant type of piping valve used as a control valve rotates a disk, ball, or cylindrical plug in a high friction seal. The actuator is often a piston originally designed for on-off service. The connections between the actuator and closure component (ball, disk, or plug) have a degree of play (looseness or gap) where a change in direction of the actuator shaft does not affect the stem or link and ultimately the closure component position in the chain of connections until the change exceeds the deadband created by the play.

When a piping valve closes, the sealing surfaces provide tight shutoff (best leakage class). Most of the change in flow occurs for a relatively small rotation from the closed position of these valves since they were originally designed for on-off service. The high friction of the sealing surfaces for tight shutoff causes excessive stick-slip and a tendency to overshoot the desired position, particularly when the valve is less than 20% open [6] [7]. The actuator shaft and stem connections and on-off actuator used by piping valves are generally not designed to make or measure small changes in intermediate positions. The result is poor flow control. The high friction near the seat (and in some designs throughout the valve travel) and a quick opening type of flow characteristic together create a large limit cycle (persistent, relatively constant amplitude oscillation) that can spread into the process. Without good flow measurements, the paths and consequences of the propagation of oscillations in the process are difficult to analyze. The details and ramifications of poor throttling performance will be discussed in the sections on Dynamics (7-5), Linearity (7-6), and Rangeability (7-7).

Sliding Stem Valves

Globe (sliding stem) control valves (throttling valves) linearly move a plug into and out of a seat. The movement is perpendicular to the flow path. The restriction to flow can be simply viewed as the annular area between the plug and the seat as shown in Figures 7-3 for a single port and 7-4 for a double port valve. The plug and seat for a sliding stem valve are called "trim" and are often supplied and replaced as a set.

The single port in Figure 7-3 is an unbalanced design because the upstream pressure is acting to either push or pull the plug stem. The double port in Figure 7-4 is a balanced design because the upstream pressure acts to both push up and pull down the plug stem. The balancing is not perfect because the areas of the plugs are not the same. The lower plug must be smaller than the upper plug to pass through the cage. Single port valves require bigger actuators but are predominantly used today because leakage is less, the flow path can be more streamlined, and the replacement of trim is easier [8].

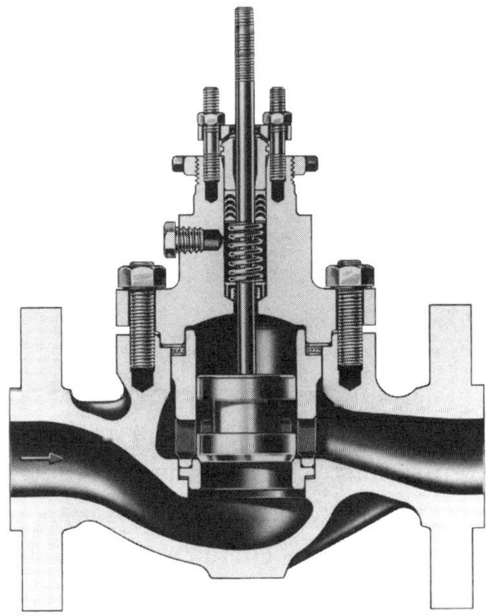

Figure 7-3. Single Port Globe Valve

Globe valves should generally be installed in the direction that is "flow-to-open" where flow acts to open the valve. The "flow-to-close" direction, where flow acts to close the valve, can cause an erratic flow and intermittent closure of the valve near the closed position. The proper direction of flow is normally noted by an arrow on the valve body.

Angle and 3-way valves are special types of globe valves. The angle valve in Figure 7-5 provides a less tortuous flow path and self-drainage, important for slurry service. The 3-way valve in Figure 7-6 is a double-ported valve that splits a flow stream into two flow streams, eliminating the need for the throttling of two control valves. However, the flexibility in the

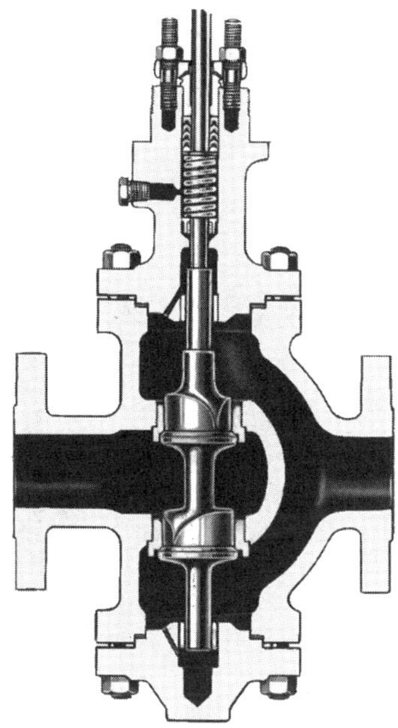

Figure 7-4. Double Port Globe Valve

stroking and characterization afforded by the use of two control valves is lost for startup, shutdown, transitions, and process control. Also, the leakage is larger because of the double-port design. The flow characteristic with stem position is more difficult to quantify and may be more affected by changes in trim clearances with temperature and wear.

Globe control valves offer the best process control except for extremely small flows, where a diaphragm type of valve may be more appropriate.

Roller Diaphragm Valves

The roller mechanism diaphragm valve shown in Figure 7-7 offers precise throttling of flow. The diaphragm eliminates the need for stem packing, provides a small, smooth path for the fine adjustment of extremely small flow rates, and meets sanitary valve requirements. The very low friction of the packless and roller bearing linkage design enables the valve to respond to incredibly minuscule changes in signal. These valves can be used for fine control of additives and nutrients in food and drug production. The flow varies to the 3rd power with stroke. The exceptional resolution and flow characteristic offer an extraordinary rangeability of better than 15,000:1 for a stroke range of 25:1. The flow is forced to be laminar. The pressure drop is small compared to the pressure available. As a result

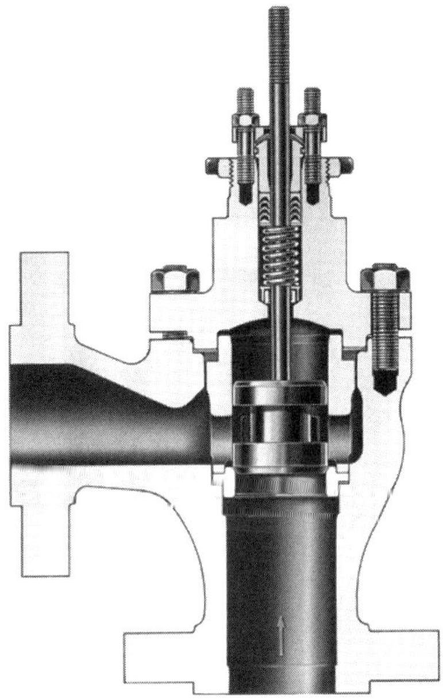

Figure 7-5. Angle Globe Valve

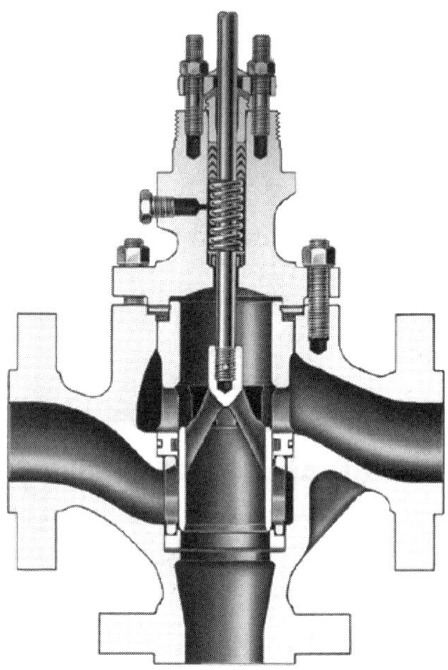

Figure 7-6. 3-Way Globe Valve

the installed characteristic is close to the inherent characteristic (more on this later), enabling an accurate linearization of the controller output in a digital positioner or in a DCS provided there is at least a 16 bit output card.

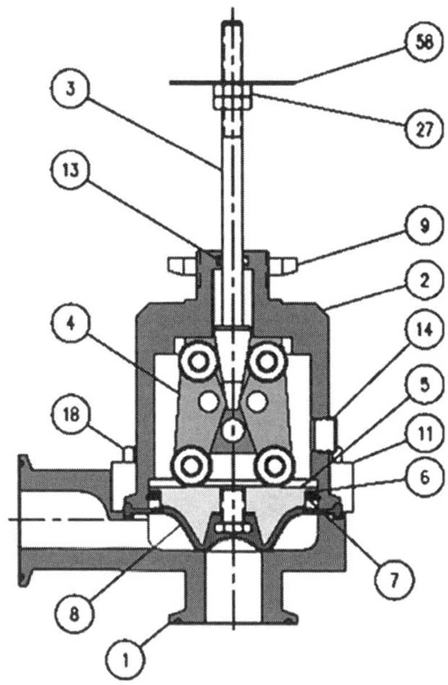

Figure 7-7. Roller Diaphragm Valve

Pinch Valves

A pinch valve (boot valve) consists of a boot that is pinched by a stem pushing on the boot or by a loading pressure surrounding the boot. Pinch or boot valves are primarily used for throttling slurries with extremely large and hard solids (e.g., poor quality limestone) that would erode and plug control valves. There is a self clearing action of rocks and debris. However, pinch or boot valves are only used if a rotary or angle valve won't work because the flow rate changes too fast when the valve opens and the variable elasticity of the boot introduces a degree of hysteresis of non-repeatability greater than seen in other valves. The result can be large flow oscillations.

Rotary Valves

Rotary valves rotate a pug, ball, or disk. The completely open position can correspond to a rotation of 60° to 90°.

The cylindrical plug valve developed by piping valve manufacturers has a rectangular opening that is rotated perpendicular to the flow path. The use of Teflon for the plug and the large opening in the plug appear to offer a low-cost solution for throttling corrosive slurries; however, high friction persists beyond the closed position because the plug is continually rotated through a sealing surface. These valves are not recommended for process control.

The contoured eccentric plug valve developed by various control valve manufacturers swings a plug into a seat at the inlet port as shown in Figure 7-8. The contour, offset, and seat are designed to provide relatively good flow control over a signal range almost as wide as a sliding stem valve.

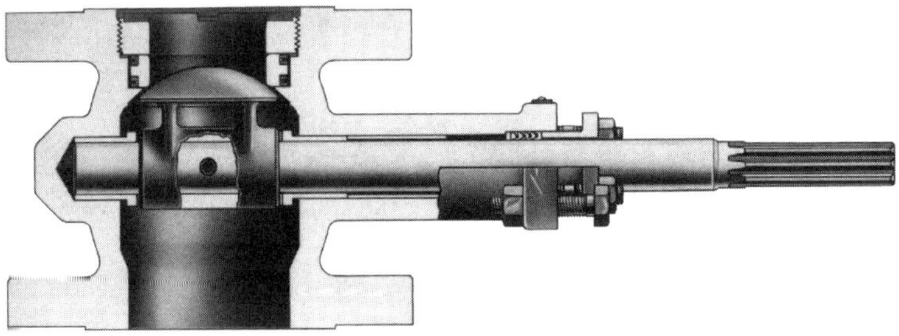

Figure 7-8. Contoured Eccentric Plug Valve

A ball valve designed for throttling service rotates a segment of a spherical surface into the inlet port. The v-notch (triangular opening) in the ball face shown in Figure 7-9 and 7-10 provides a finer adjustment of flow, particularly near the closed position, and a better flow characteristic. The segmented v-notch ball valve is designed to provide good flow control. Full bore ball valves with line size holes in the face and back of the ball that provide an unrestricted cylindrical flow path when completely open were originally designed to be piping valves and are not recommended for process control.

The butterfly valve swings a disk into and out of the inlet port. For throttling service, the disk is contoured and has a low friction seal as shown in Figure 7-11. Butterfly valves with a flat disk and tight sealing surfaces originally designed to be piping valves have a high breakaway torque and a tendency to overshoot as the disk breaks free of the seal.

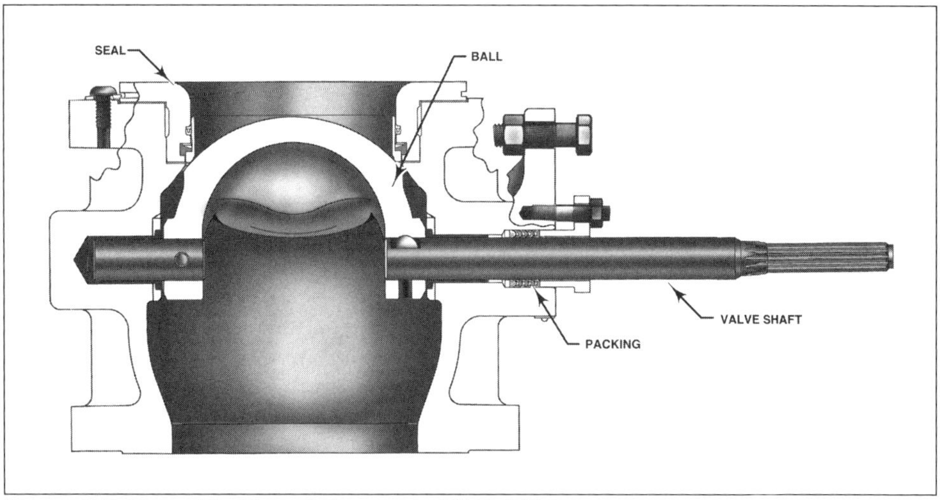

Figure 7-9. Cross Section of Segmented V-Ball Valve

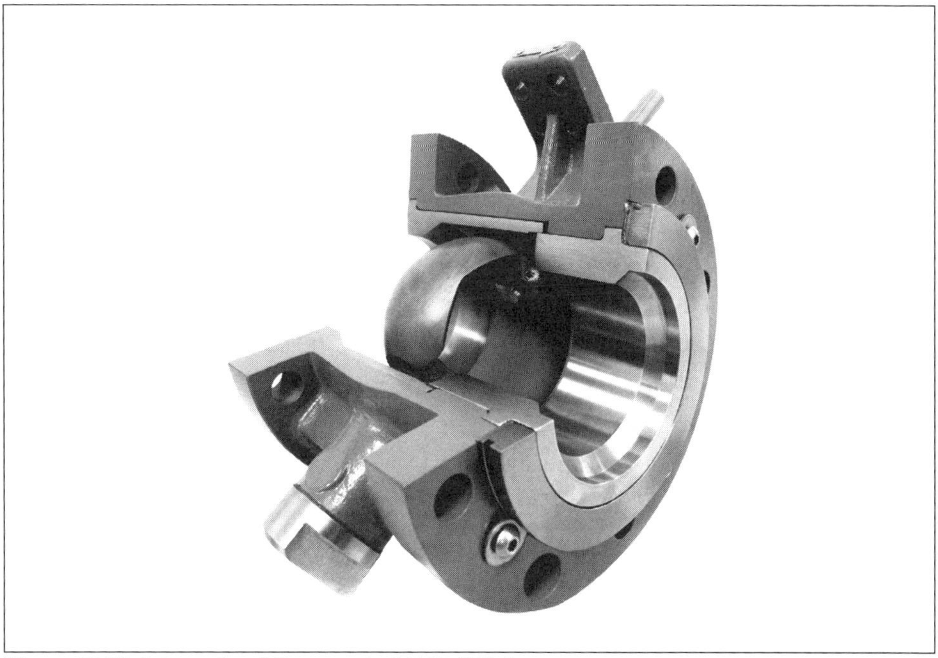

Figure 7-10. Internals of Segmented V-Ball Valve

When used for control, conventional disk butterfly valves are limited to a range of 10 to 60°. There is little change in flow between 60° and 90°. The flow rapidly drops toward zero below 10° of opening. Contoured disks and toothed disks can significantly extend the usable throttling range.

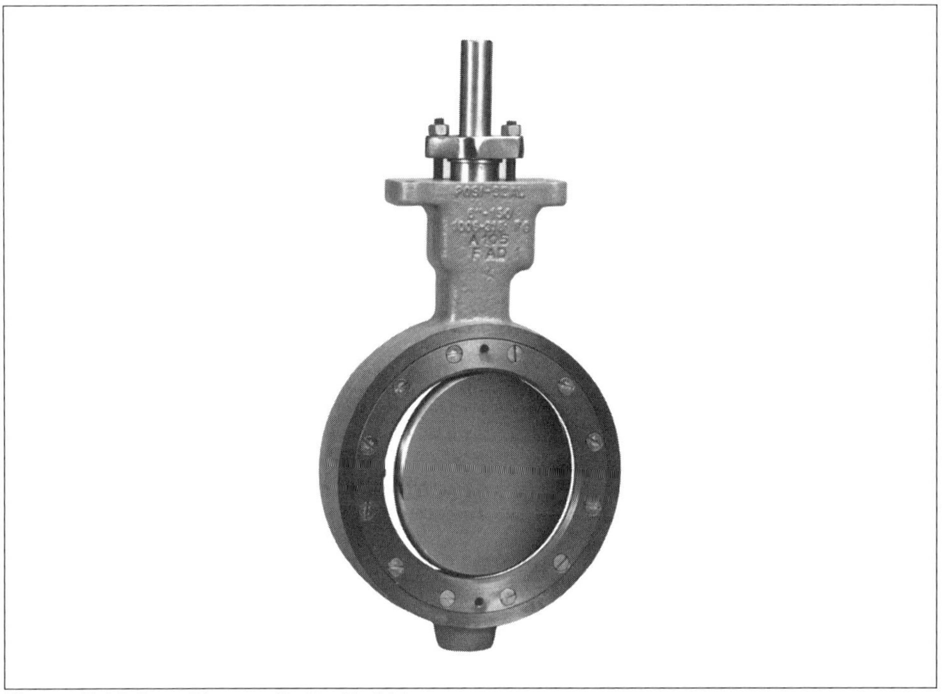

Figure 7-11. Low Torque Butterfly Valve

Dampers and Vanes

To reduce the cost of the final element for large pipes and to throttle the flow in ducts dampers are usually used to manipulate gas or vapor flow and are located at the inlet or outlet of a blower or fan, or at the inlet of a compressor. In some cases, they control vent gas flow to atmosphere or to a vent gas treatment system. The closure component may be a single wafer (Figure 7-12a), a blade, parallel blades (louvers), parallel airfoils, or rotary vanes (Figure 7-12b). The single wafer is the simplest and most rugged type of damper. A combination of parallel and opposed airfoils can be used to provide much less flow distortion (turbulence) downstream.

Dampers generally do not have as tight a seal at the closed position as do rotary control valves. Clearances between wafers, blades, airfoils, vanes, and the housing or supports allow for thermal expansion that is particularly significant in dampers for the venting of combustion gases from furnaces and boilers. The higher leakage flow is usually of no consequence because the prime mover (blower, fan, or compressor) is shut down to stop flow. In fact, the leakage flow can help protect the prime mover from being starved during operation. Where isolation is needed, a guillotine (slide gate) damper is used. Guillotines are operated in an on-off mode (flow modulation is not possible).

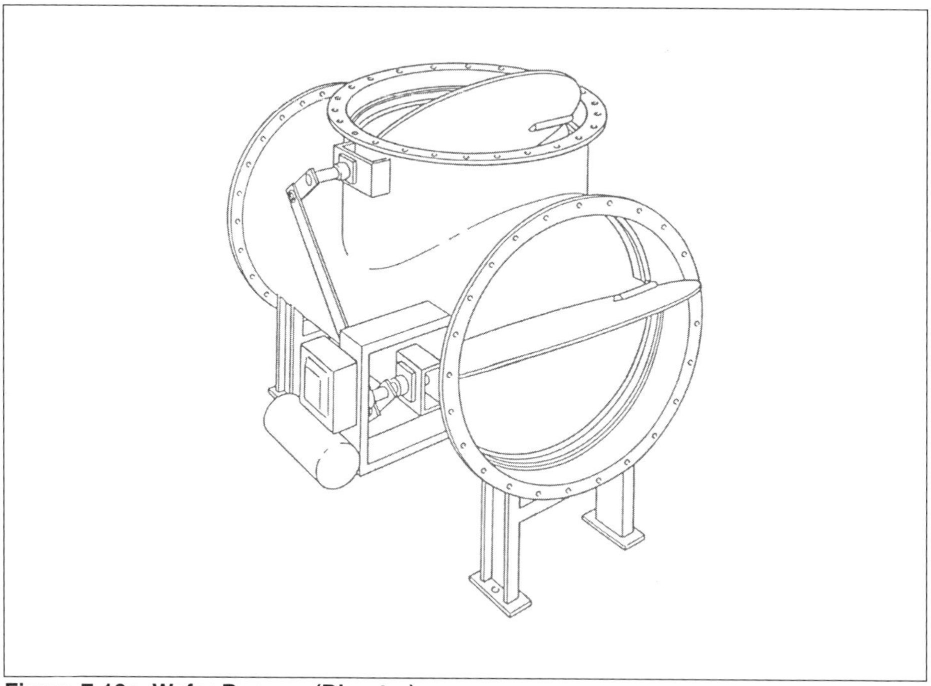

Figure 7-12a. Wafer Damper (Diverter)

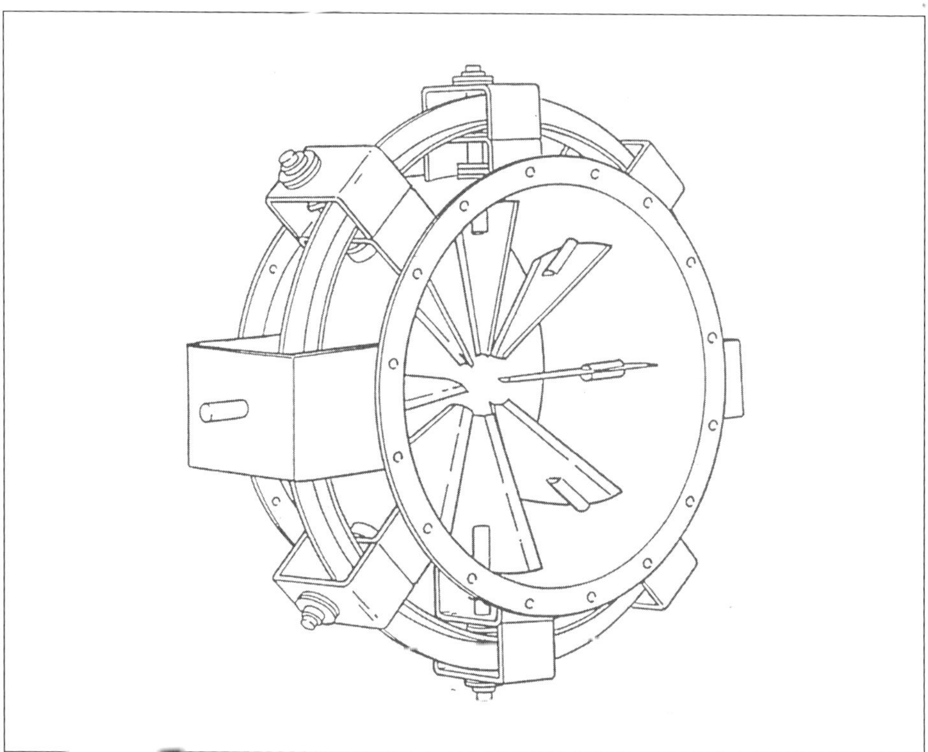

Figure 7-12b. Rotary Vanes Damper

Louver dampers have the highest leakage. The blades are so thin that the blades may flutter, buckle, or warp. The numerous blade seals, fasteners, shafts, and stops pose obstructions to flow that cause pressure drop and can catch furnace ash and scrubber slop [9].

Rotary vanes or guide vanes at the inlet of compressors and fans provide a pre-rotation of flow in a direction that matches up with compressor or fan rotation. The kinetic energy put into the pre-rotation reduces the amount of power the motor must supply, which increases the efficiency of the compressor or fan. These guide vanes are not used on induced draft fans because of potential corrosion.

Dampers may be mounted at the inlet of blowers, fans, and compressors to reduce the pressure drop and energy loss. An inlet damper, however, is more expensive because it is larger than an outlet damper and generally requires the addition of inlet ductwork.

The actuation and response dynamics for dampers are quite different than for control valves. The low friction at the closure position of most modulating dampers translates to a torque requirement to open the damper (breakaway torque) that is much lower than the requirement for rotary control valves.

The flow characteristic resembles quick opening, where most of the flow occurs in the first 40° of opening. Dampers are generally line size and have excessive flow capacity. As a result some dampers operate between 5° and 20° off the seat [10]. The result is a process gain that is extremely nonlinear. Below 20° rotation, the process gain is exceptionally high. Above 20°, the gain dramatically decreases as the rotation is increased reaching a point where the flow does not change in response to rotation.

Dampers can be incredibly large (e.g., 12 feet; 3.6 meters) [9]. The momentum (the product of mass and velocity) of a wafer damper can also be quite large. The load torque (the product of surface area and pressure drop) is large near the closed position when dampers are used on the discharge of prime movers. The resulting dynamic torque combined with the flow characteristic can cause positional instability, particularly when the damper is moving fast (high momentum) or is working near the closed high pressure drop position (high load). As a consequence, some damper control manufacturers recommend the use of flow controls (variable flow restrictors) in the two air lines from the positioner to the actuator [10]. These flow controls were particularly needed for the older design spool type positioners to prevent the damper from developing high frequency oscillations. Smart digital positioners, such as the digital valve controller with numerous tuning options and travel velocity feedback, may eliminate the need for the complications of these flow controls.

The rotation of dampers is usually limited to 60° because there is not much change in flow between 60° and 90°, and because the load torque from fluid dynamics increases dramatically beyond 60°. Dampers have linkages and connections (Figure 7-13) between the actuator and closure components with play ("slop") that creates excessive deadband from backlash. Older positioners were also poor in dealing with the torque and backlash issues. In section 7-5 on Dynamics, the nature and consequences of backlash and the limit cycles created in pressure loops and cascade loops will be discussed. The important point for now is that dampers generally provide poor flow control unless better actuator, linkage, and positioner designs are used [11, 12].

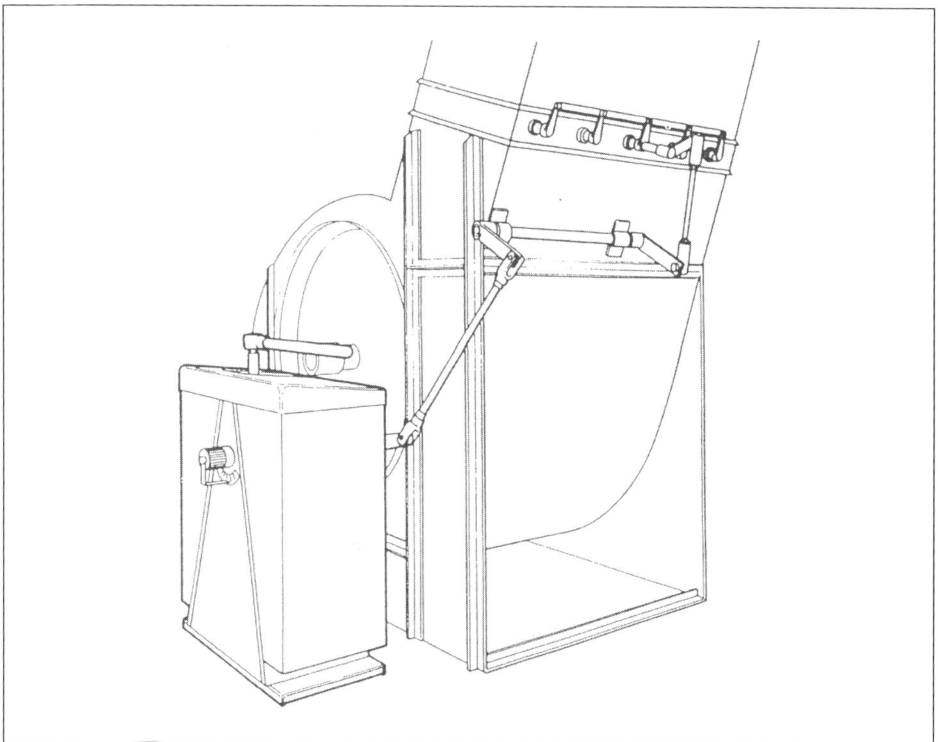

Figure 7-13. Fan Damper Linkages

Special Actuators

The signal from the control loop to a control valve or damper goes to a transducer or positioner that converts the signal to a driving force in the actuator, which positions the closure component of a valve or damper to change the restriction to flow. Most actuators use air pressure but in some cases for higher speed, greater force, or a longer stroke, an actuator may use hydraulic pressure, an electric motor, or a solenoid to move the actuator shaft.

Hydraulic actuators require supporting equipment to provide the pressurized oil and a complex trip system to provide fail safe action since there is no spring. Hydraulic actuators are exceptionally fast with moderately good resolution. The high cost, large footprint, and additional maintenance requirements have limited the use of general purpose hydraulic actuators in the process industry.

Electro-hydraulic actuators (Figure 7-14) use a stepper motor to drive a positive displacement gear pump. The pump transfers hydraulic fluid from one side of a double acting cylinder to the other. When movement stops, the piston is locked in place. The speed of movement is governed by the volumetric pumping capacity and the rotational speed of the motor. Typical speeds range from 0.1 to 0.5 inch/sec (0.25 cm/sec to 0.5 cm/sec) (4 to 20 seconds stroking time for a 2 inch (5.1 cm) stroke) [13]. The resolution is exceptional (e.g., 0.1%) [13]. Electro-hydraulic actuators are sealed and self-contained to provide a compact packaged design free of hydraulic fluid leaks. The hand wheel in Figure 7-14 enables a field operator to manually position the valve.

Electromechanical actuators use a motor and gear box to provide a rotary motion or a linear motion via a screw drive. These actuators are capable of very high torques required by large valves and high pressure drops. Upon loss of electrical power, the actuator fails to last position. It is important to provide an Uninterruptible Power Supply (UPS) for the motor and signal for applications where failure to last position is an operational or safety problem. These motor actuators historically have been slow with poor resolution.

Small electronic stepper motor driven valves are capable of modulating extremely small changes in position (0.0005 inch; 0.0013 cm) at fast speeds (100 millisecond time constant). These valves are limited to small sizes (e.g., < 1 inch; 2.54 cm) and stainless steel or aluminum construction.

Figure 7-14. Electro-Hydraulic Actuator (with Hand Wheel)

Solenoid operated valves are capable of fast action. They consist of a solenoid actuator and an integral packless sliding stem valve. Small on-off solenoid valves are commonly used on air lines to pneumatic actuators to force a valve to a preset fixed position (e.g., open or closed). Small on-off solenoid valves are also used for blasting air streams to clean filters. Modulating solenoid valves are used in non-corrosive clean streams for fast throttling. For more information, see the section on Solenoid Valves.

Pneumatic Actuators

The focus in this book will be on pneumatic actuators because hydraulic and electric actuators are used in less than 5% of the total applications in the process industry. Process plants have reliable air systems. Air may also be available upon power failure and shutdown of an air compressor because of air reservoirs. Boosters outlined in the Accessories section can be used to increase actuator speed. For all but the highest speed and force requirements, pneumatic actuators instead of hydraulic, electro-hydraulic, motorized, or electromechanical actuators are used because of the following advantages:

- Lower cost
- Fail-safe action
- Safe for hazardous areas
- Simplicity
- Reliability
- Maintainability
- Standardization
- Minimal parts inventory
- Lower energy use

Pneumatic actuators are diaphragms or pistons. The diaphragm actuator always has a spring on the opposite side of the diaphragm from the air pressure, and loaded to position the valve. The force available to position the valve is the air pressure multiplied by the diaphragm surface area minus the spring force. The specific pressure range needed to fully stroke the valve, called the bench setting, depends on the spring. The actuator size (surface area) and spring rate are chosen based on the forces required to open and close a specific valve design (e.g., in view of packing and seal friction) at operating pressures.

The diaphragm is typically a flexible fabric-reinforced elastomer with a molded convoluted shape that flexes without stretching as the valve strokes. The change in area is less than 10%. The spring, attached to a diaphragm plate, retracts or extends the closure component to the desired failure position upon loss of air pressure.

For the spring and diaphragm actuator in Figure 7-15a, the air pressure is loaded below the diaphragm to exert a force that retracts the actuator shaft. The diaphragm plate and spring act to extend the actuator shaft.

For the spring and diaphragm actuator in Figure 7-15b, the air pressure is loaded above the diaphragm to exert a force that extends the shaft. The diaphragm plate and spring act to retract the shaft.

For valves that are "retract to open" (the most common case), the actuators in Figures 7-15a and 7-15b close and open the valve, respectively, upon air failure. Fail closed actuators are used in most loops. However, there are important applications of fail open actuators. Vent valves to prevent high pressures, recycle or vent valves to prevent compressor surge, and coolant valves to prevent high temperatures often use fail open actuators. It is critical for safe operation that the correct failure position be specified for each application based on a thorough analysis.

The valve action is either "direct" ("increase-to-open"—increase in valve signal increases flow) or "reverse" ("increase-to-close"—increase in valve signal decreases flow). Fail closed valves generally but not necessarily are "direct" acting. In a DCS, where the PID (proportional-integral-derivative) block control action is set based on process action (direct or reverse), the PID algorithm assumes an increase in PID output will cause an increase in valve flow. In this case, the signal must be reversed for a "reverse-acting" or "increase-to-close" valve. This reversal of the signal can be done in the DCS AO (analog output) block or in the positioner, but not in both places. It is important that the PID and the valve action be correctly set or the PID controller output will go in the wrong direction to the PID output limit, a potentially unsafe situation.

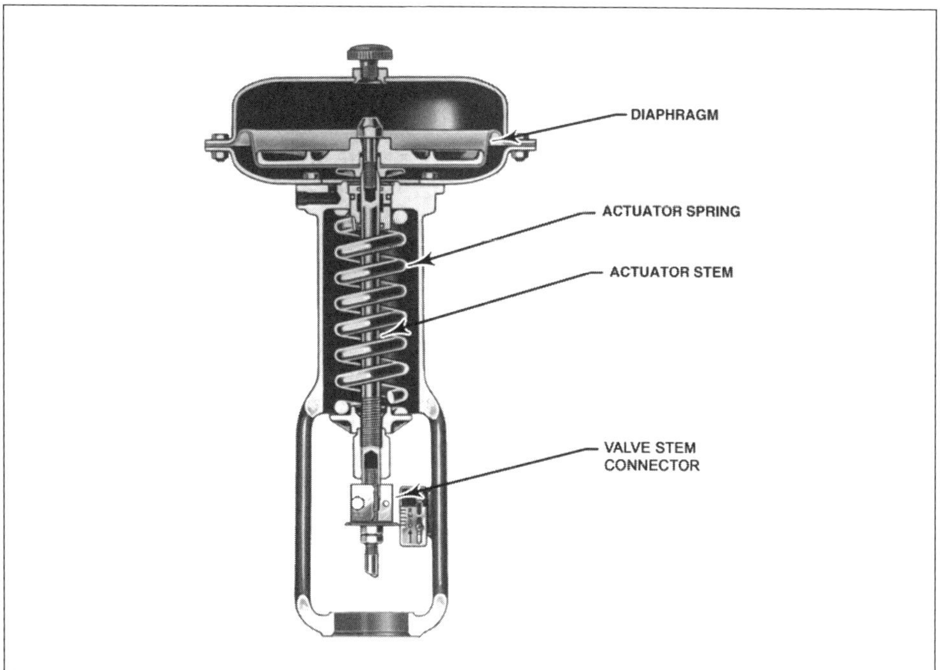

Figure 7-15a. Diaphragm Actuator with Spring to Extend (Pressure to Retract)

Diaphragm actuators offer advantages in terms of reliability, maintainability, and resolution compared to piston actuators. The diaphragm provides an air tight seal that does not wear out with travel. Since the diaphragm ends are attached and stationary, the friction associated with the movement (diaphragm flexure) is negligible. The diaphragm actuator is able to respond to small changes in air pressure.

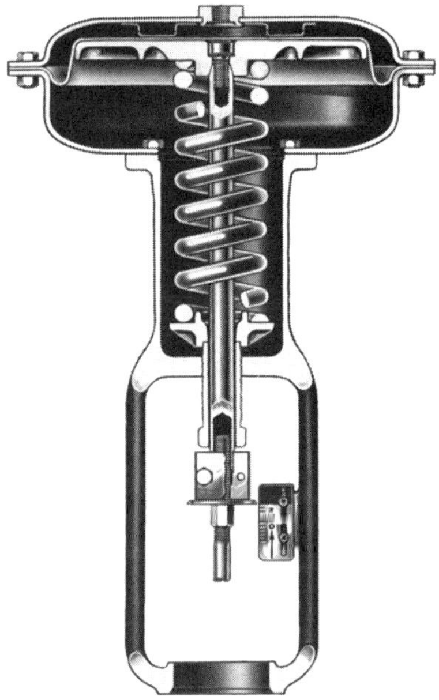

Figure 7-15b. Diaphragm Actuator with Spring to Retract (Pressure to Extend)

However, in setting up and testing a diaphragm actuator the user must be aware of the bench settings. It is important to remember that bench settings, as the name implies, are the pressure settings that are needed to fully stroke the valve on the benchtop at shop conditions (no pressure drop and no temperature extremes). The bench settings can be checked for an installed valve if the pipe is depressurized and at ambient temperature. If a positioner is used, it will drive the actuator pressure to the pressure needed to make the feedback of the actual position match the desired position. In the setup of the positioner, the maximum and minimum actuator pressures need to be specified, but the consequence of not exactly matching the bench settings is minimal if the pressure range specified is sufficient to fully stroke the valve.

The traditional spring and diaphragm actuator can be operated without a positioner. In this case, the current to pneumatic (I/P) transducer output is piped to the actuator to provide a fill and exhaust rate to achieve an actuator pressure that matches the lower bench setting at 0% signal (e.g., 4 mA DC) and the upper bench setting at 100% signal (e.g., 20 mA DC). This configuration is not recommended even for fast loops. All loops should have a digital positioner (see Digital Positioner subsection).

The diaphragm actuator pressure must overcome the spring force. The traditional diaphragm actuator was limited to about 30 psig, limiting the force that it could deliver to the valve. Diaphragm actuators have been developed that use 90 psig and a positioner to provide a more compact and powerful package (Figure 7-15c). Diaphragm actuator size and cost increase as valve size and pressure drop increase. For very large valves (e.g., > 24 inch; 61 cm) or very high pressure drops (e.g., > 1000 psi; 6.9 mPa), a pneumatic piston, hydraulic piston, or motor actuator may be needed.

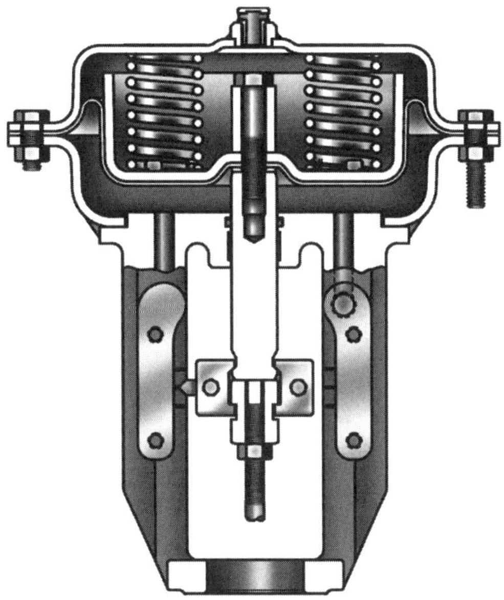

Figure 7-15c. High Pressure Diaphragm Actuator

The pneumatic piston actuator is widely used for on-off valves and for throttling large control valves and dampers. Piston actuators can be smaller than diaphragm actuators for the same load because of a higher air pressure.

Piston actuators can be single sided, where the pressure is loaded on one side of the piston to oppose the force of a spring, similar in concept to a spring return diaphragm actuator. The more common approach is a double sided (double acting) piston (Figure 7-16) where air pressure is loaded on both sides of the piston. There may be a small bias spring to directionally put the valve in the correct air fail position. However, to ensure an air fail position, a volume tank (reservoir) is needed as shown in Figure 7-27c.

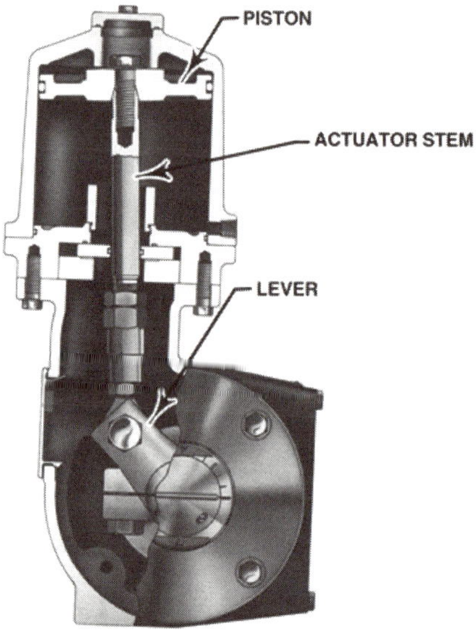

Figure 7-16. Double Acting Piston Actuator (Rotary Valve)

Motion of a double acting piston actuator depends on a difference in air pressure across the piston. Positioners are necessary to load and balance the high pressures in these piston actuators so the question of whether to use a positioner on fast loops goes away. There are no bench settings but there is a crossover pressure that is rarely set properly [8].

The crossover pressure is the balance point where piston motion ceases. The pressure on one side of the piston must differ from the crossover pressure by enough to exceed the unbalance from pressure drop and frictional forces posed by piston seals and valve stem and trim seals. If the crossover pressure is set too high, the pressure in the piston may approach the supply pressure to overcome unbalance and frictional forces. If the crossover pressure is set too low, there is not enough muscle for the piston, making the valve trim more susceptible to buffeting from fluid forces and flutter of a plug or disk [15].

Piston actuators require extremely smooth surfaces, precision machining and guiding, and lubrication of the piston O-ring to minimize the friction between the piston and cylinder walls. Wear and damage, rust, and improper lubrication can cause excessive friction and poor resolution. The O-rings also need to be periodically replaced.

A piston actuator doesn't respond to as small a change in air pressure as a diaphragm actuator. Consequently, the minimum positioning resolution is not quite as good for a piston actuator even in the best of conditions. Also for comparable sizes and the positioner capacity, the pre-stroke time delay required to make a significant pressure change is larger for a piston actuator.

Rotary vane actuators (Figure 7-17) eliminate the need to convert the linear motion of the shaft of a diaphragm or piston to the rotary motion of a rotary control valve or damper. Early design issues with finish on the inside of the chamber and O-ring friction have been addressed. The vane actuator has the same issues as the piston actuator. The addition of a spring for fail-safe action is difficult [8].

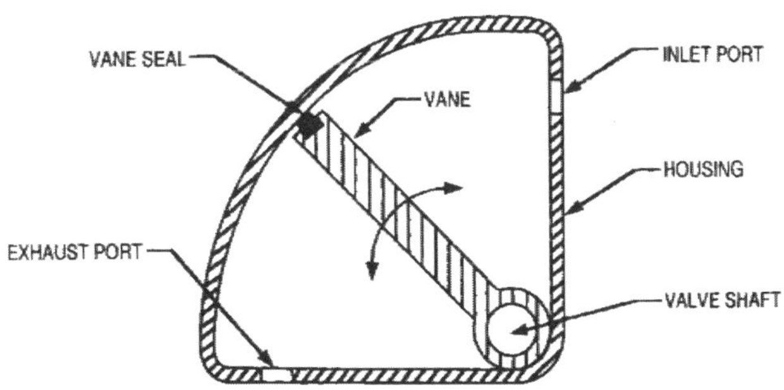

Figure 7-17. Rotary Vane Actuator [8]

Shaft and Stem Connections

The shaft of the actuator and the stem of the internal closure component (plug, ball, or disk) are normally separate. The closure component may be cast and forged with the stem or the stem may be connected during valve assembly. The actuator shaft moves the stem that moves the closure component. (While "shaft" and "stem" are more appropriate terms for the actuator and the closure component, respectively, in practice the terms "stem" and "shaft" are used interchangeably.) The amount of play (looseness or gap) in the connections between the shaft, stem, and closure component is backlash that creates deadband and determines (in part) how well the valve will respond to small changes in signal. The location and type of connection of the positioner feedback mechanism for valve travel determines whether the positioner is seeing the response of just the actuator or the actual response of the closure component.

Previous methods of testing valve response involved making much larger changes in the valve signal than would normally be made in closed loop control. Most valves will look OK with these large changes in requested position. In service, the change in controller output from scan to scan is generally small (e.g., < 0.2%) except during the start of an operation or process. For small changes in valve signals the resolution limit and deadband that prevent a good response and create sustained oscillation (limit cycle) are observable. Current test methods established by the ISA-75.25.01-2000 (R2006) standard addresses the effect of step size on response (see section 7-5 on Dynamics).

Sliding stem (globe) valves have the least amount of deadband because of the direct connection between the actuator shaft and trim stem, and low trim friction. For rotary valves, connections can be problematic since there is the need to convert the linear motion of a piston or diaphragm shaft to rotary motion and the changes in the effective lever arm length.

The direct connection design (Figure 7-18a) eliminates linkages for the translation of linear to rotary motion. This connection introduces the least amount of deadband. The actuator shaft must pivot as the diaphragm or piston strokes. The pivoting design for piston actuators is especially complex because of the effect of the pitch on the O-ring seal [8].

There are several less expensive connection designs. Unfortunately, these designs all introduce either resolution or deadband errors into the valve response that increase with wear. The link arm connection design (Figure 7-18b) has backlash from the connections at pin points 1 and 2. The rack and pinion design (Figure 7-18c) has a resolution limit from tooth spacing and backlash from play in the fit between the pinion gear and rack teeth. The slide bearing to hold the rack in contact with the gear adds friction. The scotch yoke design (Figure 7-18d) is perhaps the worst from a standpoint of control because the slot adds a lot of backlash [8]. A valve with backlash will not respond to a change in direction of a signal until the change exceeds a deadband (see Section 7-5 on dynamics).

The problems with the rotary valve connections are aggravated by high friction from the stem packing (seal) and the sealing of closure component surfaces. Some high temperature and early environmental packing

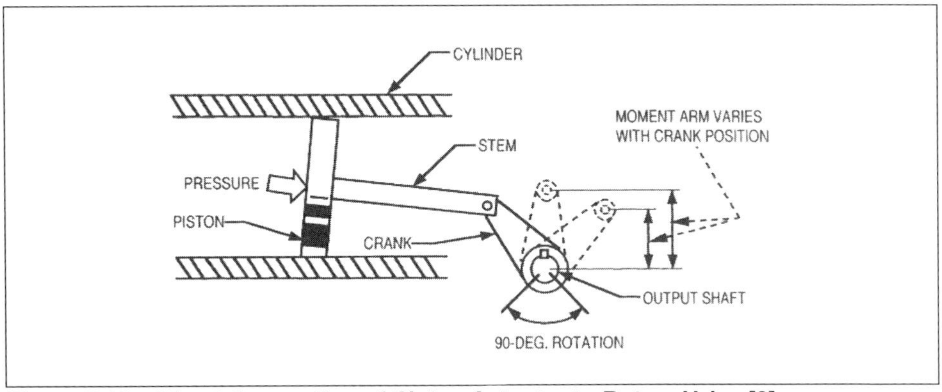

Figure 7-18a. Direct Connection of Piston Actuator to Rotary Valve [8]

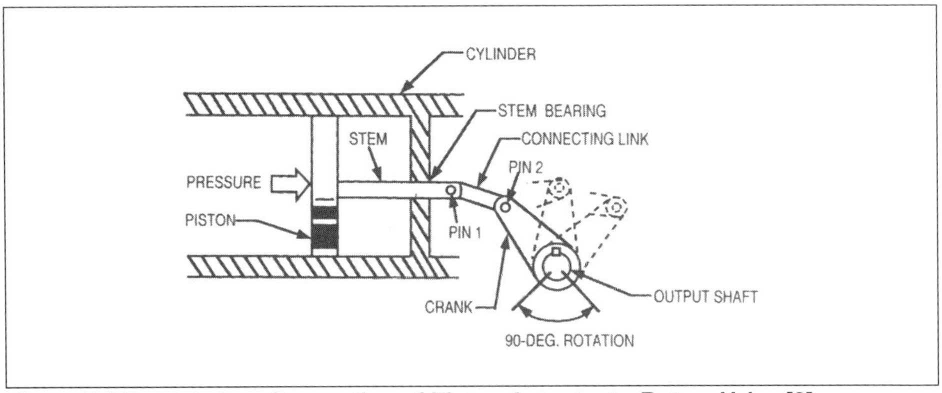

Figure 7-18b. Link-Arm Connection of Piston Actuator to Rotary Valve [8]

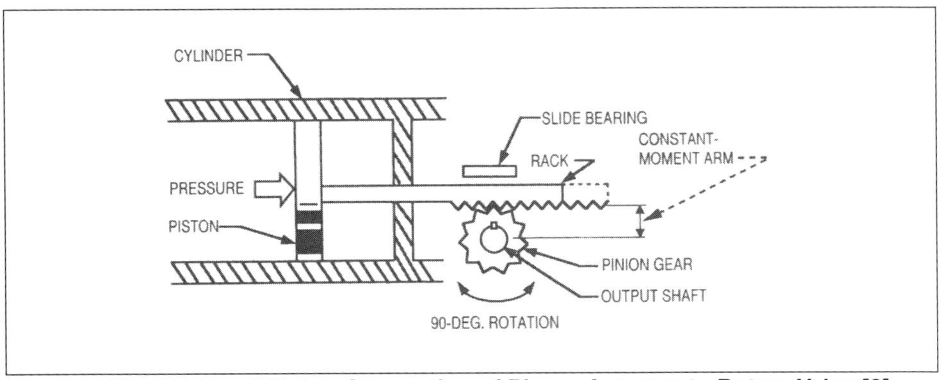

Figure 7-18c. Rack and Pinion Connection of Piston Actuator to Rotary Valve [8]

designs created excessive stem friction. Additional tightening of the packing beyond factory specifications makes the friction worse. On-off valve designs (e.g., full port ball or rectangular port plug valves) where the closure component is always rubbing against a seal designed for tight shutoff, create excessive friction. The result can be a twist in the shaft (shaft

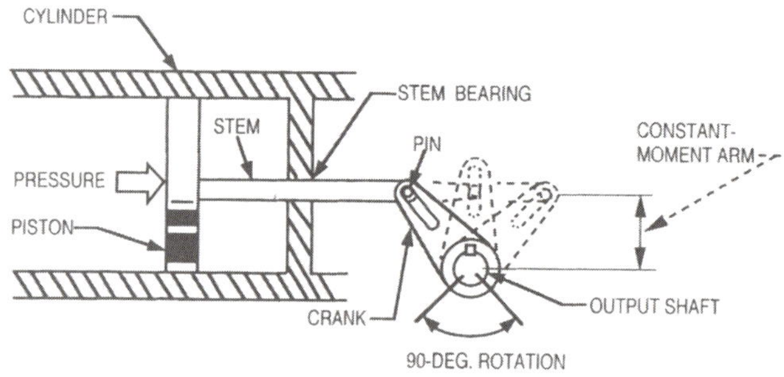

Figure 7-18d. Scotch-Yoke Connection of Piston Actuator to Rotary Valve [8]

windup), where the actuator shaft moves but the closure component remains stationary. Eventually, the closure component breaks free and jumps to the next position (stick-slip). This twist can also occur in the plug, ball, or disk stem. The windup increases as friction increases, stem or shaft length increases, and stem or shaft diameter decreases. Shaft windup occurs normally from a rest position or a change in direction of the desired stroke until the signal exceeds a deadband. However, once the stroke starts there is a jump rather than a small incremental response that is observed with true backlash. Windup can be thought of as a combination of a resolution limit and deadband (see Section 7-5 on dynamics).

Windup results in the positioner reporting the valve moved when the actual closure component position had not changed, when the positioner is on the actuator shaft rather than on the valve stem [16]. A sensitive flow measurement is needed to determine when the closure component has actually moved [17]. In the case of tight shutoff butterfly valves for compressor control and ball valves for phosphorous control, the author has witnessed a difference of 8% between the position indicated by a smart positioner and the actual position of the internal disk or ball. These and other problems can be avoided by adherence to the connections outlined in Figure 7-19. Here, a direct connection actuator design eliminates linkages, a splined shaft to stem connection eliminates the play of key-lock connections, a stem cast with the closure component (e.g., ball) eliminates the play of stem to closure member connections, a short and large diameter shaft and stem reduce shaft windup, and a positioner connection to the stem provides the best position feedback.

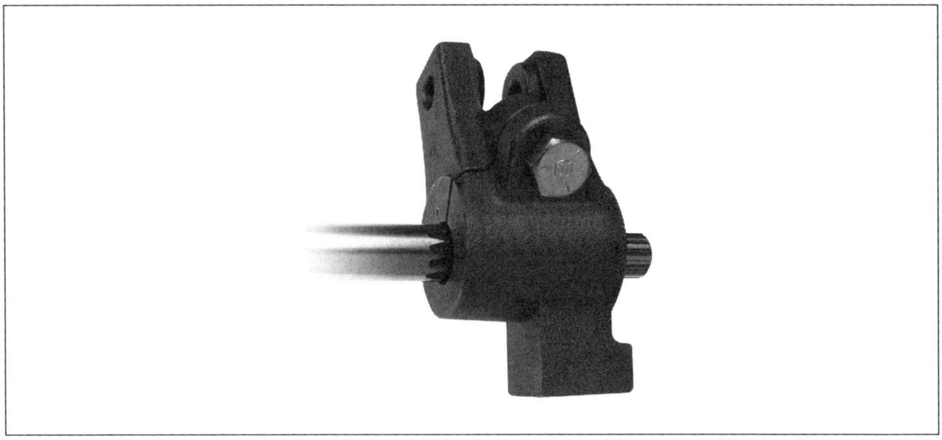

Figure 7-19. Connections for Minimal Windup and Backlash in a Rotary Valve

Solenoid Valves

Special modulating solenoid valves can be used for throttling. When the solenoid is energized, a magnetic force lifts the plunger against the force of a return spring, pulling the pilot valve off of its seat. This causes the pressure difference between the top and bottom surfaces of the main disk, allowing the main disk to follow the pilot disk by servo action [14]. The actuator and positioner are integrated into the control valve as shown in Figure 7-20. No stem packing and connections are used, which eliminates stem leakage and friction [14]. These valves are capable of fast strokes (e.g., < 1 sec) and exceptional resolution. The materials of construction are limited and the flow path is not streamlined. These valves are extensively used in nuclear submarines on utility flows.

Small two-position solenoid valves (Figure 7-21) are commonly used for on-off control. Two-ported (2-way) solenoid valves are used to turn small water and gas flows on and off. Three-ported (3-way) and four-ported (4-way) solenoid valves are used to vent and pressurize actuators. The materials of construction are often limited to brass and stainless steel. By far, the most prevalent use of these small on-off solenoid valves in the process industry is in safety instrumented systems (SIS) (see Chapter 1 for a SIS overview). The solenoid valves are put in the air lines to pneumatic actuators for on-off valves and control valves to set a fixed position demanded by a SIS or a sequence (see section on Accessories).

Pneumatic Positioners

Pneumatic positioners persisted well into the days of electronic instrumentation and the DCS because the actuator was also pneumatic. For

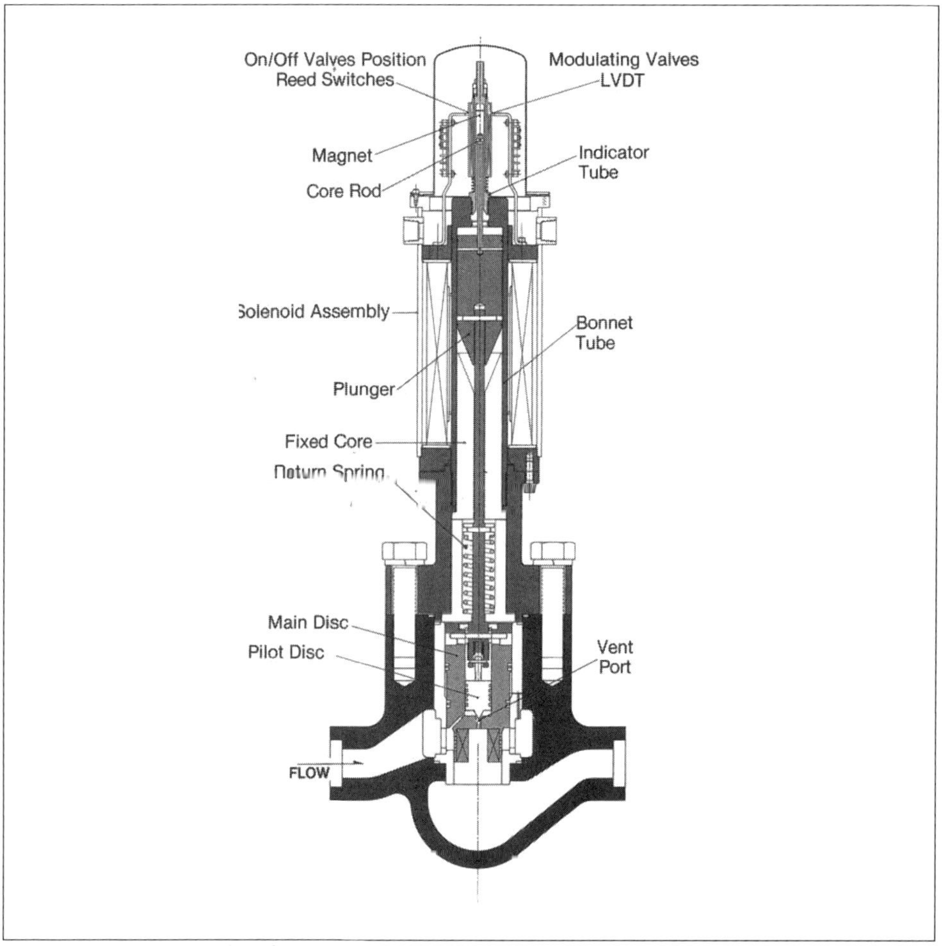

Figure 7-20. Modulating Solenoid Valve [14] ***(Courtesy of Target Rock - Curtis Wright Flow Control)***

Figure 7-21. Solenoid Valve for On-Off Control

pneumatic positioners, the analog or digital signal first has to be converted to a 3–15 psi (20.7–34.4 kPa) signal by a current to pneumatic (I/P) transducer. This type of positioner (Figure 7-22), used until the development of the electronic positioner in the 1980s, was principally limited to proportional-only by means of links, levers, bellows, flappers, and nozzles. The time required for even good pneumatic positioner designs to respond increased from 1 to 100 seconds or more for small changes in the signal (e.g., < 0.2%) (see Section 7-5 on Dynamics). For spool type positioners there was sometimes no response. Adjustment of the proportional gain was limited or nonexistent.

The proportional gain and the resolution capability depended on the pneumatic design. A two stage pneumatic relay design provided better sensitivity and a higher gain than the inexpensive spool design. Spool type positioners with poor connection designs were often used for throttling of piping valves and dampers. Pneumatic positioners were difficult to calibrate and readily lost their calibration. These positioners did not provide a valve position signal (travel readback or feedback), alerts, or diagnostics. Consequently, the user was unaware of calibration or performance problems. In order to get a position indication in the control room, a separate position transmitter had to be installed. Limit switches that indicate full open or closed position were used for batch operation, startup, and shutdown but position transmitters were rare. It was only after the advent of electronic positioners that users started to become aware of the valves as a major source of process variability next to controller tuning. Today, existing pneumatic positioners should be replaced with microprocessor based digital (smart) positioners.

Digital Positioners

The digital positioner manipulates the supply and exhaust flow to the actuator via a pneumatic relay. The exhaust flow capacity is generally higher than the supply capacity to provide faster stroking to the air failure position. The digital positioner accepts analog or digital input signals. The current to pneumatic (I/P) transducer is functionally moved from the input to the output of the controller. The I/P transducer converts the microprocessor based controller output to a pneumatic signal that is amplified by a pneumatic relay. The pneumatic relay is like a self-contained pressure controller. The relay adjusts the supply or exhaust flow to achieve an outlet pressure that matches the desired pressure set by the I/P transducer. The relay supplies air flow to increase actuator pressure and

Figure 7-22. Pneumatic Positioner

exhausts air flow to decrease actuator pressure. The relay has two outputs. In Figure 7-23, output A is direct acting (actuator pressure increases with valve signal) and output B is reverse acting (actuator pressure decreases with valve signal). For zero percent signal, relay A is at zero psig and relay B is at full supply pressure. To make the electrical signal failure position consistent with the air failure position, relay A output is connected to the diaphragm or single-sided piston. For a double acting piston, relay A output is connected to the piston end that drives the valve open and relay B output is connected to the piston end that drives the valve closed.

It is useful to think of a digital positioner as a digital valve controller where the set point is the desired position and the position feedback is the measurement. The desired position is an analog or digital signal sent from the distributed control system (DCS) or programmable logic controller (PLC). The digital valve controller output is the set point for a relay that is a high gain proportional-only fast-acting pressure controller. Since a DCS controller output is the valve signal, we have as a minimum, a triple cascade control system. Frequently, the DCS has a cascade loop where the output of a primary process controller is the set point of a flow controller whose output is the set point of a digital valve controller whose output is the set point of the actuator pressure controller whose output regulates the supply or exhaust flow(s) to the actuator. The block diagram for this triple cascade is shown in Figure 7-24 for temperature control. The primary

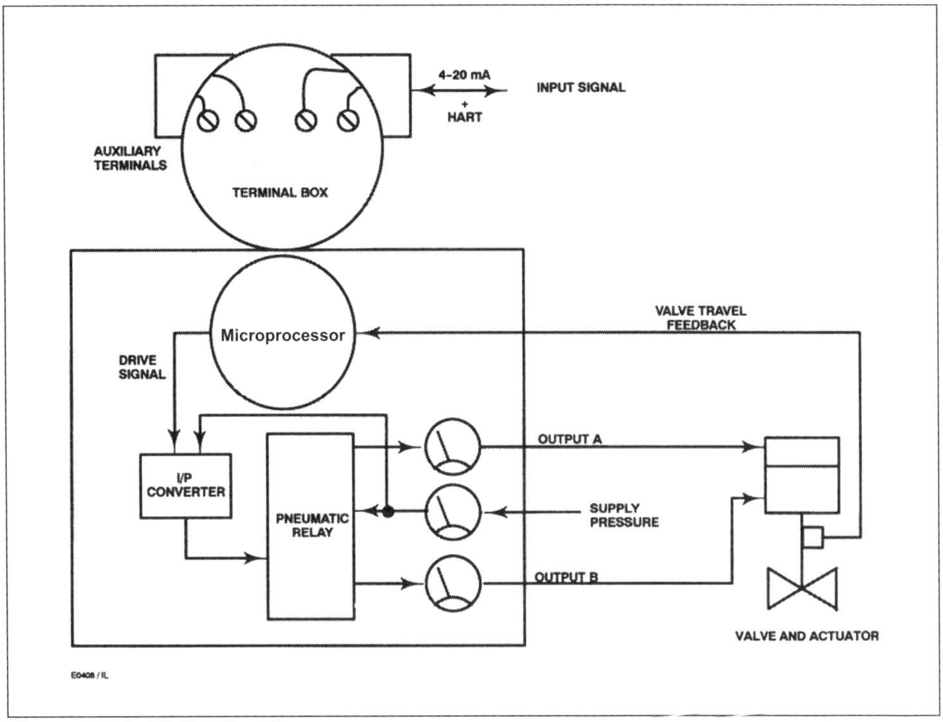

Figure 7-23. Digital Positioner

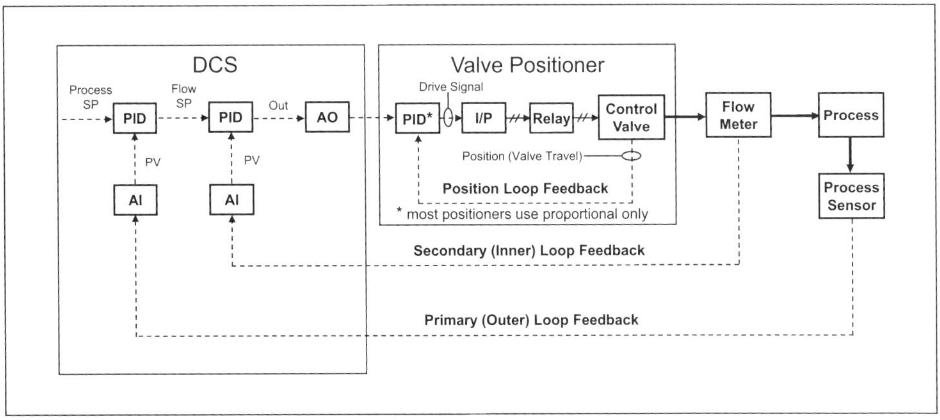

Figure 7-24. Triple Cascade Control

(master) loop is temperature. The flow controller is a secondary (slave) loop that serves the primary loop and the valve position controller is a second secondary loop that serves the flow loop. Each loop has set point and output limits that must be consistent with all loops.

Digital positioners offer the opportunity for more sophisticated control algorithms. Instead of proportional-only control, some positioners provide capabilities similar to the integral and derivative modes of the PID controller in the DCS. The derivative mode is useful for preventing overshoot and providing a faster response. Integral action can eliminate an offset between the actual and desired position. However, integral action promotes overshoot and creates a limit cycle given the inevitable resolution limits as the digital positioner searches for a match between actual position and desired position.

Proportional and derivative actions are content with an offset (bias between the desired and actual valve position). Integral action is never satisfied with anything but an exact match. An exact match between set point and measurement is rare and momentary in any loop. The use of integral action necessitates a decrease in proportional action which decreases the immediate response to load or set point change. For cascade control systems the use of integral action in secondary loops will create a limit cycle from deadband. Many digital positioners do not use integral action or offer the option to automatically turn off integral action when the offset is within an adjustable deadband.

Integral action is useful in secondary loops if there are large load changes, a noisy signal, or a need to provide ratio control. These conditions exist for flow loops, which explains the predominant use of integral action in flow controllers. When flow loops are secondary loops, a limit cycle develops for deadband. The limit cycle can be stopped (broken) by turning off the integral action when the measurement is within an acceptable band (zone) around the set point. This band is called an "Integral Deadband" or "Integral Deadzone."

The digital valve controller shown in Figure 7-25 offers improved valve position control compared to analog positioners by the using the following controller tuning settings.

- Travel gain
- Travel rate
- Pressure rate
- Minor loop feedback (MLFB) gain
- Travel integral gain

The travel gain provides the proportional control action that was the only control mode used in pneumatic positioners. Here the gain can be precisely adjusted over a wide range. The travel rate and pressure rate provide a correction to the digital valve controller output based on the rate of change of travel and pressure, respectively, similar to derivative (rate) action. The minor loop feedback (MLFB) for both travel and pressure control is the motion of the pneumatic relay. The integral gain setting has an associated integral deadzone setting to break limit cycles.

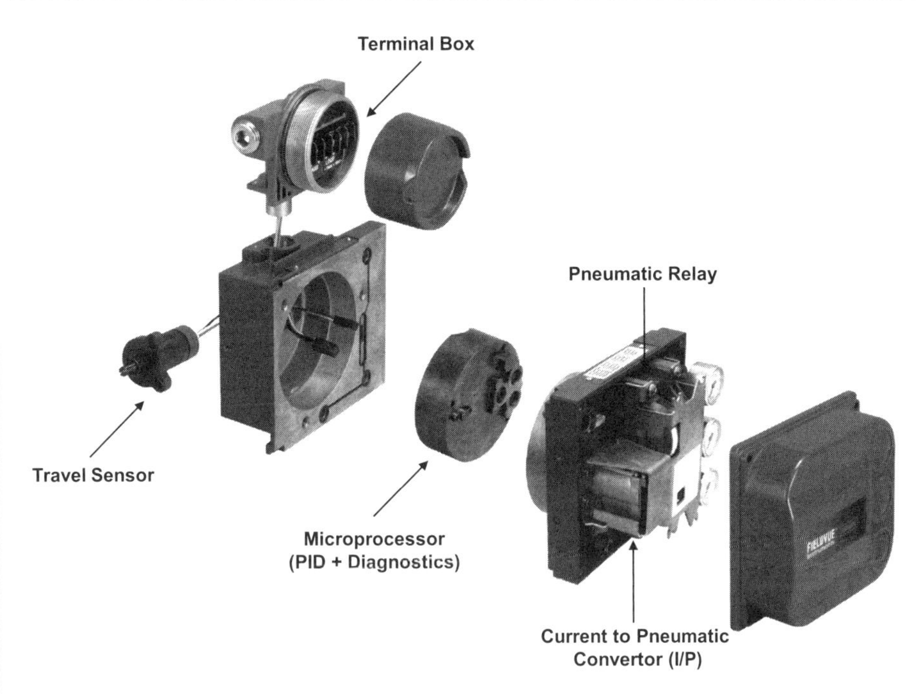

Figure 7-25. Digital Valve Controller

The digital valve controller offers a "fallback to pressure control" option for spring and diaphragm actuators. When there is a loss of the feedback measurement of valve travel, the positioner reverts to actuator pressure control. This is not suitable for piston actuators because there is no proportional relationship between actuator pressure and implied position. While similar to the classic control of a spring and diaphragm actuator via an I/P transducer where a positioner is not used, the performance is better from the use of pressure rate and MLFB gain action. In one case where a scaffold fell on the instrument and peeled off the connection between the valve stem and the instrument, the digital valve controller automatically reverted to pressure control and continued to control the process.

The digital valve controller comes with a recommended set of tuning settings for different types of actuators and valves. These "tuning sets" generally have an increasing travel gain and rate setting as the size of the valve increases. The user can stay with the recommended tuning set or go to the expert mode and enter settings for more aggressive control. For example, the gain and pressure rate settings might be increased for high friction valves to reduce stick-slip errors.

There is a wide spectrum of alerts and diagnostics in smart digital positioners that are available in the control room via HART, WirelessHART, or Foundation Fieldbus signals. Basic alerts for microprocessor problems and sensor failures may include [18]:

1. Flash ROM Fail
2. No Free Time
3. Reference Voltage Fail
4. Drive Current Fail
5. Critical Non Volatile Memory (NVM) Fail
6. Temperature Sensor Fail
7. Pressure Sensor Fail
8. Travel Sensor Fail

There are other alerts for supply pressure, calibration, and configuration. Basic alerts for travel may include [18]:

1. Travel High
2. Travel High-High
3. Travel Low
4. Travel Low-Low
5. Travel Deviation
6. Cycle Count
7. Travel Accumulator
8. SIS Valve Stuck

There are deadband settings for travel alerts (1) through (4) for determining when the alert clears. For example, if the travel high alert and deadband are 95% and 5%, respectively, the high alert would be set if the travel increased to 95% and would clear if the travel decreased to 90%. There is also a travel deviation time setting, where the deviation between actual and desired position must be equal to or larger than the travel deviation setting (e.g., 5%) for at least as long as the deviation time setting (e.g., 5 sec).

There are deadband settings for the cycle counter and travel accumulator. If the travel exceeds the deadband around the deadband reference point (Figure 7-26a), the cycle count is incremented and the deadband reference point is moved to the point of the first signal reversal. The travel accumulator (Figure 7-25b) is incremented by the change in travel that occurs between the existing deadband reference point and the new deadband reference point. An increment only occurs when the travel exceeds the deadband around the reference point.

The cycle count tells how much dithering (small rapid repetitive oscillations) there is from measurement noise or overly aggressive DCS controller gain and rate settings. The cycle count can also be used for predictive maintenance (replacement) of stem packing. The travel accumulator can be used with the cycle count to diagnose the amplitude of the cycling [19].

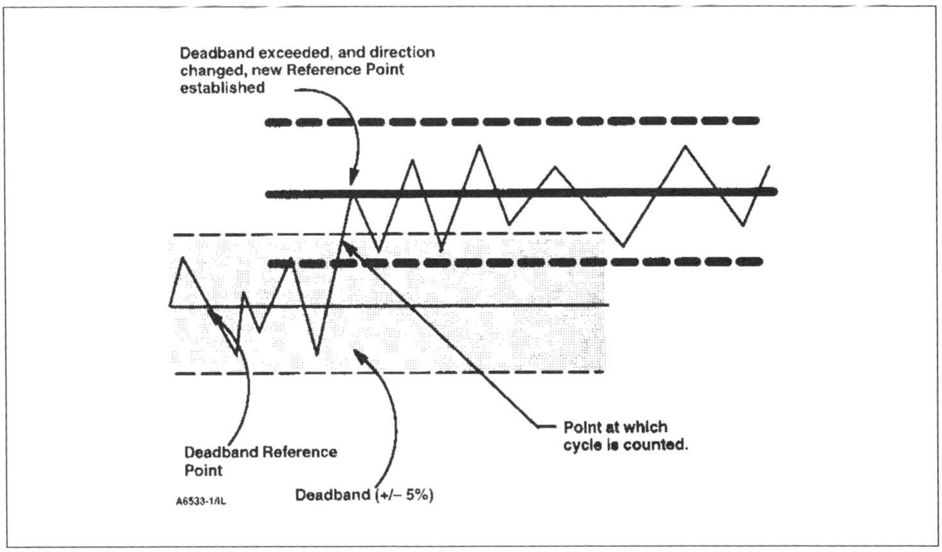

Figure 7-26a. Cycle Counter [19]

The "SIS stuck" alert function is used to detect a failure of the valve to move to the trip position commanded by a safety instrumented system (SIS).

Advanced diagnostics can be used to observe the dynamic response if the valve can be taken out of service to allow stroking (full range of motion) of

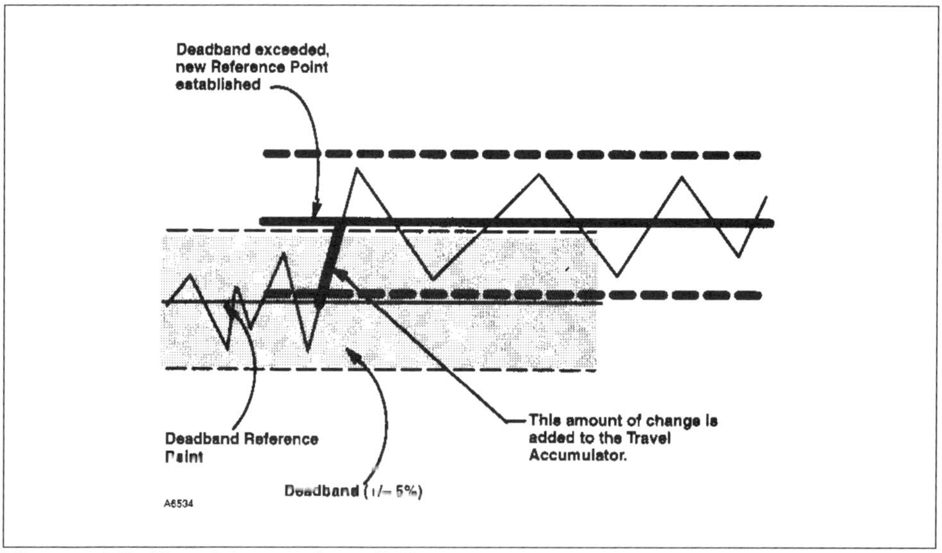

Figure 7-26b. Travel Accumulator [19]

the valve. Valve signatures (valve travel for a full scale reversal of valve signal) provide an indication of resolution and deadband. Step responses provide an indication of the valve deadtime and response time. A higher level of performance diagnostics can estimate valve friction and the performance profile from normal movement so the valve does not need to be taken out of service. Performance diagnostics can also do the following tests without interfering with valve response [19].

- I/P & Relay Integrity Test
- Air Mass Flow Test
- Relay Adjustment Test
- Supply Pressure Test
- Travel Deviation Test

Digital valve positioners should be used on all loops regardless of loop speed, because of the diagnostics and the compensation for non-ideal effects [20, 21, 22, 23, and 24]. The positioners can be tuned for faster action and the DCS controller can be tuned for slower action to adhere to the cascade rule where the secondary loop (digital valve controller) should be five times faster than the primary loop (DCS flow controller). The digital valve controller execution is faster than the DCS execution and the use of predominantly reset action in the DCS flow loop makes the cascade rule a non-issue.

In one event in the author's experience, the use of a booster instead of a positioner as advocated by the cascade rule was disastrous for surge control. The substitution of a booster for a pneumatic positioner caused the compressor surge valve to slam shut during startup. Fortunately, the safety instrumented system (SIS) shut down the compressor before it was damaged from surge. The problem was solved by putting a volume booster on the output of the positioner and adjusting the booster bypass for stability (see Accessories section) [20][24]. Also, since bench settings are often not accurately set or known for the diaphragm actuator where a positioner supposedly could be omitted, the operator and technician do not know at what controller output the valve will lift off the seat and be wide open. Piston actuators require positioners. As a practical matter HART or FOUNDATION Fieldbus digital positioners are needed to:

- Improve resolution (reduce stick-slip)
- Deal with diaphragm bench settings
- Deal with piston crossover pressure settings
- Increase actuation stiffness (prevent flutter)
- Increase or slow down speed of valve response
- Provide valve position monitoring and historization
- Provide valve alerts and diagnostic monitoring and historization

Valve Accessories

Digital positioners and air supply filter regulators are an essential part of every control valve with a pneumatic actuator. The supply regulator is set at a lower pressure (e.g., 20–40 psig; 0.96–1.91kPa) for diaphragm actuators and a higher pressure (e.g., 80–120 psig; 3.8–5.8kPa) for piston actuators. The air must be clean and dry. The air supply filter should only have to remove infrequent small particles. The air supply filter regulator is normally integrally mounted on the digital positioner.

A current to pneumatic (I/P) transducer on the controller output was extensively used before the digital control system and digital positioner became common. As we have seen, the I/P output was the input to a pneumatic positioner or the input to a diaphragm actuator. A digital positioner directly accepts a current or fieldbus input so the only reason today to buy an I/P is if you didn't want a positioner. For the reasons discussed in the sections on Positioners, a smart digital positioner is recommended for all loops. In the author's opinion, the separate I/P transducer is no longer needed for modern final elements.

The remaining accessories discussed here are used to achieve a desired throttle position or a fixed position. For the process industry, the throttle and fixed positions are generally set by an analog output and discrete output, respectively from a DCS, a programmable logic controller (PLC), or a special SIS. Prior to the advent of the DCS and PLC, electromechanical relays were predominantly used to provide the discrete contacts needed to set the fixed position.

The figures and discussion in this book for discrete positioning are meant as examples only. The actual setup for an application must be carefully designed and tested for the required functionality and the type of actuator, positioner, and valve accessories used.

The controller output that provides the signal to the positioner should be set to match the fixed position set by a discrete output for coordination of the loop and positioner with the discrete action.

The most common accessory for discrete positioning is a three-way solenoid valve. A solenoid valve is used to put a control valve or damper in a fixed position for safety or to achieve a sequence of flows for startup, shutdown, batch operation, or product transition. A solenoid valve is normally piped or tubed on each output of the positioner as shown in Figures 7-27a through 7-27c. For fail safe action, an open contact from a DCS, PLC, or SIS will de-energize the solenoid valve. The solenoid valve selection and installation are designed to ensure the valve goes to the safe position for electrical failure, wiring failure (open circuit), or air failure.

For the spring return pneumatic actuator in Figures 7-27a and 7-27b, a de-energization of the solenoid valve will exhaust the air to trip the valve and achieve the desired electrical and air failure position. A digital controller can check the performance of the solenoid valve by monitoring the pressure between the solenoid valve and the actuator. For a "pressurize to extend" diaphragm actuator and single-acting pneumatic direct relay, a connection from the solenoid valve output to Port B is used.

For the double-acting piston actuator in Figure 7-27c, the de-energization of one solenoid valve exhausts and the de-energization of another solenoid valve supplies air to the side of the piston needed to trip and achieve the desired failure position. Since a double-acting piston has no spring to

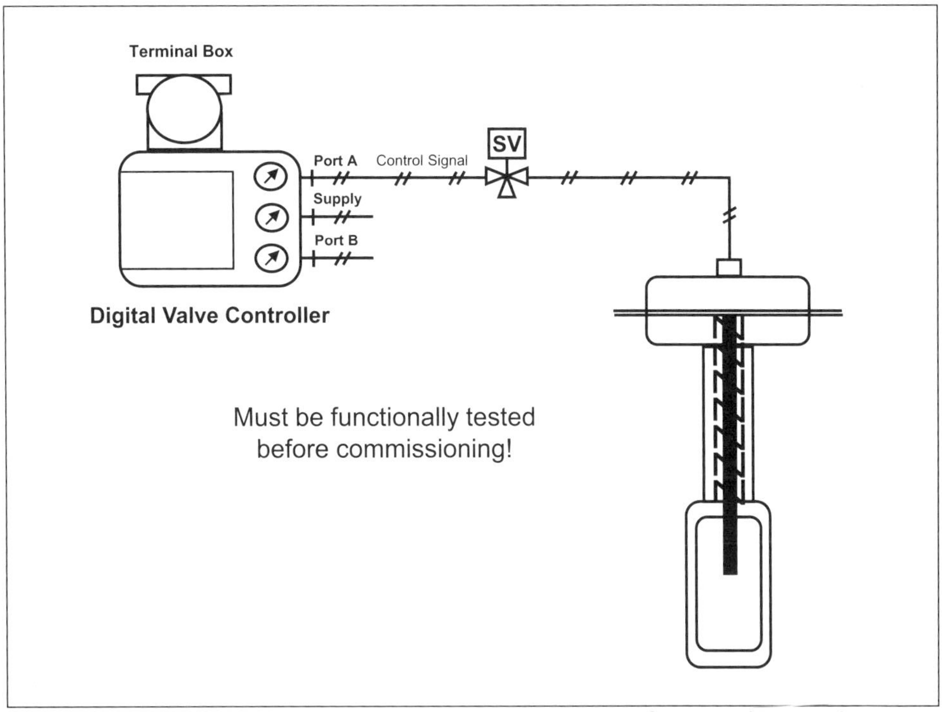

Figure 7-27a. Solenoid Valve Setup for Trip of Diaphragm Actuator (Pressurize to Extend Diaphragm Actuator)

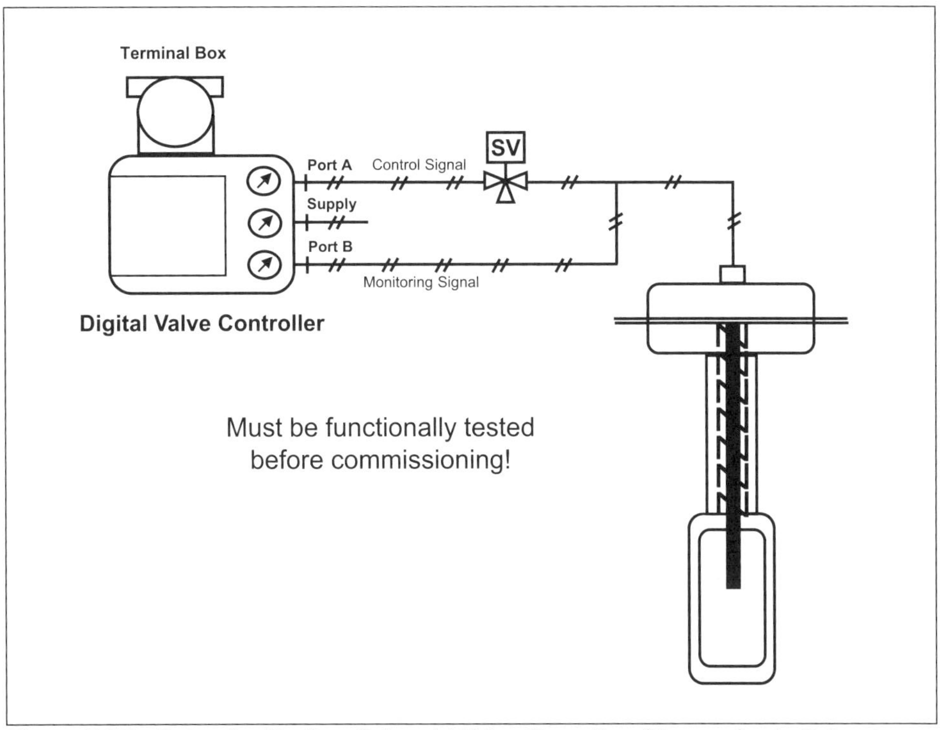

Figure 7-27b. Setup for Testing Solenoid Valve Operation (Pressurize to Extend Diaphragm Actuator and Single Acting Direct Pneumatic Relay)

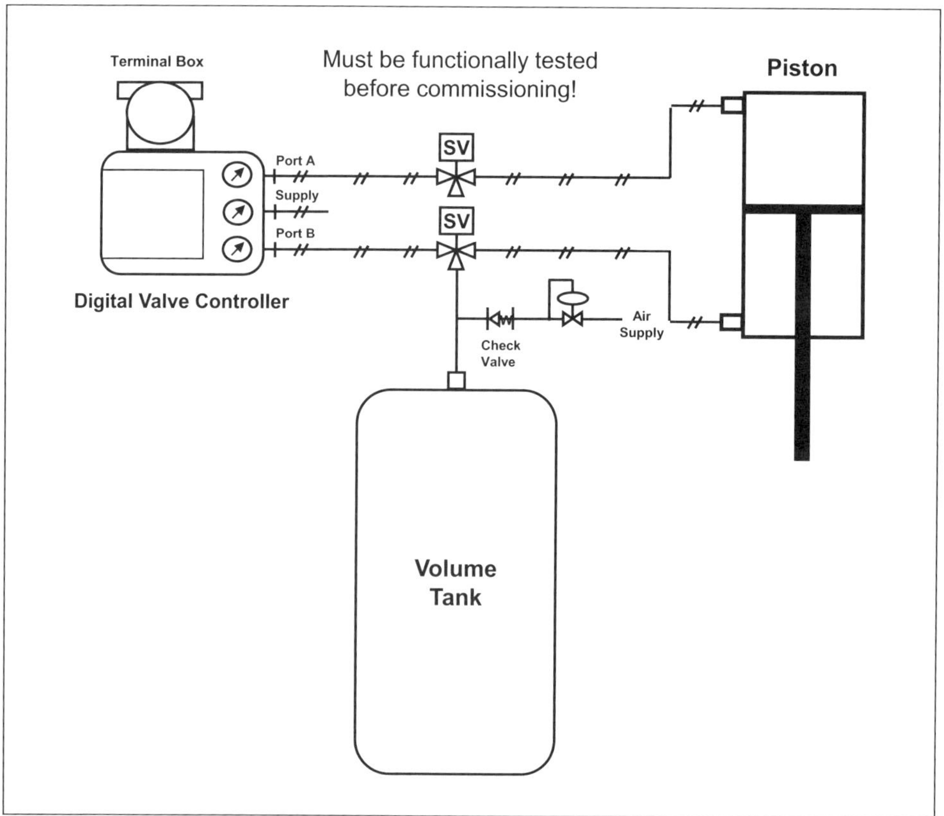

Figure 7-27c. Solenoid Valve Setup for Trip of Double-Acting Piston Actuator

ensure the valve to the safe position on air failure, a volume tank (air reservoir) is used as a backup air supply for the solenoid used to pressurize the side of the piston to achieve the trip. In the process industry, the trip and fail position are almost always the same. For on-off valves, solenoid valves are used without positioners to stroke the valve full scale upon energization and de-energization.

The next most common accessory is the limit switch to provide remote indication of a fixed position (e.g., fully open or fully closed). Proximity limit switches that magnetically sense shaft or stem position eliminate the questionable reliability of the lever mechanism used in electromechanical limit switches. In the author's experience, most indications of an improper valve position were the result of a failure of the limit switch rather than a failure of the valve. Digital positioners can be used to provide a more reliable indication of valve position. In this case, limit switches are only needed when a SIS demands a separate indication. Since an isolation valve (on-off valve) should be used instead of a control valve for a SIS, limit switches are not commonly used on new control valves.

Lock-up or trip valves are pneumatic switching devices used to lock a valve in a desired position (last, open, or closed). The common setup is to have the sensing port connected to the air supply. For air failure to last position, a two-way lock-up valve (Figure 7-28a) locks in the last air pressure from a positioner to both sides of a double-acting piston actuator when the air supply pressure drops below a minimum. The valve position may subsequently change due to changes in fluid forces. For air failure to an open or closed position, a three-way trip valve switches in full pressure from a volume tank to the proper side of the actuator when the air supply pressure drops below a minimum. Figure 7-28b is an example of a trip valve setup for failure to the closed position of an extend to close double-acting piston actuator and valve. Pneumatic lock-up and trip valves can have a 20% deadband so the set and clear switch points are not accurate. The lock-up and trip valves must be piped and maintained properly. A spring in the actuator is a much simpler and more reliable means of providing an air failure position.

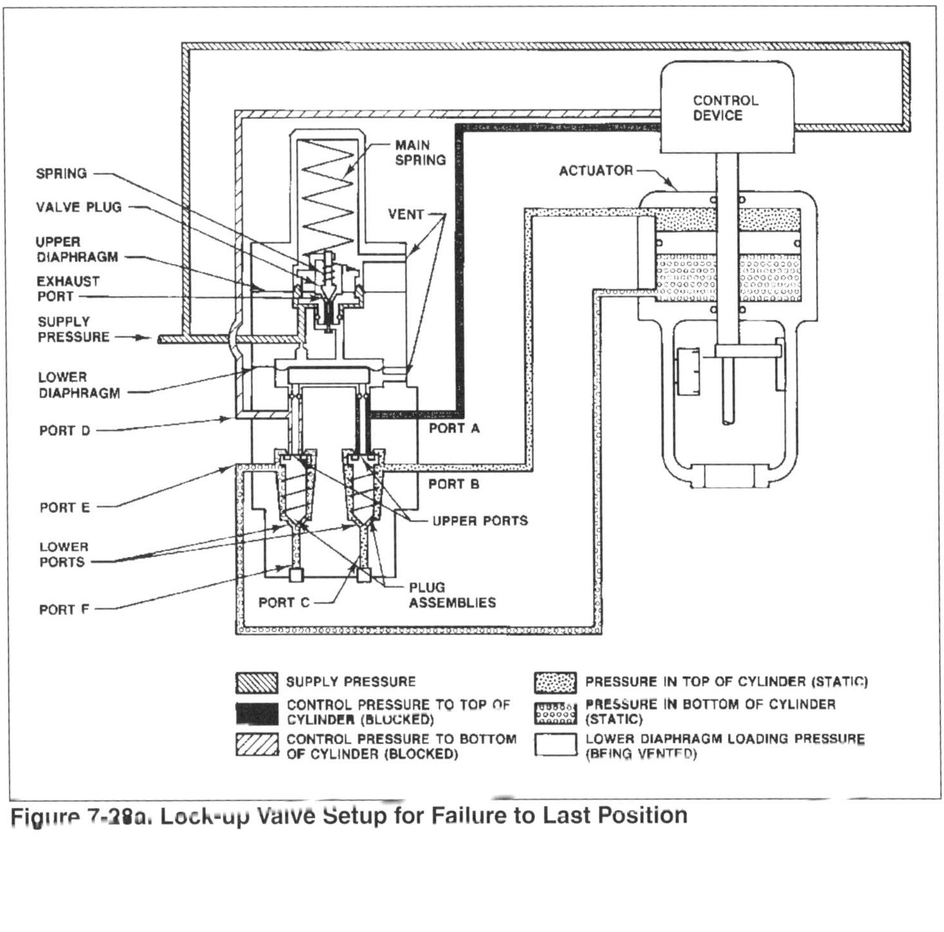

Figure 7-28a. Lock-up Valve Setup for Failure to Last Position

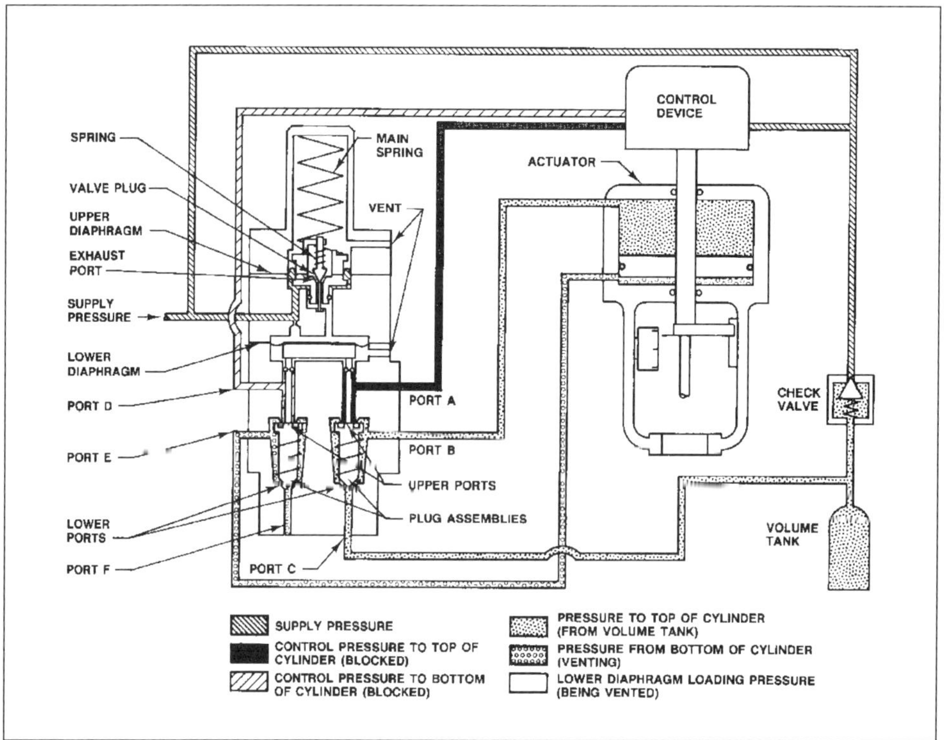

Figure 7-28b. Trip Valve Setup for Failure to Closed Position (Extend to Close)

Boosters (Figure 7-29) are used to increase the supply flow to a pneumatic actuator to make it faster. The booster has a much higher flow capacity than a I/P or positioner to achieve and maintain the pressure in the actuator. The booster is a high gain proportional-only controller that matches a change in outlet pressure to a change in input pressure that exceeds a deadband. The deadband is deliberately introduced to prevent the cycling of piston actuators. Thus, the booster will ignore small changes in input signal (coarse input port sensitivity). The booster through its high gain action tries to correct any changes in booster outlet pressure (high outlet port sensitivity). The booster offers no stiffness beyond what is provided by the actuator. In several instances, high flow rate boosters on diaphragm actuators have exhibited positive feedback (negative stiffness). This was observed in both the shop and field where a person grabbing the shaft of an 18-inch (46 cm) butterfly valve could manually move the disk when a booster was put on the output of an I/P and piped directly to an actuator. When the booster was removed (I/P output went to actuator) or was replaced with a positioner, the valve stem could not be moved. The high outlet port sensitivity of the booster caused the booster to relieve or supply pressure from an ever-so-slight compression or expansion of the diaphragm volume to provide a force in the same direction of the force applied to the stem [20][23, 24].

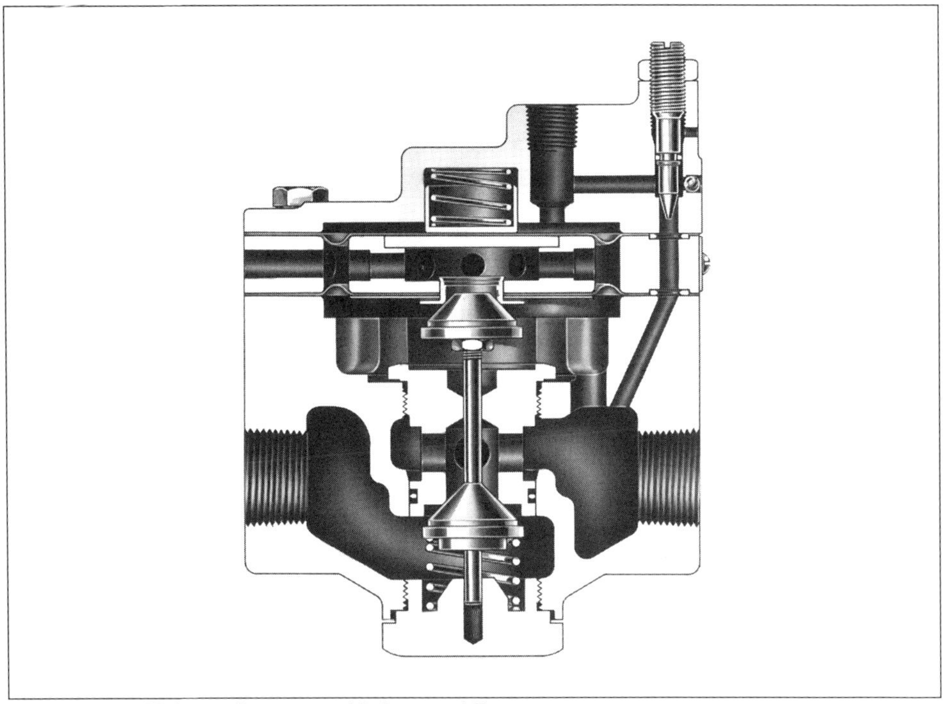

Figure 7-29. Volume Booster with Integral Bypass

Boosters have poor inlet (signal) port sensitivity. A booster will not respond to small input signal changes. For slowly changing signals, a deadtime is added that is proportional to the booster deadband or resolution limit divided by the rate of change of the input signal. Since most controller output signals make small changes from scan to scan, we have the ironic situation where a booster that is added to speed up the response of a valve actually prevents it from moving. The large step changes (e.g., > 5%) that are normally used to test valve response do not reveal the problem. The lack of understanding the implications of high outlet port sensitivity, poor inlet port sensitivity, bench settings, and fluid forces has led to the misguided rule that boosters instead of positioners should be used on fast loops.

A booster should not be used in lieu of a positioner but in conjunction with a positioner to maintain actuation stiffness, sensitivity, and a consistent input signal range for stroking the valve. A booster should be mounted on each output port of the positioner. For double-acting pistons, two boosters are required. The booster must have a bypass as noted in Figure 7-30 to prevent instability. Since the positioner is designed to be looking into the relatively large volume of an actuator, the extremely small volume of the booster inlet port will cause a rapid limit cycle. The bypass must be opened until the high frequency oscillations (e.g., 1–2 cps) stop [20] [23, 24, 25]. The bypass may be opened slightly further to provide a stability margin. Since the booster bypass slows down the speed of the actuator, it is desirable to open the bypass just enough to provide stability. If the tuning of the digital positioner is changed, the booster bypass may need to be adjusted accordingly. More aggressive tuning settings will require a larger bypass flow.

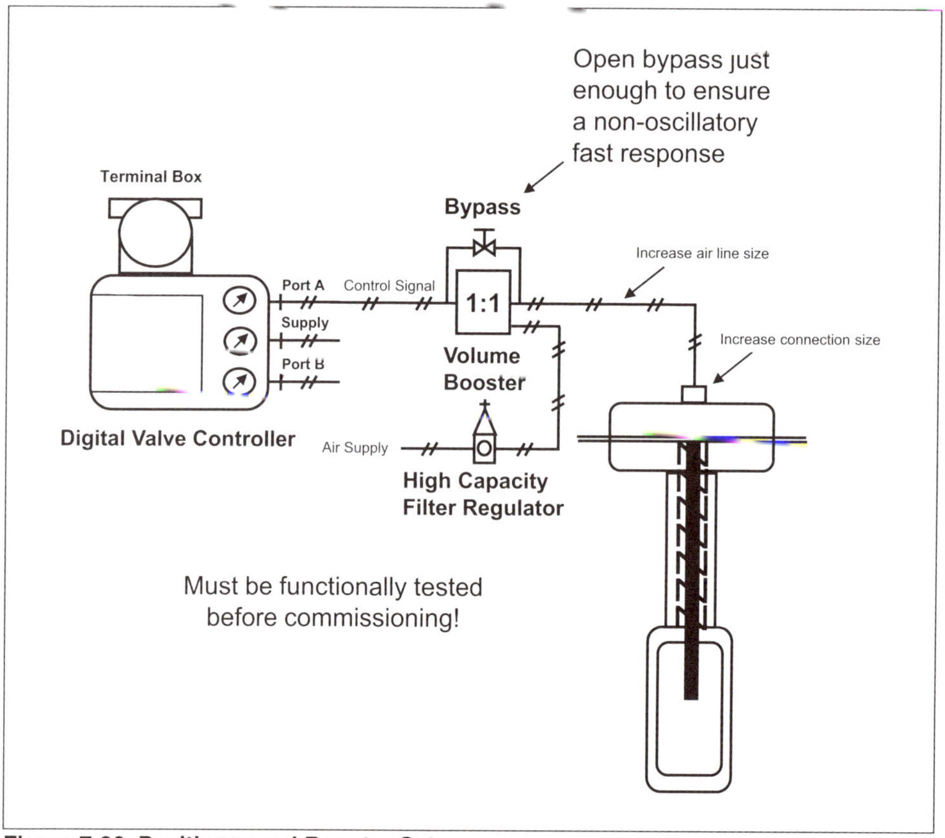

Figure 7-30. Positioner and Booster Setup

The combination of a positioner and booster can dramatically reduce the pre-stroke deadtime and stroking time of large actuators. The air pipe or tubing size must be increased and in some cases the diaphragm actuator casing connection enlarged. The flow capacity of accessories such as solenoid valves and air filter regulators must be accordingly increased. Otherwise, the true flow capability of the booster will be restricted.

Quick exhaust valves are used to provide a rapid stroke by rapidly dumping flow from the actuator. Quick exhaust valves are used to get a surge valve to stroke rapidly into the air fail open position. This discrete unidirectional action is generally not good for throttling control. Boosters are preferred for more continuous action, flexibility, and adjustability. If the rate of change of the valve position needs to be restricted in one direction (e.g., the closing of a surge valve), an adjustable directional velocity limit may be set on the valve signal in the DCS or in the digital valve controller.

Valve Best Practices

For best performance, users should consider the following during the specification of control valves:

- Actuator, valve, and positioner package from a control valve manufacturer
- Digital positioner tuned for valve package and application
- Diaphragm actuators where application permits (large valves and high pressure drops may require piston actuators)
- Sliding stem (globe) valves where size and fluid permit (large flows and slurries may require rotary valves)
- Low stem packing friction
- Low sealing and seating friction of the closure components
- Booster(s) on positioner output(s) for large valves on fast loops (e.g., compressor anti-surge control)
- Valve sizing for a throttle range that provides good linearity [4]:
    - 5% to 75% (sliding stem globe),
    - 10° to 60° (v-ball)
    - 25° to 45° (conventional butterfly)
    - 5° to 65° (contoured and toothed butterfly)
- Online diagnostics and step response tests for small changes in signal
- Dynamic reset limiting using digital positioner feedback [2]

7-4. Variable Speed Drives

Control valves change flow by changing the system resistance (restriction area) to flow. Variable speed drives on axial and centrifugal prime movers (blowers, compressors, fans, and pumps) change the discharge head (pressure) of the prime mover. The speed of peristaltic and positive displacement pumps is also manipulated but the speed control is simple. Positive displacement pumps create disruptive pressure pulses, can over pressurize downstream piping and equipment if the discharge flow is blocked, and generally require more maintenance. Consequently, positive displacement pumps are relegated to metering relatively low additive and reagent charges to vessels.

This section focuses on variable speed drives for centrifugal pumps. A liquid flow control application is used to illustrate the potential benefits and problems.

Modern variable speed drives (VSD) for pumps are variable frequency drives (VFD). The motors are AC induction motors. A DC motor has a faster response but the initial cost is much higher, as are maintenance costs due to brushes that need to be periodically replaced. DC motors are used for servomechanism control where smaller size and faster speed of response are needed.

Static Head

In non-positive-displacement pumps, flow rate is proportional to pump speed except at low speeds. The change in flow rate with speed increases at low speeds where the pump head curve is flatter, or as the pump head approaches the static head (see section 7-6 on Linearity). If the pump head becomes less than the static head, reverse flow through the pump can occur. In one disastrous case that soured a customer on the use of variable speed drives, the pump head from a VSD pump feeding a reactor dropped below the reactor pressure. Catalyst and reactants blew back through the pump into the feed tank, damaging the pump and creating a potentially hazardous situation from exothermic chemical reactions in the storage tank. This case illustrates that the pump head should not be able to drop below the highest reactor pressure excursion. On-off valves should shut to provide positive isolation and perhaps check valves should be used to prevent reverse flow.

Static head is the downstream pressure that doesn't vary with flow (e.g., liquid head and vessel pressure). The term can be misleading in that static head is not constant. Pressure excursions in vessels from batch operation or process disturbances can be fast and frequent. For example, if a pump was providing a flow that entered the top of a pressurized vessel, the static head would be the result of gas pressure in the top of the vessel plus the liquid head associated with the elevation change between the pump and vessel entry point (Figure 7-31a). The worst-case vessel pressure must be considered. The discharge head of the prime mover is proportional to the speed squared. The pump head must equal the sum of the static head and the system pressure drop. Since there is no control valve, the pump flow is at the intersection of the pump and system curves. Figure 7-31b shows how the addition of a static head shifts the intersection (operating point) to the left reducing the flow delivered.

The pump curve shifts down with speed. When the speed causes the intersection to approach the static head, the flow is no longer controllable. Hence, the magnitude of the static head determines how low in speed a VSD can go (see the subsection on turndown and the end of Section 7-7 on Rangeability).

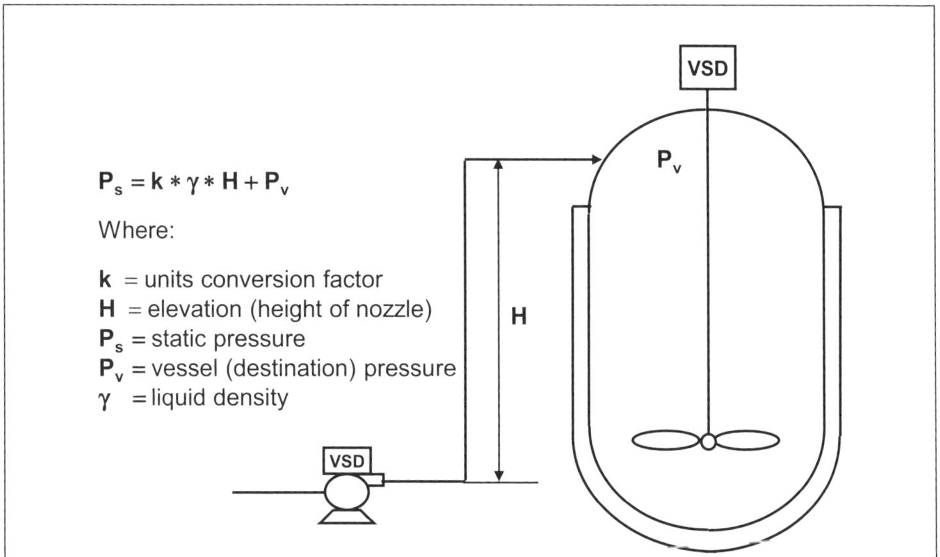

Figure 7-31a. Static Head from Elevation Change and Destination Pressure

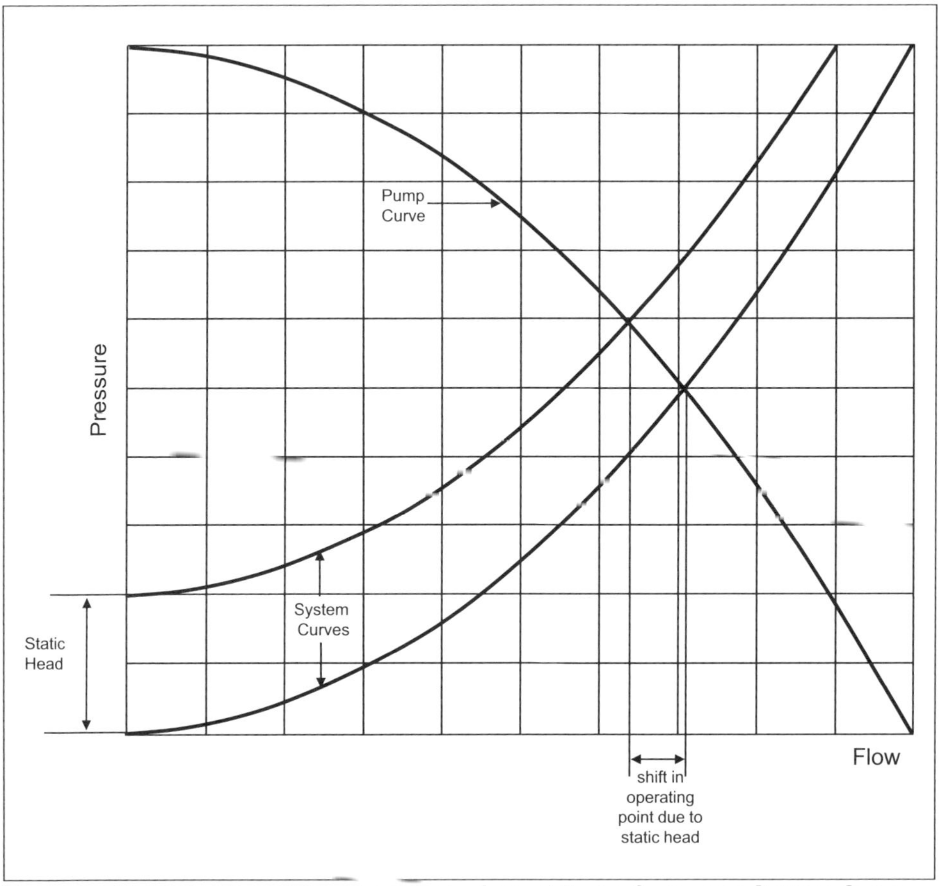

Figure 7-31b. Static Head Effect on Intersection of Pump Curve and System Curve

VFD Pulse Width Modulation

Pulse width modulation (PWM) is predominantly used today to vary the frequency of the voltage or vary the current to the pump motor. PWM introduces lower noise, has a higher input power factor, and has better low speed performance than older drive technologies. PWM also offers better rangeability from less cogging (torque pulsation) at low speeds [26]. Figure 7-32a shows the functions in a VFD inverter to convert an AC line voltage to a variable frequency voltage for speed control of an AC induction motor. Industrial motors use three phases, staggered to provide a smoother output. The AC line voltage is rectified and filtered to create a DC voltage. The DC voltage is then inverted to a variable frequency voltage whose frequency is proportional to the drive input signal by PWM. The square wave output is then filtered to create a sine wave. The rounding from the peak is more accurate for finer pulse widths. Insulated gate bipolar transistors (IGBT) are used to create a series of pulses of varying width. The dominant carrier frequency from PWM is proportional to the drive input signal. The resulting waveform from PWM is closer to a sine wave [26]. These drives come in a wide spectrum of sizes as shown in Figure 7-32b.

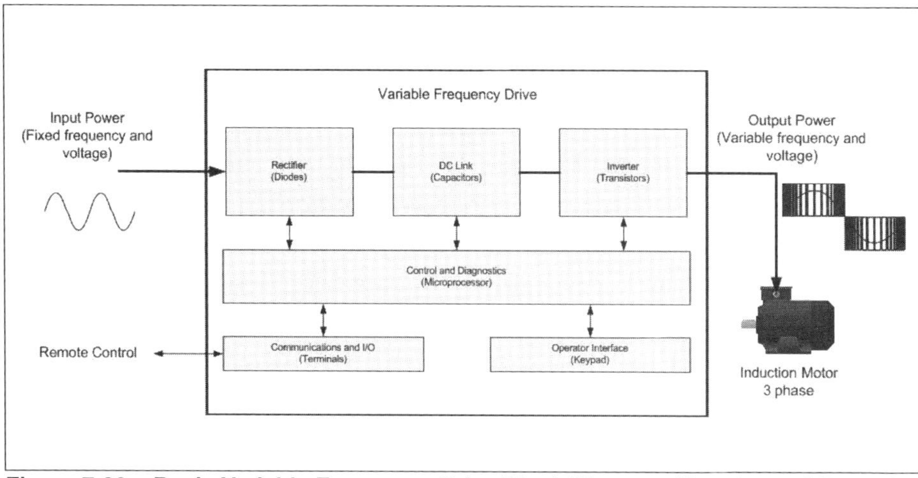

Figure 7-32a. Basic Variable Frequency Drive Block Diagram *(Courtesy of Control Techniques)*

Figure 7-32b. Spectrum of Variable Frequency Drives *(Courtesy of Control Techniques)*

PWM inverters do not have the performance problems, the harmonics, and sharp spikes seen in older drives that can damage bearings and create electrical noise in instrument signals. For example, the older six-step voltage drive technology, while inexpensive, had a number of undesirable characteristics. The motor can be pushed to its breakdown point. Shorts can cause an infinite current spike. The inverter puts out the same voltage and current at half load as at full load, which reduces efficiency. The waveform has wide and fast current variations that can damage the inverter. The inverter has numerous harmonics that increase motor losses, heat generation, and electromagnetic interference.

These invertors used an output choke to prevent damage to motor insulation; however, the input choke was optional and was often missing or insufficient. Eventually, the noise in instrument signals became bad enough that chokes were offered to meet the International Electrotechnical Commission (IEC) standards. Alternatively, isolation transformers were located close to the inverters with the power wiring between the inverter and transformer in hard pipe conduit to minimize the noise from this section of wiring [27]. While the PWM drive has lower harmonics and spikes, the rapid rise time of the pulses for precise speed control is still a source of noise and potential damage to bearings and cables.

VFD Cable Problems

Belden Inc. has studied the radiated noise from cables between the VFD and the motor. Unshielded VFD cables can radiate 80V noise to unshielded communication cables and 10V noise to shielded instrument cables. The radiated noise from foil tape shielded VFD cables is also excessive. A foil braided shield and armored cable performs much better [28]. Still a spacing of at least one foot is recommended between shielded VFD and shielded instrumentation cables. The cables should never cross. As a best practice, separate trays to isolate VFD and instrumentation cables should be used to avoid mistakes during plant expansions and instrumentation system upgrades.

All VFD systems have reflected waves from an impedance mismatch between the VFD and the motor. The amplitude of the waves depends on the voltage magnitude and rise time from the PWM drive, the distance between the VFD and motor, and the impedance mismatch. If a reflected wave gets in-phase with the radiated wave, the voltage can double and the PVC jacketed VFD cables can be damaged. XPLE jacketed VFD cables that are capable of withstanding a high voltage impulse are recommended [28]. For the "Ten Things to Consider before Selecting Your VFD" from Belden Inc., see Appendix B.

VFD Bearing Problems

The electric discharge machining (EDM) effect from inverter drives can cause arcing across the lubrication gap of a motor bearing "almost like a series of little lightning strikes" [29]. The arcing damages the bearing surface. The damage is seen as pit marks in variable speed operation and fluting in constant speed operation. The localized high temperatures from the strikes cause reactions in lubricant additives and burning or charring of the lubricant [29]. Pitting, fluting, and poor lubrication cause an increase in noise and vibration. Eventually the bearing fails.

Mechanical solutions insulate the bearing or provide a path to ground. Ceramic coatings and ceramic balls or rollers may be used for insulation. However, the insulation forces the current to go elsewhere and possibly cause damage, such as the erosion of pump impellers in prime movers.

VFD Slip

In an AC induction motor, the rotor and hence shaft speed lags behind the speed of the rotating electrical field of the stator because a difference in speed is needed to provide the rotor current and consequently the torque to balance any motor losses and the load torque from prime mover operation. This difference in speed between the stator field and the rotor of the motor is called slip. In pumps, there is dynamic slip for large changes in the pump load (e.g., static head) or desired flow rate (speed signal). There is also steady-state slip for operation at a particular load and speed. The relationship of slip to load and motor torque can be seen in the motor torque curve. National Electric Manufacturers Association (NEMA) design B motors are commonly used for pumps. The torque curve for design B as shown in Figure 7-33 has a dip on the left side of the peak but a nearly linear decrease in torque with speed on the right side of the peak. NEMA design A has a higher peak and a much sharper fall-off in torque with speed. Since variable speed motors are designed to operate on the right side of the peak, the more gradual slope of design B translates to smaller process variable errors and oscillations from disturbances and speed regulation limitations. The torque curve is shifted to the left by the inverter controls to keep the operating point on the right side of the peak as shown in Figure 7-33.

It is important to note that speed slip is not the same as valve stroke slip. In speed slip, the speed still responds smoothly to a change in drive signal, but the proportional change in speed is less. In stroke stick-slip, there is no

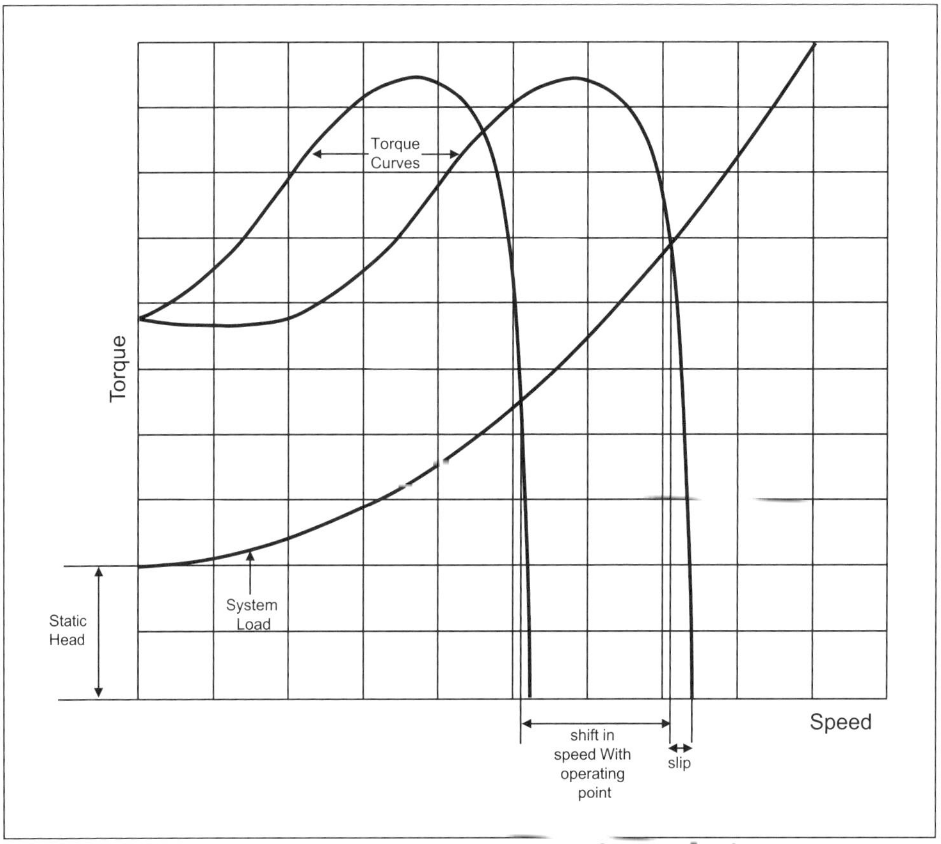

Figure 7-33. Shifting of Torque Curves by Torque and Current Control

response in the valve stroke until the change in signal is greater than the stick, at which time the valve stroke jumps to a new position corresponding to the stroke slip. Speed slip causes a change in process gain. Stroke slip introduces time delay and discontinuous action that results in a limit cycle (see section 7-5 on Dynamics).

In a synchronous motor, the rotor is designed to inherently eliminate slip so the rotor speed is at the synchronous speed of the stator. Synchronous motors are significantly more expensive and complex and are used only where inherently fast and precise speed regulation is needed. Synchronous motors have been used for ratio control of reactants or additives where small transients or offsets in the speed could cause a significant variation in the product.

VFD Turndown

Since the inverter waveform is not purely sinusoidal, it is important to select motors that are designed for inverter use. These "inverter duty" motors have windings with a higher temperature rating (Class F) [30]. Another option that facilitates operation at lower speeds to achieve the maximum rangeability offered by a PWM drive is a higher service factor (e.g., 1.15).

To help prevent motor overheating at low speeds, larger frame sizes and line powered ventilation fans are used. In the process industry, totally enclosed fan cooled (TEFC) motors are used to provide protection from chemicals and receive ventilation from a fan that runs off the same power line as the motor. The fan speed decreases as the motor speed decreases. To reduce the problem of motor overheating at low speeds, an AC line power constant speed ventilation fan and a larger frame size to provide more ventilation space can be specified. Alternately, a separate booster fan can be supplied. For very large motors (e.g., 1000 HP), totally enclosed water cooled (TEWC) designs are used to deal with the extra heat generation. For low static head pump applications, overheating at low speeds is not a problem because the torque load decreases with flow.

VFD Controls

The turndown (rangeability) of a VSD can be increased by ensuring that the pump head is large compared to the static head (see the end of Section 7-6 on Linearity and Section 7-7 on Rangeability), by using PWM inverters, and by dealing with the heating problems at low speeds. Turndown also depends upon the control strategy in the variable frequency drive. All the control strategies discussed in this section use PWM to manipulate the frequency and amplitude of voltage and current to each phase.

Open loop voltage (volts/Hertz) control has the simplest algorithm but is susceptible to varying degrees of slip. Most of the drives provided for pump control use this strategy, in which the rate of change of flux and hence speed is taken as proportional to voltage. At low speeds the motor losses are larger, making the difference between the computed and actual speed (slip) much larger. As shown in Figure 7-34, some drives make a correction to the voltage to account for estimated motor losses. Ultimately these drives depend on the DCS to correct for dynamic slip through proportional action and to correct for steady state slip through integral action in process controller(s). The turndown is normally 40:1 with 0.5% speed regulation [30].

Closed loop slip control has a speed loop cascaded to a torque loop. The torque feedback may be calculated from a current sensor. A DCS process

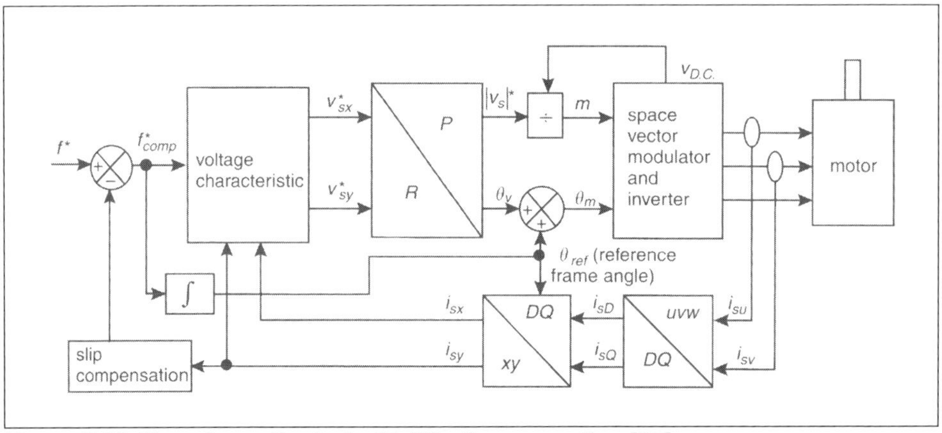

Figure 7-34. Open Loop Control with Slip Compensation [32]

controller output is the speed set point for the speed controller, whose output is the set point to a torque controller. PI rather than P-only controllers can be used since stiction and resolution limits are negligible, eliminating any concern about limit cycles from integral action.

The speed controller plays a role similar to that of the valve position controller and the torque controller serves a similar purpose as the relay controller. However, in a digital positioner the relay response is inherently much faster than the valve position response. In the VSD, the torque controller can have a relatively sluggish response. To prevent a violation of the cascade rule that requires the secondary loop (torque) to be 5x faster than the primary loop (speed), the speed loop is slowed by decreasing the speed controller gain and integral time. Since the speed set point comes from a process controller in the DCS, there is at least a triple cascade. In many cases there is a quadruple cascade control system when the DCS has a cascade control system as shown in Figure 7-24. As a result of the quadruple cascade, the detuning of the speed controller causes detuning of the flow controller, which in turn may cause detuning of the temperature controller. As a result, the ability to reject fast process disturbances may be compromised. The turndown of speed using closed loop slip control is normally 80:1 with 0.1% speed regulation [30].

Flux vector strategies offer sophisticated methods for incredibly fast multivariable motor control. Motor models can be used to provide inferential speed and torque measurements, eliminating the maintenance and response limitations of sensors. The effect of rotor temperature on rotor resistance is computed and used in the motor model. The specific strategy details are proprietary to each manufacturer. Optimum operation depends on the correct setting of numerous parameters for the application. Some of these parameters are available from the motor nameplate. Other parameters may depend on the mechanical design and process load. These strategies are used particularly for servo mechanism control systems that require a much faster and better definition of system response than process applications such as pump control systems [30]. High-performance drives that use this strategy have a rangeability of 120:1 with 0.001% speed regulation [30].

With these drives, the current or voltage state vector is controlled in an optimized orientation to maximize torque with minimal motor losses [31]. Current measurements from two of the three phases are transformed and used as feedback for closed loop control of the current state vector (Figure 7-35). Position feedback is differentiated and used as feedback for a PI torque controller whose output is the desired y-axis current (i^*_{sy}). The output of the y-axis current controller sets a y-axis voltage (v^*_{sy}) to affect torque. The x-axis current (i^*_{sx}) controller, whose set point is zero, sets an x-axis voltage (v^*_{sx}) to affect flux. A space vector modulator computes a modulation depth and angle that define the output voltage from the inverter and the angle of the conceptual voltage vector, respectively to achieve the desired flux vector as exemplified in Figure 3-35 [32].

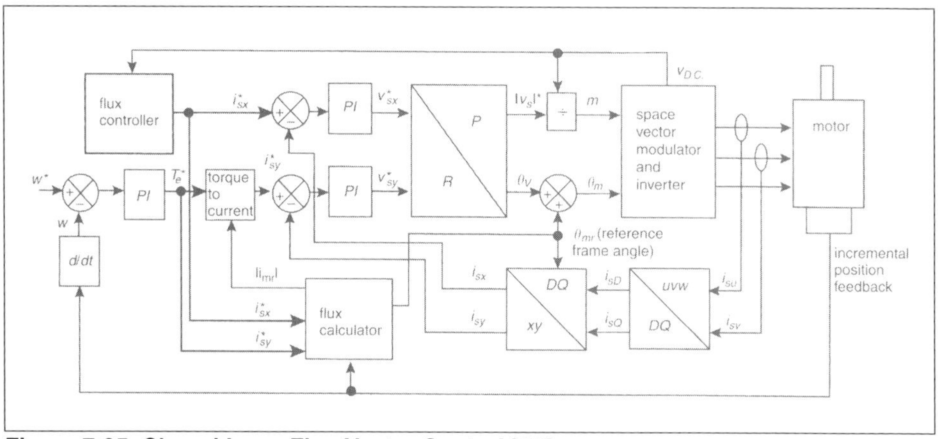

Figure 7-35. Closed Loop Flux Vector Control [32]

For liquid flow and liquid pressure control, the process response from an open loop volts/Hertz drive is so fast that slip can be corrected by the flow or pressure controller in the DCS. Inferential speed or tachometer feedback control in the DCS may require the DCS loop to be detuned because of the cascade rule and thus do more harm than good. The volts/Hertz strategy provides enough turndown for most applications. For gas pressure and flow, there may be some benefit from closed loop speed control in the drive, particularly for large volumes. For concentration, vessel pH, level, and temperature control where there is no secondary flow loop, the benefit of closed loop speed control is much more obvious because these primary loops are so slow. Flux vector control is probably not required for any of these loops but may be useful for conveyor control.

VFD Justification

As documented in the literature, the primary justification for going to variable speed pump drives is the desire to save energy. Eliminating a control valve and its pressure drop decreases the amount of head the pump must provide. Also, the pump speed can be reduced at lower flow requirements. The big incentive is seen in the relationship where the power required varies with the cube of speed [33]. Also, the cost of a motor starter can be eliminated [34]. However, the estimated energy savings often do not account for the decrease in motor efficiency with speed, the minimization of valve pressure drop by proper valve selection and sizing, the amount of time the motor is actually needed to operate at lower speeds, or the limitation imposed by static head.

For example, the motor efficiency can decrease by 40% or more at low speeds. Most continuous processes run at a relatively fixed flow rate except during grade transitions or startup. Batch processes have a fixed feed rate to reach preset total charges unless the feeds are manipulated for closed loop concentration control. Equal Percent valve characteristics allow the valve pressure drop to be decreased to 10 to 20% of system pressure drop (see section 7-6 on Linearity). Finally, a high static head will prevent the turndown of motor speed.

In the future, new microprocessor based motor controls that maintain high motor efficiency throughout the speed range may increase the potential energy savings at low speeds but decrease the energy savings at normal speeds [36].

Even in the absence of energy savings, variable speed pump drives may be the best solution for certain process applications by eliminating problems with control valves. For slurries, a VSD pump eliminates a control valve with coating, erosion, corrosion, and plugging problems. For molten metals, a VSD pump eliminates a control valve with high temperature seizing and stiction from high temperature stem packing. For highly viscous flows, a VSD eliminates a control valve with an erratic flow response from a transition between turbulent and laminar flow. For corrosive fluids that require special materials of construction, given that a pump and an isolation valve are required, the VSD eliminates an expensive control valve.

Variable speed drives are more complicated to set up and maintain than control valves. Proper implementation requires a cross-disciplinary effort drawing upon electrical, mechanical, instrumentation, and process control skills [33]. The expertise needed to set up and maintain the VSD may necessitate onsite support from a contractor or vendor.

In summary, the four main practical reasons that VSDs are not used as extensively as one might think for pump control are as follows [35].

1. **Drives are generally not built just for pumps**. They handle conveyors, extruders, etc. There are a lot of VSD menu choices and options not pertinent to pumping applications.
2. **Users don't like the complexity of the VSD**. The user must address setup, maintenance, and design issues. Special practices are needed to prevent EMI (electromagnetic interference) in instrument signals and to keep from getting harmonics back into the power supply.
3. **Someone needs to do the right calculations on dollars saved.** As mentioned, calculations typically don't take into account the drop in drive efficiency at low speeds. The duty cycle (the amount of time that speed is actually turned down) is not known (or perhaps not considered) in advance. If there is a high static head, the energy savings of a VSD disappear.
4. **It is rare to compare a VSD and a valve.** There are generally no decision points in the project for this comparison.

Which Is Faster: a Valve or a VSD?

Exceptionally fast loops (e.g., furnace pressure, liquid pressure, and compressor surge control) can ramp off-scale in milliseconds. These loops have essentially a zero process deadtime and may have a high process gain due to a narrow control range (e.g., fractional inches of water column for furnace pressure). These loops require DCS scan times of 0.05 to 0.1 seconds. Special fast scan rate digital controllers or analog controllers are needed. DCS scan time requirements of 0.2 seconds or less signify a VSD opportunity. A properly designed VSD has no measureable deadtime, while control valves and dampers take anywhere from 0.2 to 2.0 seconds to start to move. For example, an incinerator pressure and polymer pressure loop that could get into trouble in less than 0.1 second required a VSD and analog controller to stay within the desired control band [20][23][35].

A VSD has a negligible response time delay unless a deadband or dead zone is introduced into the drive electronics to slow response to process measurement noise, or if a low resolution input card is used. A control valve or damper has a deadtime that is proportional to the resolution limit (e.g., from stiction and windup) and deadband (from backlash and windup) divided by the rate of change of the process controller output. For large or fast changes in signal this deadtime disappears.

A pneumatic actuator has a pre-stroke deadtime that is the time it takes for the actuator pressure to change enough to move the actuator shaft. For large actuators, the pre-stroke deadtime can be several seconds unless a booster is added.

The inertial time constant of liquid flow response is inversely proportional to flow rate. Consequently, the process lag at low flow rates and at the start of flow can be quite long (e.g., 5 seconds) compared to the process lag at normal flow rates (e.g., 0.5 seconds). The comparison between VSD and control valve response should be at normal flow rates.

In a published comparison of the dynamic response of a control valve and a VSD-controlled pump for flow control for a system with negligible static head, the integral times were about the same for the VSD and valve loops. However, the controller gain could be increased by over a factor of six for the VSD loop. As a result, the set point response was faster [38]. In this test the valve deadband was about 8%.

In an unpublished lab test comparison between a control valve with low stiction, low backlash, and a digital positioner and a VSD with a volts/Hertz PWM drive for liquid flow control, the speed of response of the valve and VSD were similar.

Variable speed drives, control valves, and dampers have a velocity limited exponential response. The velocity limit in a pump drive depends upon the available motor torque and the inertia of the motor rotor, the pump shaft, and the pump impeller. The exponential term is generally much smaller for a VSD than for a control valve or damper. On the other hand, the velocity limit is lower for a VSD unless the actuator size is large and boosters are not used. Consequently, for small changes in signal, a well designed VSD is faster. Conversely, for large changes in signal, a small control valve is faster (see section 7-5 on Dynamics). This leads to the conflicting statements about whether a VSD or control valve is faster. Application for application, which final element is faster often depends upon the size of the change in signal.

To summarize, a VSD is most likely to offer energy savings or better loop performance as a final element for the following types of applications:

- Loops that require 0.2 seconds or faster scan time
- Valves and dampers with a 0.5% or more resolution limit or deadband
- Large utility flows
- Integrating and runaway processes without a secondary flow loop
- Low static head processes requiring frequent turndown

VSD Best Practices

With a VSD, a tachometer or inferential speed feedback signal should be sent to the process controller in the DCS that is sending the signal to the drive. The speed feedback should be used in a similar way to the position feedback from a digital positioner to prevent the process controller output from changing faster than the VSD can respond. The use of the dynamic reset limit option for the loops in the DCS can automatically prevent the process controller from outrunning the response of any type of final element (see Section 7-5 on Dynamics).

For best performance, users should consider the following during the specification and implementation of variable speed drive systems:

- High resolution input cards
- Pump head well above static head
- On-off valves for isolation
- Design B TEFC motors with class F insulation and 1.15 service factor

- Larger motor frame size
- XPLE (cross-linked polyethylene) jacketed foil/braided or armored shielded cables
- Separate trays for instrumentation and VFD cables
- Inverter chokes and isolation transformers
- Ceramic bearing insulation
- Pulse width modulated inverters
- Properly set deadband and velocity limiting in the drive electronics
- Drive control strategy to meet rangeability and speed regulation requirements
- Dynamic reset limiting using inferential speed or tachometer feedback

7-5. Dynamics

Detailed first principle models of the dynamic response of control valves require momentum, material, and energy balances, positioner dynamics, and the mechanical models of friction, backlash, and shaft windup for specific actuators, positioners, and valves. The models and the results obtained by control valve manufacturers are proprietary. Such modeling by piping valve and damper manufacturers appears to be minimal, based on the dynamic performance of the products and the guidance offered.

The results of these complex models can be simplified for control loop performance analysis to the stroking time and pre-stroke deadtime of the actuator, the positioner sensitivity, and the stick-slip from friction (stiction), and windup and the deadband from backlash and windup in the valve assembly.

Actuator Response

The actuator response can be approximated as a pre-stroke deadtime and a rate limited exponential for the stroking time. The approximated response is for an actuator not connected to a valve. It does not take into account stiction, backlash, windup, or positioner dynamics. The effect of the positioner and accessories is simplified to an effective supply and exhaust flow coefficient (C_v).

The pre-stroke deadtime is the amount of time it takes for the change in actuator pressure to be large enough to move the actuator shaft. The pre-stroke deadtime can be simply approximated as an actuator X_v factor divided by a flow coefficient (C_v) as shown by Equation 7-2. The flow coefficient is based on the source of the air flow to the actuator. Normally the air source is a positioner output. If a volume booster or quick release valve is added, the air source is the booster or quick release valve output. If there is no positioner, booster, or quick release valve, the air source is the I/P transducer. If the restriction created by a solenoid valve or by the tubing or air connection size is significant, the effective flow coefficient for a given solenoid valve or restriction flow coefficient (C_{v1}) and source flow coefficient (C_{v2}) can be approximated by Equation 7-3 [39].

$$\theta_v = \frac{X_v}{C_v} \tag{7-2}$$

$$C_v = \sqrt{\frac{C_{v1}^2 * C_{v2}^2}{C_{v1}^2 + C_{v2}^2}} \tag{7-3}$$

Where:

C_v = effective flow coefficient (dimensionless)

C_{v1} = positioner, booster, quick release valve, or I/P flow coefficient (dimensionless)

C_{v2} = solenoid valve or restriction flow coefficient (dimensionless)

X_v = actuator pre-stroke factor (dimensionless)

The stroking time is the amount of time it takes for a full scale stroke. The stroking time can be simply approximated as an actuator Y_v factor divided by a flow coefficient (C_v) as shown by Equation 7-4 [39].

$$T_v = \frac{Y_v}{C_v} \tag{7-4}$$

The X_v and Y_v factors depend upon actuator type, size, stroke, and air flow direction (supply or exhaust). Table 7-1 gives some exhaust and supply factors for diaphragm and piston actuators. Table 7-2 lists flow coefficients for a positioner, I/P transducer, and various size quick release valves and volume boosters.

Table 7-1. Control Valve Prestroke Dead Time and Stroking Time Factors [39]

Actuator, sq. in.	Travel, inches	Pressure, psig	Spring, lb/in.	Fill Factors X_v	Fill Factors Y_v	Exhaust Factors X_v	Exhaust Factors Y_v
			Diaphragm				
66	21/8	4.5–16	275	0.015	0.338	0.225	0.610
100	31/2	6–22	335	0.031	0.861	0.404	1.446
215	41/8	6–21	610	0.105	2.276	1.200	3.902
46	3/4	3–15	735	0.012	0.190	0.045	0.256
46	3/4	6–30	1470	0.016	0.226	0.031	0.290
69	3/4	3–15	1100	0.020	0.297	0.071	0.401
69	3/4	6–30	2210	0.028	0.355	0.048	0.457
105	3/4	3–15	1070	0.033	0.466	0.115	0.630
105	3/4	6–30	3320	0.046	0.574	0.078	0.727
156	3/4	3–15	2500	0.046	0.676	0.161	0.913
156	3/4	6–30	5000	0.065	0.811	0.110	1.046
220	2	3–15	1260	0.074	2.004	0.500	2.810
220	2	6–30	2520	0.104	2.243	0.390	3.038
310	2	3–15	1650	0.103	2.790	0.569	3.724
310	2	6–30	3100	0.143	3.379	0.388	4.265
450	2	6–26	4500	0.323	4.386	1.260	6.552
450	2	6–26	4500	0.380	4.586	1.353	6.870
			Piston				
17	3/4	60	—	0.085	0.050	0.024	0.050
28	3/4	60	—	0.165	0.086	0.035	0.086
56	3/4	60	—	0.296	0.169	0.050	0.169
89	2	60	—	0.715	0.719	0.106	0.719
131	2	60	—	0.995	1.060	0.272	1.060
222	2	60	—	1.730	1.800	0.738	1.800
17	4	60	—	0.020	0.278	0.024	0.278
28	4	60	—	0.051	0.460	0.035	0.460
56	4	60	—	0.099	0.901	0.050	0.901
89	4	60	—	0.181	1.453	0.196	1.453
131	4	60	—	0.227	2.144	0.272	2.144
222	4	60	—	0.603	3.600	0.738	3.600

Accessory Type	Connection Sizes, inches	Supply C_v	Exhaust C_v
Positioner	—	0.17	0.24
I/P transducer	—	0.39	0.36
Quick release valve	3/8 NPT	2.94	3.6
Quick release valve	1/2 NPT	5.46	6.49
Quick release valve	3/4 NPT	9.97	12.17
Volume booster	3/8 and 3/8 ports	3.74	2.29
Volume booster	3/8 and 1/2 ports	3.74	2.52
Volume booster	1/2 and 3/8 ports	5.32	2.30
Volume booster	1/2 and 1/2 ports	5.32	2.53
Volume booster*	1 1/2 NPT	23.1	23.1

* Fairchild 200 XLR

For a large step change in signal to the control valve, the actuator response is a ramp with some slight rounding at the beginning and at the end of the stroke (Figure 7-36a). The ramp rate is the slewing rate (rate of change of valve position). The total time required for a 100% change in signal is the stroking time (100%/slewing rate) estimated by Equation 7-4. The rounding at the start and end of the actuator stroke is from a primary (largest) and secondary (next largest) time constant, respectively in the pressure response of the actuator.

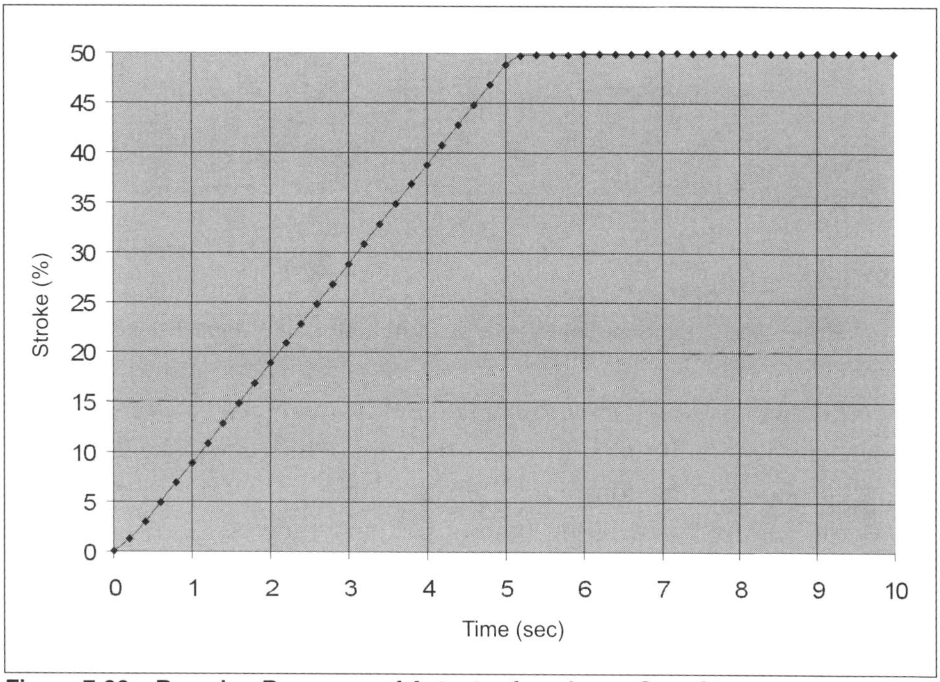

Figure 7-36a. Ramping Response of Actuator for a Large Step Change in Signal

For a small step change in signal to the control valve, the actuator response looks more like an exponential (Figure 7-36b). The rate of change of the valve position in the exponential response is limited by the slewing rate of the control valve but is essentially defined by the primary time constant with some rounding at the start by the secondary time constant. The response to a small step is the second order (two time constant) plus dead-time approximation used in the literature to describe a valve's dynamic response.

In the actuator response to both the small and large step changes there is a time during which the change in position of the actuator shaft is negligible. In particular, the change in actuator shaft position may have no effect

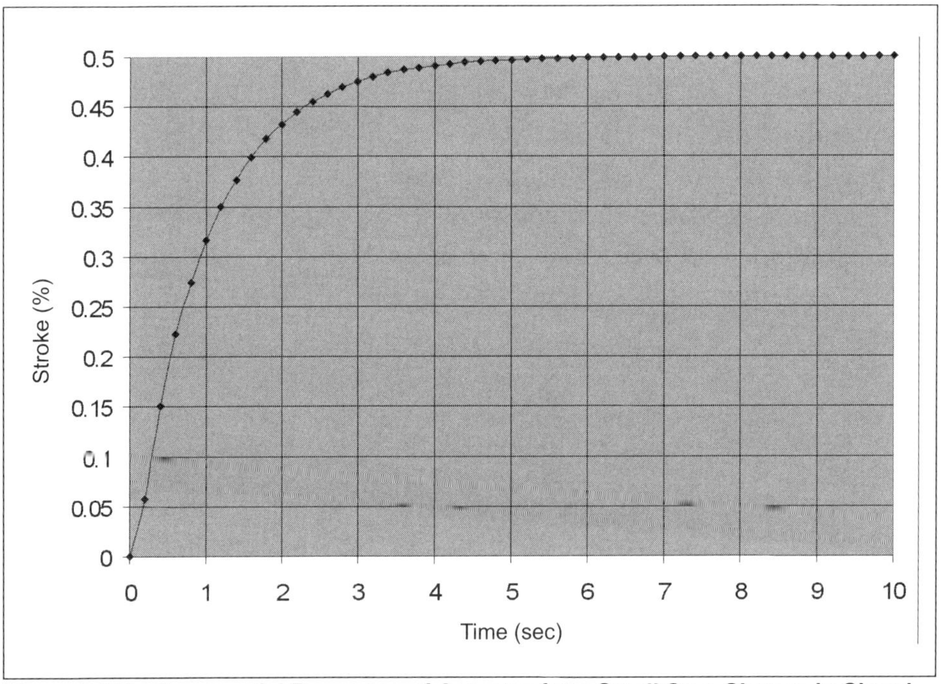

Figure 7-36b. Exponential Response of Actuator for a Small Step Change in Signal

because of backlash and shaft windup in rotary valves. This time delay before a significant response is the sum of the pre-stroke deadtime and the secondary time constant. The rest of the actuator response is effectively defined by the stroking time or primary time constant for large or small changes, respectively.

It has generally been found in industrial process control that a first order (one time constant) plus deadtime approximation (FODT) is sufficient. In this approximation all time constants smaller than the largest are taken as effectively creating deadtime [4][5]. This rule is used for all contributors to the loop's dynamic response as discussed in Chapter 1 (final element, process, measurement, and DCS). If the actuator primary time constant is much smaller than the process time constant, the actuator primary time constant is summed with the actuator pre-stroke deadtime and secondary time constant to be the dead time from the actuator in the FODT approximation of the process response. This approximation works well for small step changes in signal where the effect of the stroking time is negligible.

The effects of the stroking time and primary time constant can be combined to provide a more accurate estimation of the effective time constant for the control for a variety of step sizes. This response, as defined by Equation 7-5, is termed a "rate limited exponential." In this equation, the actuator's primary time constant is the stroking time (T_V) multiplied by the ratio factor (R_V). As a rule of thumb, the ratio factor for diaphragm actuators is about a factor of 10 smaller than for piston actuators (e.g., R_V = 0.02 to 0.1 for diaphragms and R_V = 0.2 to 1.0 for pistons) [39]. Thus, for the same stroking time, diaphragm actuators are faster, particularly for small changes in signal. This finding is consistent with the general observation that diaphragm actuators are more responsive than piston actuators to small changes in signal.

$$\tau_v = T_v * (\Delta CO / 100\% + R_v) \tag{7-5}$$

Where:

ΔCO = step change in controller output (%)

R_v = ratio of actuator's primary time constant to stroking time

τ_v = effective time constant of actuator's rate limited exponential response (sec)

T_v = stroking time (sec)

Boosters can be used to reduce the stroking time and pre-stroke deadtime by providing a large supply and exhaust flow coefficient. The booster should be mounted on the output of the positioner and the booster bypass opened enough to prevent instability as outlined in the "Accessories" subsection in section 7.3, Valves and Dampers. A larger air filter-regulator with more flow capacity is generally required so the booster is not starved for air supply. Also, tubing and solenoid valves, if any, between the booster and actuator and actuator air input connections should be increased in size so they do not present appreciable restriction to air flow as per Equation 7-3.

The response and performance of a process depends on the tuning of the process controller and response of the control valve, which in turn depends on the tuning of the positioner. The tuning of the positioner will change with actuator and valve size, type, and accessories. The supplier may initialize the tuning settings to match the valve purchased. Typically, these are conservative settings. To provide more aggressive control or to deal with changes in packing and accessories, the tuning can be changed in the field. By looking at the response to steps in the valve signal similar in size for the worst-case disturbance, judgments can be made as to whether to increase or decrease the gain and rate settings in the positioner as shown in Figure 7-37 for a digital valve controller. If the response to the final position is exponential with no overshoot or oscillation (over damped), the gain and rate settings are close to best (step response 1 in Figure 7-37). If the response overshoots the final position and then oscillates (under damped), the gain setting is too high (step response 2 in Figure 7-37). If the response oscillates before reaching the final position, the gain and rate settings are both too high (step response 3 in Figure 7-37). If the response is too slow with no overshoot and oscillation, the gain and rate settings are too low (step response 4 in Figure 7-37). For aggressive control, some overshoot and oscillation is useful for getting the process to move faster.

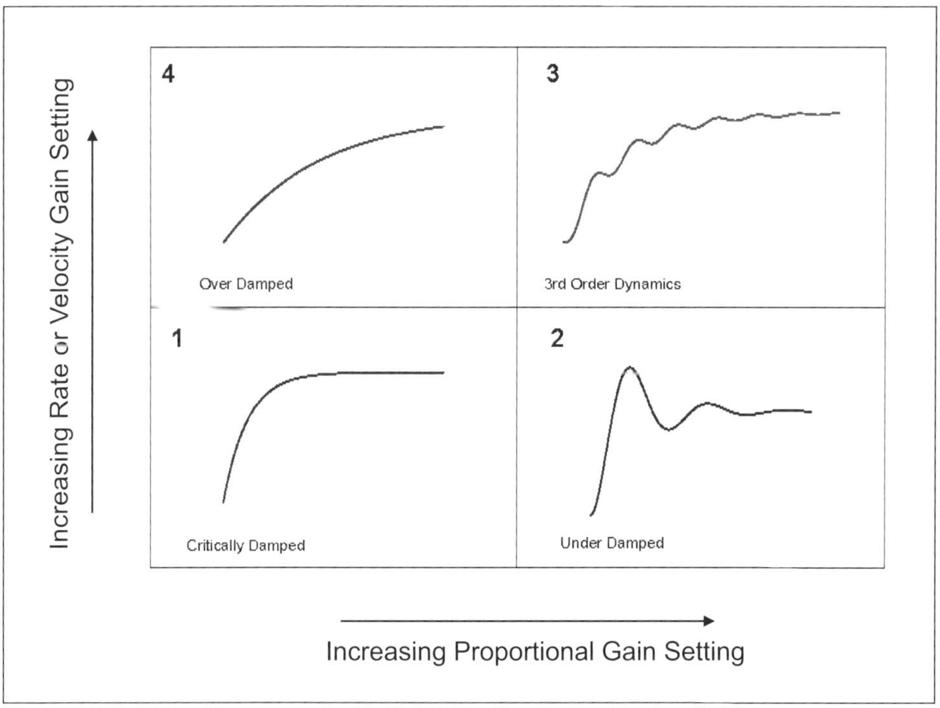

Figure 7-37. Valve Response to Digital Valve Controller Tuning Settings

A significant sensitivity problem that can dramatically increase the response time for small steps in the input signal has been identified in analog positioners. For example, the response time for a pneumatic positioner went from 1 second for a 1% step, to 10 seconds for a 0.5% step, and finally to 100 seconds for a 0.2% step [16]. The response time here is the time to 86% of the final stroke (T_{86}). Digital positioners have much better sensitivity. Little degradation in response is expected until the step size becomes less than the resolution of the relay position feedback in the digital positioner or the stick-slip of the valve or damper. Figure 7-38 shows the effect of positioner sensitivity, valve and actuator resolution limits for small and large step changes in signal.

For large step sizes, the effect of actuator size rather than valve stick-slip, backlash, and windup is seen, and all valves of the same size appear to be equal in response capability. For small step changes, there is a slight effect of valve size from pre-stroke deadtime but the major effects seen are from the valve resolution limit and deadband and positioner sensitivity. Small step sizes reveal the real dynamic capability of a control valve. Small step sizes will show the incredible deterioration in the response of piping valves being used as control valves and of inferior positioner designs. Small step sizes will show the performance advantage of sliding stem (globe) valves with diaphragm actuators and digital positioners [39, 40].

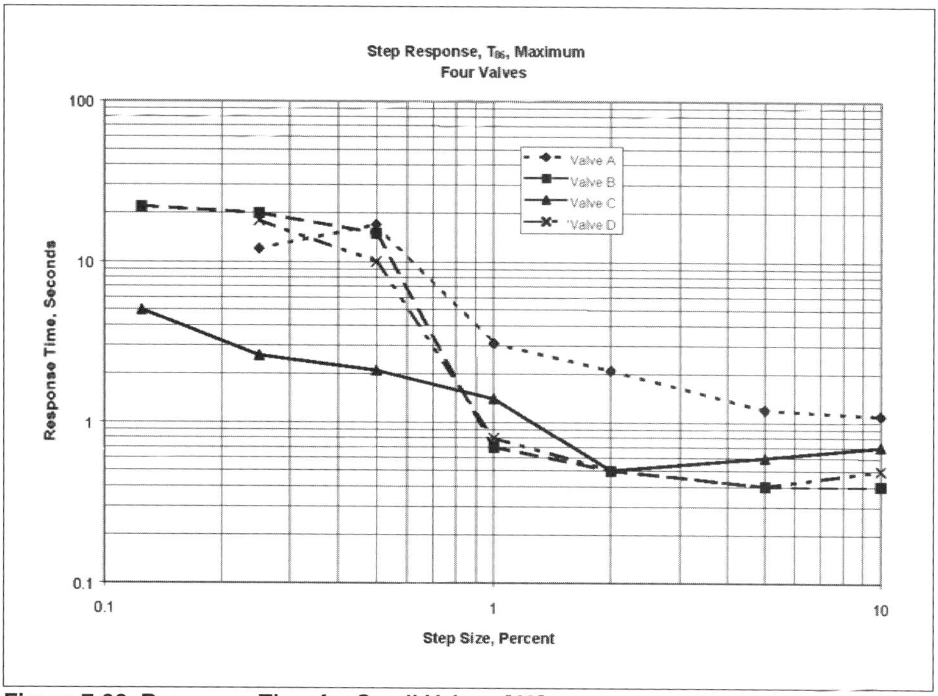

Figure 7-38. Response Time for Small Valves [42]

VSD Response

The response of variable speed drives more closely resembles a pure ramp with no rounding or time delay, provided a filter or deadband has not been added in the drive electronics to attenuate process noise in the process controller output signal. The ramp time in a VSD depends on the size of the load compared to the available torque from the motor. In general, the ramp time of a VSD is longer than the stroking time of a control valve but is shorter than the stroking time of a large damper. Longer than necessary VSD ramp times may inadvertently be imposed in the drive electronics.

There is essentially no stiction or backlash in variable speed drives for axial and centrifugal blowers, fans, and pumps but this does not necessarily mean there is no resolution limit or deadband in VSD response.

Controller outputs invariably display fluctuations that originate from process or sensor noise. These fluctuations are not representative of the actual value of the process variable and are best ignored. These fluctuations are particularly large and fast for flow and pressure loops. A deadband is sometimes introduced in the VSD electronics to prevent cycling in speed from noise. The effect may be a true deadband where the desired speed does not change upon a change in direction until the change in signal is larger than the deadband setting. The effect here is similar to backlash in a control valve. In other cases, it may be a "deadzone" setting, in that the desired speed does not change until the accumulated change in signal since the last change in speed is larger than the deadzone setting. Here the effect is similar to that of a resolution limit.

If there is no deadzone setting, the resolution limit in a VSD is largely determined by the input card. Assuming there is no sign bit, the VSD resolution limit is simply 100% divided by 2 raised to the number of bits (n) of the input card (Equation 7-6). Unfortunately, VSD manufacturers did not understand the limit cycle that would result from the resolution limit and offered until recently (e.g., 2006) a low resolution input card (e.g., 8 bit) as the standard. Higher resolution input cards (e.g., 12 bit and 16 bit) should be specified to make the VSD I/O resolution comparable to the DCS I/O resolution.

$$E_{ri} = \frac{100\%}{2^n} \tag{7-6}$$

Where:

E_{ri} = resolution error of input signal (%)

n = number of bits of input card (no sign bit) (dimensionless)

In pumps, the resolution of a speed feedback signal by pulse interval measurement is approximately 100% divided by the number of pulses per period (Equation 7-7a), which in turn is the frequency of the clock divided by the product of the pump speed and the number of pulses per revolution (Equation 7-7b). The worse case resolution is 0.15%, which corresponds to a clock frequency of 20 KHz in older drives, a pump speed of 1800 rpm, and a one pulse per revolution speed pickup. Newer drives have a clock frequency of 100 KHz or faster and most speed pickups send more than one pulse per revolution.

A single notch or embedded magnet used in keyphasor speed measurements for vibration analysis provides only one pulse per revolution. These are generally magnetic pickups. Most tachometer feedback measurements have gear teeth for magnetic pickups, and disks with holes or bands with mirrors on the shafts for optical pickups to provide many more pulses per revolution.

$$E_{rs} = \frac{100\%}{N_p} \tag{7-7a}$$

$$N_p = \frac{F_c}{S_p * \frac{P_r}{60}} \tag{7-7b}$$

Where:

E_{rs} = resolution error of speed sensor (tachometer feedback) (%)

F_c = clock frequency (cycles per second)

N_p = number of pulses (dimensionless)

P_r = pulses per revolution

S_p = speed of pump (rpm)

As we have seen, speed slip, unlike valve stick-slip, does not introduce a resolution limit. In speed slip, the drive output continues to respond to changes in the input signal. The amount of the response varies, which changes the difference between the desired and actual speed (slip). The result is a change in the process gain. In other words, slip affects the size of the response and thus the linearity of the system but does not cause a time of no response.

The net result is that the VSD resolution limit and deadband are negligible for an open loop (volts/Hertz) PWM drive with a good input card and no deadband or deadzone setting. For control valves and dampers, stick-slip and backlash always introduce a resolution limit and deadband.

Resolution, Deadband, and Hysteresis

The actuator may move, but this does not necessarily mean the closure components of the control valve (ball, disk, or plug) or the damper (wafer, blade(s), airfoils, or vanes) also move. A resolution limit and deadband are used to describe when and how the closure components respond to the stroking of the actuator.

The resolution limit from stick-slip in a control valve arises from stem packing friction and sealing or seating friction. All control valves have stick-slip. Dampers, on the other hand, primarily have backlash. The resolution limit of well designed sliding stem control valves with pneumatic actuators ranges from 0.1 to 0.2%. The resolution limit of rotary valves with shaft windup is approximately the same as the deadband. If the slip is equal to the stick, the result is a resolution limit where the response is a staircase as shown in Figure 7-39. For changes in the same or opposite direction, the closure component does not move until the change in signal from the last move exceeds the resolution limit. The valve then steps (slips) by the amount of the resolution limit. In practice, the slip is never exactly equal to the stick. Undersized actuators, and the high breakaway torque and shaft windup of piping valves can result in a slip or jump in the valve position greater than the total requested change in the valve position.

Deadband originates from play in linkages, shaft and stem connections (backlash), and shaft windup. There is a temporary disconnection and lost motion between a change in input (signal) and the resulting change in output (closure component position). In well designed sliding stem (globe) valves, deadband is negligible. In well-designed rotary control valves with direct splined shaft connections, short shafts, and minimal shaft windup, the deadband is less than 0.5% [2]. In piping valves being used as control valves and in most dampers, the deadband is 5% or more [16][20][23, 24]. Technically, there is no resolution limit associated with pure backlash in that once the closure component starts to move, the movement continues regardless of how small the change. However, shaft windup causes the valve to jump like stick-slip.

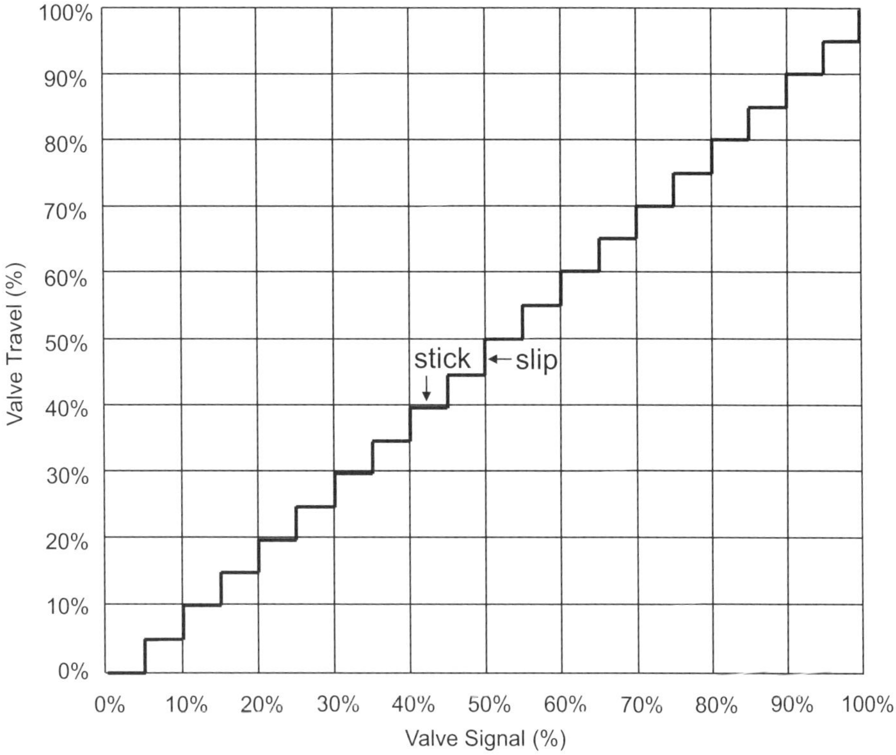

Figure 7-39. Resolution Limit from Stick-Slip (Zero Hysteresis and Zero Backlash)

Dampers in low temperature operation where stem pack friction is negligible primarily exhibit backlash (resolution limit is negligible). For rotary valves, the resolution limit is often at least half of the deadband. For pure backlash (no shaft windup), upon signal reversal once the change in signal exceeds the deadband, there is an incremental change in position that matches incremental changes in the valve signal. The formal definition of "deadband" is for a full scale reversal of signal (0% to 100% and 100% to 0%) as shown in Figure 7-40b. As practical matter deadband occurs for any reversal of signal and is defined as the signal change needed to change direction at any operating point.

The term hysteresis is often mistakenly used for deadband. In hysteresis, unlike deadband, there is no delay in the response but the slope (the amount of change in the stroke for a change in signal) is not constant, due to absorbed energy. Figures 7-40a and 7-40b show hysteresis by itself and combined with deadband, respectively. A small but negligible amount of hysteresis can exist in a diaphragm actuator's response from the flexing of the diaphragm. Pinch valves may exhibit a significant amount of hysteresis from the flexing of the boot. In general, however, deadband is the more dominant effect.

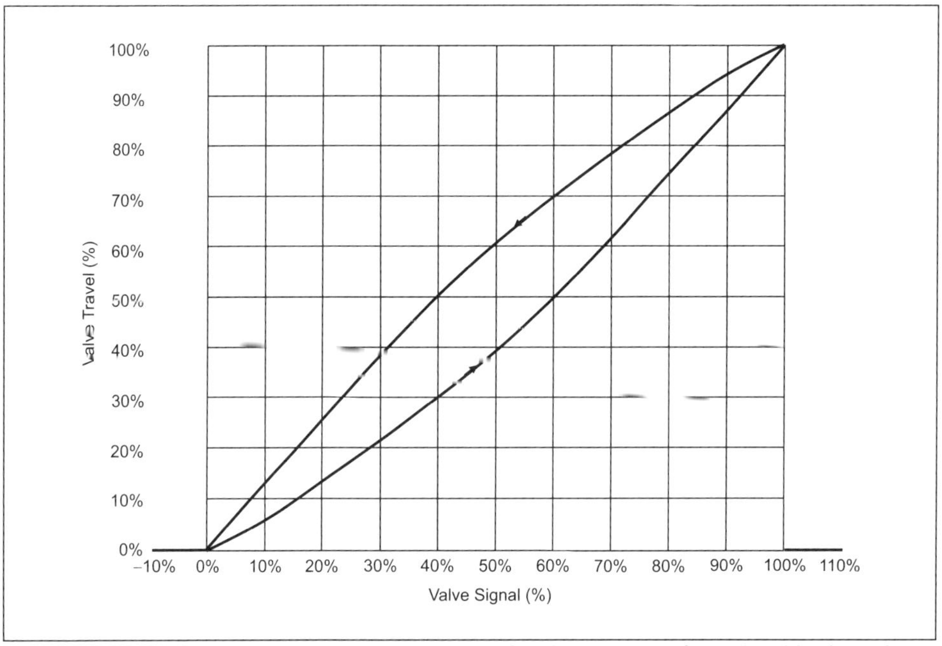

Figure 7-40a. Hysteresis from Energy Dissipation in Actuator (zero backlash and zero stick-slip)

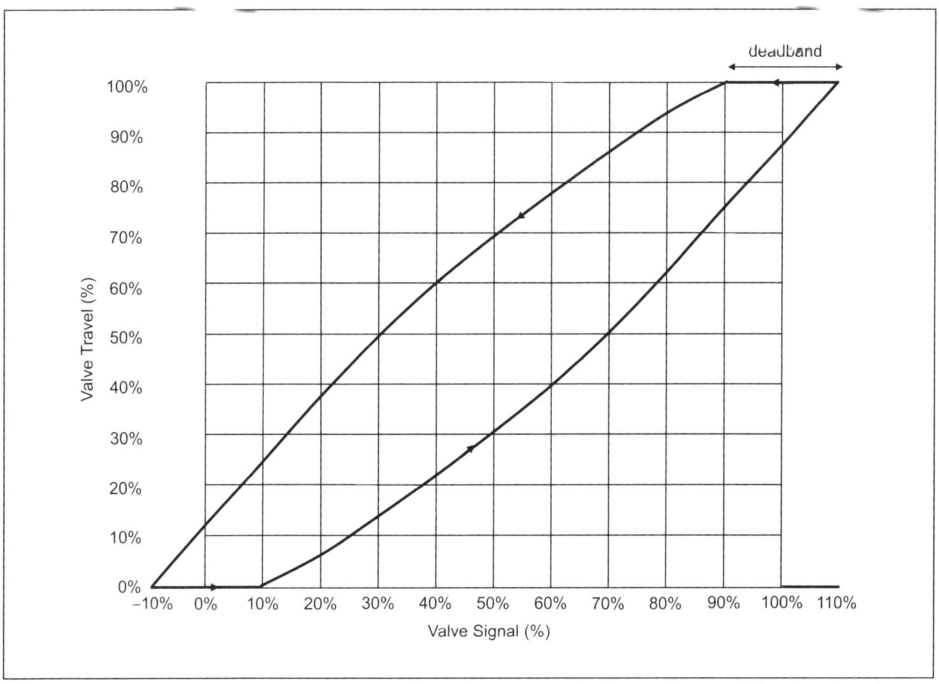

Figure 7-40b. Deadband from Backlash and Hysteresis from Energy Dissipation (zero stick-slip)

Digital positioners monitor and correct for stiction, backlash, and windup if the position feedback is in actual closure component position. The resolution and deadband for valves with digital positioners are considerably smaller, particularly when these positioners are tuned for a more aggressive response and have a position feedback that truly represents the actual position of the closure component. In cases where digital positioners have been applied to piping valves being used as control valves, the feedback of the actuator shaft position may differ from the actual ball, disk, or plug position by 5% or more because of backlash and shaft windup [16][23, 24]. Position feedback, diagnostics, and step response tests give erroneous results with these valves and there is very little real improvement. A similar problem can exist with dampers because of the backlash in linkages between the actuator shaft and the wafer, blades(s), airfoils, or vanes.

For an oscillating input, resolution limits and deadband cause a phase shift and an attenuation of the amplitude of the oscillation. Oscillating inputs are useful for determining the frequency response of a valve. However, users think in terms of the time response as seen in the trend charts (plot of process variables versus time provided by data historians integrated into and interfaced to the control system). For characterizing the time response, it is useful to look at a valve's response to ramps and steps.

For a ramping input, resolution limits create deadtime that is the resolution or deadband divided by the rate of change of the controller output (Equation 7-8a). The difference between a ramping input signal and the stroke of the closure components is the dynamic error band (Figure 7-41a). This error is a complex result of the stroking time, stick-slip, and deadband of the complete valve assembly. This error is of great practical significance because in most cases the controller output for closed loop control is ramping. However, the ISA standards for testing and quantifying valve performance use step tests [41, 42]. The contribution of the actuator lag and valve deadband to the dynamic error band can be identified by a folding back of the valve response as shown in Figure 7-41b.

For step changes in the controller output smaller than the resolution limit, there is no response. For step changes in the controller output much larger than the resolution limit, there is a response but with no deadtime from the resolution limit. There is a decrease in process gain that can be estimated by subtracting the resolution limit from the step in the controller output (Equation 7-8b). These relationships for deadtime and process gain hold true for deadband when there is a reversal in the direction of the sig-

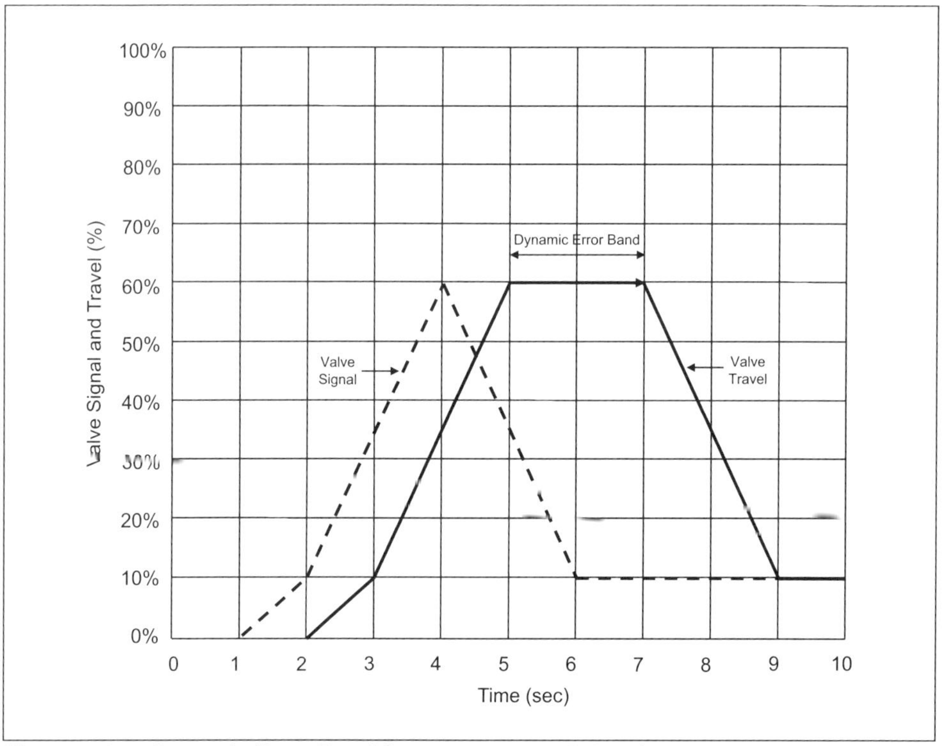

Figure 7-41a. Dynamic Error Band (as seen on trend chart)

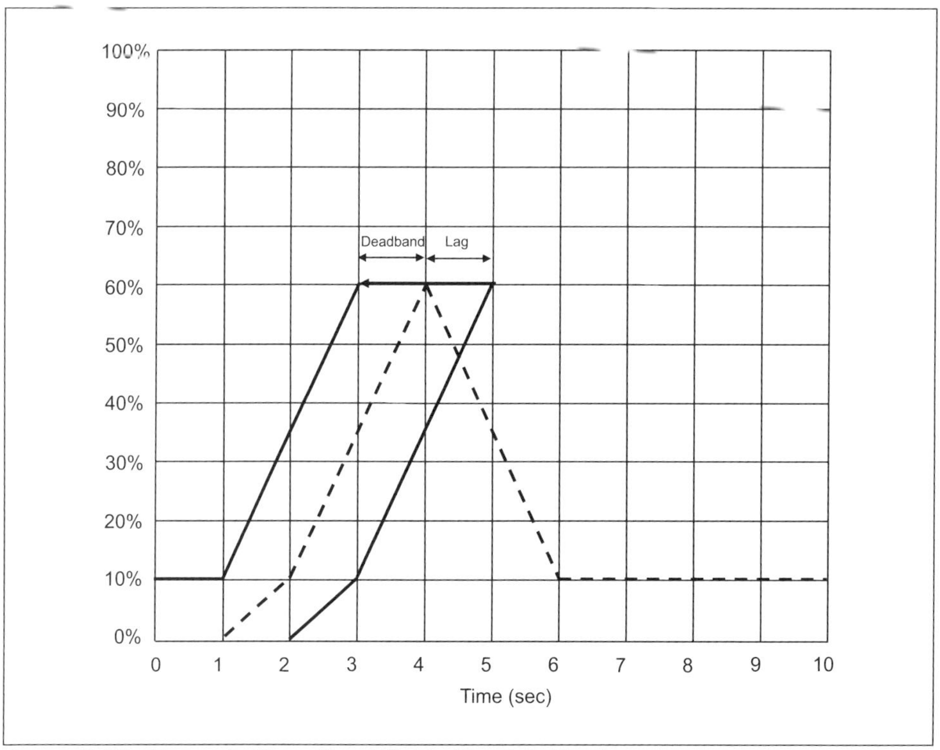

Figure 7-41b. Dynamic Error Band (folded back in time)

nal. For step changes near the resolution limit of the valve or sensitivity limit of the positioner, there may be an extremely slow response.

Open loop tuning methods identify loop dynamics from the response of the process variable to step changes in the controller output with the controller in manual or output tracking. The size of the step change is chosen to be larger than the resolution limit or deadband of the control valve. The observed deadtime will not include the valve deadtime because the valve signal will have zipped right through the resolution limit or deadband. The valve deadtime (θ_v) can be estimated from the observed deadtime (θ_o) for slow processes by Equation 7-8c, developed in the Theory section of *Advanced Control Unleashed* [4]. As a controller is detuned (detuning factor K_x is decreased), the valve deadtime is increased. Most loops have a controller gain that is less than half the maximum allowable controller gain ($K_x < 0.5$). Consequently, the use of the observed deadtime for an open loop test instead of the actual deadtime for closed loop operation usually does not cause instability. The concern is more about a decrease in process gain and an increase in deadtime that makes the performance of detuned loops with an excessive resolution limit or deadband even more sluggish.

$$\theta_v = \frac{\Delta X_v}{\Delta CO / \Delta t} \quad (7\text{-}8a)$$

$$K_p = \frac{\Delta PV}{(\Delta CO - \Delta X_v)} \quad (7\text{-}8b)$$

$$\theta_v = \frac{\Delta X_v}{K_x * (\Delta CO - \Delta X_v / 2)} * \theta_o \quad (7\text{-}8c)$$

Where:

ΔCO = step change in controller output

$\Delta CO/\Delta t$ = ramp rate of controller output (%/sec)

K_p = process gain modified for resolution limit or deadband (dimensionless)

K_x = detuning factor (e.g., 0.1 - 1.0) (dimensionless)

ΔPV = final change in process variable for step change in controller output (%)

θ_v = valve deadtime from resolution limit and deadband (sec)

ΔX_v = valve resolution limit or deadband (%)

Closed loop tuning tests identify loop dynamics from the response of the process variable to step changes in the controller set point with the controller in automatic. The identified dead time will include the additional dead time from resolution and deadband. As tuning settings change, the identified deadtime will change due to the change in the rate of change (e.g., ramp rate) of the controller output.

Limit Cycles

Resolution limits and stick-slip cause continual hunting in a controller with integral action on any process because the measured process variable never exactly matches the set point [43]. Integral action will ramp the controller output for an offset of the process variable from the set point. For a fast-changing process variable, such as flow, the controller output ramps back and forth (sawtooth wave), and the process variable steps up and down (square wave) from stick-slip (Figure 7-42). If there are no disturbances, the result is a perpetual constant amplitude oscillation called a limit cycle. The limit cycle amplitude is the stick-slip multiplied by the open loop gain (Equation 7-9a) [3]. The open loop gain is the change in PV in % of scale divided by the change in controller output in %. The open loop gain is the product of the valve gain (per section 7-6), the process gain, and the measurement gain (per Equation 7-9d). The limit cycle period is proportional to the integral time and is inversely related to the controller gain (Equation 7-9b) [3]. Detuning the controller (decreasing the controller gain and decreasing the controller integral time) increases the limit cycle period. The limit cycle amplitude is independent of tuning.

$$A_o = S_v * K_o \qquad (7\text{-}9a)$$

$$T_o = 4 * T_i * [1/(K_o * K_c) - 1] \qquad (7\text{-}9b)$$

$$K_o = K_v * K_p * K_m \qquad (7\text{-}9c)$$

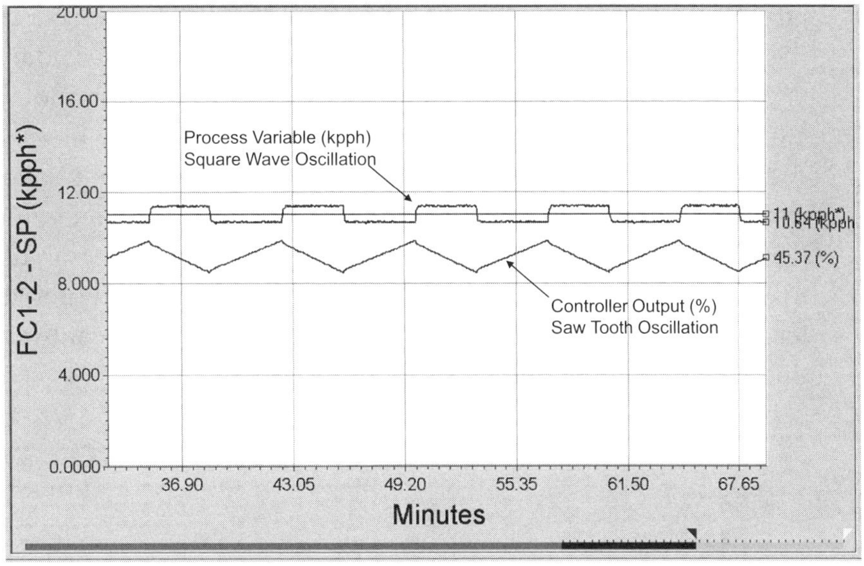

Figure 7-42. Limit Cycle in Flow Controller from a Resolution Limit

$$K_m = \frac{100\%}{S_m} \tag{7-9d}$$

Where:

A_o = amplitude of limit cycle (%)

K_c = controller gain (dimensionless)

K_o = open loop gain for self-regulating processes (dimensionless)

K_v = valve gain (Flow e.u. / CO %)

K_p = process gain (PV e.u. / Flow e.u.)

K_m = measurement gain (CV % / PV e.u.)

S_m = span of measurement scale (PV e.u.)

S_v = valve stick-slip (%)

T_i = controller integral time (sec/repeat)

T_o = period of limit cycle (sec)

If there are two or more integrators in the process and control system that affect the final element, deadband will cause a limit cycle. For an integrating process, such as level, the controller output is a sinusoidal oscillation and the process variable ramps with some rounding of the peaks (smoothed sawtooth) from deadband (Figure 7-43). The flow is a clipped oscillation. If there are no disturbances, the result is a limit cycle. The limit cycle amplitude is the deadband divided by the controller gain (Equation 7-10a). The limit cycle period is proportional to integral time and is inversely related to controller gain (Equation 7-10b). Detuning the controller (decreasing the controller gain and decreasing the controller integral time) increases both limit cycle amplitude and gain.

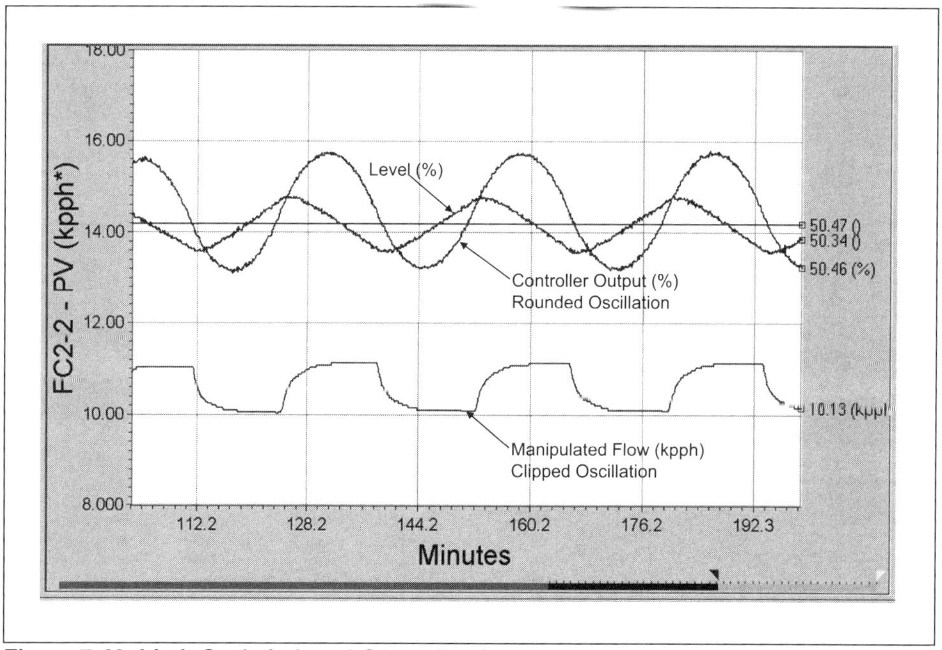

Figure 7-43. Limit Cycle in Level Controller from Deadband

$$A_o = B_v / K_c \tag{7-10a}$$

$$T_o = 5 * T_i * [1 - 2/(K_c^{0.5})] \tag{7-10b}$$

Where

A_o = amplitude of limit cycle (%)

B_v = valve backlash (deadband) (%)

K_c = controller gain (dimensionless)

T_i = controller integral time (sec/repeat)

T_o = period of limit cycle (sec)

The limit cycle in process variables for a valve with a minimal resolution limit and deadband (e.g., 0.2%) is often not noticeable because the oscillation is filtered (smoothed or washed out) by process volumes. Sensor and signal filter time constants have a similar effect. The smoothed amplitude is proportional to the product of the limit cycle amplitude and period and is inversely proportional to the residence time (volume/flow) (Equation 7-11a and 7-11b) [3]. Technically, the volume should have some back mixing from circulating streams, agitation, or boiling.

To minimize the filtered amplitude of a limit cycle, the controller should use fast but smooth tuning. Moderately aggressive tuning corresponds to a detuning factor of about 0.25 (K_x = 0.25) or a Lambda factor of 4.0 (λ_f = 4) for flow loops. The transition to a new set point should be fast but smooth with minimal overshoot (discounting non-ideal valve action). As the tuning is made faster, the limit cycle period (from resolution and deadband) and the limit cycle amplitude (from deadband) are minimized. If the tuning is too aggressive, there will be increased variability in the manipulated flow from the use of piston actuators and spool positioners in fast processes.

For inline process equipment (e.g., blender, desuperheater, heat exchanger, static mixer, extruder, and sheet line), the increased variability in a feed flow from a poor valve response propagates unfiltered into downstream temperature and composition because there is no smoothing from back mixing. Process variability studies that concentrate on flow and an inline process may conclude that slow tuning is best for poorly responding valves. The flow labs at valve manufacturers have inline systems (no back mixed volumes). However, even for this case, globe valves with diaphragm actuators and digital positioners will show a reduction in variability as the tuning is made faster as seen in Figure 7-44 [44].

The practice of slowing down the tuning to reduce variability is popular because users are taught that fast tuning causes oscillations and interactions. Studies that concentrate on the variability of temperature and concentration in columns, evaporators, crystallizers, and reactors conclude that fast tuning is best for all valves since fast fluctuations and fast limit cycling are washed out.

$$A_f = A_o * \frac{T_o}{2\pi * T_r} \tag{7-11a}$$

$$T_r = V / F \tag{7-11b}$$

Where:

A_f = filtered amplitude of limit cycle (%)

A_o = original amplitude of limit cycle (%)

F = process flow (cubic meters per sec; cubic feet per sec)

T_o = period of limit cycle (sec)

T_r = process residence time (sec)

V = process volume (cubic meters; cubic feet)

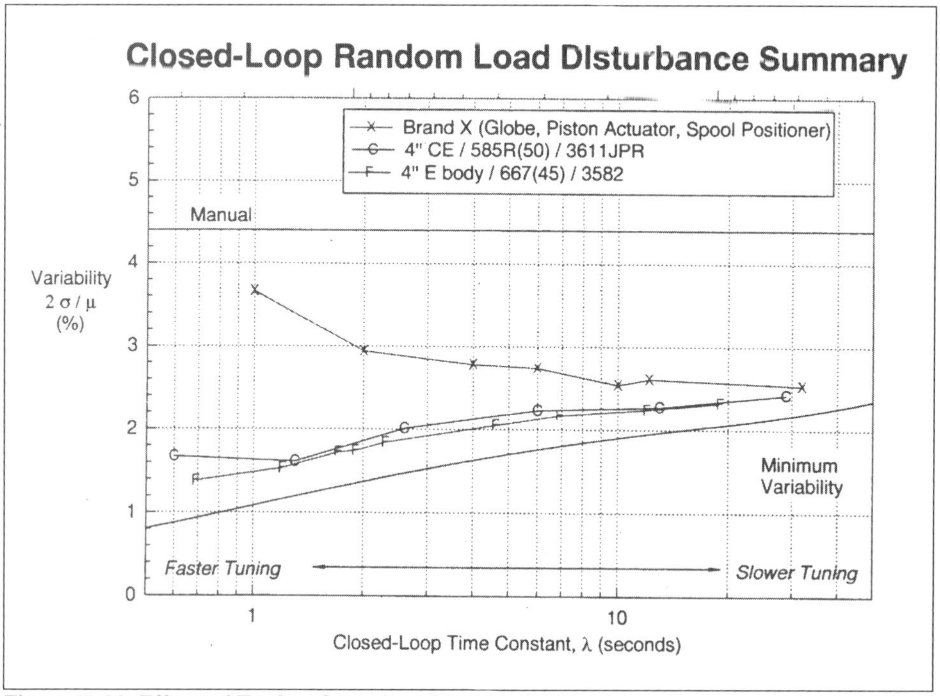

Figure 7-44. Effect of Tuning Speed on Flow Variability for Different Valves [44]

In the literature, the source of limit cycles is usually stated as stick-slip. When limit cycles from deadband are addressed, the case is stated to be a level loop. The use of integral action is not stated as a necessary condition. The rules on the source of limit cycles that lead to a deeper understanding are:

1. A limit cycle from a resolution limit develops when there is at least one integrator in the process or control system.
2. A limit cycle from deadband develops when there are at least two integrators in the process or control system

The integrators can be from an integrating process response (e.g., level or batch temperature and composition) or from an integrating action in the feedback controllers in the DCS or in the positioner. The use of two or more controllers with integral action in the triple cascade control system shown in Figure 7-24 will result in a limit cycle from deadband even if the temperature has no integrating process response.

The limit cycle from a resolution limit and deadband can be stopped until the next disturbance by turning off the integral action when the process variable is close to set point. The offset from set point that triggers the suspension of integral action should be about equal to the limit cycle amplitude. This feature exists in DCS controllers and digital positioners.

7-6. Linearity

In linear systems the dynamics (gain, deadtime, and time constant) are constant. We have already seen that the deadtime and time constant for valves and variable speed drives are variable and difficult to quantify. We will focus here on valve gain and VSD gain in pumps, which is what most practitioners consider when judging the linearity of a control system. Both the valve and VSD gain are the slope at the operating point on a plot of flow rate versus signal taking into account the effect of the pump curve, static head, and system frictional loss. As we have seen, the frictional loss is the pressure drop from the resistance to flow of everything in the flow path (e.g., piping, manual valves, and process equipment) except for the control valve. Some publications include the control valve pressure drop in the system loss. The system frictional loss is proportional to the flow rate squared.

Control Valves

From Equation 7-1 for liquid flow through a control valve, we see that the flow is proportional to the flow coefficient and the square root of the pressure drop. For a flow coefficient that is proportional to the stroke (linear trim) and a constant pressure drop and specific gravity, a plot of flow versus stroke is a straight line (constant slope). The valve and process gain for a flow loop are linear since the valve gain is constant. While this case exists, it occurs in less than 5% of the installations. The more general case is a flow coefficient that is a nonlinear function of stroke and a pressure drop that varies with operating point. The result is a variable slope and hence a nonlinear valve gain.

The pressure drop available to the control valve is the pressure difference between the upstream source and the sum of the downstream static head and system friction loss. Figure 7-45 shows how the pressure drop available for the control valve considerably decreases with flow rate for a system where maximum flow is large enough that the drop in pump head and the rise in system pressure drop are significant. For old plants pushed beyond nameplate capacity, the pump and piping size may be too small, causing the control valve to run out of available pressure drop. In these cases, increasing the valve size or capacity does not increase the flow appreciably because the new valve still has no pressure drop (driving force). Per Equation 7-1, if the pressure drop approaches zero, it doesn't matter how much the valve's flow coefficient is increased. As a practical matter the pressure drop must not cause an excessive loss of sensitivity at the upper end of the stroke range (see section on Installed Valve Characteristic) and be greater than the fluctuations in upstream and downstream pressures.

If the flow through a control valve is extremely small, the reduction in pump head and the increase in system pressure drop from friction loss are extremely small. The pressure drop available to the control valve is constant. These conditions commonly occur for pH control valves because the reagent flow is extremely small. For small utility flows from a large header that is pressure controlled, the changes in valve pressure drop may also be negligible. For cases where the valve pressure drop is constant, the installed flow characteristic is the same as the inherent flow characteristic. The three major types of control valve inherent flow characteristics are quick opening, equal percentage, and linear. The term "trim" is used for the closure component. In a sliding stem valve, the plug, seat(s), and cage

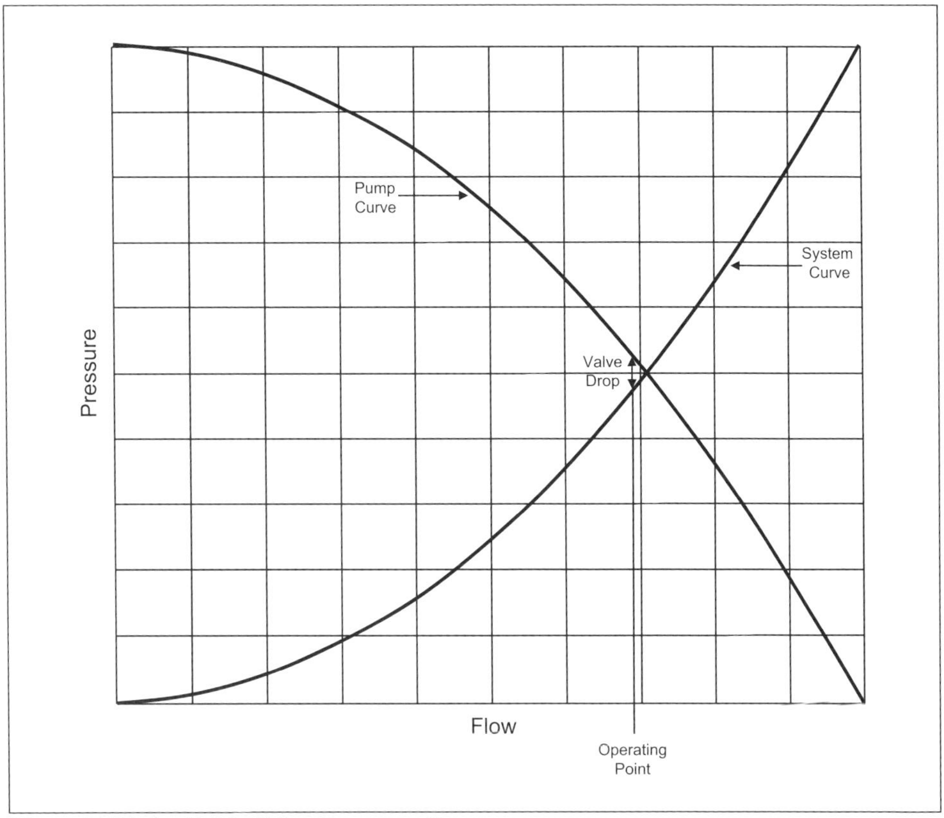

Figure 7-45. Available Valve Pressure Drop

are provided as a matched set (trim set). The term "trim" is often used in conjunction with the specification of the "inherent flow characteristic" provided by the closure component. The commonly stated choices in control valve specifications are quick opening trim, linear trim, and equal percentage trim.

Inherent Valve Characteristic

Figure 7-46 shows the theoretical inherent flow characteristic for the three major types of trim. The actual inherent flow characteristic will deviate from the theoretical inherent flow characteristic. The user should use plots of flow coefficients versus stroke for an individual valve from sizing catalogs or programs provided by the supplier. The distinguishing feature of these flow characteristics for process control is the slope (change in flow per change in stroke), because this slope is the valve gain. The plot is commonly labeled as percent flow versus percent stroke. In reality it is the flow coefficient expressed as a percent of the maximum flow coefficient versus percent stroke. The inherent flow characteristic does not take into account various application (installed) effects, such as pipe reducers, pump curve, system frictional loss, high fluid viscosity, and flashing (formation of bubbles).

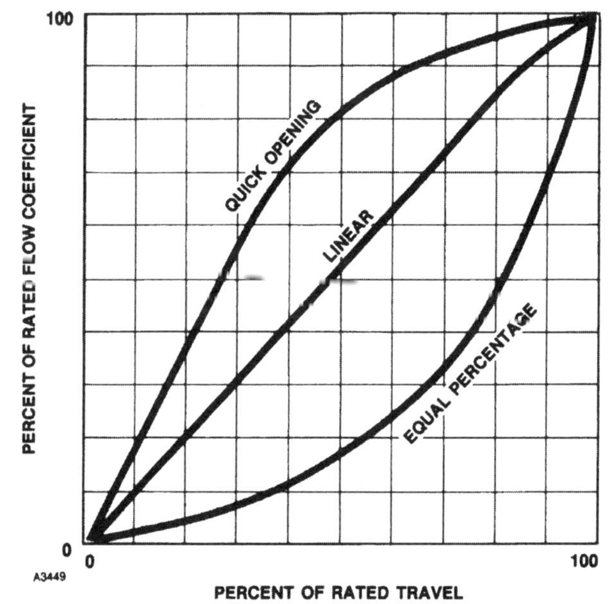

Figure 7-46. Theoretical Inherent Valve Characteristic

The inherent characteristic for quick opening trim has a constant slope from 0 to 40% stroke. At 40% stroke about 70% of the flow capacity is attained. Above the 40% position, the slope drastically decreases. Above the 60% position, there is very little additional flow. Quick opening trim characteristics are used for pressure relief and surge control where a rapid change in flow rate is needed. Very few applications benefit from a quick open characteristic.

The inherent characteristic for linear trim has a constant slope throughout its stroke range. The fractional flow coefficient is the fractional stroke (Equation 7-12). The linear flow characteristic is useful for vessel pH and header pressure control systems where the valve pressure drop is relatively constant. The next section, Installed Valve Characteristic, shows how a variable pressure drop causes a linear inherent characteristic to distort to a quick opening installed characteristic.

The inherent characteristic for equal percentage trim has a slope that is proportional to flow rate. The change in flow is equal to the flow on a percentage basis. The slope is almost flat near the closed position and exponentially increases with stroke. The fractional flow coefficient is an exponential function of the fractional stroke (Equation 7-13). The range (ratio of maximum to minimum stroke) over which the equal percentage is valid is normally taken as 50:1 (R = 50) [45]. Equal percentage trim is the most popular choice because of many practical advantages.

The equal percentage trim design minimizes the effect of stick-slip near the closed position where stiction is greatest. This decrease in stick-slip increases the practical rangeability of the control valve.

The control valve pressure drop at rated capacity normally only needs to be about 5 to 15% of the system friction loss for equal percentage trim, which reduces the pump head and hence the pump size and energy requirement. The equal percentage characteristic is more forgiving of under-sizing because of higher slope near the wide open position and of over-sizing because of the lower slope near the closed position.

The next section, on Installed Valve Characteristics, shows how a variable pressure drop causes an equal percentage inherent characteristic to approach a linear installed characteristic. A constant valve gain is desirable for flow and liquid pressure loops and for temperature and pH loops on well-mixed vessels.

For cases where the valve pressure drop is constant, the equal percentage characteristic compensates for a process gain that is inversely proportional to feed flow rate. Examples are pipeline concentration control by the direct manipulation of reagent and additive flows, and for heat exchanger control by the direct throttling of coolant valves.

If a secondary flow loop is installed, there is no compensation of the process gain by the valve gain. For cascade control systems, where the secondary loop throttling the valve is a flow loop, the primary loop (e.g., pH, concentration, or temperature) does not see the nonlinearity of the installed characteristic since it is manipulating flow rate rather than valve position. While cascade control is generally beneficial, there are cases where the cascade rule that requires the secondary loop to be 5x faster than the primary loop may necessitate detuning a fast primary loop to slow it down. For inline systems, if feedforward or ratio control is not needed, the faster response and nonlinear compensation from direct manipulation of a sliding stem (globe) valve with a digital positioner may provide better rejection of fast disturbances.

For linear trim:

$$C_x = \frac{X_v}{X_{v\max}} \tag{7-12}$$

For equal percentage trim;

$$C_x = R^{\left[\frac{X_v}{X_{v\max}} - 1\right]} \tag{7-13}$$

Where:

C_x = flow coefficient expressed as a fraction of maximum (dimensionless)

R = range of the equal percentage characteristic (e.g., 50)

X_v = actual valve stroke (%)

$X_{v\,max}$ = maximum valve stroke (%)

The equal percentage inherent flow characteristic offers these practical advantages:

- Minimum effect of stick-slip near the closed position
- Maximum rangeability
- Minimum pressure drop and hence pump size and energy use
- Maximum forgiveness of under-sizing and over-sizing
- Maximum linearity for valve pressure drop that significantly decreases with flow
- Beneficial process gain compensation for some inline control systems

The inherent flow characteristic of a modified parabolic (modified percentage) trim has a slope that resembles the equal percentage characteristic for the lower portion of the stroke range (e.g., stroke < 40%). The slope becomes relatively constant for the upper portion of the stroke range (e.g., stroke > 60%). This trim offers some of the advantages of equal-percentage trim in terms of stick-slip and over-sizing and provides a more linear installed flow characteristic for a given valve pressure drop and flow relationship. Signal characterization in the DCS offers more flexibility than a special trim for linearization.

The actual inherent valve characteristic does not exactly match the theoretical inherent flow characteristic. The difference is particularly noticeable for rotary valves and dampers. The flow coefficients from the valve sizing catalog should be plotted versus stroke to get the actual inherent flow characteristic. Often the actual characteristic shows a decrease in the slope above 70% stroke. This "bending over" of the characteristic is accentuated by application effects, especially for rotary valves and dampers. The result is a decreasing valve gain and an eventual loss of sensitivity (negligible change in flow) as the valve approaches the wide open position.

Installed Valve Characteristics

The installed characteristic of control valves takes into account the effect of a variable valve pressure drop. The pressure drop available to a control valve is seen in Figure 7-45 as the difference between pump head (upstream pressure) and the sum of the static head and system frictional loss (downstream pressure). The variation in pressure drop due to frictional loss makes an equal percentage trim much more linear.

During operation within the proper throttle range of a properly selected trim (equal percentage for most applications), valve nonlinearity does not cause a noticeable deterioration in performance because process controllers are usually detuned by a factor of four or more. However, to develop a deeper understanding of the consequences of improper design and the potential to do better for difficult and demanding loops, a detailed analysis is offered.

At the maximum rated flow, a pressure drop that is 10% of the difference between pump head and static head may be used to conserve energy. This corresponds to a pressure ratio of 0.1 as defined by Equation 7-14 as the ratio of the valve pressure drop for a fully open valve to the valve pressure drop for a completely closed valve [45]. The maximum rated flow corresponds to a 100% position. In the valve sizing, the maximum required flow may be chosen to be at a position in the more linear throttle range of the control valve (e.g., < 75% for a sliding stem valve, < 60° for a ball valve, and < 45° for a conventional disk butterfly). Sizing programs that allow you to specify the pump and systems curves and static head will give the pressure drop and flow at other valve positions.

$$\Delta P_R = \frac{\Delta P_{100\%}}{\Delta P_{0\%}} \tag{7-14}$$

Where:

ΔP_R = valve pressure drop ratio (dimensionless)

$\Delta P_{0\%}$ = valve pressure drop at completely closed position (kPa; psi)

$\Delta P_{100\%}$ = valve pressure drop at fully open position (kPa; psi)

The installed characteristic (Equation 7-15) is a function of the pressure drop ratio and the fractional flow coefficient (C_x) that depends on the inherent flow characteristic. A plot of Equation 7-15 for a linear and equal percentage trim shows how the installed characteristic becomes more "quick opening" and linear, respectively, as the pressure drop ratio decreases (Figures 7-47a and 7-47b). For both trims, the flow characteristic bends over more at the top end of the stroke and the flow at the low end of the stroke is higher as the pressure drop ratio decreases. The result is a decrease in rangeability. Thus the desire to save energy by decreasing the valve pressure drop at maximum flow must be tempered by the need for rangeability.

$$Q_x = \frac{C_x}{\sqrt{\Delta P_R + (1 - \Delta P_R) * C_x^2}} \tag{7-15}$$

Where:

C_x = flow coefficient expressed as a fraction of maximum (dimensionless)

ΔP_R = valve pressure drop ratio (dimensionless)

Q_x = flow expressed as a fraction of the maximum rated flow (dimensionless)

For a linear installed flow characteristic, the fractional flow should be equal to the fractional valve position (Equation 7-16). The desired inherent characteristic to provide a linear installed characteristic can be computed

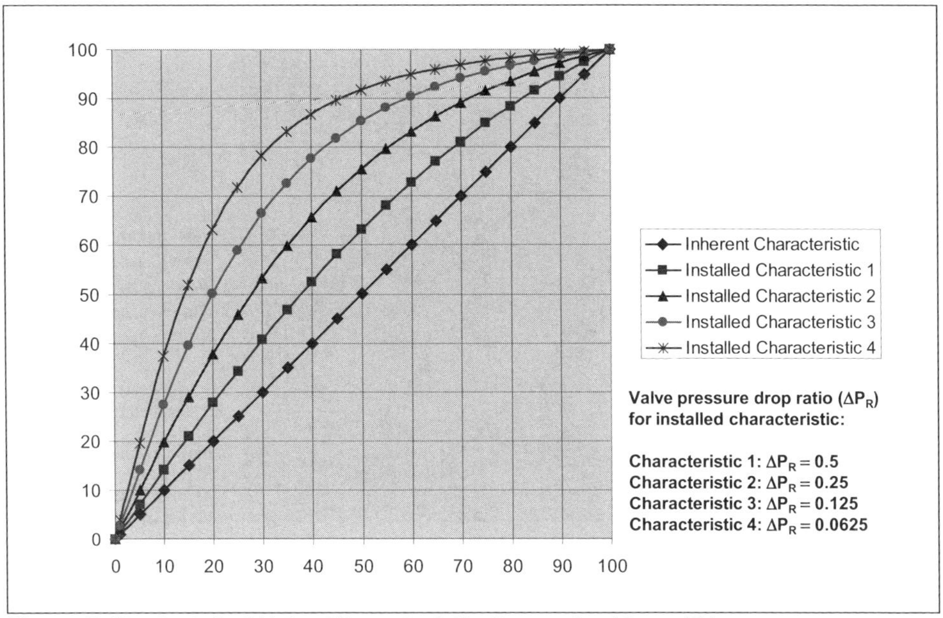

Figure 7-47a. Installed Valve Characteristic Curves for Linear Trim

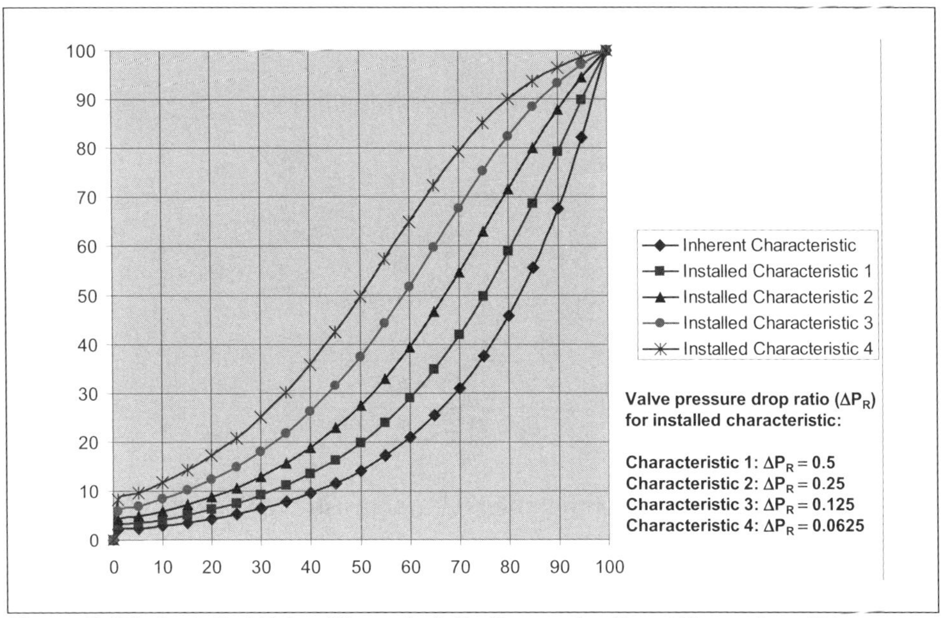

Figure 7-47b. Installed Valve Characteristic Curves for Equal Percentage Trim

by substituting Equation 7-16 into Equation 7-15 and solving for the fractional flow characteristic (Equation 7-17) [45].

The modified percentage trim is based on Equation 7-17 with a pressure drop ratio of 0.33. If this pressure drop ratio ($\Delta P_R = 0.33$) is substituted into

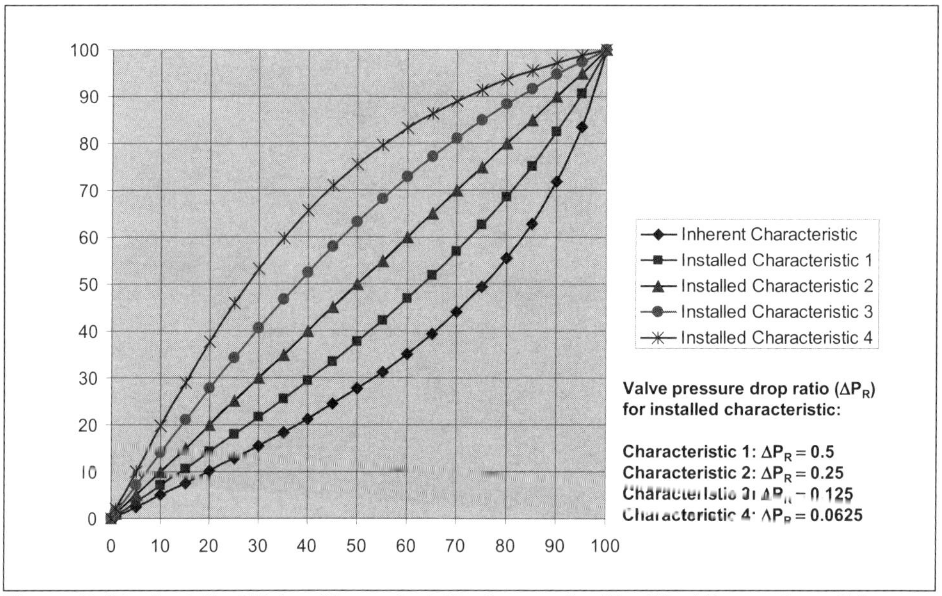

Figure 7-47c. Installed Valve Characteristic Curves for Modified Percentage Trim

Equation 7-17, we have an inherent flow characteristic expressed by Equation 7-18.

The desired linear installed flow characteristic:

$$Q_x = \frac{X_v}{X_{v\max}} \tag{7-16}$$

The necessary inherent flow characteristic:

$$C_x = \sqrt{\frac{Q_x^2 * \Delta P_R}{1-(1-\Delta P_R) * Q_x^2}} \tag{7-17}$$

The modified percentage flow characteristic ($\Delta P_R = 0.33$):

$$C_x = \frac{Q_x}{\sqrt{3-2 * Q_x{}^2}} \tag{7-18}$$

Where:

C_x = flow coefficient expressed as a fraction of maximum (dimensionless)

ΔP_R = valve pressure drop ratio (dimensionless)

Q_x = flow expressed as a fraction of maximum (dimensionless)

X_v = actual valve stroke (%)

$X_{v\,max}$ = maximum valve stroke (%)

The modified percentage characteristic provides better linearity and rangeability for a variable pressure drop as shown in Figure 7-47c. The same effect can be accomplished more exactly in the DCS. Equation 7-17 would be written into a calculation block between the controller (PID) output and the analog output (AO). The inputs to this calculation are the desired fractional flow (controller output) and the pressure drop ratio. The output is the fractional flow coefficient. For linear trim, the fractional flow coefficient is the fractional stroke. For equal percentage trim, Equation 7-13 (solved for the fractional stroke) must be added into the calculation block. The inputs to this additional calculation are the fractional flow coefficient and range of the equal percentage characteristic. The output is fractional stroke, which multiplied by 100% is the set point for the AO block.

Everything we have presented is based on theoretical inherent flow characteristic, pump curves, and system curves. As we have seen, the actual inherent characteristic of control valves deviates from the theoretical one, particularly at the low and high ends of the stroke. Also, many rotary valves do not follow the equal percentage characteristic. In general, the inherent characteristics of rotary valves fall somewhere between the theoretical equal percentage and linear characteristics. Figures 7-48a through 7-48c show the installed characteristic curves and valve gains for a conventional disk butterfly, v-ball, and sliding stem (globe) valve. The throttling range suggested by control consultants to keep the changes in valve gain less than a factor of four (0.5 > valve gain < 2.0) is marked on the plot of valve gain for each type of valve.

The DCS controller is tuned for smooth control at the valve position that has the maximum valve gain. Typically, the maximum gain occurs at around 25° for a rotary valve and 55% stroke for a globe valve. The control loop response is slower at other valve positions since the process gain is lower but the controller gain is the same.

Figures 7-48a through 7-48c provide an example of the installed characteristic and valve gain for the major types of control valves. The use of signal characterization or gain scheduling in the DCS can make the control loop response more consistent and extend the throttle range. Both methods slow down the change in signal to the valve on the steeper (high gain) portion of the installed characteristic, which makes the deadband and stick-slip greater in terms of controller output. Conversely, the methods speed up the change in signal to the valve on the flatter portion of the curve, which makes the deadband and stick-slip smaller in terms of controller output.

This speedup of the change in signal is particularly helpful near the closed position, where the stick-slip is the greatest. There is a practical limit to how much the flow range can be extended. In Figure 7-48a, we see that the installed characteristic curve is essentially flat above 60°. The flattening can occur at lower positions in valves in old plants pushed beyond nameplate capacity, where the piping and pump sizes are too small. The loss in sensitivity will cause a controller output to wander since the controller sees no real change in flow.

The controller output limit should be set to prevent the controller output from wandering on the flat part of the curve, but some plants may be reluctant to give up that last 0.5% of capacity. In these applications, signal characterization may be a temporary solution. A better solution would be to increase the pump head or decrease the system frictional loss.

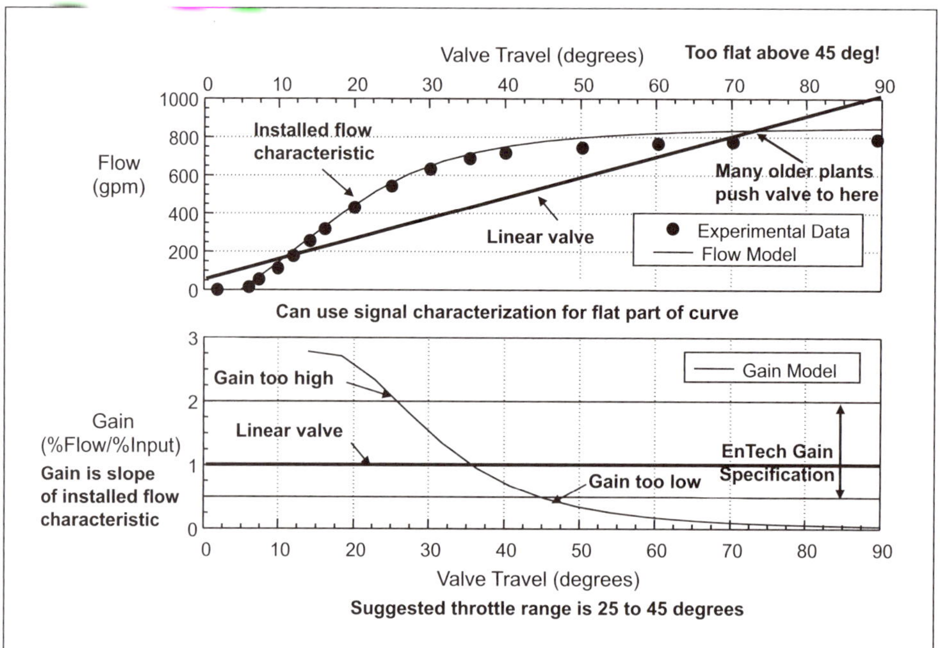

Figure 7-48a. Installed Conventional Disk Butterfly Valve Characteristic

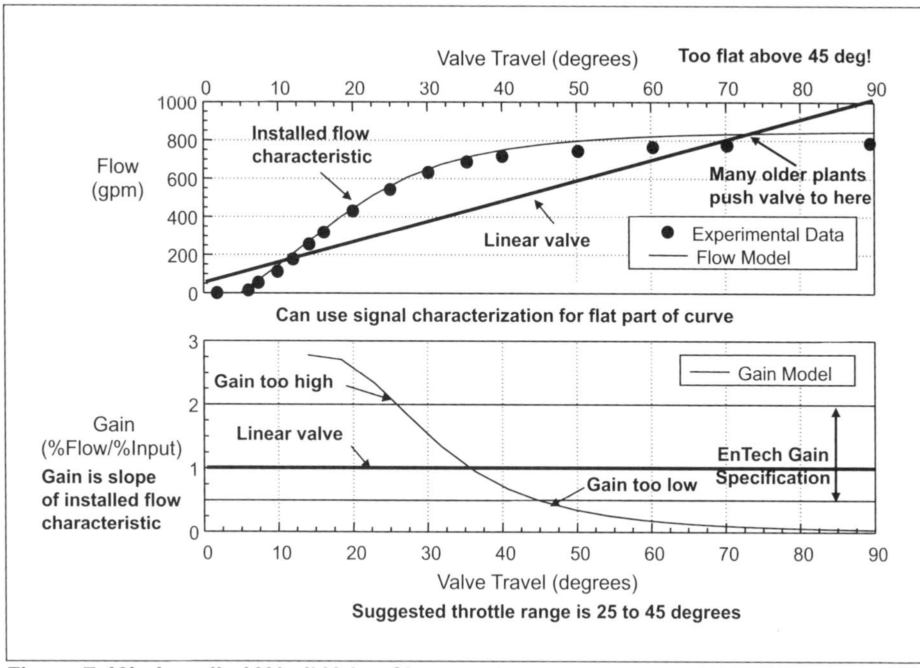

Figure 7-48b. Installed V-ball Valve Characteristic

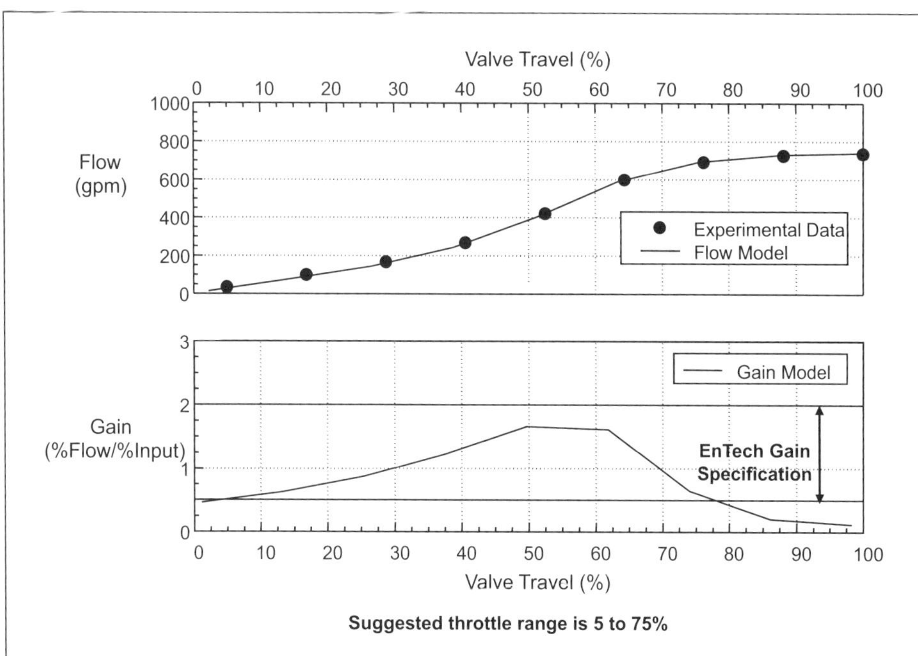

Figure 7-48c. Installed Sliding Stem (Globe) Valve Characteristic

The valve flow characteristic is best captured online with a high rangeability, low noise flow meter (e.g., magmeter or Coriolis meter) and by a ramp in the valve signal at normal operating conditions. A digital positioner is essential for this test. For small sliding stem valves, the whole ramp may be done in less than 20 seconds (20 points at a 1 second update time). The valve gain can be computed online by a simple calculation in the DCS to divide the filtered change in flow by the change in stroke. A ramp that is too slow can make the calculation too noisy by making the true change in flow small relative to the size of the noise. Alternately, an adaptive controller can be used to identify the process gain at various operating points for gain scheduling.

Variable Speed Drives

In a variable speed drive for liquid flow, the pump characteristic curve shifts with pump speed. Since there is no control valve, there is no valve drop and the flow is at the intersection of the pump curve and the system frictional loss curve. Figure 7-49a and 7-49b show the intersections for a system with zero static head and significant static head, respectively.

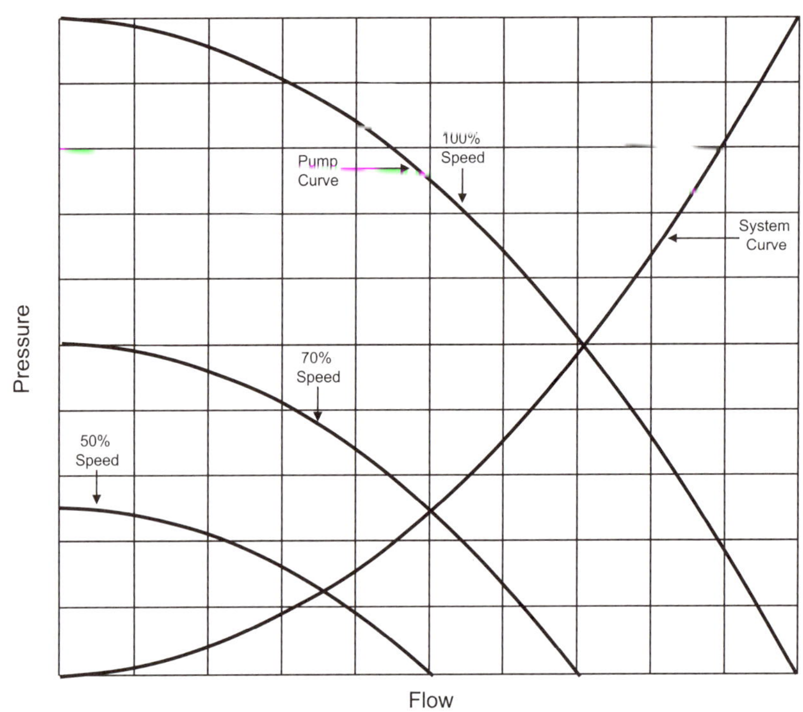

Figure 7-49a. Pump and System Curves for VSD Flow Control (zero static head)

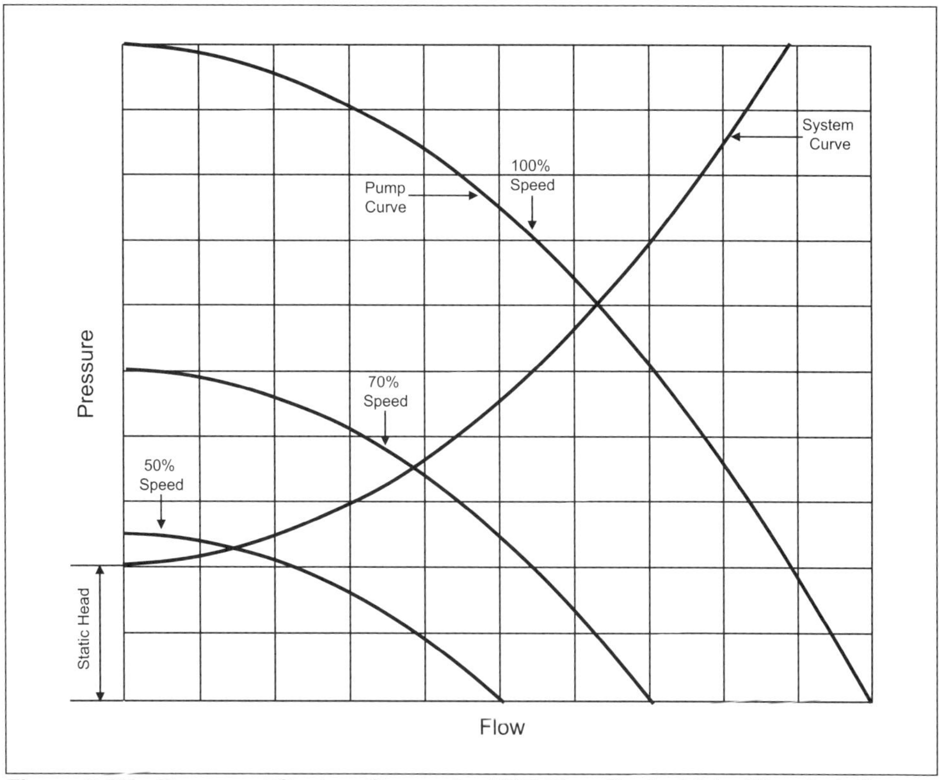

Figure 7-49b. Pump and System Curves for VSD Flow Control (significant static head)

For a negligible static head and an idealized pump, motor, and VSD, the change in flow with speed is linear until it reaches the point where the loss in pump efficiency and the increase in slip at low speed causes the flow to gradually decline to zero (Figure 7-50a).

If we ignore the loss in pump efficiency and increase in slip, a pump curve that approaches the static head will show a sharp bend downward to zero flow at low speed. The plummet of the speed depicted in Figure 7-50b at 50% speed causes a significant increase in gain, a noisier flow at low speeds, and a huge reduction in turndown [46].

A flat pump curve will cause almost a quick open type of flow characteristic. The high gain (sensitivity) at low speed can cause cycling [46]. Operation on a relatively flat pump curve can occur from improper pump selection or over-sizing.

Forward curved blades are used in fans to reduce size and cost. Unfortunately these blades cause a significant flattening at both low and high speeds making a highly nonlinear S-shaped plot of flow versus speed.

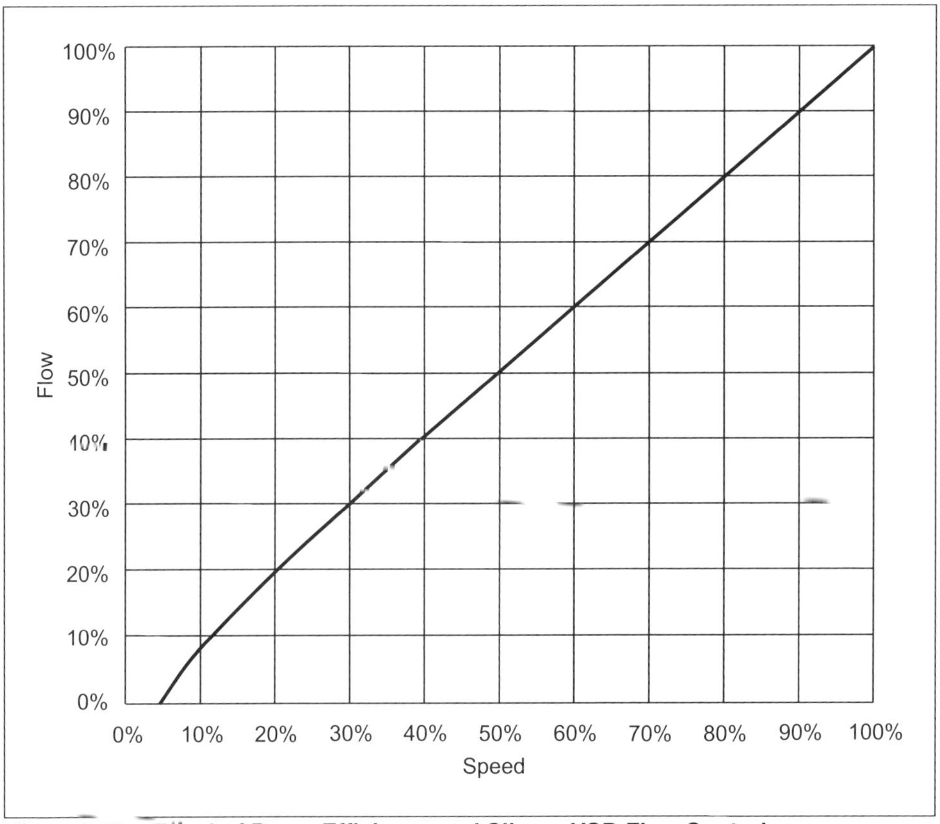

Figure 7-50a. Effect of Pump Efficiency and Slip on VSD Flow Control

Due to the complexities and unknowns in the piping system, the prime mover's characteristic curve, and the VSD's drive performance, it is difficult to get a plot of installed flow versus speed and the installed gain for a VSD. As we have seen, the emphasis on VSD applications to date has been in terms of energy savings. References [38] and [46] provide details of the positive and negative aspects, respectively, of VSD flow control for a given set of application conditions. Much more needs to be studied and published in terms of process control and loop performance. Until more data and guidance are available, it is important that the installed VSD gain be captured by a ramp or adaptive controller as discussed at the end of the Installed Valve Characteristics section.

7-7. Rangeability

The rangeability of a final element is the ratio of the maximum to minimum controllable flow. For pumps, the maximum controllable flow is near the intersection of the system frictional loss curve with the pump curve for maximum speed. The definition of the minimum controllable flow

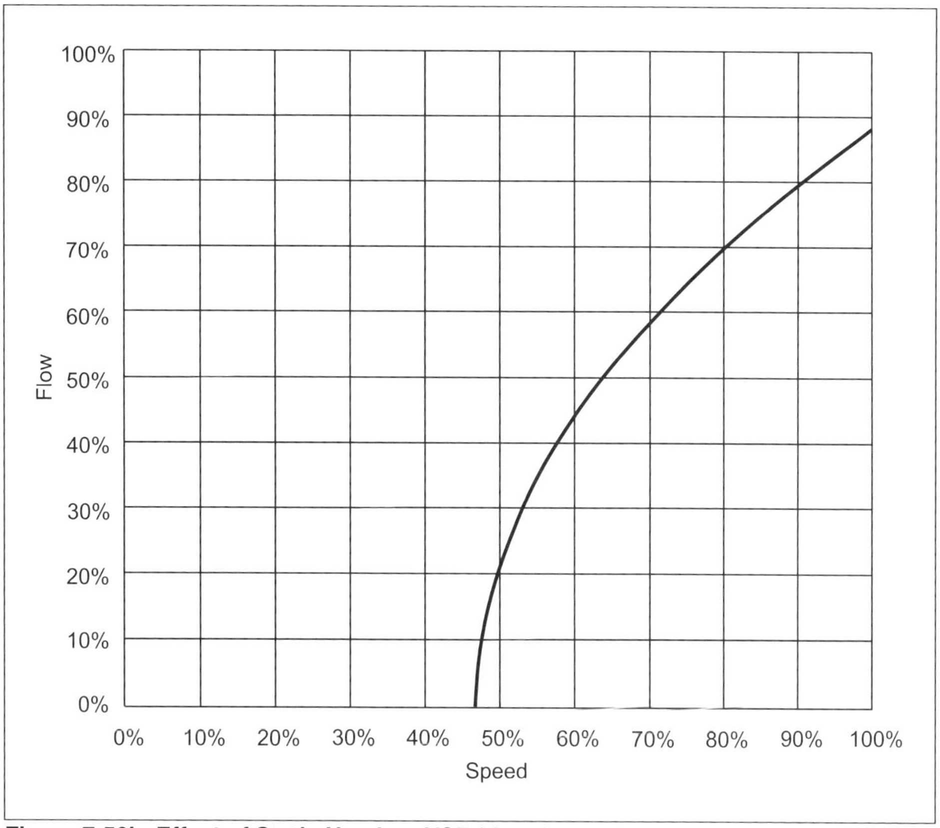

Figure 7-50b. Effect of Static Head on VSD Flow Control

depends on whether the final element is a valve or a VSD and the viewpoint of the supplier, user, or contractor.

Control Valves

Some valve manufacturers define the minimum controllable flow as the point where the inherent flow characteristic exceeds some specified tolerance. This definition has very little practical significance. Process control consultants tend to specify the minimum controllable flow to be where the valve gain becomes unacceptably small (e.g., valve gain < 0.5). The author prefers to base the minimum flow on the valve stick-slip (S_V) near the closed position. In this case the minimum practical controllable flow is the flow on the installed characteristic at a valve position equal to the stick-slip. The use of stick-slip provides a rangeability that is much larger than what control consultants might suggest but still much smaller than what piping valve manufacturers might state.

The minimum practical controllable flow can be estimated by first substituting the slip-stick near the closed position into the equations for the linear trim (Equation 7-19a) or equal percentage trim (Equation 7-19b). The resulting minimum flow is then substituted into the equation for the installed characteristic to get the minimum practical controllable flow (Equation 7-19c). Note that fractions instead of percentages are used. The rangeability of the control valve is simply the inverse of the minimum practical controllable flow expressed as a fraction (Equation 7-20d).

Minimum fractional flow coefficient for a linear trim and stick-slip:

$$C_{x\min} - \frac{S_v}{X_{v\max}} \tag{7-19a}$$

Minimum fractional flow coefficient for an equal percentage trim and stick-slip:

$$C_{x\min} = R^{[\frac{S_{vv}}{X_{v\max}} - 1]} \tag{7-19b}$$

Minimum controllable fractional flow for installed characteristic and stick-slip:

$$Q_{x\min} = \frac{C_{x\min}}{\sqrt{\Delta P_R + (1 - \Delta P_R) * C_{x\min}^{\ 2}}} \tag{7-19c}$$

$$R_v = \frac{1}{Q_{x\min}} \tag{7-19d}$$

Where:

C_x = minimum flow coefficient expressed as a fraction of maximum (dimensionless)

ΔP_R= valve pressure drop ratio (dimensionless)

$Q_{x\ min}$ = minimum flow expressed as a fraction of the maximum (dimensionless)

R_v = rangeability of control valve (dimensionless)

R = range of the equal percentage characteristic (e.g., 50)

$X_{v\,max}$ = maximum valve stroke (%)

S_v = stick-slip near closed position (%)

Variable Speed Drives

For variable speed drives, estimating rangeability gets trickier. The decrease in process gain from speed slip offsets the increase in process gain as the pump discharge head approaches the static head. If there are no overheating or cogging problems as suggested in the case for a pump and valve system with a well designed open loop (volts/Hertz) PWM drive, high resolution input card, and negligible static head, the rangeability is normally 40:1. When the pump head is operating near the static head, the minimum controllable flow is set by rapid changes in the static head and frictional loss. These rapid changes could be due to process noise and sudden or large disturbances. The speed can not be turned down below the amplitude of these fast fluctuations.

The rangeability of a VSD could drop to 4:1 for the following systems:

- Older VSD technologies such as six-step voltage (excessive slip at low speed)
- Systems with a high static head (flow plummets to zero at a low speed)
- Operation on the flat portion of the prime mover curve (cycling at low speed)
- Exposure to hot process or flue gases (motor overheats at low speed)

7-8. Questions

1. What do most control loops end up manipulating in the process to deal with disturbances or reach a new set point?
2. What are some of the drawbacks of using a regulator instead of a DCS loop for temperature or pressure control?
3. Why are piping valves not recommended for throttling service?
4. Which type of valve and actuator tends to have the best sensitivity (ability to respond to small steps in desired valve position)?

5. For small flows, what type of control valve offers the greatest rangeability?
6. For moderate flows, what type of control valve offers the greatest rangeability?
7. For large flows, what type of control valve offers the greatest rangeability?
8. For slurries, particularly pulp and bark, what type of control valve is best?
9. What types of final elements tend to have the greatest deadband?
10. When does deadband cause a limit cycle?
11. When does a resolution limit cause a limit cycle?
12. What accessory is used to position the valve to a fixed position from a discrete signal?
13. What accessory is used to make a valve with a large pneumatic actuator respond faster?
14. What is the principal limit to the turndown of a variable speed drive?
15. Why is the energy savings from a variable speed drive overestimated?
16. Why are variable speed drives not replacing control valves in projects?
17. Is a drive or a valve faster?
18. Why is an equal percentage inherent flow characteristic so popular?
19. How does the trend of a valve response change with step size?
20. How does the trend of a valve response change with ramp rate?

7-9. References

1. Baumann, H. D. *Control Valve Primer—A User's Guide, 4th Edition.* Research Triangle Park: ISA, 2006.
2. McMillan, G. K. "What Have I Learned? – The Cost and Source of Oscillations – Part 3." http://www.modelingandcontrol.com/, May 4, 2009.
3. McMillan, G. K. "What Is Your Flow Control Valve Telling You." *Control Design*, www.controldesign.com/articles/2003/164.html, May 2004.

4. Blevins, T. L., McMillan, G. K., Wojsznis W. K., and Brown, M. W. *Advanced Control Unleashed*. Research Triangle Park: ISA, 2004.

5. McMillan, G. K. *Good Tuning—A Pocket Guide, 2nd Edition*. Research Triangle Park: ISA, 2005.

6. McMillan, G. K. "A Fine Time to Break Away from Old Control Valve Problems." *Control*, Nov. 2005.

7. McMillan, G. K. and Weiner, S. "Downturn Turndown" in "Control Talk," *Control*, http://www.controlglobal.com/articles/2009, Jul. 2009.

8. Fitzgerald, B. *Control Valves for the Chemical Process Industries*. McGraw-Hill, Inc., 1995.

9. O'Keefe, W. "Reliable Dampers: Vital for Today's Powerplant." *Power*, Oct. 1981.

10. Peters, M. E. "Damper Control." *Automax Inc. white paper*, Feb. 1997.

11. Anderson, J. I., Aparicio, A. B., and Rutkoski, S. T. "Better Damper Control Delivers Big Savings." *Chemical Processing*, http://www.chemicalprocessing.com/articles/2004/62.html, Apr. 2004.

12. Nelson, D. "Better Damper Control Nixes NOx." *Chemical Processing*, http://www.chemicalprocessing.com/articles/2004/154.html, 2004.

13. *Rexa Electraulic™ Actuators and Drives*. Rexa Bulletin X2L-1.

14. *Electronically Actuated Control Valves*. Target Rock Corporation Bulletin EAC V-1, Oct. 1990.

15. Rinehart, N. and Ingram, N. "Smart Control Valve Diagnostics: Predictive Maintenance and Beyond." *Valve Magazine*, Volume 16, No. 4, Fall 2004.

16. McMillan, Gregory, K., "Improve Control Loop Performance." *Chemical Processing*, http://www.chemicalprocessing.com/articles/2007/200.html, Oct. 2007.

17. Coughran, M. T. "Valves: Testing for Peak Performance." *InTech*, Volume 41, No. 10, Research Triangle Park: ISA, 1994.

18. *FIELDVUE® Instruments – A Planning Guide for Effective Use – HART® Implementations*. Emerson, Dec. 2007.

19. *FIELDVUE® DVC6000 Series Digital Valve Controllers, Instruction Manual*, Form 5647. Emerson, Jul. 2007.

20. McMillan, G. K. "A Funny Thing Happened on the Way to the Control Room." http://www.easydeltav.com/controlinsights/FunnyThing/default.asp.

21. Arant, J. B. "Control Valve Notes and Positioner Myths – Part I." *Control*, Nov. 1993.

22. Arant, J. B. "Control Valve Notes and Positioner Myths – Part II." *Control*, Dec. 1993.

23. McMillan, G. K. "Valves That Can Turn on a Dime or at Least a Quarter – Part I." http://www.modelingandcontrol.com, Apr. 9, 2007.

24. McMillan, G. K. "Valves That Can Turn on a Dime or at Least a Quarter – Part II." http://www.modelingandcontrol.com/, Apr. 19, 2007.

25. Mamzic, C. L. "Improving the Dynamics of Pneumatic Positioners." *ISA Journal*, Research Triangle Park: ISA, Aug. 1958.

26. Parsons, B.W. "The Growing Use of AC Variable Speed Drives." *Automation*, Aug./Sep. 1982.

27. McMillan, G. K. "What Have I Learned – Wiring and VFD Problems." http://www.modelingandcontrol.com/, May 19, 2009.

28. Shuman, B. "Building a Reliable VFD System." Belden Inc., VFDWP, 2009.

29. Spear, M. "Get a Fix on ASD Problems." *Chemical Processing*, http://www.chemicalprocessing.com/articles/2005/597.html, Nov. 2005.

30. Heider, R. and Clay, L. "Make the Most of Variable Frequency Drives." *Control,* http://www.chemicalprocessing.com/articles/2009/043.html?page =print, Jan. 2009.

31. Bartos, F. "Field Oriented Control for Servos." *Control Engineering*, Feb. 2004.

32. Drury, B. *The Control Techniques Drives and Controls Handbook.* IEE Power and Energy Series 35. Cambridge: Cambridge University Press, 2001.

33. Spitzer, D. "Getting up to Smart Speed." *Control*, Sep. 2008.

34. Spitzer, D. *Variable Speed Drives: Principles and Applications for Energy Cost Savings, 3rd Edition*. Research Triangle Park: ISA, 2008.

35. McMillan, G. and Weiner, S. "Deal or No Deal" in "Control Talk," *Control*, http://www.controlglobal.com/articles/2008/063.html, Feb. 2008.

36. Hurst, J. and Dominguez, H. "Raising the Standard for Electric Motor Efficiency." *InTech*, Research Triangle Park: ISA, Apr. 2007.

37. McMillan, G. K. "Analog Control Holdouts." http://www.modelingandcontrol.com/, Apr. 2, 2007.

38. Shinskey, G. "Flow and Pressure Control Using Variable Speed Drives." *Proceedings of Control Conference*, pp 161-167, Chicago, 1980.

39. McMillan, G. K. *Tuning and Control Loop Performance, 3rd Edition*. Research Triangle Park: ISA, 1994.

40. Coughran, M. "Performance Influences in Globe Control Valves." *InTech, Engineer's Notebook – Valve Performance*. Research Triangle Park: ISA, Aug. 1996.

41. ISA-75.25.01-2000 (R2006) "Test Procedure for Control Valve Response Measurement from Step Inputs." Research Triangle Park: ISA, 2006.

42. ISA-75.25.02-2000 (R2006) "Control Valve Response Measurement from Step Inputs." Research Triangle Park: ISA, 2006.

43. Ruel, M. "Stiction: The Hidden Menace." *Control*, Nov. 2000.

44. Reinhart, N. and Jury, F. "The Impact of the Control Valve on Process Optimization." Hydrocarbon Processing, Process Optimization Conference, Houston, Apr. 1997.

45. Keith, G. "Linear Installed Characteristics Simplify Control Valve Selection." *Control Engineering*, Nov. 1977.

46. Smith, C. "Watch Out with Variable Speed Pumping." *Chemical Processing*, http://www.chemicalprocessing.com/articles/2008/069.html May 2008.

8

Control Valves

8-1. Learning Objectives

A. Appreciate the large variety of valves available to meet application requirements.

B. Be introduced to different valve and actuator constructions.

C. Understand how actuator, positioner, and valve design affect performance.

D. Learn what information is needed for selection and sizing.

E. See how process conditions and flow range affect valve selection.

F. Gain some insight into how valve and piping system design affect flow characteristics.

G. Find out what features are important for process control.

H. Learn terminology and issues to intelligently discuss industrial applications.

I. Visualize how valve components are integrated into a valve package.

8-2. Introduction and Overview

In the process industry, nearly every control loop relies upon on a final element to change a flow to affect the process. A control valve is used as the final element in 95% or more of the applications in chemical and petrochemical plants. A control valve must precisely and reliably regulate small and large flow rates despite exposure to solids, corrosive chemicals, and extremes in process temperature, pressure, viscosity, and velocity. The task of the control valve is more challenging than that of any other field component in an automation system. The wide range of process conditions and requirements has led to a large variety of control valves.

In Chapter 7, we covered the concepts and functionality important for process control and other final elements, such as dampers and variable speed drives. This chapter provides a cross-section of modern control valves that can be used to meet cost and performance objectives. Significant technological advances have been made in the last 15 years to improve the response and minimize the life-cycle cost of the control valve. Many of the improvements were triggered by studies in pulp and paper in the 1990s that concluded 35% of the variability in control loops was due to valve response problems.

In this chapter, the focus is more on the details necessary for selection, specification, installation, and maintenance of the most extensively used types of control valves. The coverage is necessarily limited by space and time and is not meant to be an all-inclusive survey of control valves. The amount of detail provided is intended to provide a perspective on what is important in the design, implementation, and maintenance of a control valve in the process industry. For actual applications, the publications, catalogs, software, and application specialists offered by the control valve supplier should be used. There is a long tradition of expertise at the local business partners and representatives of the major control valve manufacturers. With the decline in the number of experienced personnel and specialists in the process industry, it is more important than ever to take advantage of the technical information and assistance offered by the control valve supplier.

8-3. Basic Construction and Operation

The closure component in a valve changes the area provided for flow. The positioning of the closure component to regulate flow is called throttling. For sliding stem (globe) valves, the element (plug) is linearly moved into and out of a seat ring. For rotary valves, the closure component (e.g., ball, disk, butterfly, or eccentric plug) is rotated within a circumferential seal.

A more general name for the closure component is "closure member." For sliding stem valves, the plug and seat ring come as a set called "trim." An actuator is used to provide the force to move the closure component. Positioners provide feedback control of the position of the closure component.

The basic operation of valves and actuators is quite straightforward. Most control valves are supplied as an integrated package that includes an actuator, positioner, valve, and various accessories. Figure 8-1a shows an example of a modern control valve package for tight process control. This control valve package consists of a pneumatic diaphragm actuator, digital positioner, and sliding stem (globe) valve body.

The pneumatic actuator provides the force to position the closure component of the valve. The digital positioner converts the digital signal from the process control system (e.g., a distributed control system) to air pressure in the actuator. The positioner supplies or exhausts air to the actuator to provide enough force to make the actuator's shaft position match (in percent of stroke) the input signal. The closure component in its simplest form in the globe valve is a plug with a stem and a seat ring. In some designs, a cage is used to guide the plug (Figure 8-1b). The actuator shaft is attached to the stem. As the stem slides up and down, the plug moves in and out of the seat ring to vary the area for flow.

Pneumatic actuators are normally diaphragm or piston actuators. With a diaphragm actuator, operating pressure is applied to one side of a flexible diaphragm. Pressure on the diaphragm forces the diaphragm, shaft, and attached valve stem toward or away from the attached valve. Often, travel in the other direction is provided by the force of a spring. (Figure 8-1c). The diaphragm actuator provides excellent resolution.

Figure 8-1a. Diaphragm Actuator, Positioner, and Sliding Stem (Globe) Valve

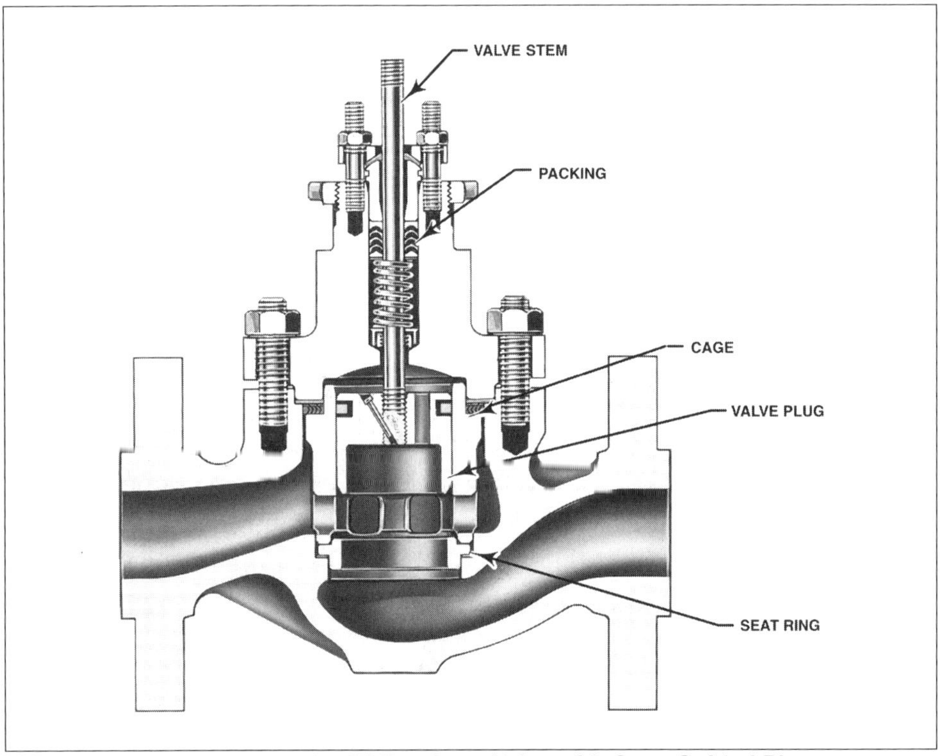

Figure 8-1b. Internal Parts of a Sliding Stem Valve with Cage Guided Plug

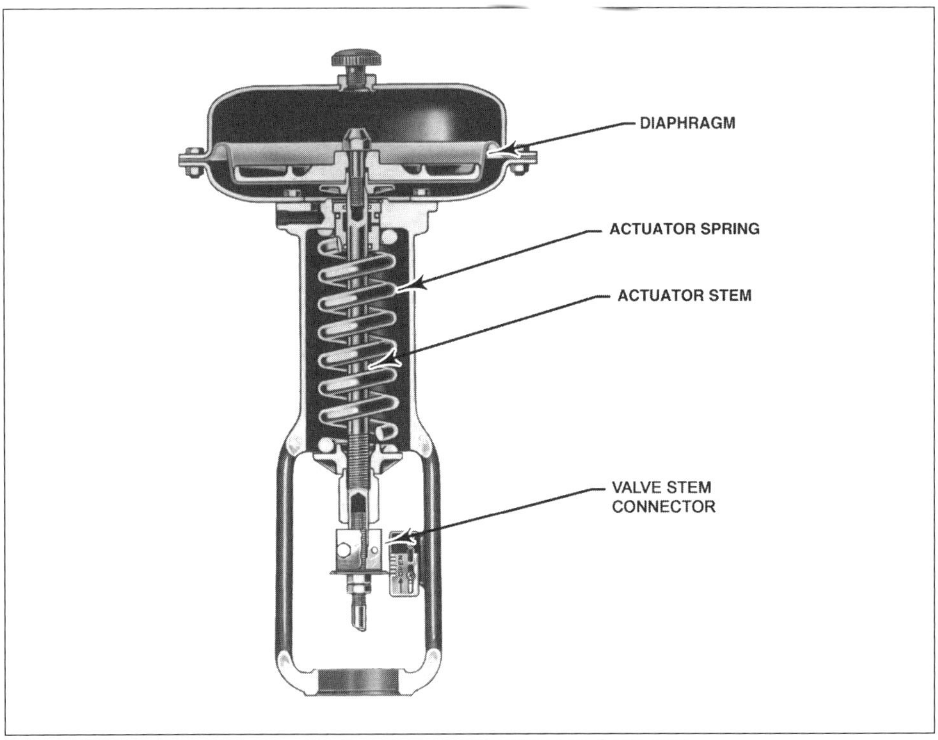

Figure 8-1c. Internal Parts of a Pneumatic Diaphragm Actuator

There are methods of regulating flow through sliding stem valves other than with a valve plug and seat ring. Some designs use a clapper that is forced against a seat ring or a diaphragm that is forced against a port.

Because the stem moves in relation to the valve body, it must be sealed to minimize leakage of process fluid. A packing of graphite, TFE, or composition materials often provides the seal between the stem and the valve body or bonnet. Elastomeric O-rings are an alternate means of sealing the valve stem. In the past, high friction packing was used to prevent deformation at high temperature or to provide a seal tight enough to prevent emissions of hazardous materials. In modern valves there is an increased emphasis on minimizing the friction of the packing to reduce stick-slip.

Figure 8-2a shows a piston actuator, a digital positioner, and a ball valve (a type particularly suitable for good control of high flow rates and slurries.) Pneumatic piston actuators operate in a manner similar to that of the diaphragm actuator, except the operating pressure is applied to a piston. The piston is held in a cylinder. Piston actuators normally use pressure applied to both sides of the piston, but spring-return piston actuators also are available. As the actuator shaft moves up and down, a lever arm attached to the actuator shaft rotates the stem of a ball segment inside the valve to vary the area for flow (Figure 8-2b). The position of the ball is expressed in terms of 0 to 90° rotation. The ball may have a v-notch to provide a better flow characteristic (plot of flow rate versus position).

Figure 8-2a. Piston Actuator, Positioner, and Rotary (Ball) Valve

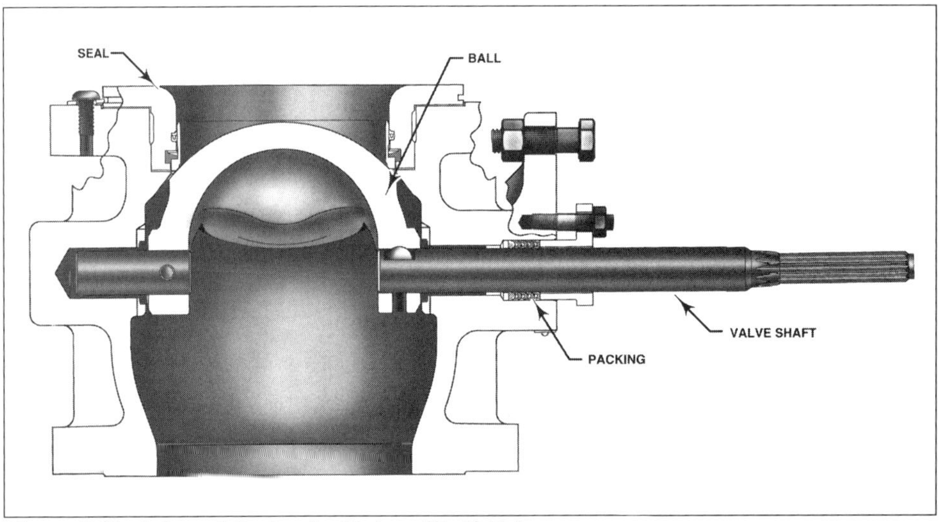

Figure 8-2b. Internal Parts of a Rotary (Ball) Valve

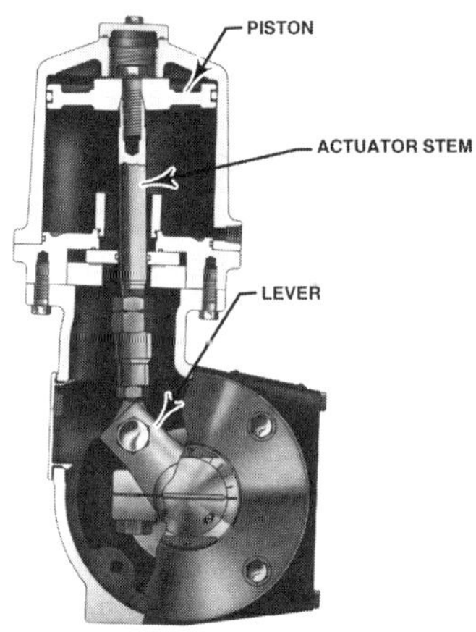

Figure 8-2c. Internal Parts of a Pneumatic Piston Actuator

For rotary valves some actuator models use a rack-and-pinion gear set instead of a lever to convert the linear motion of the actuator to rotary motion (Figure 8-3). However, the gear teeth impose a resolution (minimum change in position) that is not fine enough for tight control. The gear teeth can also become worn, causing some slip between positions. Section 7-5 of chapter 7 discusses how coarse resolution and stick-slip can cause deadtime and limit cycling. Rack and pinion actuators should only be used in applications where coarse control and limit cycling are permissible.

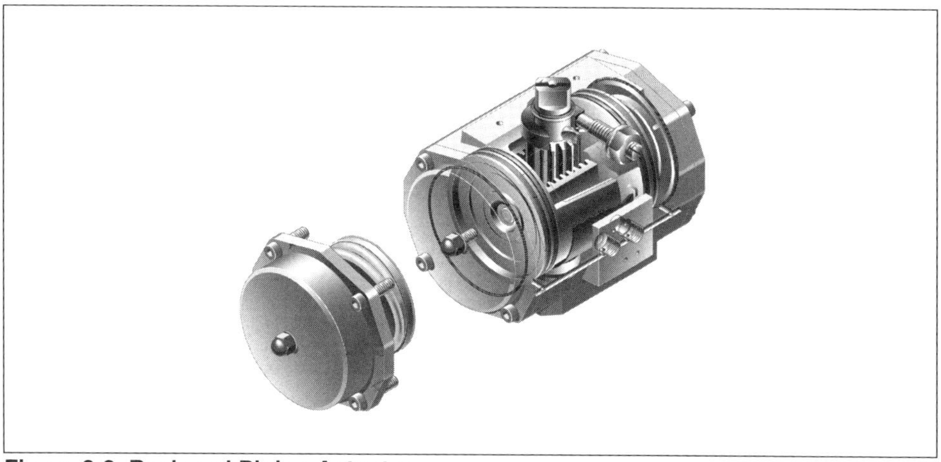

Figure 8-3. Rack and Pinion Actuator

If there is no compressed air available to operate an actuator, electric actuators can be used. There are models that use an electric motor to operate a valve through mechanical means, and those that use an electric motor-driven pump to provide hydraulic fluid to a cylinder and piston. In another design, a stepper motor turns a lead screw and nut assembly. The nut pushes or pulls the valve stem with it. Other models operate with a rack-and-pinion gear output for rotary valves (Figure 8-4).

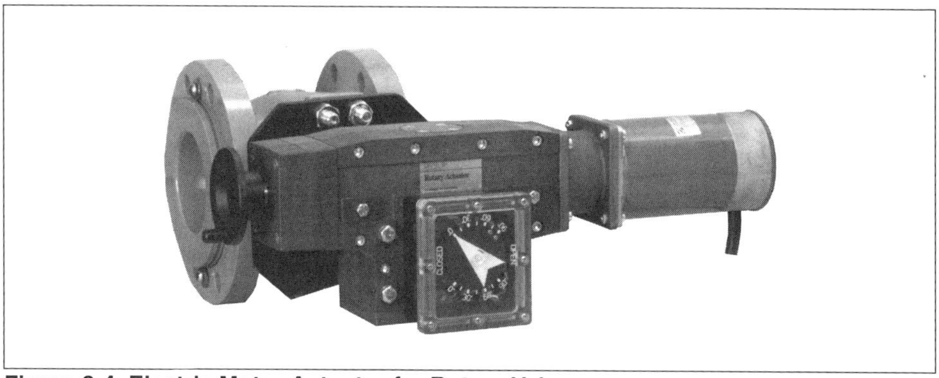

Figure 8-4. Electric Motor Actuator for Rotary Valves

Positioners

Many positioners operate with a nozzle and flapper arrangement that senses the pneumatic output from a controller and in turn, operates a pneumatic relay to provide the required output to the valve actuator. While there are other means of operation in other types of valve positioners, the nozzle-flapper principle is described here.

If the positioner is an electro-pneumatic positioner, the direct-current input signal is applied to an electro-pneumatic converter in the positioner. The converter then provides the pneumatic input signal pressure used by the pneumatic positioner. Electro-pneumatic positioners were predominantly used in the 1970s and 1980s (Figures 8-5a and 8-5b).

Supply pressure is connected to the relay. The relay limits flow to the nozzle so that when the flapper is not restricting the nozzle, air can bleed out faster than it is being supplied. The pneumatic input signal from the control device or the output signal of the converter in electro-pneumatic positioners is connected to the bellows.

When the input signal increases, the bellows expands and moves the beam. The beam pivots about the input axis, moving the flapper closer to the nozzle. The nozzle pressure increases and through relay action, increases the output pressure to the diaphragm or piston actuator.

The increased output pressure to the actuator causes the actuator stem to move. Stem movement is fed back to the beam by means of a cam. As the cam rotates, the beam pivots about the feedback axis to move the flapper slightly away from the nozzle. The nozzle pressure decreases and reduces the output pressure to the actuator.

Stem movement continues, backing the flapper away from the nozzle, until equilibrium is reached. When the input signal decreases, the bellows contracts (aided by an internal range spring) and the beam pivots about the input axis to move the flapper away from the nozzle. Nozzle pressure decreases and the relay permits the release of diaphragm casing pressure to the atmosphere. The actuator stem moves upward. Through the cam, stem movement is fed back to the beam to reposition the flapper closer to the nozzle.

When equilibrium conditions are obtained, stem movement stops and the flapper is positioned to prevent any further decrease in diaphragm case pressure.

Digital valve positioners were developed in the 1990s to provide better position control and diagnostics. Digital valve positioners receive an electrical input signal that provides electrical power to the unit as it communicates the set point. The signal is routed through the terminal box to the printed circuit board assembly, where a microprocessor runs a digital con-

Figure 8-5a. Electro-Pneumatic Valve Positioner on a Rotary Valve

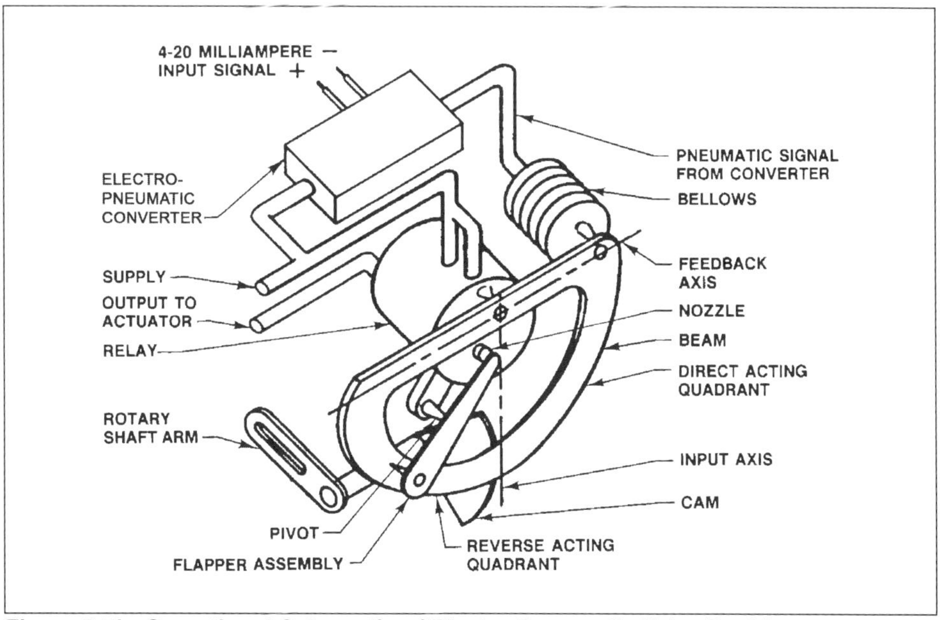

Figure 8-5b. Operational Schematic of Electro-Pneumatic Valve Positioner

trol algorithm resulting in a drive signal to the current-to-pneumatic converter (Figures 8-6a and 8-6b).

The converter is connected to supply pressure and converts the drive signal into a pressure output that is sent to a pneumatic relay. The relay, also connected to supply pressure, amplifies the small pneumatic signal into a single larger pneumatic output signal used by the actuator.

For double-acting actuators, the valve positioner provides two pneumatic output signals. A change in relay output pressure sent to the actuator causes the valve to move.

Valve position is provided by feedback linkage to the positioner travel sensor. The travel sensor is electrically connected to the printed circuit board to provide a travel feedback signal used in the control algorithm. The valve continues to move until the correct position is reached.

Figure 8-6a. Digital Valve Controller

8-4. Sizing and Selection

While valve sizing and selection processes are related, the required valve type may be known before sizing activity begins and before a selection of materials and options is made. However, if after calculating the valve size it is determined that the desired style of valve is not available in that size, it will be necessary to select another type and repeat part of the selection process.

Selection of a valve type is often a matter of tradition, personal preference, or company policy. For some applications, there is a valve specifically

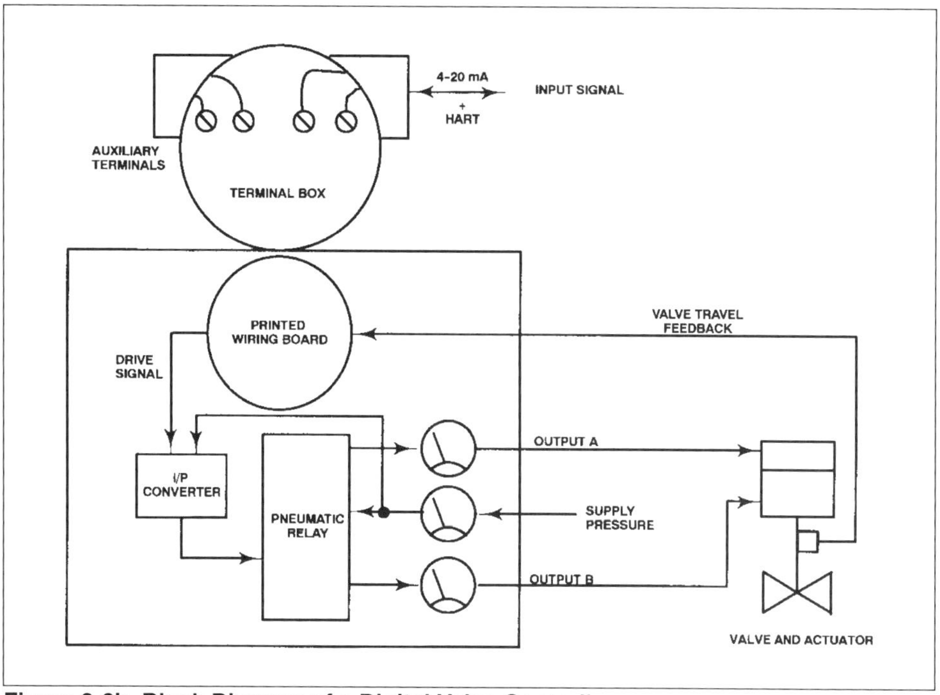

Figure 8-6b. Block Diagram of a Digital Valve Controller

designed to meet the need. As an example, sanitary-service applications require self-draining valves that can be cleaned in place and can meet other sanitary-service requirements (Figure 8-7).

Variations of basic valve types also are available. For diverting (flow-splitting) or converging (flow-mixing) service, where a flow stream is to be split into two flow streams, where two streams are to be combined, or where one flow is to be switched between two outlets, a three-way valve can eliminate the need for separate valves (Figure 8-8). To accommodate plant piping or where a self-draining valve is desired, an angle-pattern valve is the choice (Figure 8-9).

Performance Based Selection

Control valve performance is recognized as an important factor in efficient plant operation. High control valve performance helps ensure low process variability, which in turn reduces raw material costs and scrap costs. However, there has been some disagreement as to what is meant by "high performance." Piping valve manufacturers termed their rotary valve offerings high performance valves because they had high temperature ratings and/or extremely low leakage flows (tight shutoff). These valves often had high stiction near the closed position, excessive backlash in the linkage, and

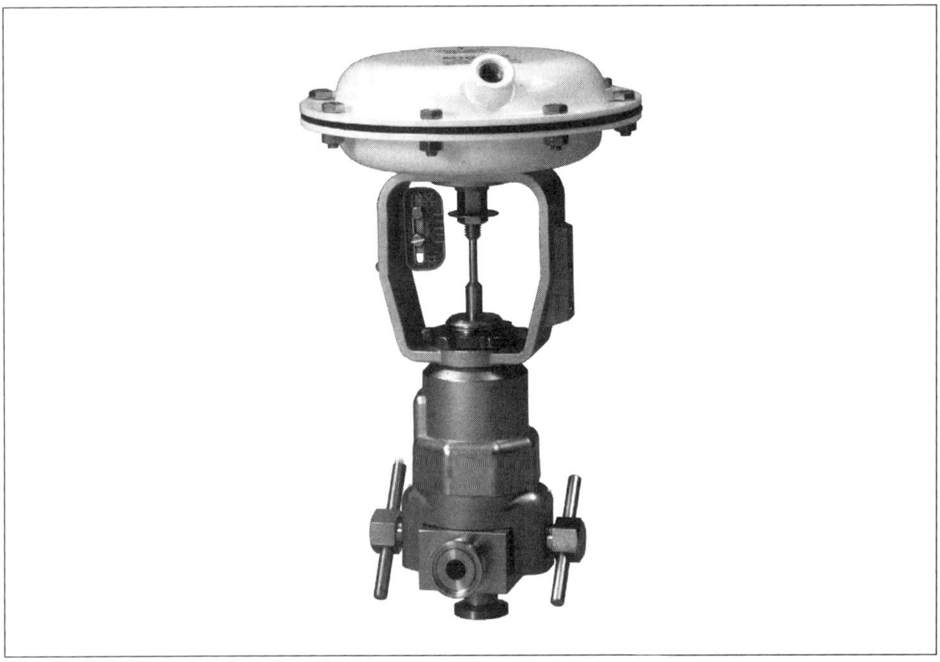

Figure 8-7. Sanitary-Service Valve

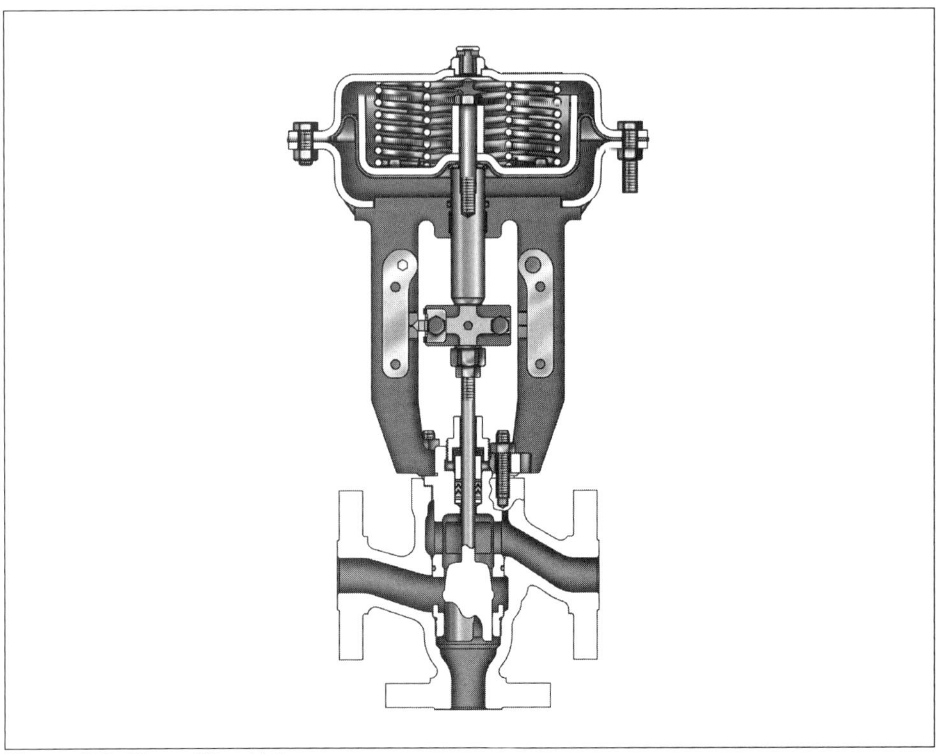

Figure 8-8. Three-Way Valve

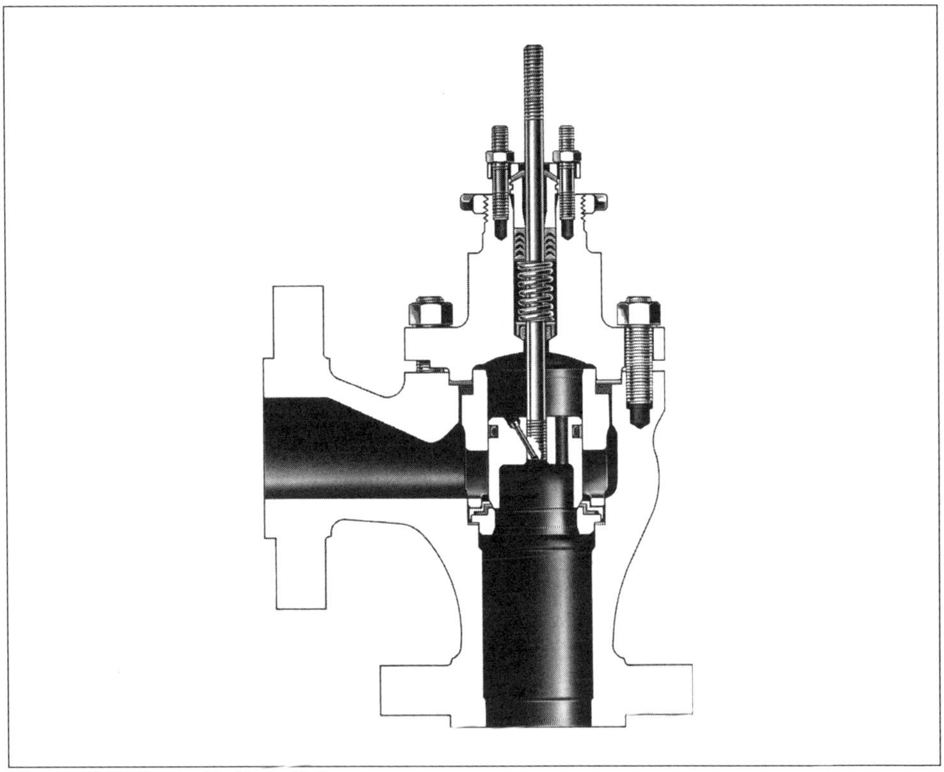

Figure 8-9. Angle-Pattern Valve

actuators with a poor resolution limit. The result was actually high process variability. Control valve manufacturers have been able to reduce the stiction and backlash in rotary valves that can provide low leakage. The term "high performance" is valid for these valves.

For tight process control, control valves must be able to respond to changes in the signal from scan to scan. For most control loops the change in controller output, even over the course of several successive scans, is less than 0.5% except for set point changes. The ability to respond to small changes is achieved by [1, 2, 3, 4, 5]:

- Minimizing actuator sensitivity
- Maximizing positioner sensitivity
- Minimizing valve resolution limit (stick-slip or stiction) from friction
- Minimizing valve deadband (backlash) in shaft and stem connections

- Minimizing rotary valve shaft windup from friction and long shafts

Stick-slip and shaft windup originates from packing, seating, and seal friction. Shaft windup occurs for the start or reversal of stroke when the actuator twists the shaft of a rotary valve but the ball, disk, or plug does not move until the signal exceeds a deadband. Control valves with resolution limits and deadband create deadtime and sustained equal amplitude oscillations called limit cycles (see Section 7-5 on Dynamics).

The nonlinearity (changes in valve gain with stroke) of a control valve is important but not as critical as resolution and deadband. The valve gain is the slope of the installed characteristic of the control valve. Changes in valve gain can be compensated by gain scheduling or signal characterization. Adaptive controllers can find and provide this gain scheduling automatically [6]. Even rough approximations and compensation of the valve gain provide some benefit. For resolution limits and stick-slip there is no technique that corrects for the steps in the valve position. For backlash, a deadband compensator can bias the output upon a change in signal direction but an incorrect setting can readily cause stick-slip and oscillations (limit cycling). The most practical action to help prevent oscillations is to turn off integral action in the process controller when the error (difference between set point and the process variable) is acceptable. The suspension of integral action stops limit cycling at the cost of an offset of the process variable from the set point.

Before the advent of online adaptive controllers that could identify the process gain and schedule the controller gain as a function of controller output, guidelines such as 0.5 to 2.0 (% flow per % signal) were developed to specify the permissible change in valve gain. The throttle range was defined as the region of the installed valve characteristic where valve gain stayed within the specified limits.

Each type of valve has a different throttle range based on this criterion. Part of the complication in setting this range is that a change in pressure drop and thus the piping system design affects the installed characteristic. For a given decrease in valve pressure drop with flow, typical throttle ranges are stated. The exact range is not as important as the relative size of the range. Globe valves and some new high-range contoured disk valves offer the widest throttle range (60% or more of total travel). Ball valves follow closely with a slightly narrower range. Next are rotary eccentric plug valves, followed by conventional disk butterfly valves.

A more practical definition of throttle range is based on the stick-slip and backlash near the closed position and where the control valve characteristic becomes essentially flat (see the Linearity (7-6) and Rangeability (7-7) sections of Chapter 7). The flow characteristic of a conventional disk butterfly valve typically becomes too flat for rotations greater than 60°. Rotary valves with high seal stiction can exhibit a stick-slip of 10% or more near the closed position (see the dynamics Section 7-7 of Chapter 7). [1, 2, 3, 4, 5]. The stick-slip is measured by making small incremental changes in the valve signal until the valve moves. The accumulated steps is stick. The movement is slip. If position feedback is not representative of closure component movement (often the case for rotary piping valves), a low noise and high rangeability flow measurement (e.g., magnetic or Coriolis flowmeter) is needed to detect valve movement.

Valve Selection

There is a wide range of valve types and styles to meet size, process flow variability, process fluid, temperature, durability, and pipe connection requirements. The globe valve offers the widest material selection and lowest process variability. However, particles and fibers can plug the passageways of globe valves. For slurries, V-notch ball valves are preferred to provide a clear flow path. For large flows, rotary valves (e.g., ball, butterfly, and eccentric plug) are significantly more economical. For line sizes above 24 inch (61 cm), the only choice is a butterfly.

Actuator Selection

Pneumatically operated actuators are in common use where compressed air is available to power the assembly. For applications requiring relatively short travels and low operating force or torque, spring-opposed diaphragm actuators should be considered since they generally offer the best sensitivity. Many of these operate from standard pneumatic signals of 3 to 15 or 6 to 30 psig (0.2 to 1.0 or 0.4 to 2.0 bar) and valve travel (linear) up to approximately 4 inches (102 mm). New high pressure actuators that use a 6 bar air (90 psi) supply are capable of much higher torque loads (Figure 8-10).

Table 8-1. Sample of Valve Types

Valve Size, NPS (nominal pipe size)	Type Available
1/4 to 24	Globe
1 to 12	Eccentric plug
1 to 24	Ball, high-range rotary
3 to 72 and larger	Butterfly
Feature or Other Characteristic Desired	**Type Available**
Compact; low weight.	Butterfly
Efficient low-flow trim	Globe
High capacity	Butterfly
Low pressure drop	Ball, butterfly, and eccentric plug
Minimum process variability	Globe
Slurry service	V-notch segmented ball
Severe-service capabilities	Globe and ball
Ruggedness	Globe and eccentric plug
Flangeless or lug-type (single flange)	Butterfly
Flanged or flangeless	Eccentric plug or ball
Flow characteristic options	Globe
Material options	Globe

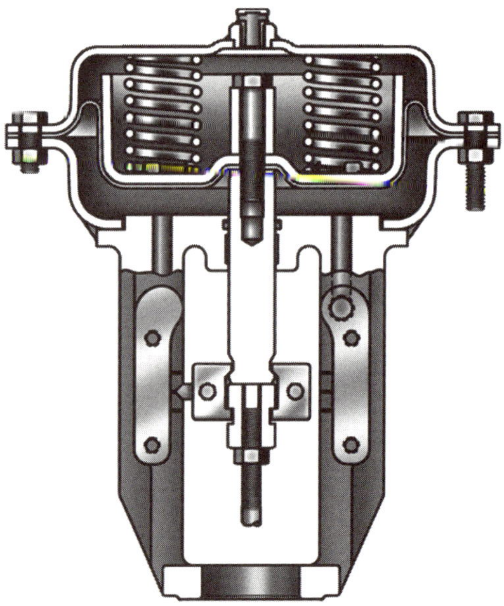

Figure 8-10. High Pressure and Torque Diaphragm Actuator

Pneumatic piston actuators are typically used for higher valve pressures and pressure drops and for longer shaft travel. As has been mentioned, pneumatic piston actuators are often double-acting (loading pressure is applied to both sides of the piston), but spring-return piston actuators are available. Piston actuators require position feedback control for balancing the pressures to both sides of the piston. If no supply air is available, electric actuators can be considered.

By selection of actuator type or changing internal parts, the actuator spring can provide spring-close or spring-open action. This capability provides the desired failure action on loss of air to the actuator.

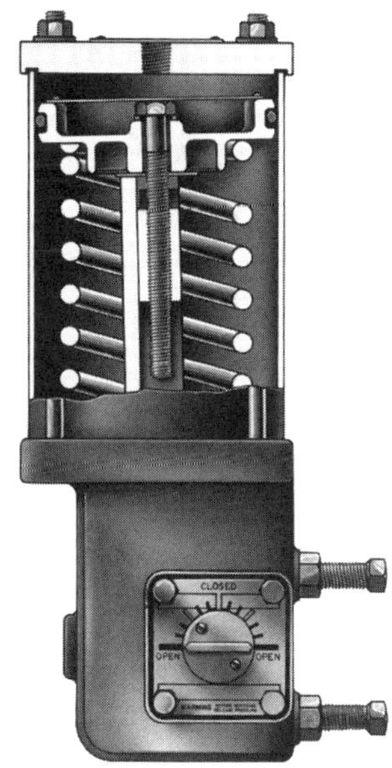

Figure 8-11. Pneumatic Spring-Return Piston Actuator

While a diaphragm actuator does not necessarily need position feedback control like a piston actuator, the use of smart digital positioners (e.g., digital valve controllers) is recommended for all types of actuators for improved valve response and diagnostics.

Positioner Selection

Smart digital positioners (e.g., digital valve controllers) are communicating devices with a feedback position controller that is conceptually analogous to a proportional only (P-only), proportional-derivative (PD), or proportional-integral-derivative (PID) controller in a process control system. The controller monitors valve position, compares it to the control signal and, in pneumatic actuators, increases or decreases pressure to the actuator as needed to correctly position the valve. Smart digital positioners offer the highly valued advantage of providing communication with the valve and diagnostic data on valve and instrument operating conditions (Figures 8-12a and 80-12b). Automatic calibration saves time over traditional methods of setting zero and span (starting point of valve travel and amount of travel). The advanced feedback control algorithm design and flexible tuning in a smart digital positioner provide tighter position control for a wide variety of valves.

An important consideration in positioner and valve controller selection is electrical and hazardous area classifications. Typical ratings include explosion proof, intrinsic safety, and flameproof. Manufacturers offer equipment that meets approval agency guidelines such as these for products used in areas with hazardous atmospheres. There are several approval agencies. Some ratings have become accepted worldwide, and some are for specific geographic areas.

Accessory Selection

In addition to controllers and positioners, there are accessories to be selected. Some of these are designed specifically to work with control valves and actuators:

- Supply pressure regulator—Supplies positioners, controllers, and other devices with a regulated supply pressure (Figure 8-13)
- Position transmitter—Transmits a pneumatic or electronic signal that is proportional to valve stem position (HART or Fieldbus positioners or smart digital positioners automatically offer position readback to the control system)
- Limit switch—Operates alarms, signal lights, relays, solenoid valves, or motors when travel (stroke) reaches pre-set points
- Three-way pneumatic switching valve—Switches pneumatic pressure on or off; can be used to open and close valves by a pneumatic signal (Figure 8-14)
- Three-way electrical solenoid valve—Switches pneumatic pressure on or off; can be used to open and close valves by energization or de-energization of a solenoid (Figure 8-15)

Setup Parameter	Factory Default Alert Settings	Recommended Initial Alert Settings	Notes
Control Mode	Analog	Analog	
Restart Control Mode	Resume Last	Resume Last	
Burst Mode Enabled	No	No	Set to Yes if a Tri-Loop is used
Burst Mode Command	2	2	
HART Tag	As Specified On Order	Fill in Plant Information	Tag from DCS, P&ID, etc.
Message	Blank	Fill in Plant Information	
Descriptor	Blank	Fill in Plant Information	
Date	Factory Calibration Date	Set to Current Date	
Valve Serial Number	Blank	Fill in Valve Serial Number	
Polling Address	0	0	
Max Supply Pressure	20	20	Should be 5psi above upper benchset or Actuator Maximum or DVC Maximum.
Feedback Connection	Rotary – All	Rotary – All	Determined by Setup Wizard
Zero Power Condition	Open	Open	Determined by Setup Wizard
Travel Sensor Motion	Clockwise	Clockwise	Determined by Setup Wizard
Analog Input Units	mA	mA	% for Delta V and Provox with HART Smart Cards
Analog In Range High	20 mA	20 mA	100 for Delta V and Provox with HART Smart Cards
Analog In Range Low	4.0 mA	4.0 mA	0 for Delta V and Provox with HART Smart Cards
Travel Range High	100%	100%	
Travel Range Low	0%	0%	
Pressure Units	PSI	PSI	
Temperature Units	F	F	
Tuning Set	F	F	Determined by Setup Wizard
Input Characteristics	Linear	Linear	
Set Point Filter Time	Filter Off	Filter Off	
Travel Limit High	125%	125%	End User to Determine
Travel Limit Low	−25%	−25%	End User to Determine
Travel Cutoff High	99.50%	99.50%	This parameter is dependent on service and trim type eg. CAVIII or standard. For example in BFP Recirc with a Reverse Acting valve we might want this set at 92% whereas for a microflute or Baumann™ trim 99.5% might be acceptable
Travel Cutoff Low	0.50%	0.50%	This parameter is dependent on service and trim type ie CAVIII or standard. For example in BFP Recirc with a Direct Acting valve we might want this set at 8% whereas for a microflute or Baumann trim 0.5% might be acceptable.
Minimum Opening Time	0 secs	0 secs	
Minimum Closing Time	0 secs	0 secs	
Integral Enabled	Enabled(1)	Enabled(1)	
Integral Gain	9.4 repeats/minute	9.4 repeats/minute	
Integral Deadband	0.25%	0.25%	
Travel Hi/Lo Alert Enabled	No	No	End User to Determine
Travel Hi Hi/Lo Lo Alert Enabled	No	No	End User to Determine

–continued–

Figure 8-12a. Typical Diagnostic Setup a Digital Valve Controller (Part 1)

Setup Parameter	Factory Default Alert Settings	Recommended Initial Alert Settings	Notes
Travel Alert High Point	125%	125%	End User to Determine
Travel Alert Low Point	−25%	−25%	End User to Determine
Travel Alert High-High Point	125%	125%	End User to Determine
Travel Alert Low-Low Point	−25%	−25%	End User to Determine
Travel Alert Deadband	1%	3%	
Travel Deviation Alert Enabled	No	Yes	
Travel Deviation Alert Point	125%	5%	
Travel Deviation Time	5 sec	5 sec	Alternatives are to use the step response test to determine the time it normally takes for a 5% step and then set the Travel Deviation Time to 2 times that value. Another option would be to use twice the T63 time
Pressure Deviation Alert Enable	Yes	Yes	
Pressure Deviation Alert Point	5 psi	5 psi	
Pressure Deviation Time	10 sec	10 sec	
Cycle Counter Alert Enable	No	No	This is a good tool to track problems. The end user needs to keep track of problems such as packing leaks, travel sensor failures, actuator diaphragm failures, etc. Upon the first problem showing up, this number needs to be recorded in the CMMS or AMS and reset. Repair the problem and record the number when the next problem occurs. This will give a history of good data that will allow us to enable this alarm and then generate a work order to address the problem before it fails.
Cycle Counter Alert Point	2,147,483,646	2,147,483,646	
Cycle Counter Deadband	3%	3%	
Cycle Counter	0%	0%	Verify 0% on initial setup
Travel Accumulator Alert Enable	No	No	Same comments as Cycle Counter Alert Enable
Travel Accumulator Alert Point	2,147,483,646	2,147,483,646	
Travel Accumulator Deadband	3%	3%	
Travel Accumulator	0%	0%	Verify 0% on initial setup.
Auxiliary Input Alert Enable	No	No	
Auxiliary Input Alert Point	Closed	Closed	
Supply Pressure Alert Point	0 psi	3 psi above upper benchset.	7 psi below nominal for piston actuators.
Drive Alert Enable	No	Yes	
Flash ROM Fail	No	No	
No Free Time	No	No	
Ref Voltage Fail	No	No	
Drive Current Fail	No	No	
Critical NVM Fail	No	No	
Temperature Sensor Fail	No	No	
Pressure Sensor Fail	No	No	
Travel Sensor Fail	No	No	

1. Firmware 7 only. Firmware 6 default is disabled.

Figure 8-12b. Typical Diagnostic Setup a Digital Valve Controller (Part 2)

- Volume booster—Increases stroking speed when mounted on the output of valve positioner by increasing air supply and exhaust rate to a pneumatic actuator
- Loading regulator—Provides manual control of actuators and other pneumatic devices
- Supply line filter—Filters supply air
- High-pressure instrument air supply system—Filters, protects against overpressure, and reduces and regulates high supply pressure
- Shutdown trip system—In pneumatic systems, provides a failure mode that opens, closes, or locks the valve in position in the event of a loss of supply pressure

Figure 8-13. Supply Pressure Regulator

Sizing

After valve type been selected, determining the required valve size, actuator size, materials, and options follows. There are valve and actuator packages that make sizing and selection easier. Conversely, there are valve and actuator product lines that allow custom selection from many materials, sizes, and options.

Figure 8-14. Three-Way Pneumatic Switching Valve

Figure 8-15. Three-Way Electrical Solenoid Valve

Valve sizing is based on separate sets of equations for liquids and gases. The equations have been slightly modified over the years, and now are documented in international standards. Today, the equations are included in computer-based valve and actuator programs that make the process relatively simple.

In the operation of a computer-based sizing program, the user enters piping information and the service conditions (e.g., upstream and downstream pressures, flow, temperature, density, viscosity, and vapor pressure) for minimum, normal, and maximum operating points. The program returns a value called the required flow coefficient, C_V. The required C_V is then used with manufacturers' published data to arrive at the size of valve needed. Many of the programs include the published flow data, so that the program will return the required valve size rather than just the required C_V. The same computer programs also can determine the expected noise level and required actuator size.

Like valve size, actuator size selection can be determined with equations and published data or by computer sizing programs.

Other Selections

Materials subjected to temperature extremes or in contact with potentially corrosive process fluids must be carefully selected. Valve bodies are available in a range of materials such as bronze, cast iron, stainless steels, high nickel alloys, and special alloys or materials. Internal parts are available in many different materials to suit the application. Valve suppliers, materials suppliers, and others publish relative compatibility charts to assist in selection.

End connection style is yet another selection to be made. Threaded, flanged, and welding-end connections are all available. For compatibility with piping systems, these connections comply with industry-wide standards. There are several standards for valves used in different world areas, and some end connection standards are unique to a specific industry.

With many sliding-stem valves there is a choice of flow characteristic, while some valve types offer only one. The characteristic you need will determine which valve model is best for your application. The inherent flow characteristic is the plot of percent flow rate versus percent stroke with a constant pressure drop across the valve. The most common inherent flow characteristics are quick-opening, linear, and equal percentage (Figure 8-16). The best characteristic depends upon the application (see 7-6, Linearity and 7-7, Rangeability in Chapter 7). As has been discussed, the equal percentage characteristic provides the best rangeability and control loop performance for a valve pressure drop that decreases with flow due to a significant frictional system loss (e.g., piping pressure drop). Frictional system losses alter the installed flow characteristic of linear trim to quick opening and the installed flow characteristic of equal percentage trim to linear.

A linear installed flow characteristic is generally the most desirable. The exception is pressure relief and surge prevention where a high initial flow rate is required. The equal percentage characteristic is also the most forgiving in terms of errors in valve sizing. For these and other practical reasons, more than 90% of the control valves in a typical process plant have an inherent flow characteristic that is similar to an equal percentage characteristic in that the valve gain increases with stroke up to its high throttle range limit. Control valve manufacturers publish guidelines to help with the selection of the valve type and inherent characteristic.

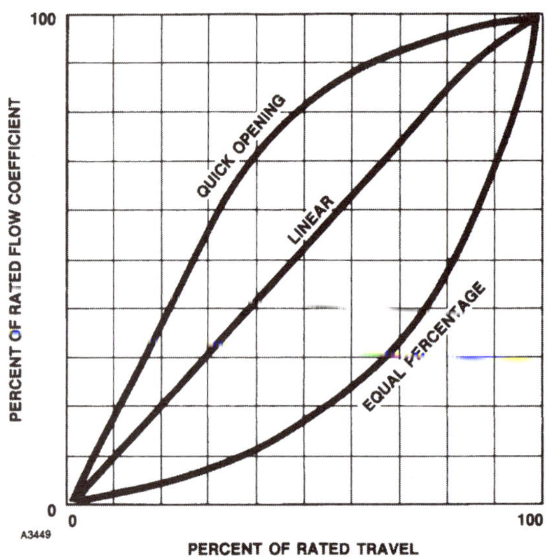

Figure 8-16. Inherent Flow Characteristics

Required Process Information

Before consulting a supplier or supplier's literature to begin sizing and selection, gather the following information about your process:

- Fluid to be controlled
- Fluid temperature
- Fluid viscosity
- Fluid specific gravity

- Maximum and minimum required flow rate
- Maximum and minimum inlet and outlet pressures
- Pressure drop at shutoff and in normal flow conditions
- Maximum allowable noise level
- Connecting pipe size and desired connection type (flanged, threaded, etc.)
- Failure mode (fail open or closed, e.g., on loss of supply pressure)
- Supply pressure available at installation site
- Instrument signal (e.g., analog, Foundation Fieldbus, HART, and WirelessHART)

Final selection can be an iterative procedure. The desired valve might not be available in a material that is compatible with the process fluid. Substitution of a valve type often requires repeating part or all of the selection process.

8-5. Installation

Valves and Actuators

Correct valve and actuator installation is critical for safe, efficient operation and to ensure compliance with environmental regulations and with plant safety requirements and operating procedures. Read and follow the manufacturer's instructions before beginning installation. Especially important is the safety information in the instructions (Figure 8-17).

If the equipment is to be stored before installation, be sure it is stored in a safe location. Contact the manufacturer for storage standards for the type of equipment you have purchased.

When unpacking the equipment, be sure to remove all packaging materials such as shipping stops, bracing, and packaging materials.

Often, valve and actuator openings will be fitted with covers or plugs that must be removed before installing. Do not remove these until you are ready to install.

Figure 8-17. Follow All Safety Instructions when Installing and Maintaining Valves

Inspect the equipment for any damage that might have occurred in transit or storage. If there are any signs of damage, do not install the equipment until the damage is evaluated and corrected. Also, make sure that no foreign material has collected in the inlet or outlet openings of valves or in the connections of pneumatic actuators. Foreign material that enters the flow stream or pneumatic supply can damage parts such as seats and seals. It also can block flow through the valve, instruments (e.g., sensors), or connected process equipment (e.g., static mixers, heat exchangers, and filters).

Check the location where the equipment is to be installed. Make sure there is enough clearance to move the equipment into place and to allow easy access for adjustment and maintenance. Compressed air or electricity must be made available to power the assembly, and a signal pressure or electronic signal must be brought in through tubing or wires.

There must be enough space to remove actuators from valve bodies because actuators must often be lifted clear of a valve stem before they can be removed.

Some valves may be installed in any position, but check the instruction manual to find the preferred position. Most often, the actuator is positioned vertically above the valve.

There are many safety precautions that must be considered before beginning installation. Failure to observe these precautions and to take appropriate action could result not only in damage to the process and the equipment but in severe injury to people installing or using the equipment. Many process fluids are inherently dangerous, and you must be prepared and equipped to contain and handle them safely. Also, many control valve assemblies are large and heavy and require special equipment to move and install the valve. Refer to the manufacturer's instructions for safety information.

A few of the more obvious points are mentioned below. Also refer to your plant's safety procedures. Check to see if any government regulations apply to your installation.

- Install auxiliary safety devices, such as relief valves and emergency shutdown systems if needed.
- Provide and use protective gloves, clothing, and eyewear when performing any installation operation.
- A remote vent line might be required if equipment is capable of venting flammable or hazardous gases. Electrical grounding is often required.
- Consider any other devices or processes that are connected to the equipment you are going to install. Take precautions to avoid over-pressuring connected devices.
- If hoisting the valve, take care to prevent injuring people or equipment in case of slipping.
- Check for leakage at packing, joints, and tubing connections when you pressure test the assembly.
- Monitor and readjust valve packing when the valve is put into operation.

In addition to following all safety procedures, apply good practices when making the actual installation. Flange gaskets are required on some flanged valve body styles (Figure 8-18 left); check the manufacturer's instructions for recommendations. With any flange type, uniform tightening to specified torque values of flange bolts and nuts is important. Follow accepted piping practices.

Valves intended to be welded into the pipeline have welding ends for this purpose (Figure 8-18 right). Dimensions of welding ends conform to industry standards. When a valve is welded into a pipeline, there is danger of welding heat damaging internal valve parts. Check the manufacturer's recommendation to determine if the internal parts must be removed before welding.

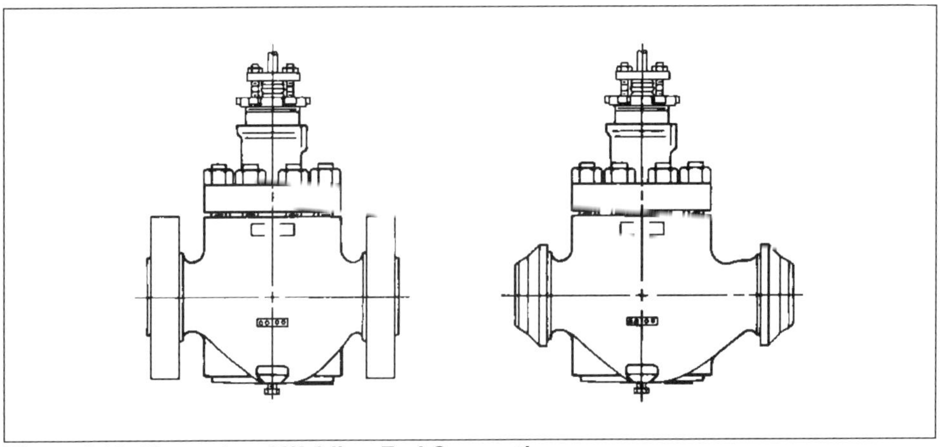

Figure 8-18. Flanged and Welding-End Connections

Positioners, Controllers, and Accessories

If the valve and actuator are part of a packaged control valve assembly, the positioner and accessories (e.g., air supply regulator, solenoid valves, and volume boosters) are attached to the valve and ready for installation with the valve.

If they are separate components, the manufacturer's instructions should provide information on attaching the positioner and accessories to the control valve. The installation site must have available electrical power, compressed air, required plant piping, and cabling, output cards, or WirelessHART communication for signals. As with valves, refer to the manufacturer's instructions, plant practices, and applicable government regulations for safety information.

The safety precautions for control valves and actuators apply to instruments and accessories. In addition:

- Supply pressure must never exceed the maximum safe working pressure of the instrument or of any connected equipment. High-pressure regulators in the supply line might be necessary.

- If natural gas or other flammable or hazardous gas is to be used as the supply pressure medium, the gas must be vented properly. A remote vent might be needed.
- The supply pressure medium must be clean, dry, and oil-free. Use a supply pressure filter if necessary.
- Tubing and instrument components that contact the supply pressure medium must be of suitable material that will not be corroded by the supply medium.
- The electrical and enclosure classifications for the instrument must be correct for the installation. All equipment must be properly grounded.

For valve positioners, calibration and alignment are often required during installation.

Alignment might require the positioning and re-positioning of physical parts, such as pneumatic bellows, links, and connecting arms. Calibration usually requires setting adjustments in response to applied signals and observing the output of the device. Refer to the manufacturer's instructions for details.

8-6. Adjustments

Valves and Actuators

Control valves and actuators require little adjustment. They are covered briefly here.

Many, but not all, spring-and-diaphragm actuators require spring adjustment. The spring is adjusted when the valve and actuator are first put into operation. There should be little need for periodic adjustment unless service conditions change or if the initial setting proves not to be optimum. This adjustment is often called the "bench set" or "initial wind-in."

Spring adjustment sets the starting point and range of diaphragm pressure over which the valve will complete full travel (Figure 8-19). The goal is to be sure that the valve plug or disk will begin to travel from the fully closed position and will reach full travel over the desired range of diaphragm pressure. A few actuators have adjustable travel stops that limit valve travel at one or both ends of travel.

Figure 8-19. Actuator Spring Adjustment

On many valves the packing can be adjusted (Figure 8-20). The adjustment applies force on the packing to create a better seal between the packing and the valve stem and between the packing and the packing box wall. This adjustment must take into account both leakage and friction.

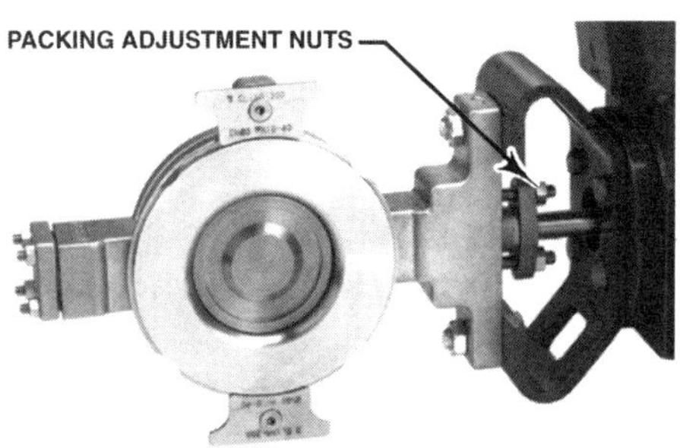

Figure 8-20. Packing Adjustment

Digital Positioners

Adjustments for these instruments vary with the type of instrument and the manufacturer. Adjustments are often required to set the starting point of valve travel (zero) and the span (amount of travel). Also, there may be gain and rate tuning adjustments. Refer to suppliers' instructions for information on how to adjust a particular instrument. See Dynamics Section 7-5 in Chapter 7 for more details on tuning a digital positioner.

8-7. Maintenance

Valves and Actuators

Control valve and actuator assemblies are not complex devices, but actuator forces and some process fluids, temperatures, and pressures they control can be dangerous if released accidentally or unexpectedly during maintenance. Care must be taken to avoid injury to personnel and property. Maintenance should not be attempted by persons who are not trained and equipped to work on the equipment.

The manufacturer's maintenance instructions must be available and understood before proceeding. Also, plant maintenance safety guidelines and applicable government regulations must be observed.

The valve assembly and the process it controls should be monitored continually. Monitor performance of the process that the valve is controlling. If the process is within normal tolerance but is nearing tolerance limits, it could be that preventive maintenance is required before serious problems develop. Regular preventive maintenance is much more effective than merely reacting to problems once they become problems.

A much better approach is to anticipate or predict problems via predictive maintenance. Computer-based valve instruments, such as smart digital positioners with in-service diagnostics software, offer a significant improvement over reactive or preventive maintenance. On-screen displays show a highly visible indication of several valve parameters. You can check variables such as instrument pressure leakage, supply pressures, travel deviation, instrument supply pressure quality, friction, and trends in friction.

Maintenance is a critical activity for any valve that handles erosive flow that can wear away even the hardest materials. For such valves, a predictive maintenance program is important to reduce maintenance costs, downtime, and process variability from damaged internal components in the flow path.

The best computer diagnostics cannot reveal every possible problem. A schedule of visual inspection of valves and actuators should also be established and followed. Check for signs of packing leakage, leakage at joints, and new or unusual vibrations or noise. Any of these could indicate the need for maintenance.

Often, plants have regularly scheduled shutdowns. If it is determined that a valve or actuator will require maintenance during shutdown, spare parts that might be needed, but are not normally stocked, should be obtained beforehand. The equipment can also be updated during a plant shutdown.

Taking large valves or welding-end valves out of the line requires considerable time and expense. For these valves, maintenance scheduling is even more critical. Check other components in the system to be sure the valve must be removed. If some other component in the system is causing the problem, removing, disassembling, and re-installing the valve would be a wasted effort.

A few of the typical maintenance procedures required by valves and actuators are:

- Packing lubrication. Packing forms a seal with a moving stem or rotating shaft. Some packing materials and process fluids require periodic lubrication. If so, a packing isolator and lubricating valve will make lubrication easier (Figure 8-21).
- Adjusting and replacing packing. Because there is continual movement between the valve stem or shaft and its packing, loading force on the packing must occasionally be increased. Innovative live-loaded packing systems compensate for wear automatically and greatly reduce packing maintenance (Figure 8-22).
- With any packing system, the packing will eventually need to be replaced. In addition, scratches on the valve stem or shaft can damage the packing, making packing replacement and maintenance to the stem or shaft necessary.

- Replacing gaskets and seals. Gaskets and seals are often called "soft parts" and should be kept on hand at all times.
- Seating parts. With metal seats, the valve plug and seat ring, or other seating parts in rotary valves sometimes require maintenance. A lapping process is effective to remove minor scratches in some designs, and it improves shutoff capabilities. In other valves, seat rings, seal rings, elastomeric seating parts, or valve disks might require replacement to improve shutoff.
- Damage from cavitation, vibration, erosion, or corrosion normally requires parts replacement.
- With actuators, replacement of worn O-ring seals, piston rings, diaphragms, and gaskets are typical maintenance operations.
- Actuator springs can be replaced if it is necessary to change the operating range of some actuator models.
- If an actuator is removed and disassembled, it might be necessary to re-adjust the spring compression when re-assembling.

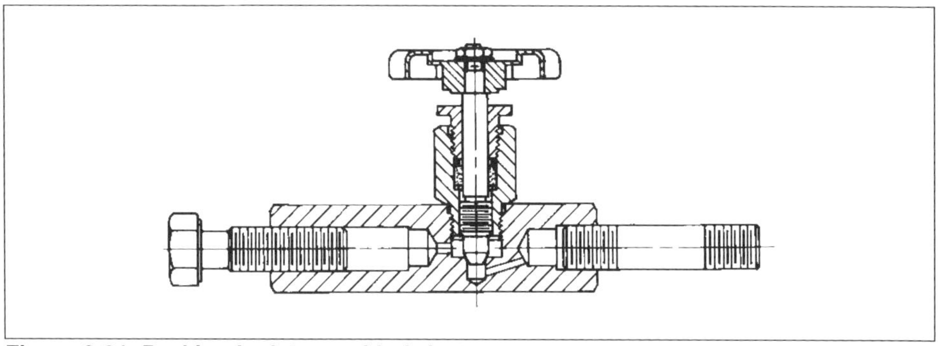

Figure 8-21. Packing Isolator and Lubricator

Manufacturers' instruction manuals contain several aids to maintenance. Normally, there will be cross-sectioned or exploded-view drawings to help you identify the parts referred to in the maintenance steps (Figures 8-23 and 8-24). Some manuals offer explanatory drawings to help you understand, use, and maintain the product (Figure 8-25).

As with installation, there are many safety precautions that must be considered before beginning maintenance. Failure to observe these precautions and to take appropriate action could not only result in damage to the process and the equipment, but also cause severe injury to people main-

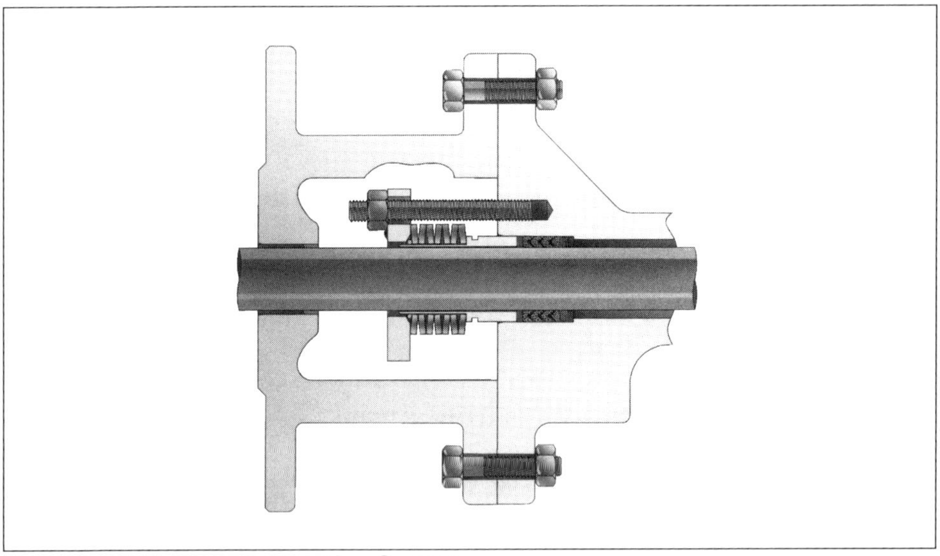

Figure 8-22. Live-Loaded Packing System

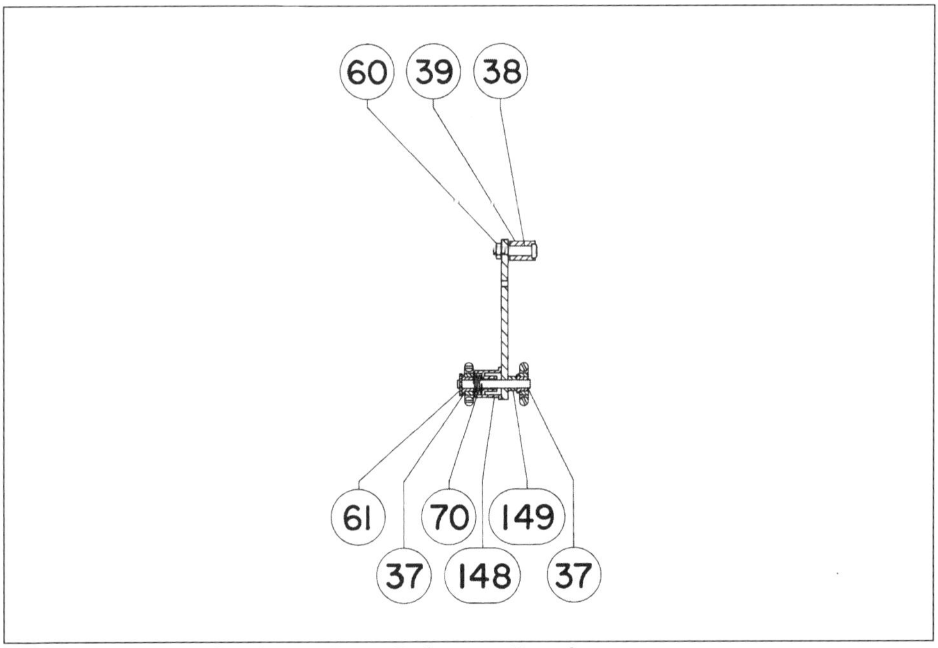

Figure 8-23. Cross-Sectioned Parts Reference Drawing

taining or using the equipment. Refer to the manufacturers' instructions for safety information. A few of the more obvious points are mentioned below. Also refer to safety procedures that your plant might have. Your plant also might have warning signs posted in areas where you must be aware of hazards. Products often have warning tags attached to alert the

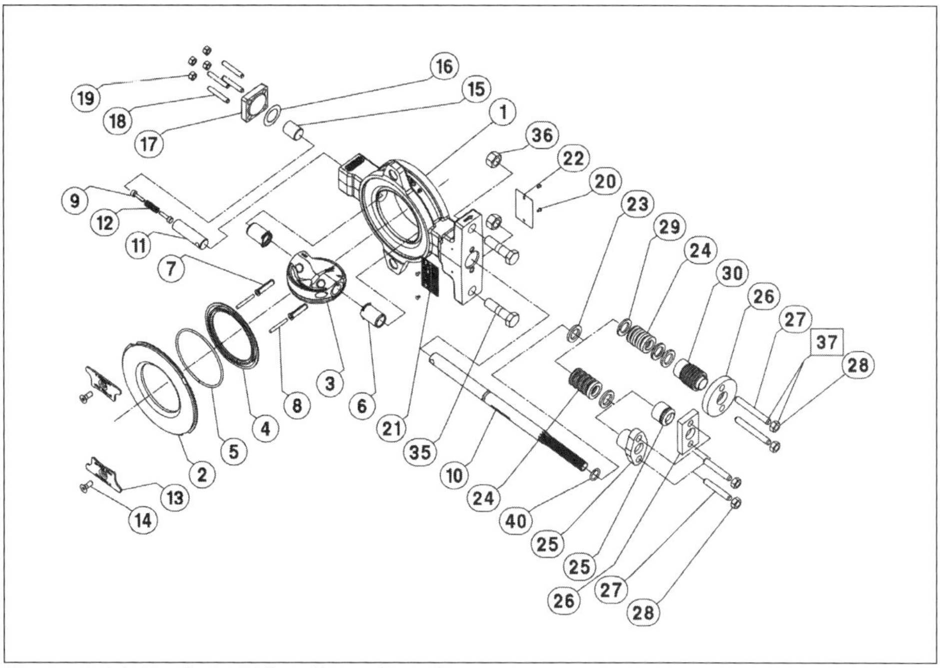

Figure 8-24. Exploded-View Parts Drawing

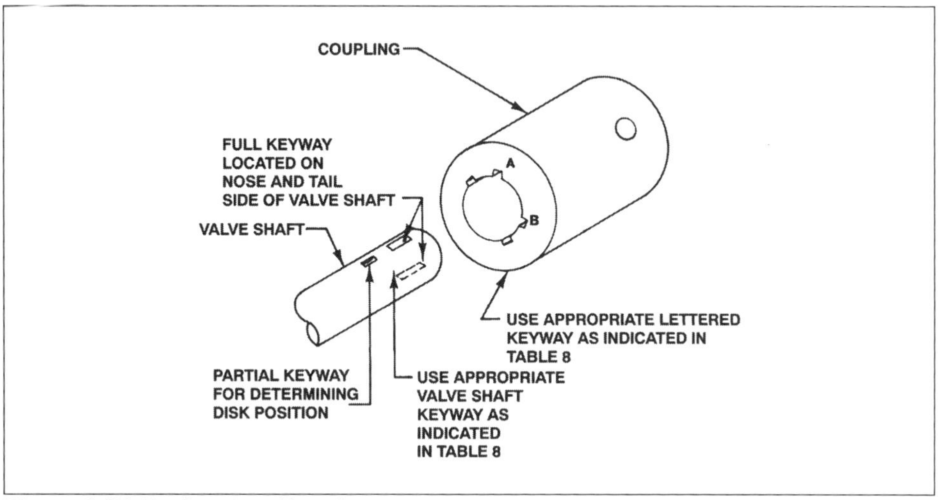

Figure 8-25. Explanatory Drawing

user (Figure 8-26). Check to see if any government regulations apply to your maintenance procedures.

- Provide and use protective gloves, clothing, and eyewear when performing any maintenance operation.

- The valve packing box might contain process fluid that remains pressurized, even when the valve has been removed from the

Figure 8-26. Product Safety Tag

pipeline. Cautiously remove packing box parts (e.g., packing follower and retainer) so that fluid escapes slowly and safely.

- If hoisting the valve, take care to prevent injuring people or damaging equipment in case of slipping.
- Stay clear of any moving part and stay clear of any part that might drop or shift during disassembly.
- Do not apply pressure to a partially assembled valve.
- Do not remove a stuck bonnet by pulling on it with equipment that can stretch the bonnet or otherwise store energy. The sudden release of stored energy can cause uncontrolled movement of the bonnet.

Smart Digital Positioners

Most maintenance procedures for instruments require that the valve and actuator be taken out of service before working on the positioner.

Refer to the manufacturer's instructions, plant safety practices, and applicable government regulations for safety information.

The safety precautions given above for control valves and actuators apply to instruments and accessories as well. Also, for instruments using electric power or electronic signals, be sure to shut off and remove electric power before opening any instrument cover in hazardous areas.

8-8. Diagnostics

Performance diagnostics can be provided through smart digital positioners mounted on a valve-actuator assembly. Control system operators can communicate with the valve using a hand-held field communicator or using a control system console over the same two-wire control signal loop or with a wireless communication system. In addition to performance diagnostics, the communications allow access to calibration, configuration, maintenance, and additional process data.

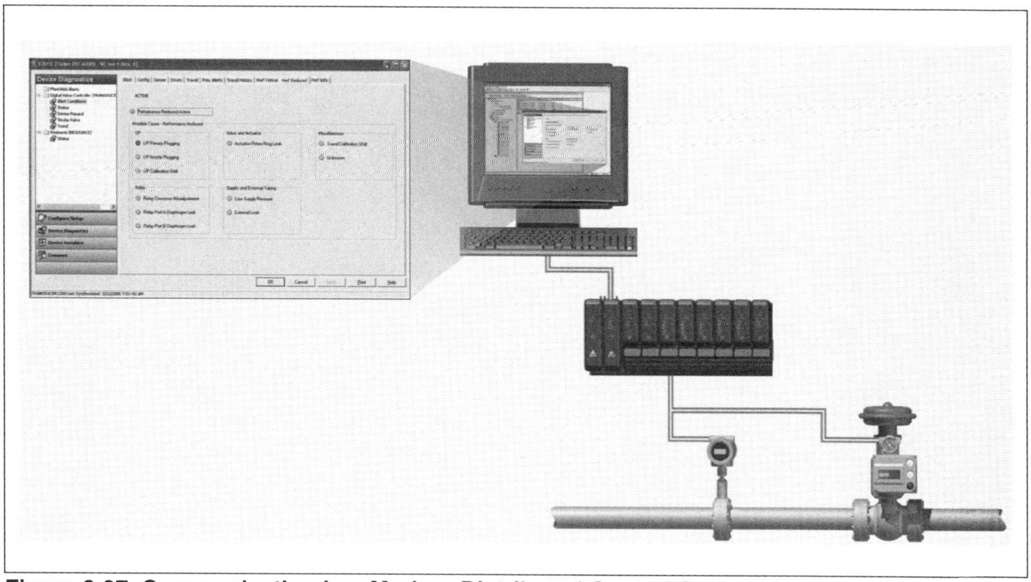

Figure 8-27. Communication in a Modern Distributed Control System

For both preventive and predictive maintenance, performance diagnostics detect problems and help the operator correct conditions that could lead to valve failure, preventing unplanned shutdowns. While valves are in service, the controller helps diagnose, analyze, and show trends in critical variables. On-line alerts can tell operators when (and what) action is needed. The system can email or send a paging signal to an operator when needed.

Some examples of problems that performance diagnostics can detect are:

Detected Problem	Test
Blocking or binding of the valve	Valve friction
Broken or loose linkage	Deadband
Dirty instrument air	Current-to-pneumatic and relay integrity
Actuator and tubing leaks	Air mass flow
Poor calibration	Travel deviation
Insufficient air supply	Supply pressure test
Instrument elastomer wear (e.g., seals)	Current-to-pneumatic and relay integrity

Other features of smart digital positioners with diagnostics capabilities are auto calibration, performance tuning, travel control, and partial-stroke testing. Valve-actuator signatures (bench set, seat load, friction, etc.) can be compared against previously stored signatures to discover performance changes that might indicate the need for adjustment or repair.

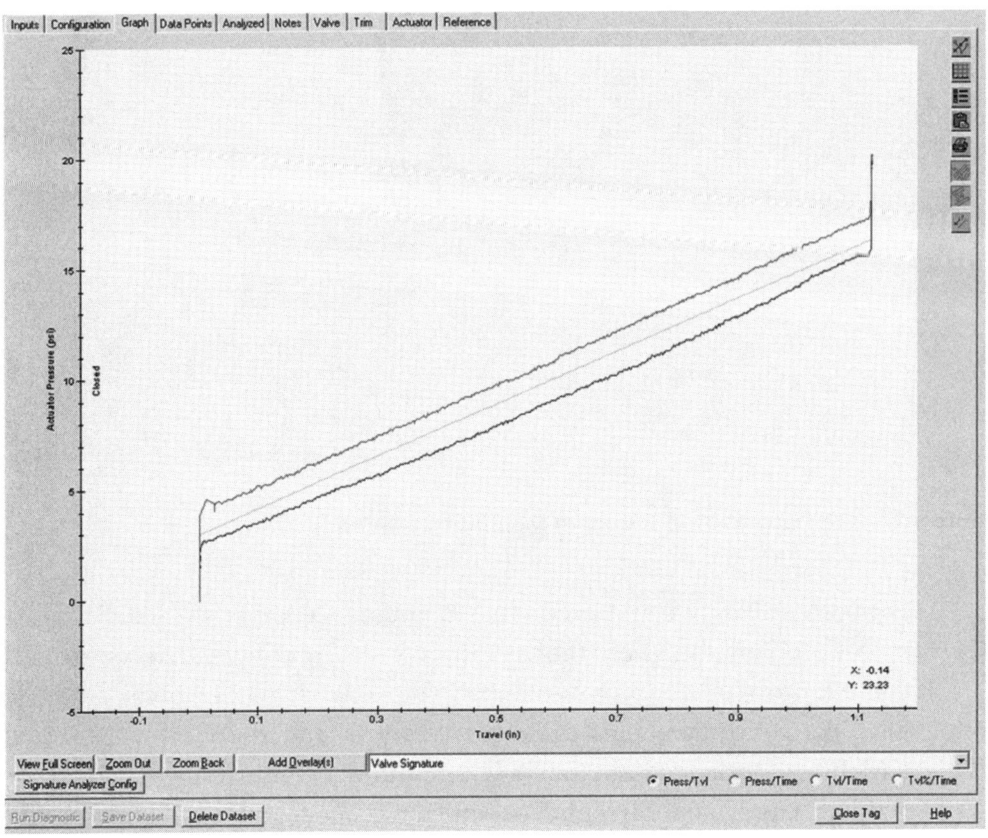

Figure 8-28. Valve Signature Display (Actuator Pressure versus Valve Travel)

Performance

Each of the components in a process control loop is important to plant productivity. Each must perform accurately and reliably. But the overall performance of the control loop must not be overlooked. It is well understood that the goal of any process plant is to make a profit through the production of a quality product that meets customer needs. A quality product conforms to a set of specifications. Any deviation from the specifications means customer dissatisfaction, lost profit due to excessive material use, reprocessing costs, or wasted product. Also, process plants must often closely control their processes to meet corporate and regulatory policies.

Process variability is a measure of how well a process is being controlled to a desired value, the set point. Process variability is expressed in percent. Mathematically, it is typically defined as $2s/m$, where m is the set point and s is the standard deviation of the process variable.

Reducing process variability through better process control allows optimization of the process and the production of products "right the first time." In a loop performance study in the 1990s, it was found that as many as 80% of the loops did not do an adequate job of reducing process variability. Furthermore, the control valve was found to be a major contributor to this problem.

Manufacturers in the process industries now test entire control loops to verify variability performance. The loop is tested under dynamic process conditions. Tests like these are the only true measure of process variability.

Some of the more important control valve dynamic considerations that are the major sources of process variability (see Dynamics Section 7-5, in Chapter 7) are:

- Resolution (magnitude of signal change necessary to move valve) (%)
- Deadband (magnitude of signal change necessary to reverse direction of valve) (%)
- Response time (e.g., T_{85} = time to 85 percent of final stroke for step input) (sec)
- Pre-stroke deadtime (time delay before actuator shaft starts to move) (sec)

Figure 8-29. Flow-Loop Performance Testing

- Stroking time (time for full scale stroke) (sec)
- Gain (slope of installed valve characteristic) (% flow rate change per % travel)

The response time will greatly depend upon the size of the step change in control system output. Some of the major design features that determine the above dynamics are:

- Actuator resolution
- Actuator size
- Positioner sensitivity
- Positioner tuning
- Positioner flow coefficient for flow rate and exhaust rate
- Accessory flow coefficient for flow rate and exhaust rate
- Shaft and stem connection backlash
- Packing and seal friction

- Inherent valve characteristic
- Source and destination (upstream and downstream) pressure
- Piping system pressure drop

Valve positioners allow for precise positioning accuracy and faster response to process upsets when used with a conventional digital control system. It is important to consider the positioner and actuator together because the combination of these two components affects the static performance (e.g., offset between desired and actual position) and the dynamic response of the valve assembly.

With the increasing emphasis upon economic performance of process control loops, positioners should be considered for every valve application where process optimization is important. When a change in the process variable has been detected and a control signal transmitted, the positioner must rapidly supply a large but closely controlled volume of air to the actuator in order to make the actuator and valve respond promptly. For many processes, it is important that the valve reach a specific position quickly. A quick response to small signal changes (0.5% or less) is important.

The ISA-75.25.02-2000 standard specifies a response time T_{85} that is the time measured from initiation of the input signal step change to when the valve travel reaches 85% of its final value [7, 8, 9].

Valve type and sizing can have a large impact on the performance of the control valve assembly in the system. A valve must be of sufficient size to pass the required flow under all possible conditions, but a valve that is too large for the application is a detriment to process optimization.

Flow capacity of the valve is also related to the type of valve through the inherent flow characteristic of the valve. The inherent flow characteristic is the relationship between the valve flow rate and the valve travel when the differential pressure drop across the valve is held constant.

With a constant pressure drop across the valve, the valve flow rate becomes only a function of the valve travel and the inherent design of the valve trim. Typical valve characteristics defined in this manner are named linear, equal percentage, and quick opening. With some valve designs, these flow characteristics are a selection to be made. With other designs, only one characteristic is available.

More important than the inherent characteristic is the installed flow characteristic (plot of flow rate versus travel for the valve installed in the plant at process operating conditions). The installed valve gain is the slope of this curve but with the change in flow rate and travel expressed as a percent so the valve gain is dimensionless. The installed flow characteristic can be estimated by computer programs if the prime mover (e.g., blower, compressor, fan, or pump) characteristic curve or header pressure, piping system resistance, system static heads, and range of process operating conditions are known. An installed flow characteristic and valve gain can be simulated in a flow lab to show the effects of the valve's inherent characteristic and the piping system (Figure 8-30).

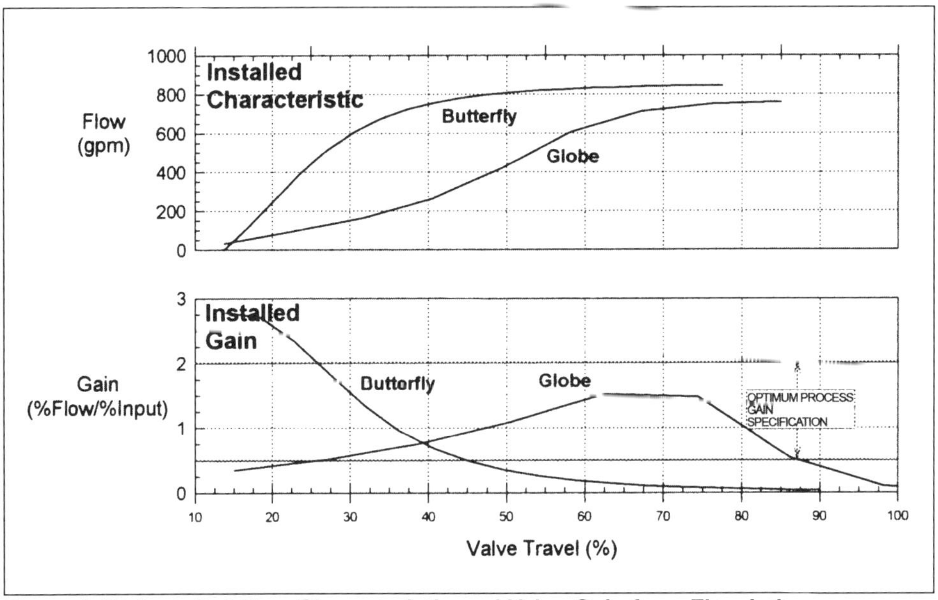

Figure 8-30. Installed Flow Characteristic and Valve Gain from Flow Lab

8-9. Valve and Actuator Capabilities

This section illustrates several types of valves and actuators found in the process control industry. Typical specifications are given for most types covered. Suppliers also offer versions of these designs that are available in other materials, ratings, and flange specifications to international standards.

Valve sizes are shown with the NPS (nominal pipe size) designations and with ASME pressure rating class designations. Depending on the supplier, versions with other size designations, end connection standards, and pressure ratings are normally available.

General-Purpose, Heavy-Duty Globe Valves

This style of valve is available with a wide range of options and finds significant application in most process industries. Heavy-duty globe valves are rugged, yet simple in design (Figure 8-31). These valves are used in general control applications and are suited to erosive, corrosive, cavitating, flashing, and noisy applications. A broad range of construction materials allows use with many process fluids.

Features commonly offered are:

- Choice of soft or metal seats in balanced or single-seated designs.
- Choice of flow characteristics and end connections
- Angle versions, also valves with expanded outlets
- Hydrodynamic noise and cavitation abatement options

Sizes: NPS 1/2 to 24

End Connections: Threaded, flanged, welding-end

Wetted Materials: Cast iron, steel, stainless steel, high-nickel alloys; TFE and elastomeric soft seats, and TFE, composition, and graphite packing

Pressure Ratings or Maximum Pressures: Class 125 through Class 600 with some models available to Class 2500

Temperature Capabilities: to 1000°F (538°C)

Compact Globe Valve and Actuator Package

Valves, actuators, positioners, and accessories can be integrated into a compact control valve package to simplify selection and virtually eliminate the need for actuator sizing (Figures 8-32a and 8-32b). Many different levels of integration are available. One common benefit of these packaged valves is simplified selection. Once the valve is selected, actuator selection is automatic.

Such integrated valves are often used as utility or general-purpose valves. Additional uses include more severe chemical and petrochemical applications such as polysilicon, ethylene, and other specialty chemicals. There are models available for a broad variety of services.

Features commonly offered by suppliers are:

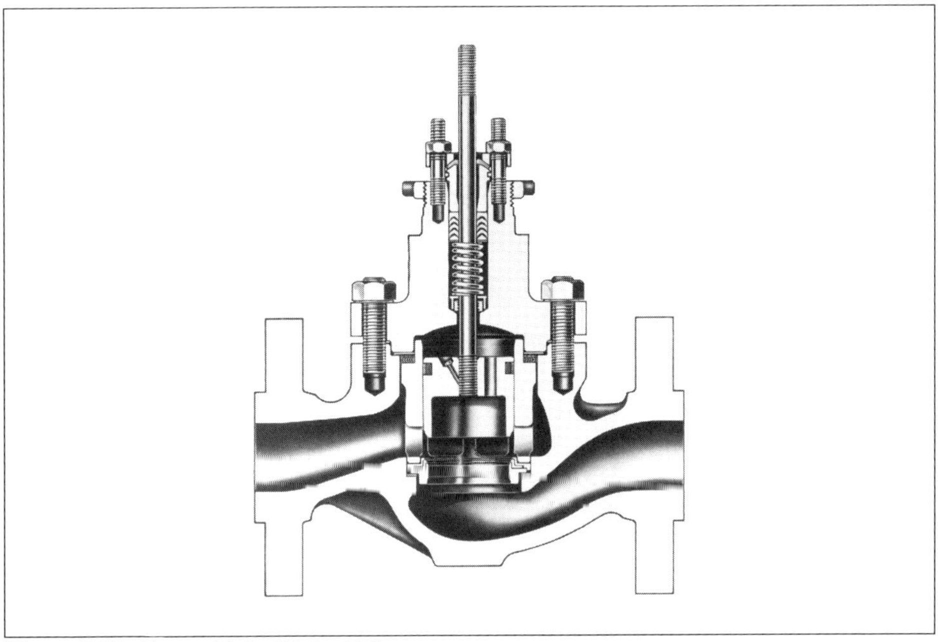

Figure 8-31. Heavy-Duty Globe Valve

- Simple, modular design
- Compact, multiple-spring actuator portion for lower overall height
- Integrated digital valve positioners
- Choice of materials and ratings

Sizes: NPS 1/2 through 6

End Connections: Threaded or flanged

Wetted Materials: Steel, stainless steel, and alloys; TFE and composition packing

Pressure Ratings or Maximum Pressures: Class 150 and Class 300

Temperature Capabilities: to 700°F (371°C)

Lightweight Utility Valves

These valves are used for general utility service in many process industries such as chemical and food and beverage and for heating, ventilating, and air conditioning (Figure 8-33).

Common lightweight utility valve features include:

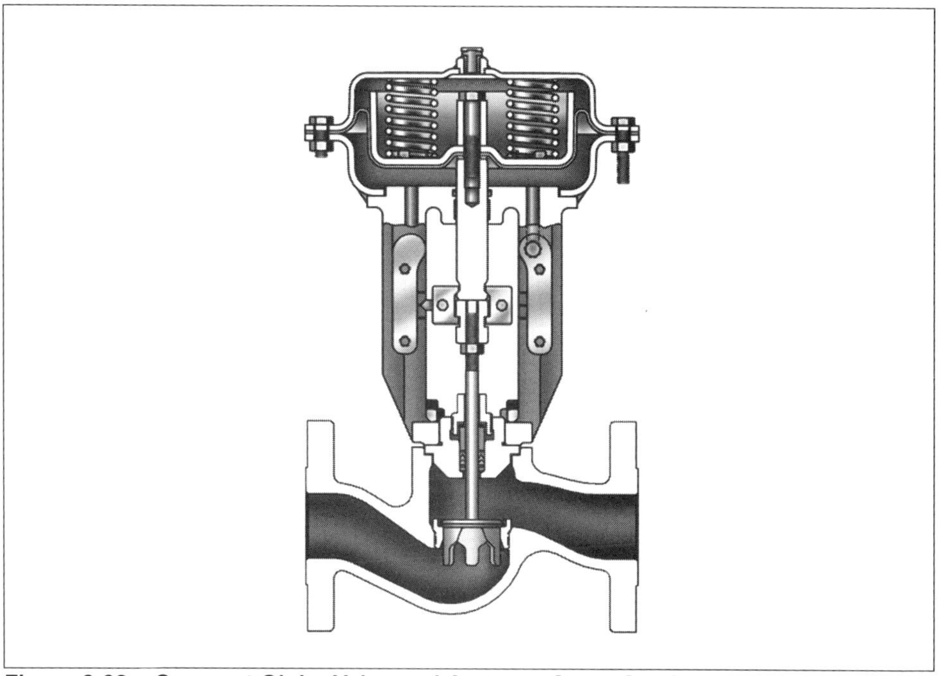

Figure 8-32a. Compact Globe Valve and Actuator Cross Sectional View

Figure 8-32b. Compact Globe Valve and Actuator External View

- Small and lightweight for easy installation
- Multi-spring, field-reversible diaphragm actuator

- Bronze, iron, steel, and stainless steel bodies available
- Good control when teamed with a good actuator and positioner
- Tight shutoff

Sizes: NPS 1/2 to 3

End Connections: Threaded or flanged

Wetted Materials: Bronze, iron, steel or stainless steel body with stainless steel internal parts; TFE and graphite packing

Pressure Ratings or Maximum Pressures: To 400 psig (28 bar)

Temperature Capabilities: to 400°F (204°C)

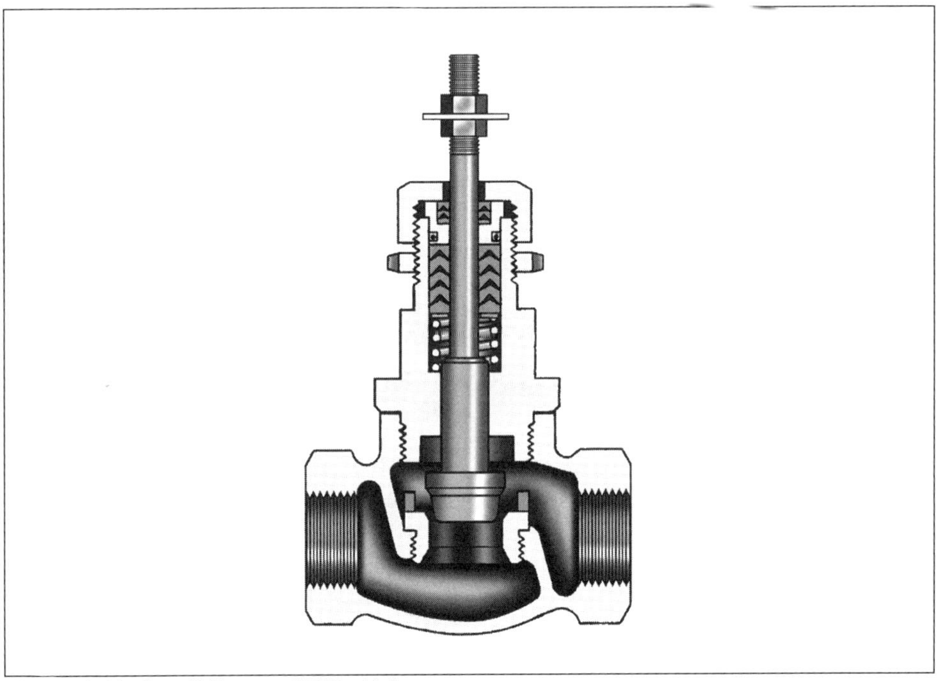

Figure 8-33. Small Bronze Utility Valve

Low-Flow Valves

Low-flow valves are compact sliding-stem valves intended for accurate control of small flow rates. These valves can use conventional valve plugs and seat rings or closure diaphragms that are forced against a valve orifice to regulate or shut off flow (Figure 8-34).

The maximum C_V (valve sizing coefficient) values for these valves range from 0.00013 to 2.5. They find application in various industries but are often used in laboratories, paper and textile production, and specialty chemical production.

Features commonly offered by suppliers are:

- Compact, low overall height
- Can be used for viscous fluids and in corrosive atmospheres
- Highly polished internal surface finishes
- Models without packing are available
- Self-draining

Sizes: NPS 1/4 or 1/2
End Connections: Threaded NPT or metric
Wetted Materials: Stainless steel, nickel alloys, TFE, fluoroelastomers
Pressure Ratings or Maximum Pressures: to 3000 psig (207 bar)
Temperature Capabilities: to 450°F (232°C)

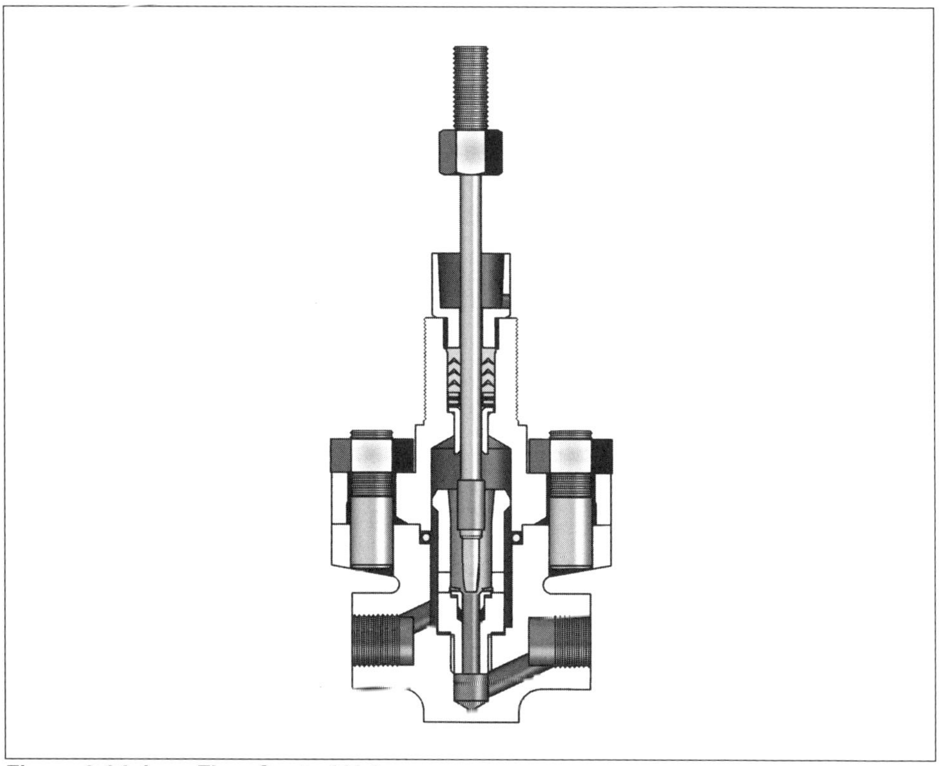

Figure 8-34. Low-Flow Control Valve

Sanitary-Service Valves

Sanitary-service valves are used in biotechnology, pharmaceutical, food and beverage, cosmetics, and many other industries where cleanliness and sterility are required (Figure 8-35).

Features commonly offered by suppliers are:

- All stainless steel for metal parts in contact with the process fluid
- The ability of the valve to be cleaned or sanitized in place
- Conformance to industry and agency guidelines, such as FDA (U.S. Food and Drug Administration) and USP (United States Pharmacopeia)
- Highly polished internal surface finishes
- No valve packing required
- Self-draining

Sizes: NPS 1/2 through 6

End Connections: Welding ends, industry-specific clamped ends

Wetted Materials: Stainless steel, compatible elastomers

Pressure Ratings or Maximum Pressures: to 275 psig (18.9 bar)

Temperature Capabilities: to 300°F (149°C)

High-Pressure Valves for Oilfield Applications

This type of valve is used for many high-pressure, on-off or throttling applications in the gas and oil industry where the flowing media can be gritty or sticky or tend to build up on internal valve parts (Figure 8-36). They can be used with separators, scrubbers, and other processes. Also, they find application as shut-down valves.

Features commonly offered by suppliers are:

- High-strength valve plug guiding for high pressure drops
- Threaded-on bonnet for maintenance convenience
- Trim materials for oil and gas flow media

Figure 8-35. Sanitary-Service Valve

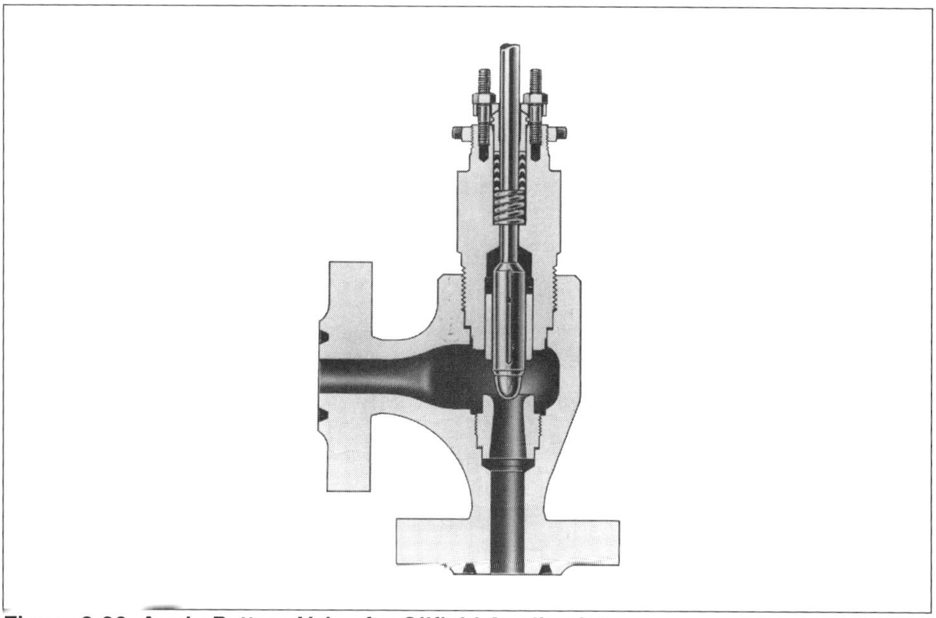

Figure 8-36. Angle-Pattern Valve for Oilfield Applications

- Angle valve configuration is available where a self-draining valve is desired

Sizes: NPS 1 or 2

End Connections: Threaded or flanged

Wetted Materials: Steel, stainless steel, ceramic trim option, TFE and composition packing

Pressure Ratings or Maximum Pressures: to Class 2500 and API 10,000

Temperature Capabilities: to 450°F (232°C)

Segmented High-Capacity Ball Rotary Valves

Segmented ball valves are rugged valves that offer the flow capacity and the low pressure drop capability of other rotary valves. These valves can be used to control gases, steam, liquids, and fibrous slurries. Segmented V-Notch ball valves are used widely in the pulp-and-paper and chemical industries (Figure 8-37).

There is a selection of sealing options: composition, metal, and flow-ring.

Features commonly offered by suppliers are:

- Slurry version available with highly wear-resistant internal parts
- Shearing action of the ball promotes smooth, non-clogging opening and closing
- High rangeability (ratio of maximum to minimum controllable flow)
- Noise attenuating closure component (e.g., trim) is available

Sizes: NPS 1 through 20

End Connections: Flanged or flangeless for installation between pipeline flanges

Wetted Materials: Steel, stainless steel, and alloy bodies; steel, stainless steel, alloy steel closure components; TFE and graphite packing

Pressure Ratings or Maximum Pressures: to Class 600

Temperature Capabilities: to 800°F (427°C)

Figure 8-37. Segmented V-Notch Ball Valve

High-Range Rotary Control Valves

The equal percentage installed flow characteristic of this valve allows throttling over a signal range almost as wide as a sliding stem valve with minimal sealing friction (see Rangeability Section 7-7 in Chapter 7). Although similar to a butterfly, the special design of the disk puts it into a separate category. This valve finds applications in pulp-and-paper, chemical, and many other industries where tight control is needed (Figure 8-38).

Features commonly offered by suppliers are:

- Equal percentage flow characteristic for increased rangeability
- High capacity
- Lower operating torques required
- Splined shaft-lever connection for reduced lost motion (backlash)

Sizes: NPS 2 through 12 and larger

End Connections: Wafer (flangeless) valve for installing between pipeline flanges

Wetted Materials: Steel, stainless steel, alloy bodies; stainless steel or alloy internal parts; TFE and metal seal, TFE and graphite packing

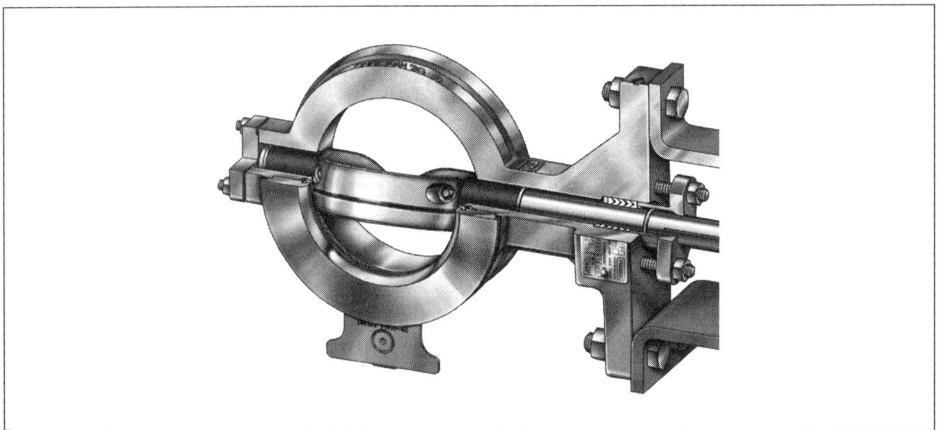

Figure 8-38. High-Range Rotary Valve

Pressure Ratings or Maximum Pressures: Class 150/300

Temperature Capabilities: to 800°F (427°C)

Throttling Butterfly Valves

Butterfly valves are high-capacity valves. Many types of throttling butterfly valves are available, including lined. One design uses a contoured disk and teeth for low noise, low torque, and better throttling at low flow rates (Figure 8-39). Eccentrically mounted disks and offset disks and a variety of sizes are offered by suppliers. Large sizes are often available. Butterfly valves find applications in nearly all industries.

Features commonly offered by suppliers are:

- Bi-directional flow capability on some models
- Soft, metal, or pressure-assisted seals
- Splined shaft-lever connection for reduced lost motion (backlash)

Sizes: NPS 2 through 24 and larger

End Connections: Wafer (flangeless) valve for installing between pipe flanges

Wetted Materials: Steel, stainless steel bodies; stainless steel or alloy internal parts; a variety of materials for internal parts, TFE or graphite packing

Pressure Ratings or Maximum Pressures: Class 150/300

Temperature Capabilities: to 800°F (427°C)

Figure 8-39. Butterfly Valve with Low-Noise Low-Torque Disk

On-Off Butterfly Valves

On-off butterfly valves are high-capacity valves normally with low leakage (tight shutoff) specifications. If these valves are used for throttling by the addition of a positioner, the stiction and backlash are excessive due to seal, linkage, and actuator design. On-off butterfly valves can use traditional actuators or on-off actuators sometimes called quarter-turn (90°) rotation. These valves are commonly positioned by discrete outputs from batch, startup, and shutdown sequences and safety systems.

Features commonly offered by suppliers are:

- Tight shutoff
- Soft, metal, or pressure-assisted seals
- Variety of actuator connections available (splined, square, and keyed).

Sizes: NPS 2 through 24 and larger

End Connections: Wafer (flangeless) valve for installing between pipe flanges

Wetted Materials: Steel, stainless steel bodies; stainless steel or alloy internal parts; a variety of materials for internal parts, TFE or graphite packing

Pressure Ratings or Maximum Pressures: Class 150/300; through Class 2500 are available

Temperature Capabilities: to 800°F (427°C)

Figure 8-40. On-Off Butterfly Valve with Rack-and-Pinion Actuator

Dampers

Dampers can be used to control the flow of air and other gases in almost every industry including power, refining, utilities, and chemical. They are used to control draft, vents, air for burners, flue gas, and many utility gas flows.

Similar to control valves, dampers can use rotary actuators for actuation or use linear-output piston devices that operate through a lever arm. Actuator types often used are pneumatic vane (quarter-turn) actuators with or without spring return and manual operation, rack-and-pinion pneumatic piston actuators, electro-hydraulic actuators, and electric-motor actuators.

The actuators can be fitted with some of the same positioners and accessories that are used with control valves.

Dampers vary widely in size and application. Suppliers offer custom-made mounting and integration of components. The backlash from linkages is often excessive (see Valve and Damper Section 7-3 and Dynamics Section 7-5 in Chapter 7).

Pneumatic Spring-and-Diaphragm Actuators

Pneumatic spring-return diaphragm actuators can be used on many sliding-stem and rotary valves. Actuator sizes can be denoted by the number of square inches of effective diaphragm area or by the amount of thrust and travel they make available to control valves, or by the size of valve on which they are used (Figures 8-41 through 8-43).

Sizes: Actuators with an effective area as small a 6 in^2 (37 cm^2) for use on NPS 1/4, 1/2, and other small valves. Actuators with an effective diaphragm area up to 450 in^2 (2902 cm^2) can be used on valve sizes up to NPS 24. Output thrust can be as high as 6300 lb (28,000 N). Typical torque output from rotary actuators is 12,100 lbf·in (1370 N·m).

Actuator Stem Travel: Up to 4 inches (102 mm) for linear output and 0 to 90° for rotary output.

Materials: Diaphragms are typically a fabric-reinforced nitrile, silicone, or other elastomeric material; other parts are steel, cast iron, or aluminum.

Pressures: Operating pressures are normally between 0 and 30 psig (0 and 2.0 bar) or 0 and 60 psig (0 and 4.0 bar). New high pressure compact actuators are capable of using supply pressures of 90 psig (6 bar) to provide greater force (Figure 8-38).

Accessories: Actuators are provided with mounting bosses and holes to mount valve positioners, electronic-to-pneumatic transducers, and supply pressure filter regulators.

Pneumatic Piston Actuators

Pneumatic piston actuators can be used on many sliding-stem and rotary valves. Actuator sizes can be denoted by the number of square inches of effective piston area, by the amount of thrust and travel they make available to control valves, or by the size of valve on which they are used (Figures 8-44 through 8-46).

Sizes: Actuators with an effective piston area up to 129 in^2 (961 cm^2) can be used on valve sizes up to NPS 24. Long-stroke piston actuators can be used on valves with connection sizes up to NPS 36.

Output thrust can be as high as 25,000 lb (111,000 N). Typical maximum torque from rotary actuators is 55,700 lbf·in (6290 N·m).

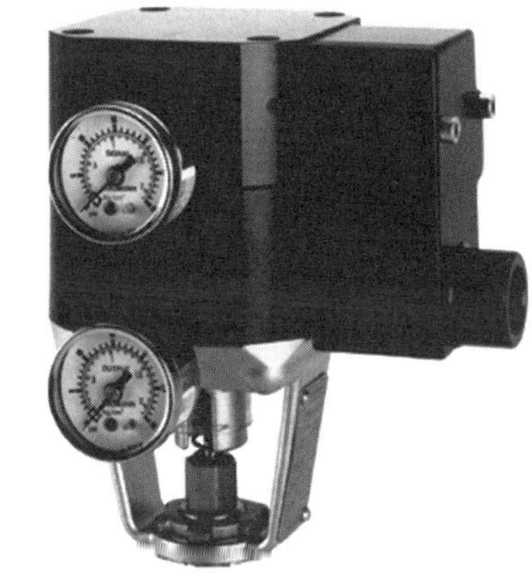

Figure 8-41. Small Diaphragm Actuator (6 in^2; 37 cm^2 Diaphragm Area)

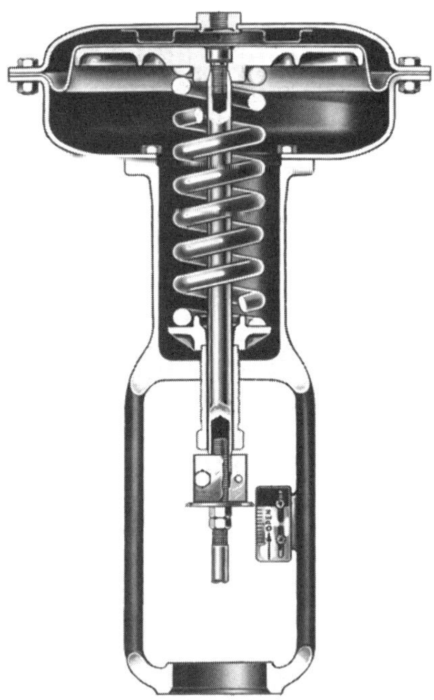

Figure 8-42. Diaphragm Actuator Configured to Extend the Actuator Stem with Increasing Loading Pressure

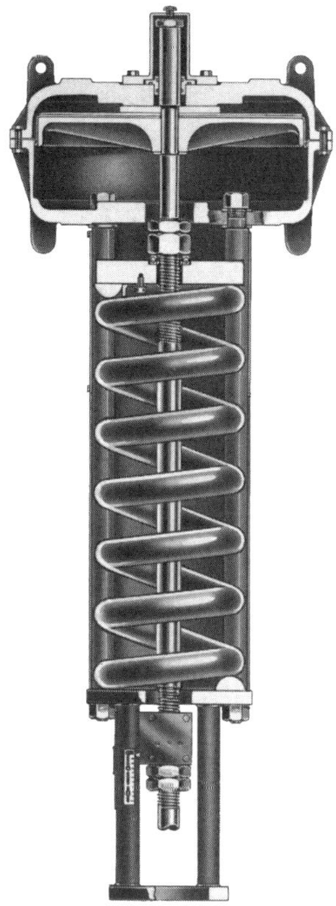

Figure 8-43. Large Diaphragm Actuator (450 in^2; 2902 cm^2 Diaphragm Area)

Actuator Stem Travel: Up to 8 in (204 mm) for linear output and 0 to 90° for rotary output. Special constructions can provide up to 24 in (610 mm) travel.

Materials: Seals are nitrile or other elastomeric materials and seal bushings are frequently brass; other parts are steel, cast iron, or aluminum.

Pressures: Operating pressures up to 150 psig (10.3 bar) are possible.

Accessories: Actuators are provided with mounting bosses and holes to mount valve positioners, electronic-to-pneumatic transducers, and supply pressure filter regulators.

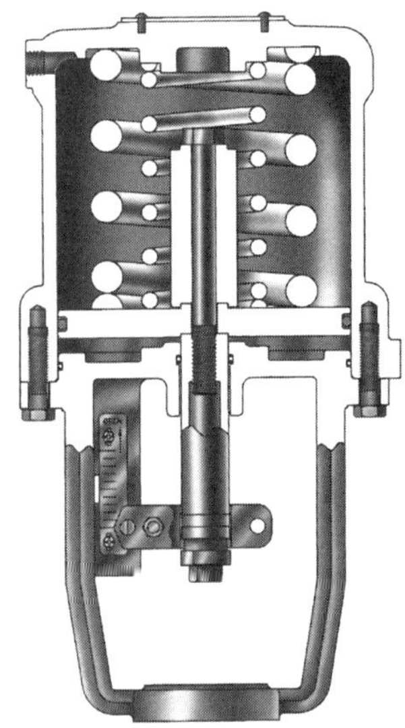

Figure 8-44. Spring-Return Piston Actuator Configured to Retract the Actuator Stem with Increasing Loading Pressure

Figure 8-45. Piston Actuator for Use on a Rotary Valve

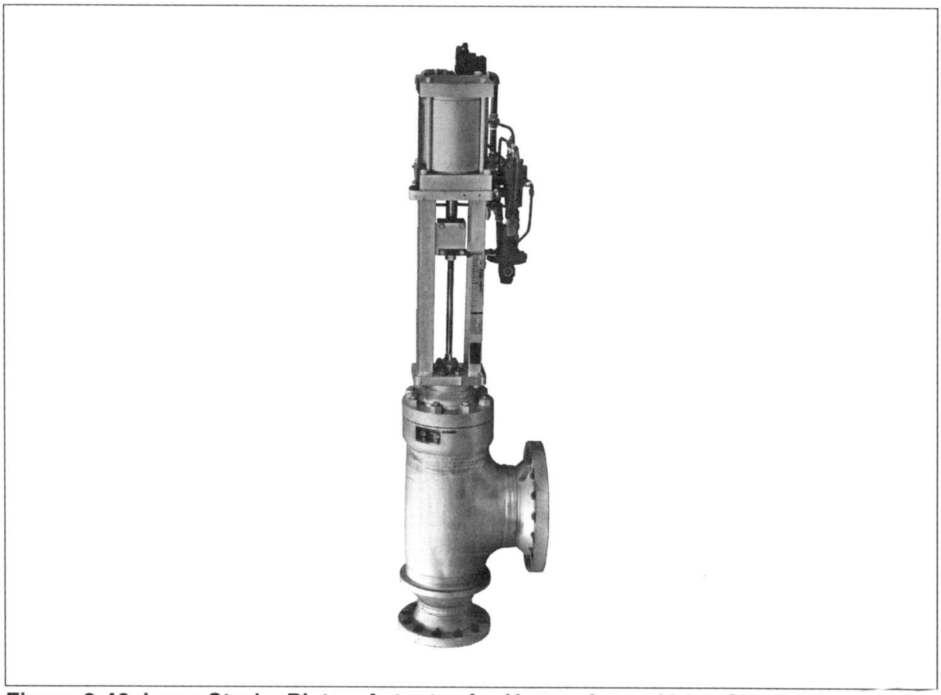

Figure 8-46. Long-Stroke Piston Actuator for Use on Large Valve Sizes

8-10. Questions

1. Two methods of converting linear actuator output to rotary input for a rotary valve are a lever and ______________.

2. Two common sealing methods for valve stems and valve shafts against process fluid are packing and ______________.

3. For flow-splitting applications, which valve is an obvious choice?

   a. Rotary valve

   b. Globe valve

   c. Three-way valve

   d. Plug valve

4. The globe valve exhibits a desired gain over a narrow range of travel. True or False?

5. Typical pneumatic signal ranges are 3 to 15 psig (0.2 to 1.0 bar) and ______.

6. Advantages of a digital valve controller include:

    a. Communication with the final control element

    b. Auto calibration

    c. Diagnostic capabilities

    d. All of the above

7. Common valve flow characteristics are quick opening, linear, and __________. __________

8. Setting the spring compression in a spring-and-diaphragm actuator is called "initial winding" or "__________."

9. Installing a control valve assembly often includes the step of setting the zero and ________ before putting the unit in operation.

10. Monitoring equipment and process operation in order to anticipate and prevent equipment failures is called __________ maintenance.

11. A study found that as many as ____ % of control loops did not do an adequate job of reducing process variability.

12. Two of the most common causes of deadband in process control equipment are __________ __________ and backlash.

13. The recommended range of gain in a control valve is:

    a. 30% to 60%

    b. 1 to 4

    c. The higher the better

    d. 0.5 to 20

14. The angle-pattern valve is a variant of the __________valve.

8-11. References

1. McMillan, G. K. "What Have I Learned? – The Cost and Source of Oscillations – Part 3." http://www.modelingandcontrol.com/, May 4, 2009.

2. McMillan, G. K. "What Is Your Flow Control Valve Telling You?" *Control Design*, www.controldesign.com/articles/2003/164.html, May 2004.

3. McMillan, G. K. "Valves That Can Turn on a Dime, or at Least a Quarter – Part I." http://www.modelingandcontrol.com/, Apr. 9, 2007.

4. McMillan, G. K. "Valves That Can Turn on a Dime, or at Least a Quarter – Part II." http://www.modelingandcontrol.com/, Apr. 19, 2007.

5. McMillan, G. K. "Improve Control Loop Performance" *Chemical Processing*, http://www.chemicalprocessing.com/articles/2007/200.html, Oct. 2007.

6. McMillan, G. K., Sowell, M., and Wojsznis, P. "Adaptive Control Takes a Leap Forward." *Chemical Processing*, Sep. 2004.

7. ISA-75.25.01-2000 (R2006) "Test Procedure for Control Valve Response Measurement from Step Inputs." Research Triangle Park: ISA, 2006.

8. ISA-75.25.02-2000 (R2006) "Control Valve Response Measurement from Step Inputs." Research Triangle Park: ISA, 2006.

9. Coughran, M. T. "Valves: Testing for Peak Performance" *InTech*, Volume 41, No. 10. Research Triangle Park: ISA, 1994.

For greater detail on these topics, there are many printed and on-line sources. A few are listed below. Manufacturers and suppliers offer product catalogs, selection guides, and technical papers; many of them are available on-line. Industry standards organizations also publish helpful guides.

10. ANSI/ISA-75.05.01-2000 (R2005) "Control Valve Terminology." Research Triangle Park: ISA, 2005.

11. ANSI/ISA-75.01.01 (IEC 60534-2-1 Mod) – 2007 "Flow Equations for Sizing Control Valves." Research Triangle Park: ISA, 2007.

12. Fitzgerald, B. *Control Valves for the Chemical Process Industries.* New York: McGraw-Hill, Inc., 1995.

13. Baumann, H. *Control Valve Primer — A User's Guide, 4th Edition.* Research Triangle Park: ISA, 2009.

14. *Control Valve Handbook, 4th Edition.* Fisher Controls International LLC, 2005.

15. *Control Valve Sourcebook — Power and Severe Service.* Fisher Controls International LLC, 2003.

16. *Control Valve Sourcebook — Refining*. Fisher Controls International LLC, 2004.

17. McMillan, G. K. *Process/Industrial Instruments and Controls Handbook*. New York: McGraw-Hill, Inc., 1999.

18. IEC 60534-1 "Industrial Process Control Valves — Part 1: Control Valve Terminology and General Considerations." 3rd Edition. Geneva: IEC, 2005.

9

Wireless

9-1. Learning Objectives

A. Learn about the latest technology advances in wireless networks.

B. Understand the importance of key features, such as security, which are essential for every wireless sensor network.

C. Learn terminology and issues to intelligently discuss wireless industrial applications.

D. Understand the impact of enhanced PID technology when performing control across wireless sensor networks.

E. Recognize the advantages of wireless smart measurements.

F. Understand the importance of exception reporting when supporting loops with fast dynamics.

G. Learn the importance of mesh technology when considering installation requirements.

H. Understand how wireless mesh networks can be used in monitoring and control applications.

I. Understand how wireless mesh networks can be used to improve overall plant operations.

9-2. Introduction and Overview

Smart devices are an essential part of nearly every control system. These smart devices provide greater process insight, reduce engineering costs, and otherwise contribute to improving the overall operational performance of the plant. Smart devices offer advanced diagnostics that can report the health of the devices and in many cases, the health of the process that the devices are connected to. It is not uncommon for smart devices to include diagnostics that can detect plugged sensing lines, burner flame instability, loss of agitation, wet gas, orifice wear, leaks, or cavitation. These devices tell the user how well they are operating and when they need maintenance. Improvements in sensor technology and improved diagnostics have led to a large variety of smart instruments. Wireless technology can allow users to more readily migrate to smart instruments and valves by eliminating the need to add smart input and output (I/O) cards for the distributed control system (DCS) and reducing the time for commissioning. For a more fundamental understanding of sensor and transmitter roles and capability, see Chapter 1.

Smart devices incorporate features such as status on measured values, time stamps, event latching and confirmation, and block data transfer. The availability of additional variables, such as density and temperature for Coriolis mass flowmeters, provides additional information useful for understanding changes in the process. Using these smart devices results in fewer trips into the plant, a much better controlled process, and significantly less downtime.

The advantages of wireless networks go well beyond the obvious initial savings in the design, installation, and commissioning of new smart instrumentation. The bigger benefits are seen in the lifecycle cost of smart instrumentation and the more effective and expansive use of smart instrumentation capabilities. As discussed in Chapter 1, smart instrumentation installation errors and drift are dramatically reduced by advances in sensing element design and the compensation of installation conditions and sensor response. This reduction in onsite process calibrations requirements combined with faster troubleshooting by the use of online diagnostics and, with wireless networks, the elimination of wiring noise and connection problems translates to significant maintenance cost savings [2].

Unfortunately, utilizing the features of these smart devices is not always possible with the installed base of control systems. What is needed is an infrastructure that can be used with both new and existing system installations. Wireless sensor networks, in particular WirelessHART™, offer a low-cost, effective, and rapid deployment of smart devices in both new and existing installations.

WirelessHART (Highway Addressable Remote Transducer) was designed from the ground up to support the capabilities of new systems while at the same time not leaving behind the large installed base of wired HART devices. Wireless communication can be easily added to existing wired HART transmitters by the simple addition of a Thum^{TM} communicator (a device that connects to and is powered by the existing wiring in the transmitter and communicates via an antenna, additional process variables and diagnostics not previously available due to a lack of intelligence in the DCS I/O). Additionally, WirelessHART transmitters can replace non-smart transmitters or provide new smart measurements. The replacement or addition of wiring and DCS I/O cards is not needed. WirelessHART supports existing applications such as pressure loops while at the same time supporting new applications such as vibration monitoring.

Good control performance requires that the overall control loop execute four to ten times faster than the process time constant plus deadtime. When calculating overall control performance, the amount of time required to read input signals, execute control, and drive outputs must be taken into consideration. These times are referred to as *latency*. Variation in input, output, and control execution times is referred to as *jitter*. Too much latency or jitter can severely degrade overall control performance [1][4]. Managing latency and jitter in the input and output operations in wireless networks is challenging. Wireless signal strength varies in time; wireless nodes can become unavailable; data may be routed dynamically on different paths. Oversampling burns batteries out prematurely. Control performance is degraded if process data is missing or delayed in the network [1][3].

Wireless sensor network technology can be viewed as a platform for new kinds of applications, as well as a platform for smart devices in existing plant environments [22]. WirelessHART targets unit-level process operations and supports the following applications:

- Equipment and process monitoring
- Control
- Asset management
- Diagnostics/predictive maintenance
- Location tracking
- Nomadic, "wireless worker" applications
- Portability for optimization of installation location
- Research and development in labs and pilot plants

Background

HART technology has been in existence since the late '80s. In its initial release the HART Field Communications Protocol was superimposed on a conventional 4- to 20-mA signal, providing two-way communications with smart field instruments without compromising the integrity of the measured data. During its 20+ years of existence, the HART protocol has evolved from a simple 4- to 20-mA-based protocol to the current wired and wireless-based technology with extensive features supporting security, unsolicited data transfers, event notifications, block mode transfers, and advanced diagnostics (Figure 9-1) [5]. As has been mentioned, diagnostics now include information about the device, the equipment the device is attached to, and in some cases the actual process being monitored.

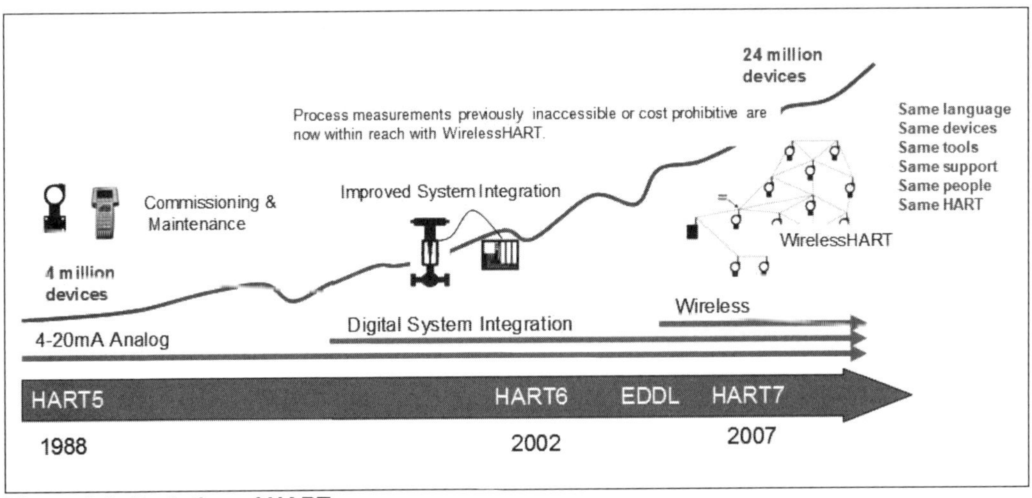

Figure 9-1. Evolution of HART

Compared to wired HART, WirelessHART incorporates many new features providing improved performance, enhanced diagnostics, and better maintenance capabilities [5]. WirelessHART is a secure networking technology operating in the 2.4-GHz industrial, scientific and medical (ISM) radio band. It utilizes IEEE 802.15.4–compatible direct-sequence spread spectrum (DSSS) radios with channel hopping on a packet by packet basis [5, 6]. WirelessHART communication is arbitrated using the WirelessHART Network to schedule link activity. A given communication link may be dedicated to communication between a network pair or the link may be shared. WirelessHART networks are organized as a mesh topology. In a mesh topology, each device is connected to several other devices and all devices are capable of routing communications traffic. Devices can be connected to each other through several hops. A wireless device network (Figure 9-2) supports a wide variety of devices from many manufacturers.

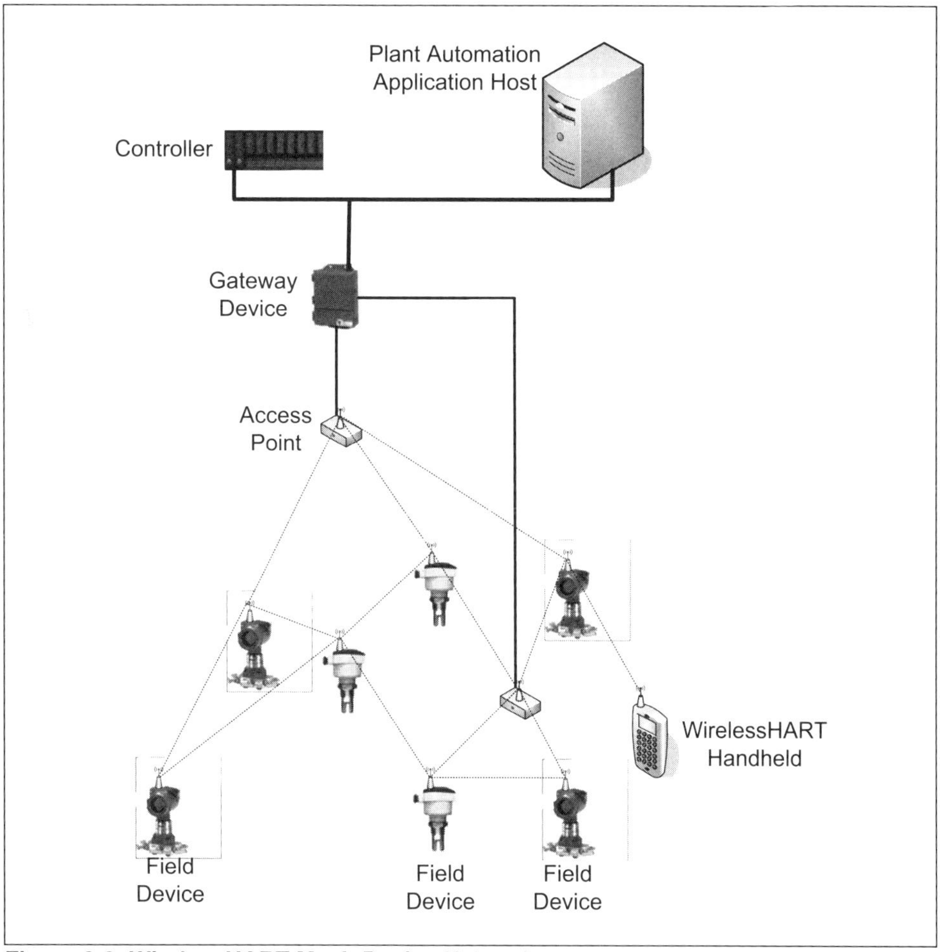

Figure 9-2. WirelessHART Mesh Device Network

Mesh networks are self-healing: the network can still operate even when a device breaks down or a connection goes bad. As a result a mesh topology is very reliable.

Improved Measurement and Control

Wireless device networks are used in a wide variety of applications including monitoring and controlling tank levels, monitoring emission levels and water quality, monitoring equipment health, and a wide variety of control applications. Measurements are often used to generate reports for the FDA, EPA, and other agencies. These measurements are also used by plant personnel to make decisions about the operation of the process, plan maintenance activities, schedule production runs, and validate the quality of the finished product. In many systems, these wireless measurements are part of feedforward and feedback control strategies.

There are many kinds of final control elements including valves, agitators, blowers, conveyors, etc. These control elements receive a signal and perform some action. In the case of an on-off valve, the valve will attempt to open or close; valve status must then be communicated. In the case of a regulating valve, the read-back of the actual valve position is important in order to determine if there are problems with the final control element itself that is impacting control performance. In the case of an agitator, it's important to know if the agitator is moving and, perhaps, at what speed. Newer final control elements are engineered to include a radio as part of the device; older devices can be connected to a wireless network using an adapter [5].

An important factor for DCS and host applications is the device status that is communicated with measurements. Device status provides an indication of the measurement quality ("level of goodness"), whether the device making the measurement is healthy, and the timeliness of the measurement. Devices perform checks on hardware and software associated with the input or output and, in turn, report this through the status value.

The status of an output parameter is calculated to specifically indicate the quality of the value: good, poor, bad, or fixed. A "good" signal may be used for control. A "poor" value is suspect and may not reflect the true measurement or calculated value. A "bad" value means that the parameter value does not reflect the true measurement, calculation, or control value. "Fixed" means that the parameter value is constant and is not being updated in a periodic fashion. When a status other than "good" is received, the control system must reflect the status value in its diagnostics, and in the case where the value is used in control, the control strategy must either take the status into account or switch the control loop to manual.

Status also reports additional information such as configuration changed, cold start, loop current fixed (in wired systems), loop current saturated, non-primary variable out of limits, and primary variable out of limits.

All WirelessHART devices also include an Extended Device Status capability. The Extended Device Status value provides information on whether the device has malfunctioned (Maintenance Required), a device variable (e.g., temperature or pressure variable) is in an alarm or warning state (Device Variable Alert), or when power is critically low (Critical Power Failure). Both device status and Extended Device Status are included as part of burst mode communications. Burst mode communications is featured whereby devices send digital measurements and computed values on a periodic or exception basis. Burst mode with exception mode turned on results in a significant reduction in communications.

Another feature of WirelessHART is the time stamp that is communicated with all measurement values. The addition of the time stamp allows applications to determine how current the value is and whether or not there is jitter in the measurement, and in some cases, to use the time value to determine appropriate control actions using the measurement value.

Many wireless devices, such as analyzers, vibration monitors, and more complex devices, generate multiple digital measurement values. WirelessHART allows each burst communication to communicate up to eight digital values. If more than eight digital values need to be communicated, multiple burst communications can be sent.

Improved Operations

Operators provide a key function in plant operations. How well the plant operates is directly correlated to how well an operator is able to assess the state of the process (i.e., is the plant operating at its design set point), interact with equipment, respond to process conditions, and communicate with other plant personnel. Operators are responsible for starting up and shutting down equipment, adjusting the equipment to meet production schedules, and responding to unplanned situations, such as equipment failures, power changes, etc. When things aren't working correctly, they are the first line of defense.

Consider the following scenario in which the pH of the process fluid in a tank is controlled by a valve and the pH measurement is oscillating about its set point. As a first step in investigating this problem, the operator would likely put the valve position target, the actual valve position, the pH reading, and the pH set point on a trend. After reviewing the trended oscillations, the operator may decide to adjust the controller gain. In this scenario, the operator notes that the oscillations remain the same regardless of whether there has been an increase or decrease in controller gain. The operator also notes that the read-back value of actual valve position indicates that the minimum change in valve position is 0.5% and notes that a step in the actual valve position change always precedes the pH change. The operator suspects that the oscillation is being caused by the resolution (stick-slip) of the control valve. Following through on this the operator requests a valve signature from the valve and compares what he or she is seeing with the actual process.

In order to support this scenario, several key functions must be supported. First, the sampling and reporting must be fast enough to not to miss the oscillations. Second, the actual valve position and signal status must be reported along with the target position. Third, the control valve needs a standard way to report more complex information such as the valve signature.

WirelessHART has extensive support for all of these requirements. In this case, the operator would set the sampling interval to a fast rate and enable exception reporting. With exception reporting, even though devices are configured to report at a fast rate, they only communicate if the measured value deviates by more than a set amount. To support the second requirement, multiple values can be communicated in the same packet. To ensure that both upstream and downstream communications get through, graph routing is used in both directions (graph routing is a set of directed paths that connects one device to another device). The third requirement, accessing the valve signature, is addressed using Block Data Transfer. Block Data Transfer is used to create a connection between the DCS or host and the remote device. Block data transfer is designed to maximize the utilization of the communication bandwidth while performing the transfer.

Operators are responsible for a large number of control loops, many pieces of equipment, and in many cases, more than one process or utility area. To support this large scope, operators rely on a well engineered alarm system. Since smart devices analyze themselves and report their health on an ongoing basis, they are ideal components of an alarm system. As we have seen, smart devices also have the ability to provide additional information about the process they are inserted into—for example, they can detect plugged sensing lines, burner flame instability, loss of agitation, wet gas, orifice wear, leaks, bubbles in the line, and cavitations. Both wired HART and WirelessHART support these capabilities through event notifications. In the case of WirelessHART, communication priorities and buffers have been optimized to support event notification.

"Event notifications" include the device's status byte, Extended Device Status byte, and the device's device-specific status information. It is possible to specify a limited set of bits that will trigger event notification. To prevent spurious event notifications a De-bounce Interval is configured. This defines the amount of time that a condition must persist before the event notification is time stamped and sent out. Once an event has been latched (once a condition has been detected the corresponding bit is set and remains set until it is acknowledged), it is transmitted repeatedly at the rate indicated by the Retry Period until the event has been acknowledged. Event notifications include a time stamp representing the first time a notification occurred.

Configuration, Installation, and Checkout

WirelessHART builds on HART. The HART infrastructure provides an extensive set of features supporting configuration, calibration, installation, and checkout. Devices are described in Electronic Device Description Language (EDDL). EDDL describes the devices' resource (hardware, electronics, and diagnostics information) and transducer information, methods, and user interfaces. Since new devices and new types of measurements will be added over time it is important that WirelessHART maintains both forward and backward compatibility. WirelessHART configuration conforms to the IEC 61804-3 standard [23].

As part of the installation process, field devices are configured and calibrated for the specific process where they are being used. As part of the configuration, each device is given a tag and scaled (device scaling includes high scale value, low scale value, engineering units, and decimal places), and signal conditioning is applied (e.g., in the case of a valve, it is important to know which direction the valve moves when the signal value is increased). A field device may be commissioned and calibrated at the factory, on site in the instrumentation shop, or after the device has been installed.

The set of parameters a device supports and the methods available to the device are described by the device's description. Users usually have standard parameter value templates for different types of devices and use these templates to customize devices for use in their plants. Tools such as the AMS 375 field communicator can be used to configure and calibrate field devices.

Device measurements are communicated to control systems or host applications. The parameters to be communicated and the rate that they are communicated depend on the process that the device is being used in. Configuring control systems and host applications requires a different set of information. In the case of the control system what is important are measurements that the device supports and the signal information that is associated with those measurements. Signal information includes information such as signal tag (since devices can be multivariable they can have many signal tags associated with them), scaling information, alarm information, linearization, and description. Control system configuration is performed off-line and downloaded into the control system at various points in the configuration cycle.

9-3. Principles

Supporting process automation applications requires consideration of each of the following: latency, energy use, robustness, security, and performance. Managing latency, while at the same time conserving power, is a difficult problem. To minimize latency, communications need to be organized so that packets are not delayed en route from source to destination. To minimize energy use, devices should be kept in a low-power mode as much as possible [7][13]. Deployments must be easy to set up, and once deployed, devices must run unattended for long periods of time, usually years.

At the communication distances typical in wireless sensor networks, listening for information on a radio channel costs about the same as data transmission. Furthermore, the energy cost for a device in idle mode is approximately the same as for a device that is in receive mode. Thus, the biggest single action to save power is to turn the radio and the device itself off during idle times, i.e., put the device to sleep. Turning the device off implies advanced knowledge about when the device will be idle. The approach taken by WirelessHART is to configure the device with knowledge about when it should wake up, perform some function, and go back to sleep; this configuration is performed by a network manager and is called scheduling [5][8].

In WirelessHART, communications are precisely scheduled using an approach referred to as Time Division Multiple Access (TDMA). The vast majority of communications are directed along graph routes. As mentioned, scheduling is performed by a centralized network manager that uses overall network routing information in combination with communication requirements that devices and software applications have provided. The schedule is subdivided into links and transferred from the network manager to individual devices; devices are only provided with the links where there is a transmit or receive requirement. The network manager continuously adapts the overall network graph and network schedule to adjust for changes in network topology and communication demands.

Several key features of the WirelessHART Network, such as self-healing, self-organization, and redundant routing, combine to make the wireless network robust. The network manager is responsible for the creation of this schedule and the associated connections. It is also responsible for the distribution of this schedule to the individual network devices. This scheduling function may be broken into the following phases [5]:

1. Support devices joining the network. As part of this the network manager is responsible for authenticating and orchestrating the join process.

2. Establish routes. As part of this the network manager is responsible for the creation of routes that can be used by plant automation hosts, gateways, other devices, and the network manager itself to perform communications with the application layer in network devices (the application layer in a device is everything above the network communication stack, for example in pressure transmitter this pressure measurement, device diagnostics, etc.).

3. Schedule communications. As part of this the network manager is responsible for the establishment of superframes and links that the user layer application of a network device may use to transfer process data, alerts, diagnostics, and other traffic to a gateway device for access by the plant automation host. The superframes also include links for network management and the join process.

4. Schedule control functions. For network devices that are actuators, interlocks, or any device that affects the process,

the network manager is responsible for the establishments of routes, superframes, and links that the plant automation host may use to send set points and outputs to the user layer application in field devices.

5. Adapt the network. The network manager continually collects data from devices on the health of connections and traffic patterns and uses this information to adjust routing and scheduling.

The effectiveness of the overall network ultimately boils down to a combination of routing and scheduling. The network manager develops a set of routes and then uses these routes in combination with requests for network resources to schedule the network. The schedule is then distributed to each device using network management commands. Each device contains device management, which is responsible for communicating with the network manager and storing network configuration in the device. The overall communication stack (Figure 9-3) follows the International Organization for Standardization (ISO) Open System Interconnection Reference Model (OSI Reference Model or OSI Model) [12].

The wireless network achieves very high reliability through the use of several mechanisms: multiple paths to devices, multiple RF channels, and multiple communication tries. If improved reliability is required, more paths can be inserted by the network manager. As additional devices are added to the mesh network the strength of the overall mesh gets better and better. Additional devices improve path diversity. Additional access points, and devices in general, increase throughput, reduce latency, and can be used to route around areas of potential interference [7, 8].

All communication on the wireless sensor network is time-synchronized. The basic unit of measure is a time slot, which is a unit of time of fixed (10 ms) duration commonly shared by all devices in the network. The duration of a time slot is sufficient to send or receive one packet and an accompanying acknowledgement, including guard-band times for network-wide synchronization (a guard band is a frequency that insulates one signal from another).

At the lowest level, communications are divided into time slots. Since WirelessHART supports up to 15 channels, 15 time slots may be in operation at the same time. To improve reliability, the physical channel is alternated at run time using an algorithm described in the WirelessHART

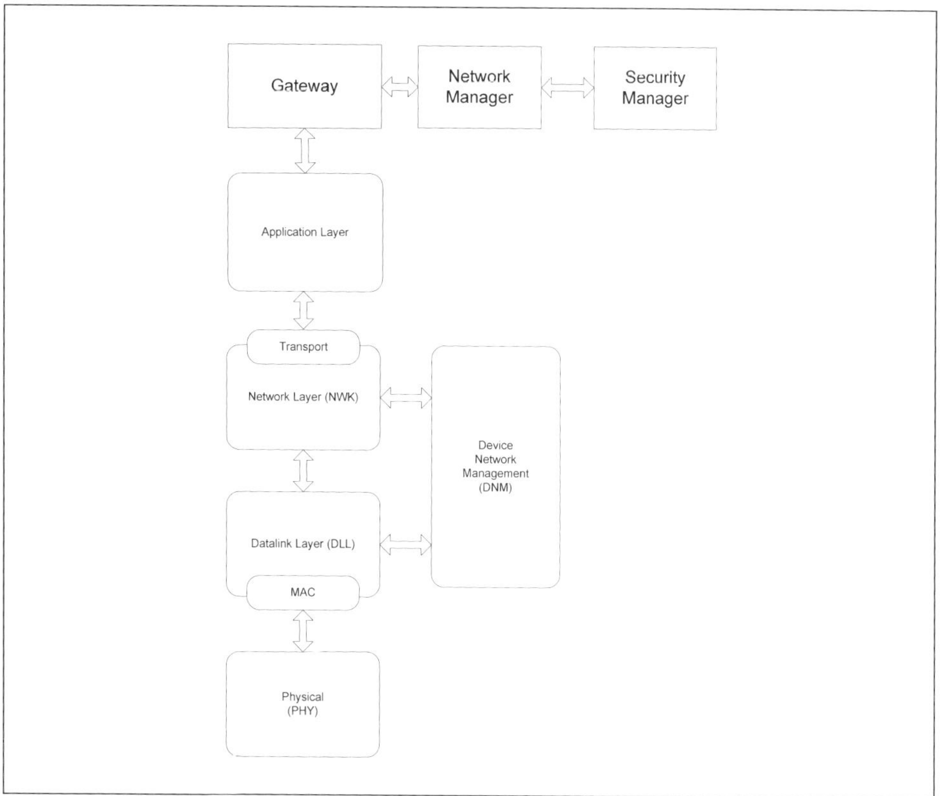

Figure 9-3. WirelessHART Communication Stack

standard. Within each time slot, devices are scheduled to transmit or receive. If there is only one transmitting and one receiving device the time slot is designated as a dedicated slot (a slot is also referred to as a link—the term *link* is used to define the association between devices during a specific time slot). If more than one transmitter can transmit into the link, the link is designated as shared. In this shared case, the transmitters must perform a clear channel assessment (CCA) before they attempt transmission. Should simultaneous transmissions (collisions) occur, devices use a back-off algorithm to manage communications. If more than one receiver is enabled to receive, the link is designated as a broadcast time link.

As has been mentioned, precise time synchronization is critical to the operation of the sensor network. Since all communication happens in time slots, devices must have the same notion of when each time slot begins and ends, with minimal variation. Time propagates outwards from the gateway (a gateway is a device that is used to connect wireless network control systems and higher level business applications) (Figure 9-4).

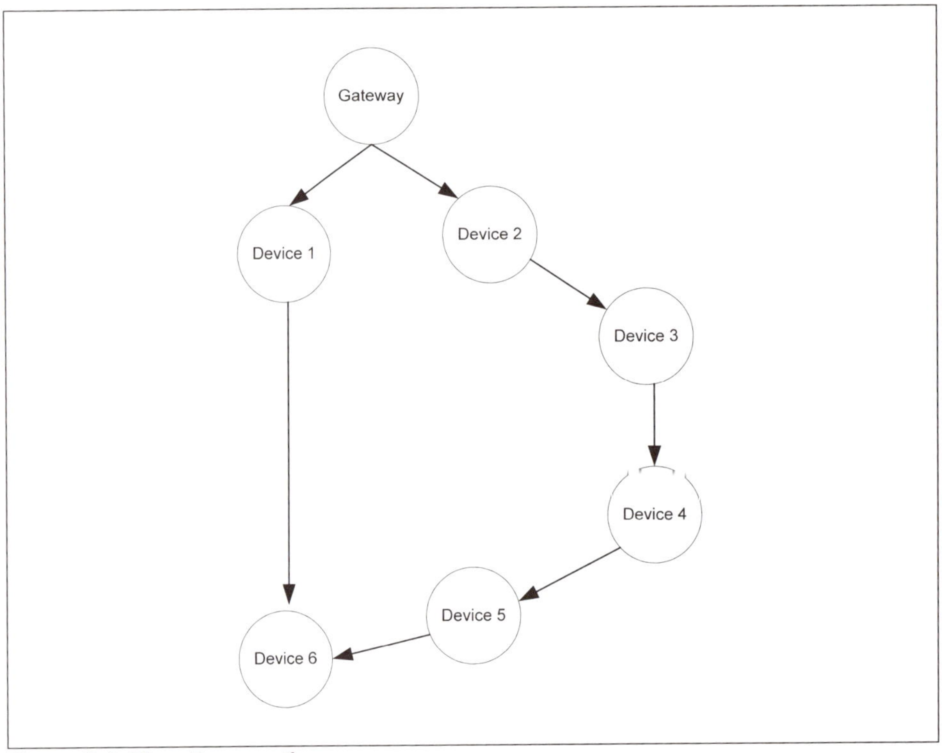

Figure 9-4. Time Propagation

Communications are organized by superframes and time slots. As shown in Figure 9-5, in each time slot there must be one transmitter and one or more receivers (one for dedicated links; multiple for shared or broadcast). In this example, devices A and B communicate during link 0; devices B and C communicate during link 1; and link 2 is not being used. Every three links, the schedule repeats. An overall schedule may have many superframes. Real superframes would typically be hundreds of time slots long.

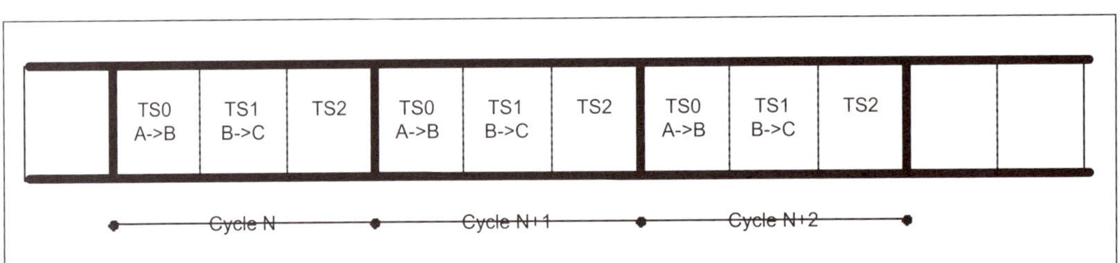

Figure 9-5. Example of a Three-slot Superframe

9-4. Security and Reliability

Communications across a wireless mesh sensor network must be both secure and reliable. Security is inherent in the design. To join a WirelessHART network, devices must pass a rigorous join process during which they authenticate themselves with the Network manager and in turn receive session keys [11]. All communications in WirelessHART are encrypted.

Communications are secured by ensuring that only devices approved to join the network join, that communications are not tampered with, and that bad guys cannot spoof the network through the use of replay attacks (a replay attack occurs when the same packets are resent on the same network). Communications are made reliable by retrying on different channels, retrying at different points in time, and communicating on different paths.

A security manager works with the network manager to secure the network from threats to its operation. The security manager generates and manages the cryptographic material used by the network. It is responsible for the generation, storage, and management of keys (Figure 9-3).

All communications in the network are encrypted and enciphered using the Advanced Encryption Standard (AES) [9, 10]. The standard comprises three block ciphers, AES-128, AES-192 and AES-256, adopted from a larger collection originally published as Rijndael. AES was announced by National Institute of Standards and Technology (NIST) as U.S. FIPS PUB 197 (FIPS 197) on November 26, 2001 [10]. A mesh network uses AES-128. The AES ciphers have been analyzed extensively and are now used worldwide. Many communication chips now contain a built-in AES coprocessor [20].

Joining the Network

As was mentioned above, devices that wish to join the sensor network must be authenticated through the join process. As part of the join process, devices are initialized with a join key and a network ID. Devices that are part of the existing network advertise for new neighbors; devices wishing to join the network listen for these advertisements. Devices wishing to join the network use their join key to encrypt a join request that they send to the network manager. The network manager uses the unique ID for the device to look up the device's join key, decrypts the join request, and if everything checks out, allows the device to join the network. The network manager then completes the join process by providing the joining device with a set of superframes, links, routes, and graphs. Join keys can be unique for each device or shared—the strategy is determined by the user and set in the gateway. Once a device joins the network the network manager can change the device's join key.

Securing Communications

All communications across the sensor network contain a security header (Figure 9-6); the security header provides a bridge between the security sub-layer in the source and the destination. The security header is designed to ensure that the payload is not tampered with. The security header starts with a security control byte that specifies the type of security that is being used. The counter is part of a nonce counter. Nonce counters are used to detect and handle reply attacks. The Message Integrity Code (MIC) ensures that the enciphered payload has not been tampered with.

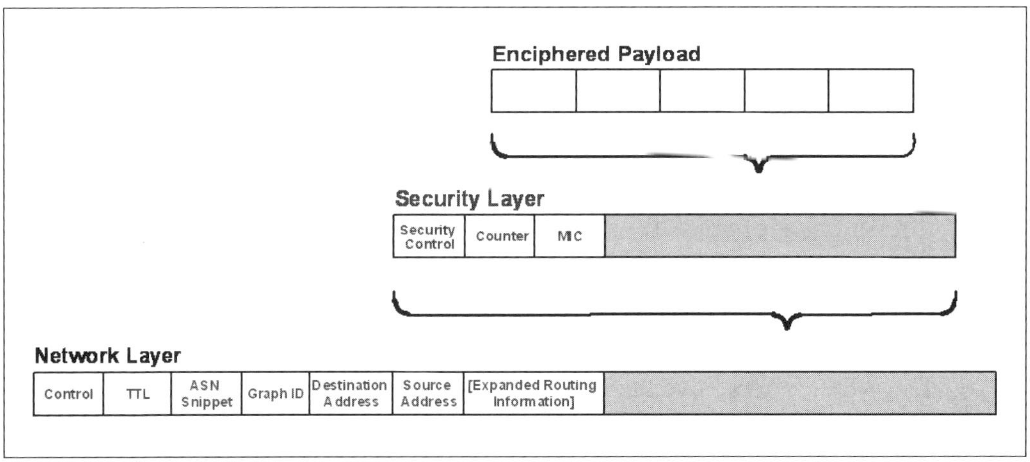

Figure 9-6. Security Sub-Layer

Sessions

Wireless sensor networks must seamlessly support services for a wide range of applications, including real-time data, alarms and events, block data transfers, configuration, calibration, and diagnostics. Sensor networks support communication traffic sourced by the network manager, by the gateway, by host applications, by devices, by location devices, and by users through host level applications and handhelds. To support these applications, session-oriented communications are used. Through these sessions applications have access to network resources (network resources are the channels and bandwidth provided by the network). Each device supports the following four sessions:

- Unicast network manager communications
- Broadcast network manager communications
- Unicast gateway communications
- Broadcast gateway communications

Additional sessions must be added for handhelds and peer-to-peer communications. Only the network manager can add or remove sessions from a device.

Reliability

Wireless sensor networks must be reliable. Reliability is achieved through:

- Secure communications
- Adapting to different types of interferers
- Retries on different paths, on different frequencies, and at different points in time
- Self organizing

The first point, secure communications, ensures that communications are not tampered with. The second point, adapting to different types of interferers, is what makes the network robust. In the wireless world interferers can be things like a truck parked in front of a transmitter, a cell phone, a walkie-talkie, a BlueTooth device, etc. To handle interferers WirelessHART is designed to retry on different frequencies, on different physical paths, and at different points in time. Since there is no way to predict all of the sources of interference using static site surveys, the sensor network must be self-organizing and must adapt. The only topology robust enough to support long term operations is a mesh. As more devices are added to the sensor network the mesh gets stronger and more robust. As we have seen, additional devices improve path diversity. Additional access points, and devices in general, increase throughput, reduce latency, and can be used to route around areas of potential interference.

9-5. Communication Rules

The network manager is responsible for managing routes, graphs, and allocating communication resources. Once these communication resources have been distributed to devices they are utilized by devices to send and receive communications. When contention for resources occur, priority is used to decide which messages get serviced first. The rules for communications are summarized in the following sections.

Network Routing

Three routing techniques are used to route packets: graph routing, superframe routing, and source routing. When using graph routing, a device sends packets with a Graph ID in the network layer header along a set of paths towards the destination. All devices on the way to the destination must be pre-configured with graph information that specifies the neighbors to which the packets may be forwarded. In a properly configured network, all devices will have at least two devices in the graph through which they may send packets.

Superframe routing is a special case of graph routing. In superframe routing packets are assigned to a superframe. Packets are instructed to follow the superframe en route from the source to the destination. With superframe routing the Graph ID is set to the Superframe ID. Since the packet follows the superframe, it is not necessary to explicitly configure graph edges.

With source routing, pre-configuration of the forwarding devices is not necessary. To send a packet to its destination, the source device includes in the network layer header an ordered list of devices through which the packet must travel. As the packet is routed, each routing device utilizes the next device address from the packet header to determine the next hop to use. Since packets may go to a destination without explicit setup of intermediate devices, source routing requires knowledge of the network topology. Only the network manager has enough information to construct source routes.

The network manager contains a complete list of routes, graphs, and devices. When devices are initially added to the network, the network manager stores all neighbor entries including signal strength information as reported from each device. The network manager uses this information to build a complete network graph. The network graph is optimized for reliability, hop count, reporting rates, power usage, and overall traffic flow (Figure 9-21).

Every graph in a network is associated with a unique Graph ID. To send a packet on a graph, the source device includes a Graph ID in the packet's network header. The packet travels along the paths corresponding to the Graph ID until it reaches its destination, or is discarded. In order to be able to route packets along a graph, a device needs to be configured with a graph table. The graph table contains entries that include the Graph ID and neighbor address. Redundant paths may be set up by having more than one neighbor associated with the same Graph ID. The device routing the packet must perform a lookup in the graph table by Graph ID, and send the packet to any of the neighbors included in the table. Once any neighbor acknowledges receipt of the packet (Media Access Control [MAC] level acknowledgement), the routing device may remove the packet from its transmit buffer. If an acknowledge is not received, the device will attempt to retransmit the packet at its next available opportunity. The packet will be retried until it is successfully sent or is timed out.

The network manager generates a set of graphs covering communications from each device to the gateway, from the gateway to each device, and broadcasting from the gateway to all devices in the network. Routes and graphs are allocated using a set of rules as follows:

- If possible one of the neighbors should be an access point (i.e., a single hop).
- Route through powered devices if they are available.
- Minimize the number of hops.
- Use a combination of weighted signal strengths to select between alternative routes (the weights need to be applied to all of the nodes in a particular set of hops towards the destination).
- Keep the number of hops to four or less.
- Prune the number of neighbors to four or less.

Network Scheduling

Network scheduling involves allocating superframes, links, routes, and graphs. During scheduling the network manager ensures that all the devices in the network are given sufficient network resources to meet their data reporting needs. As part of doing this the network manager also pays attention to latency, as well as power consumption. Managing latency, while at the same time conserving power, is an important consideration. To minimize latency, communications need to be organized so that packets are not delayed en route from source to destination. To minimize the energy use, devices should be placed into sleep mode as much as possible. Power consumption limits the utility of wireless sensor networks, which must operate unattended on the order of years. Replacing batteries is a laborious task and extremely difficult in some environments. Conserving energy is therefore critical for prolonging the lifetime of the sensor network.

WirelessHART takes the approach that applications and devices specify their communication requirements and a centralized network manager, in turn, allocates communication resources. The centralized network manager then distributes the schedule to each of the effected devices. When devices are not scheduled to transmit or receive they sleep.

Each wireless network contains exactly one overall schedule that is created and managed by the network manager. The schedule is subdivided into superframes. Each superframe is further subdivided into superframe relative links that repeat as the superframe cycles.

Requests for communication resources are directed to the network manager through network management commands. For example, a software application may request a device to schedule a burst mode communication every second. The device in turn would make a request to the network manager for the communication resources that it needs to meet this request. This kind of network-centric approach allows a sensor network to collect and process data in a very energy-efficient manner.

Over the lifetime of the wireless network it is important that the network manager be able to automatically adapt the schedule. For example, applications like asset management are started and stopped, valve and device diagnostics are run, devices are added and removed, and interferers such as scaffolding or other communication devices come and go. As part of the adapt algorithm, the network manager continually receives feedback. Feedback is provided through periodic health reports, network level alerts, and network statistics. Adapting includes adding and removing superframes, links, graphs, and routes. Summarizing the key points that have been presented:

- Scheduling reduces contention and ensures that all devices are provided with the opportunity to communicate. This is one reason why network scheduling yields such high end-to-end packet reception.

- Scheduling must be power aware.
- In many situations broadcasting is used to improve the overall effectiveness of the network. The network manager must set up routes for broadcasting.
- Time synchronization is critical—the entire network is dependent on a common understanding of time.
- To minimize latency the network manager needs to allocate links in sequence from the source to the destination.
- The sensor network must adapt to changing conditions in the plant and in the software applications.

Priority

Being able to handle sudden increases in traffic is an important characteristic of all networks. In wireless networks one of the key mechanisms used to do this is message priority. Message priority helps mitigate situations where devices need to perform information flow control. An example of this is when a piece of process equipment trips, for example a boiler. In these situations it is important that measurement and control communications get through while some of the less time-critical information is delayed. Priorities are set so that network management commands are the highest priority (this allows the network manager to temporarily increase communication resources, re-route, etc), followed by real-time published messages, followed by alarms and alerts. In the case of alarms and alerts, since they are time-stamped at the device, no critical information is lost if an alarm or alert is delayed a few time slots.

Optimizing Communication Resources

Scheduling is performed by a centralized network manager which uses overall network routing information in combination with communication requirements that devices and applications have provided. The schedule is subdivided into links and transferred from the network manager to individual devices; devices are only provided with the links for which they have transmit or receive requirements. The network manager continuously adapts the overall network to changes in network topology and communication demand.

In order to meet the requirements for latency and jitter, process measurements must be communicated often enough to not miss oscillations in the signal. To support doing this in a very efficient manner, measurements can be published periodically or periodically-by-exception—when reporting by exception a maximum time interval can be specified [2]. In addition, the reporting rates can be adjusted to report more frequently when the signal value is higher or lower than a configured threshold value. Measurements also include their sampled time stamp. Multiple variables can be communicated at the same time.

Monitor and Control over a Wireless Sensor Network

Wired digital communications have been in use for many years in supporting monitor, control, and diagnostic applications. Most plants today use digital communications for a significant portion of their control applications. Wireless digital communications are relatively new in process plants. One of the key questions that is often asked is, "Is wireless ready for control applications [2]?"

WirelessHART is effectively HART and as such was designed from the beginning to support monitor and control applications. Key factors such as sampling intervals, jitter, and latency are either easily managed or are well within the requirements of typical process control applications. To minimize communications, exception reporting was added. The rule of thumb is that feedback control should be executed four to ten times faster than the process response time, where response time equals the process time constant plus deadtime. Because wired measurement systems are often unsynchronized with the control system (Figure 9-7), measurement values are oversampled by a factor of as much as two to ten times.

With wireless systems, because of the inherent time-synchronized nature of the mesh network, it is possible to reduce or eliminate this oversampling. Sampling can be scheduled to occur just before communications are scheduled. It is also possible to turn on exception reporting and only send values that have changed (Figure 9-8).

A control system's ability to meet its control performance requirements is affected by both delays (latency) and variation (jitter) in when the information is available. Latency and jitter are often introduced by the communication system—for example, when data is transmitted from a transmitter to a gateway. Latency is the time it takes for a communication packet to make its way from the source to the destination, while jitter is variation in latency between different communication cycles (Figure 9-9). Excessive latency (which effectively adds deadtime to the process) and jit-

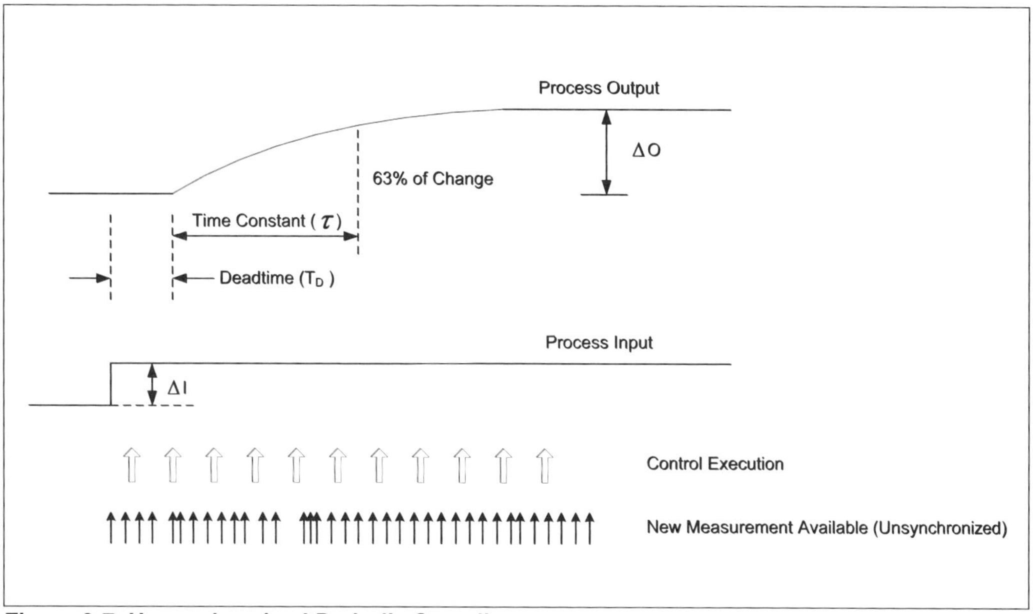

Figure 9-7. Unsynchronized Periodic Sampling

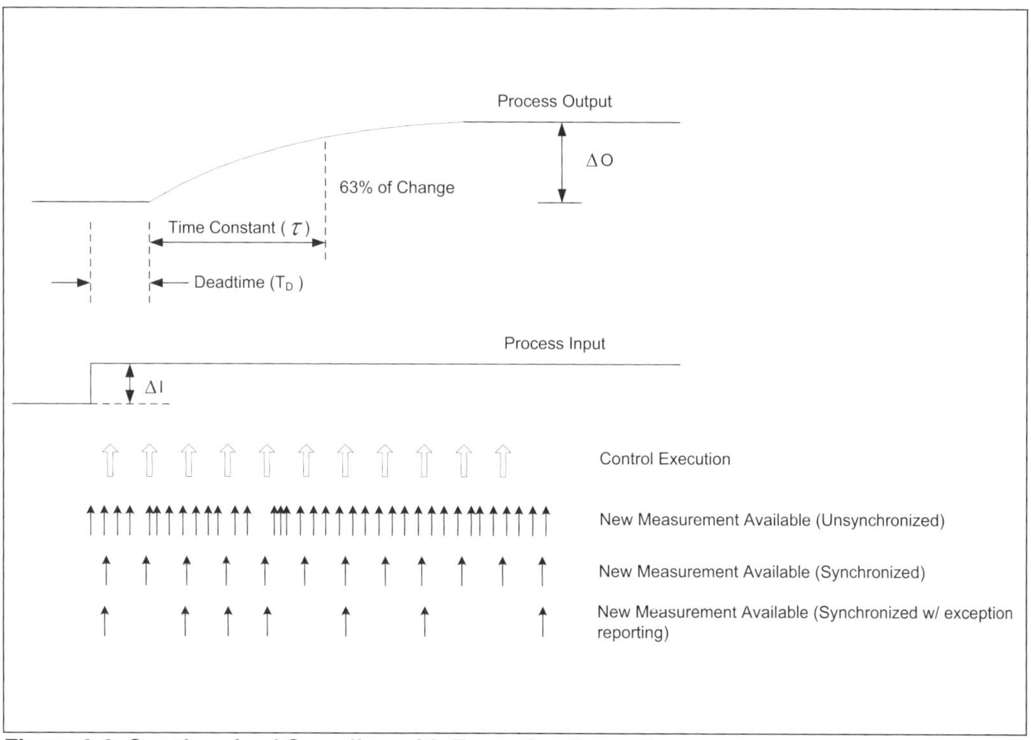

Figure 9-8. Synchronized Sampling with Exception Reporting

ter (which adds error into the control calculations) can lead to significant degradation in control performance.

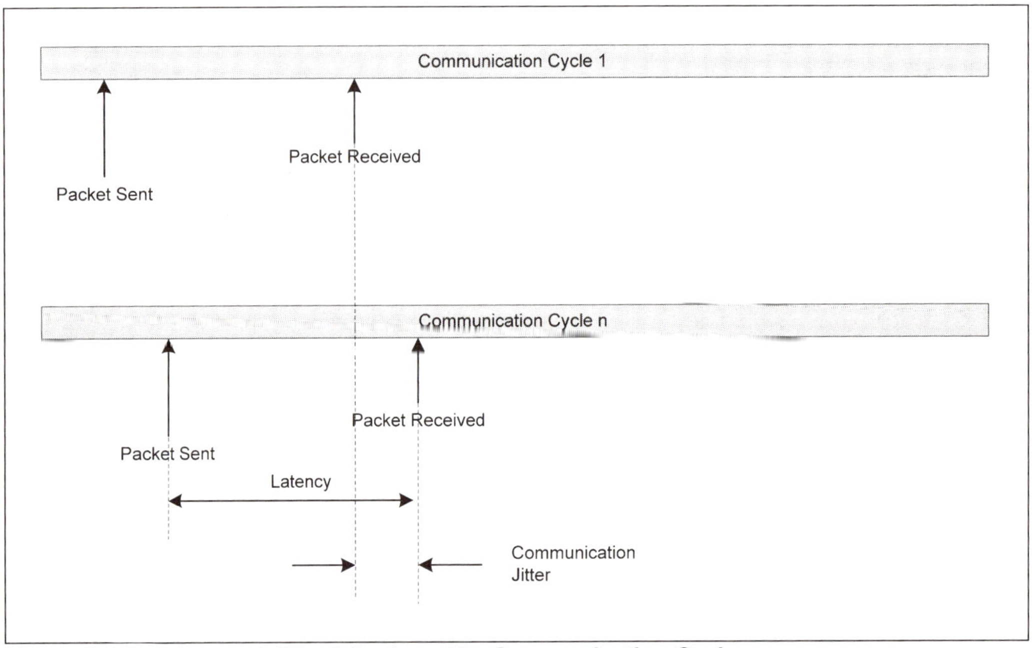

Figure 9-9. Latency and Jitter Introduced by Communication Cycles

Comparing wireless to wired, wireless transmission rates are faster than some traditional wired fieldbus technology. For example, if the communication rate is 31.25 kilobits/second, the communications delay will be 32 microseconds/bit. Wireless has a much faster communications rate—250 kilobits/second—so the delay introduced by the communications rate is only 4 microseconds/bit. Since communications occur in 10-ms time slots (Figure 9-10), the time delay for a single hop will be 10 ms.

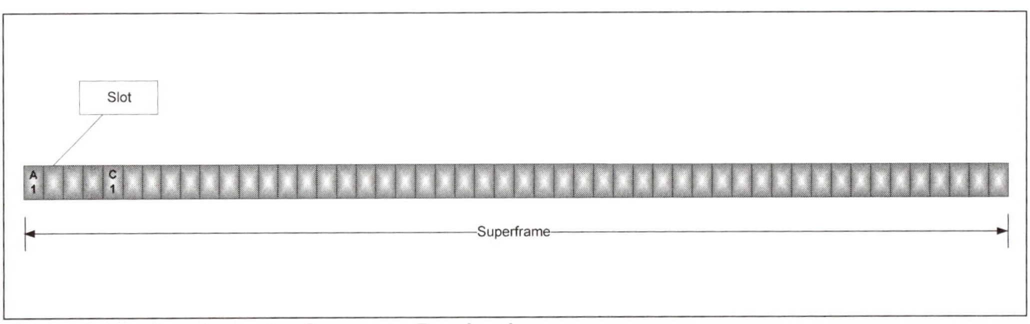

Figure 9-10. Time Links for Source to Destination

In many scenarios communications require more than one time link for a message to travel from the source to the destination (Figure 9-11). If a communication can't reach its destination directly, it can "hop" from device to device. This ability to route around physical obstacles or interference is a core feature of mesh technology.

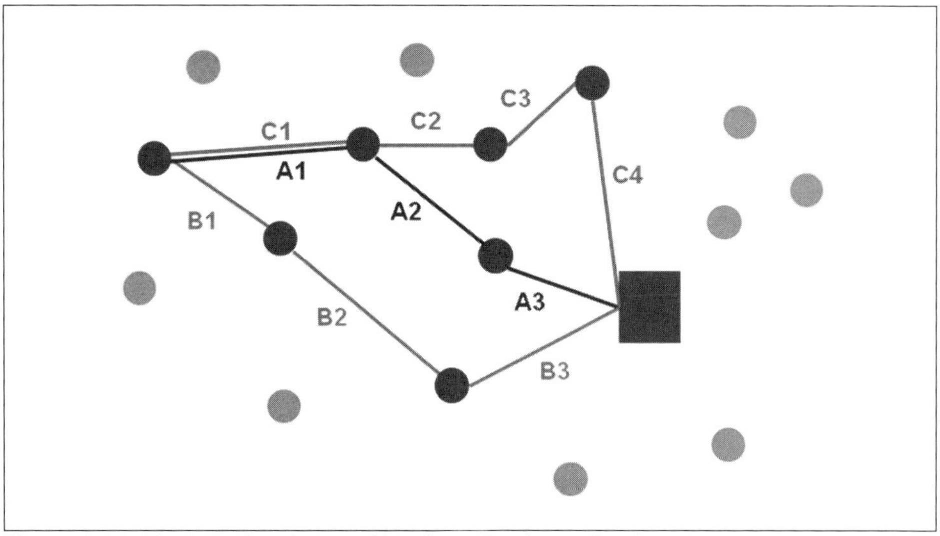

Figure 9-11. Mesh Technology to Hop from Device to Device

Changing the route the data travels can contribute to variations in communication time (jitter). Although each additional hop increases latency, in typical applications the average delay is well within the requirements for control.

In most cases, a wireless network will be able to retry a failed message in the next time slot or the one following. For example, we'll assume it takes 10 ms to process a message and assign it to another time slot. Path A in Figure 9-11 could therefore produce as much as 50 ms of total latency (10 ms + [10 ms + 10 ms] + [10 ms +10 ms]). Path B has the same number of hops and thus the same communications latency. But Path C has an additional hop, bringing total communications latency to 70 ms. This timing difference introduces a 20-ms jitter in the communications. (In many cases, the routing device will be able to retry in the very next link, which would reduce the total latencies to 30 ms for Paths A and B and to 40 ms for Path C.)

Experience in hundreds of wireless field device installations shows that communication latency on average is much lower than in this example. In real plant settings, typically 30% of the devices communicate directly with the gateway or network access point (10 ms) and about 50% are one hop away (30 ms). The rest are three to four hops. Using these numbers from existing plant installations, the average latency time will be about 30 ms.

Example – Scheduling Monitoring and Control for a Bioreactor Process

For actuators, interlocks, or any device that affects the process, the network manager is responsible for the establishments of routes, graphs, superframes, and links that the control system may use to send set points and outputs to the user layer application in field devices. In this example the wireless transmitters and actuators associated with a bioreactor process are used to demonstrate the operation of the wireless mesh network (Figure 9-12).

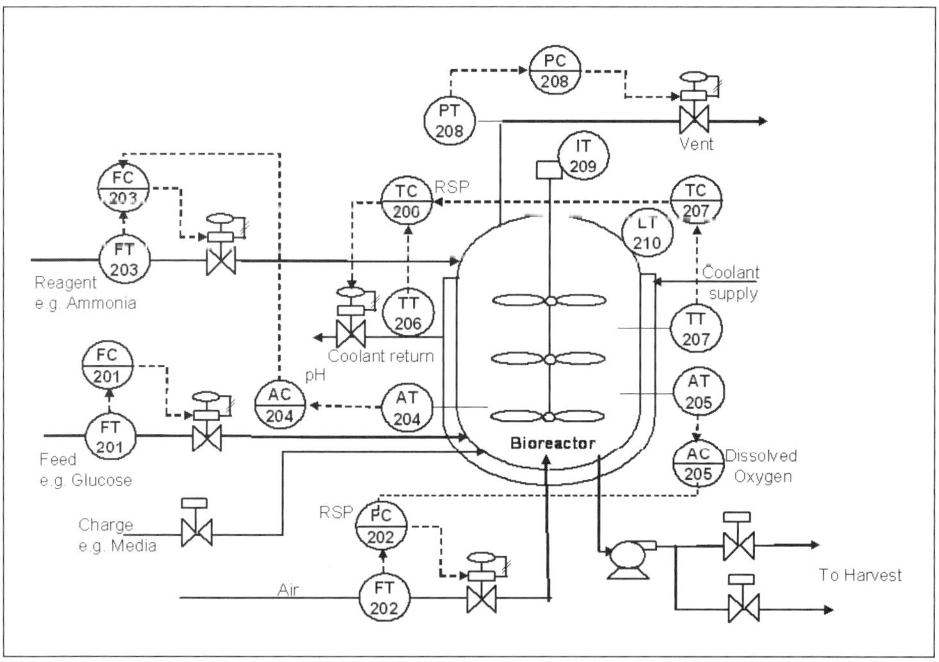

Figure 9-12. Bioreactor

The measurements, regulating valves, and blocking valves included in this example are summarized in Figure 9-13.

Network Management and Host Request

The first step in the formation of the wireless network is to allocate communication resources for network management. Devices wishing to join

Category	Device	Measurement
Measurement	D2	Reactor Level (LT210)
	D3	Feed Flow (liquid –FT201)
	D4	Reactor Gas Pressure (PT208)
	D5	Reactor Temperature (TT207)
	D6	Agitator Amps (IT209)
	D7	Return Water Temperature (TT206)
	D8	Reagent Flow (FT203)
	D9	Air Flow (FT202)
	D10	Dissolved Oxygen (AT205)
	D11	pH (AT204)
Regulating Valve	D12	Feed Flow (FV201)
	D13	Reagent flow (FV203)
	D15	Coolant Flow (FV206)
	D16	Vent Flow (FV208)
	D17	Air flow (FV202)
Blocking Valve	D18	Charge Flow (FZ211)
	D19	Harvest Flow – FZ212
	D20	Harvest Flow – FZ213

Figure 9-13. Instrument and Valve List for a Bioreactor

the wireless network listen for advertisement packets and then use information in the packets to generate join requests. When a join request is received by the network manager, the network manager verifies the device's join key [5] and either completes or rejects the join process. As part of this process, the network manager creates dedicated links in one of the superframes for device management and advertisement functions. A link will also be defined for the newly added device to send advertisement packets. This new link information is transferred into the network and reflected in all of the effected devices.

It is expected that any new device will listen for a period of time for advertisement packets and select the device with the strongest signal for joining the network. If two new devices happen to transmit a join request at exactly the same time using the advertised link and channel, then the messages will collide. In this event, the collision will be detected and the transmission re-attempted after a backoff time. Thus, the join requests will be spaced and will most likely be successful on the next attempt.

The overall network routing will be determined by the network manager based on signal strength as determined by the physical layout of the plant, number of hops, and traffic flow. The overall routing will be further adapted to reflect diagnostic information, retries, etc.

Example – Bioreactor Process Measurements

The bioreactor generates 10 process measurements. These measurements are communicated at several different update (scan) rates (Figure 9-14). The frequency at which this information is required by the control system is specific to the process equipment and the measurement type, for example, pressure, temperature, flow, level, and analytical. Thus, as part of control system configuration, the user will configure the following information for all devices that are accessed through the wireless gateway:

- Device Tag – which uniquely identifies the device (e.g., HART Tag)
- Measurement value(s) that are to be accessed in the device
- How often each measurement value is to be communicated to the gateway.

Device	Measurement	Update Rate
D2	Reactor Level (LT210)	16 sec
D3	Feed Flow (liquid –FT201)	1 sec
D4	Reactor Gas Pressure (PT208)	1 sec
D5	Reactor Temperature (TT207)	4 sec
D6	Agitator Amps (IT209)	8 sec
D7	Return Water Temperature (TT206)	16 sec
D8	Reagent Flow (FT203)	1 sec
D9	Air Flow (FT202)	1 sec
D10	Dissolved Oxygen (AT205)	4 sec
D11	pH (AT204)	4 sec

Figure 9-14. Measurement and Scan Rates for the Bioreactor

To support the configuration of multiple superframes for the transfer of process information at different rates, the measurement scan rates used by the field device and the associated communication should be configured as integer multiples of the fastest update time that will be supported by field devices (an example of a field device is a pressure transmitter). For this example, the supported scan rate will be defined as 2^x where x is positive or negative integer values; for example, scan rate selections of 1 sec, 2 sec, 4 sec, 8 sec, 16 sec, and 32 sec.

To avoid introducing latency into the measurement value that is communicated to the gateway, it is important that processing by the sensor be coordinated with the link configured for the measurement transmission (Figure 9-15).

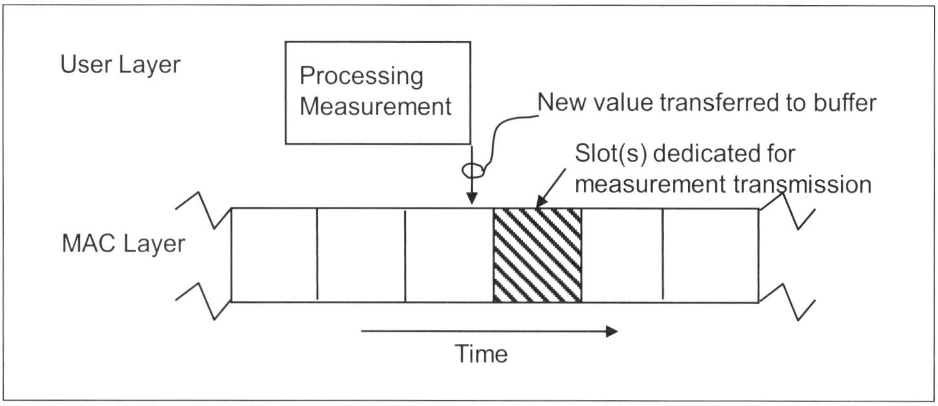

Figure 9-15. Synchronization of Measurement Processing and Transmission

The scheduling of communications associated with process measurements included in a network can be simplified by defining a superframe for each scan period and developing the schedule by allocating links for transmission of measurement data starting with the fastest to the slowest scan rates. In the schedule, a device may only occur once within a link time since at any given time a device is either transmitting or receiving on one channel (i.e., since each device has only one radio, it can only use one channel at the same time).

Scheduling Example – Single Hop

As an initial example of scheduling the most simple case will be used: all devices are able to communicate to the gateway through a single hop.

Using the recommended approach of defining superframes for each scan period and allocation of links from the fastest to the slowest measurement, the graph associated with the batch bioreactor measurements would appear as a hub-and-spokes diagram (Figure 9-16). Routing information would be shown as a set of single hops (Figure 9-17).

To support immediate re-transmission after a failed transmission to the gateway, additional links would be added in the schedule immediately after each transmission. The schedule for this configuration is illustrated

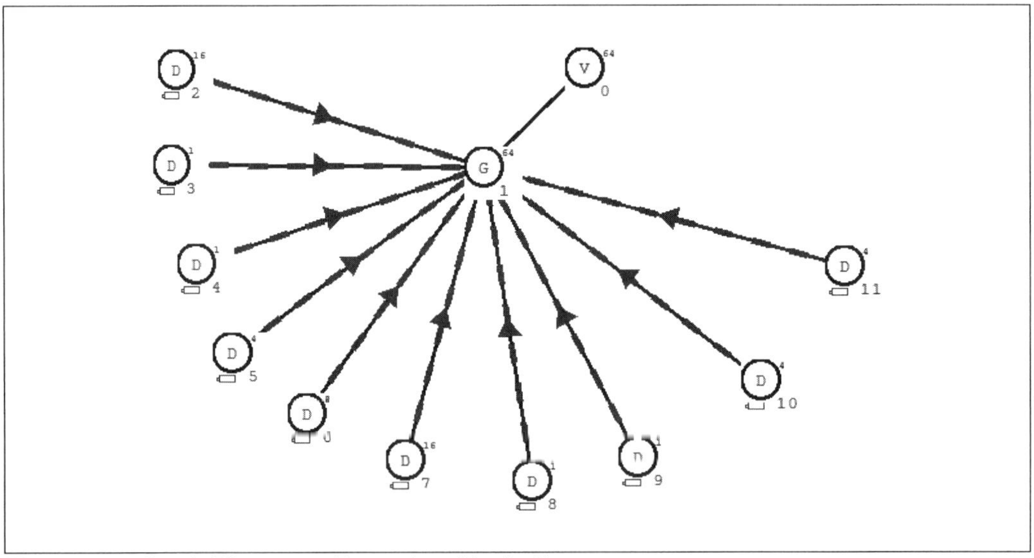

Figure 9-16. Batch Bioreactor Example – Single Hop

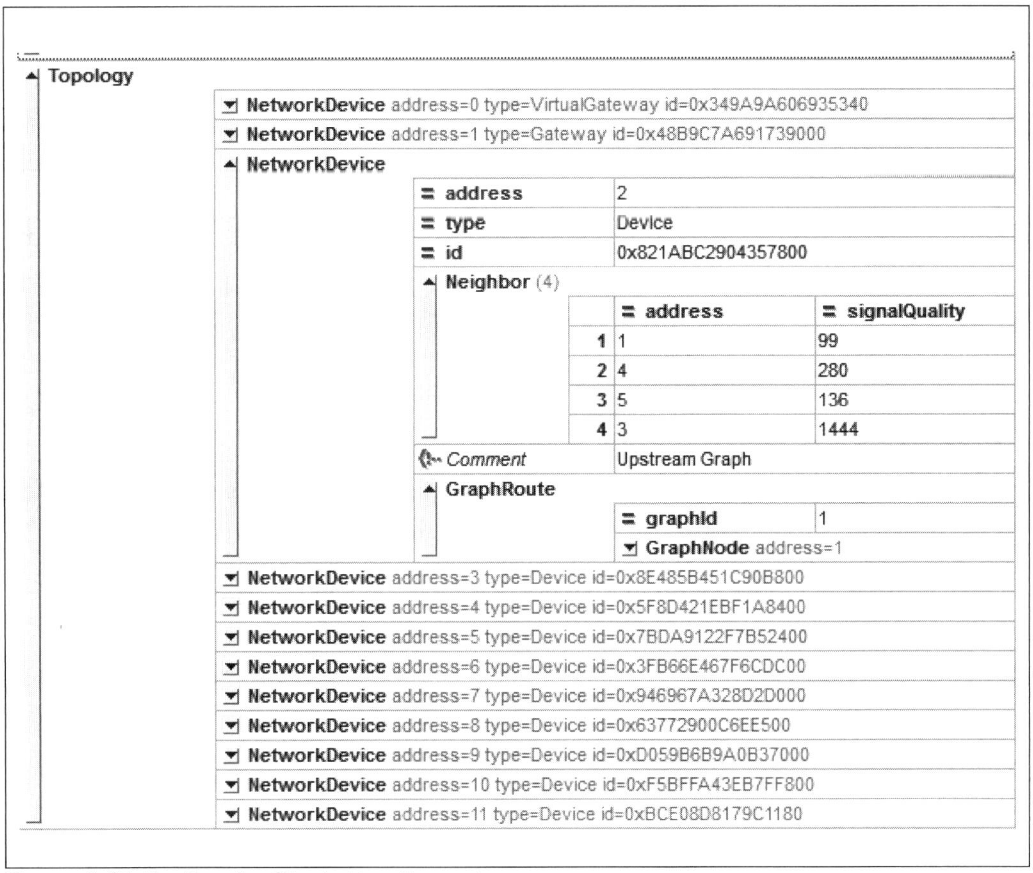

Figure 9-17. Routing for Single Hop Example

in Figure 9-18. The schedule shows radio utilization in the top portion and device communications in the bottom portion.

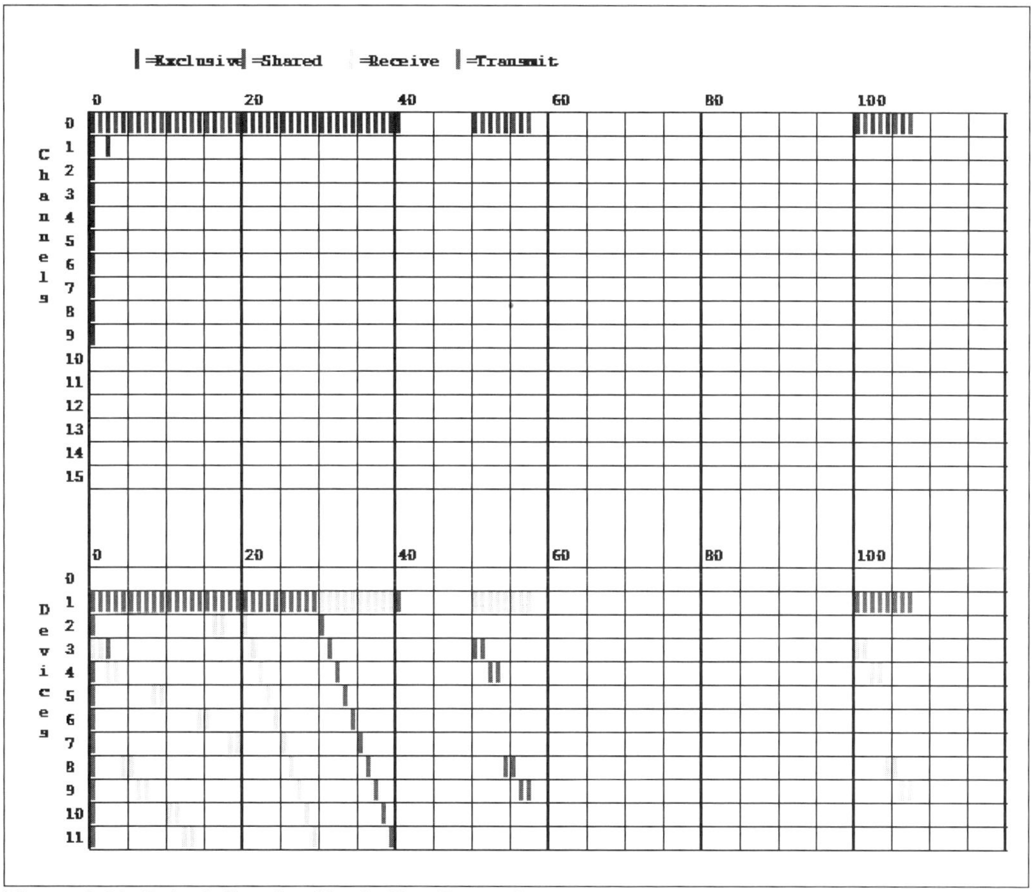

Figure 9-18. Scheduling for Single Hop Example

Scheduling Example – Multiple Hop

The previous example presented the ideal case where all network devices were directly connected to the gateway. In most networking scenarios multiple hops will be used (Figure 9-19).

In multi-hop mesh networks, the routing is through two or more neighbors. At the point of communication, the device decides which neighbor to send to based on the link that is available. For example, device 2 can send packets through either device 4 or device 5 (Figure 9-20).

The schedule for this multi-hop example is shown in Figure 9-21.

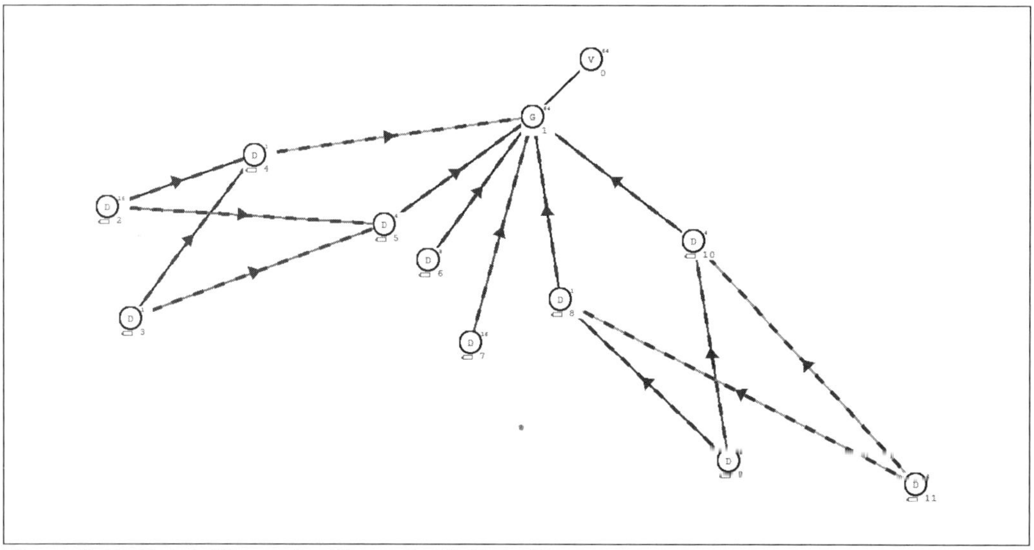

Figure 9-19. Batch Bioreactor Example Multiple Hop

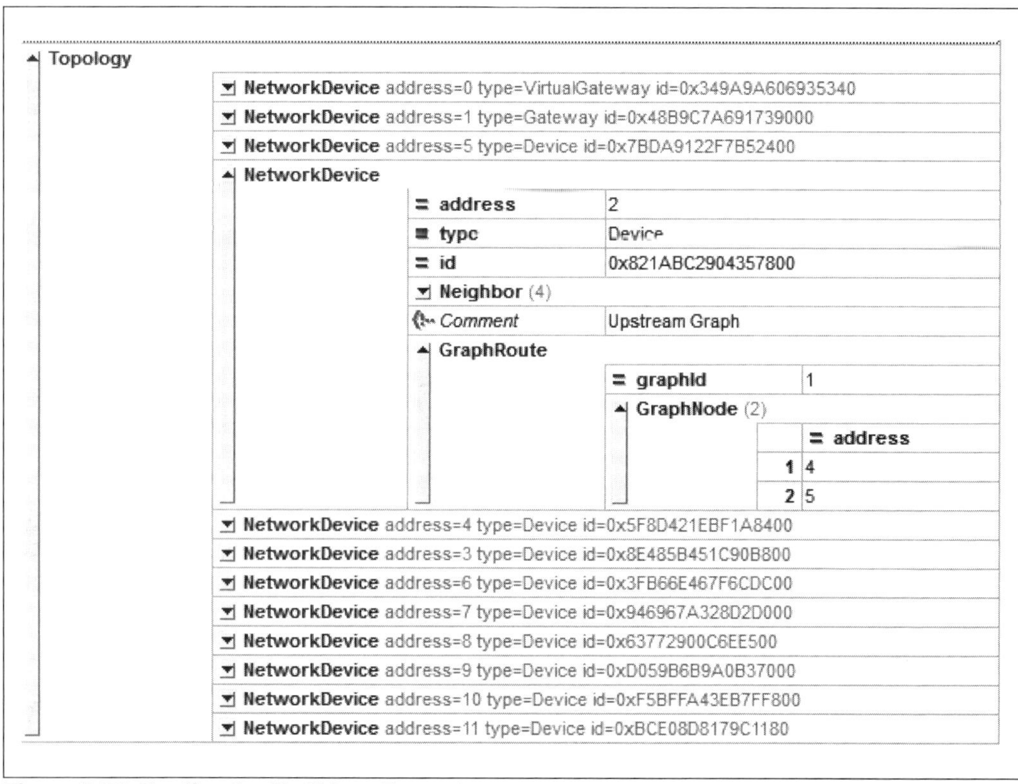

Figure 9-20. Routing for Multi-hop Example

Scheduling Example – Incorporating Control Devices

The example up to this point has only shown measurement devices. As a final step in this example, all the throttling valves are included at 1s scan

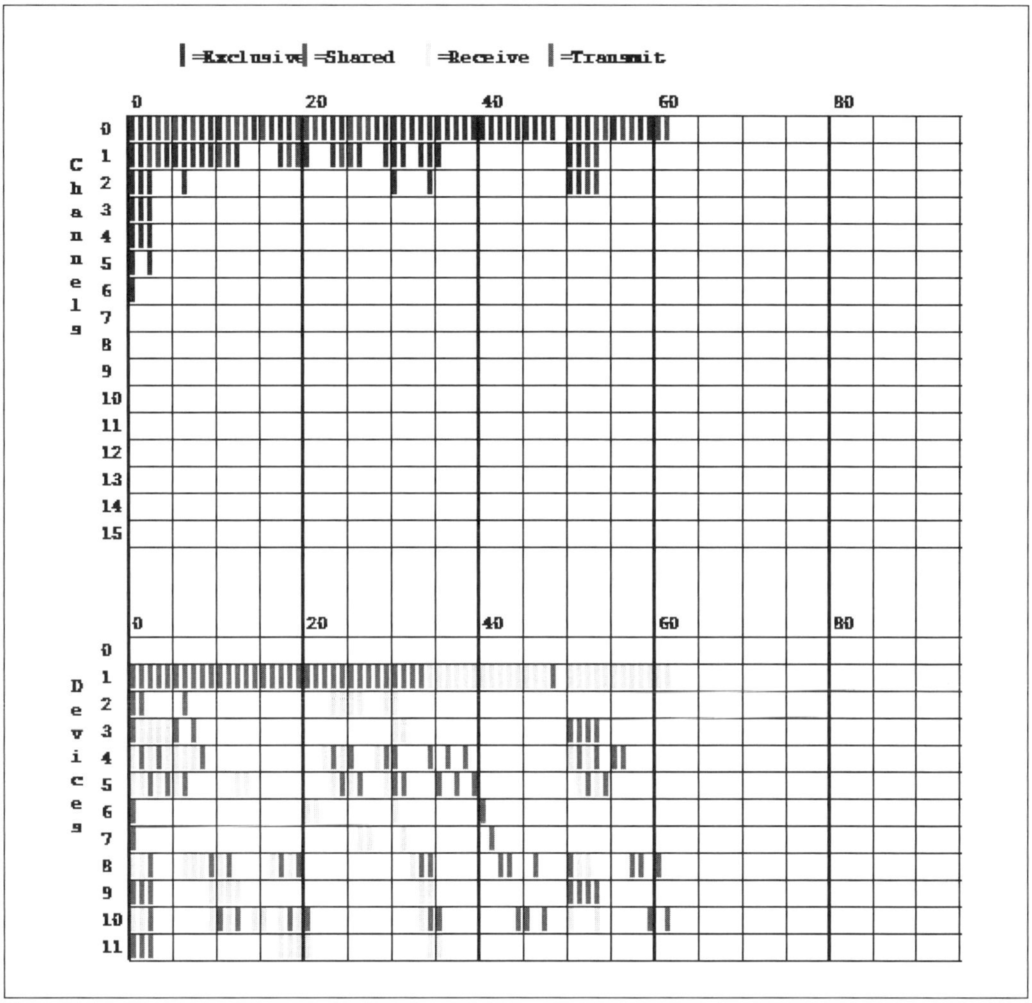

Figure 9-21. Scheduling for Multi-hop Example

rates, and all the blocking valves are included with 4s scan rates (Figure 9-22).

The example also shows the effects of adding a second access point (shown as G1 and G14 in the diagrams). The overall schedule is shown in Figure 9-23.

Over time the network topology will change—for example a truck could drive into the middle of the mesh. When this happens the network manager will restructure the network—this is illustrated in Figure 9-24.

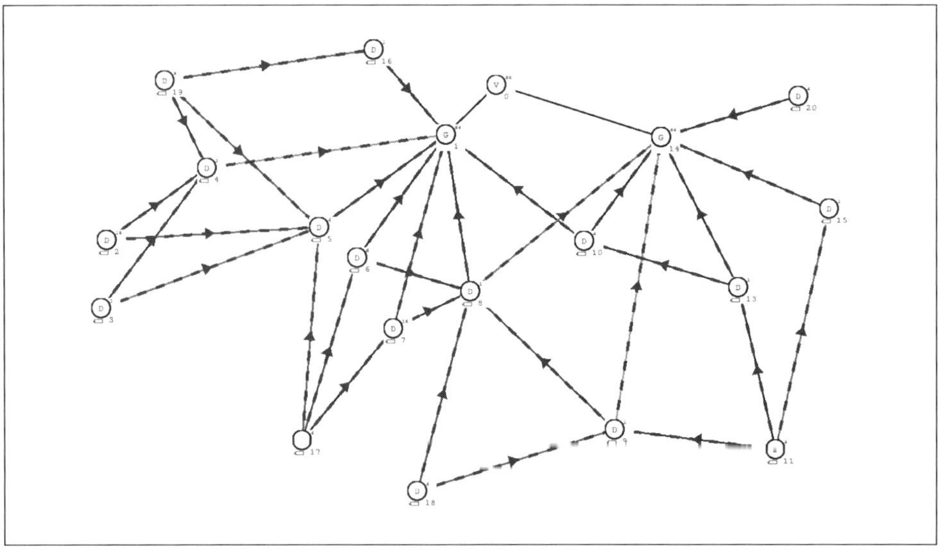

Figure 9-22. Batch Bioreactor Example – All Valves and Devices Included

9-6. Process Control

Utilizing wireless communication to provide measurements that will be used in closed loop control was illustrated earlier in Figure 9-7 and Figure 9-8. As discussed, power consumption can be reduced by eliminating oversampling and by only communicating values when they change.

A modern wired process control system is often structured as shown in Figure 9-25 [21][24]. As illustrated in this figure, sensors and actuators are connected to controllers via control networks, such as Fieldbus [15] and Profibus [17]. A sensor provides measurements and status of a physical property; for example, the flow in a pipe associated with the process. Based on the measurement from sensors, the controller determines any adjustment in the actuators that is needed to maintain the process at a target value, i.e., the set point. The control loop is executed periodically at a rate fast enough to correct any unwanted deviations in the process.

In contrast to non-wireless control systems, communications in a wireless control system tend to have more variation in communications. This variation may be caused by interferers, power failures, and environmental factors such as lightning storms. Current control strategies address this by oversampling. In standard process control systems, a missed I/O communication is considered an error. Usually, a control loop is configured with a maximum number of lost I/O communications, after which the loop declares failure and the values are set according to the fail-safe configuration.

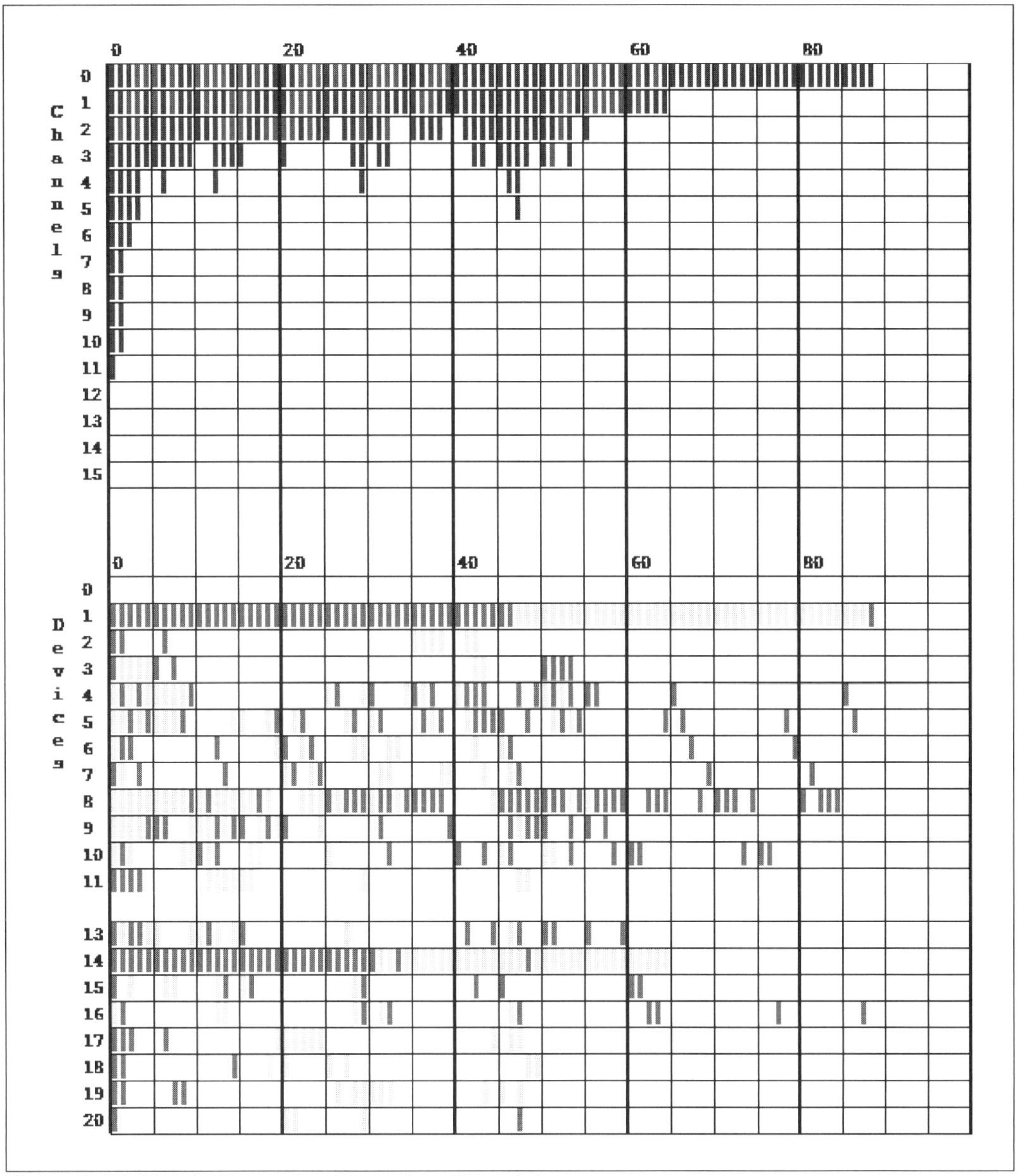

Figure 9-23. Batch Bioreactor Example – Schedule

Therefore, communication variation presents a challenging problem for standard control paradigms. Consider a PID block with inputs from a wireless channel. Suppose the inputs are lost at time t_1 and re-established at time t_2. The derivative component of the PID would cause a spike in the output at t_2. Also, from t_1 to t_2, the reset component may wind up based on the error that existed at time t_1.

Modifying the calculation of the integral and derivative components of PID algorithms by detecting missed communications and compensating

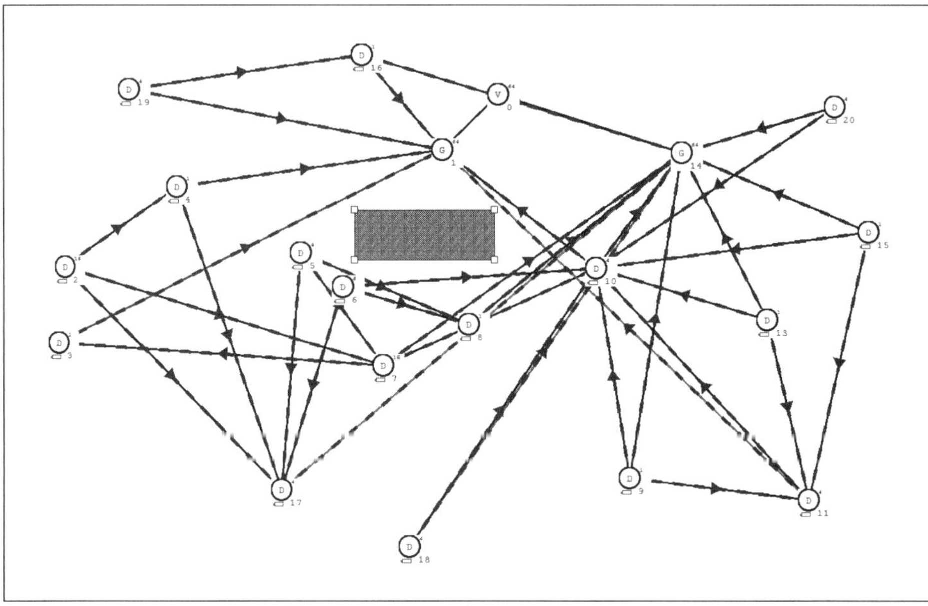

Figure 9-24. Adapting the Network to Interferers

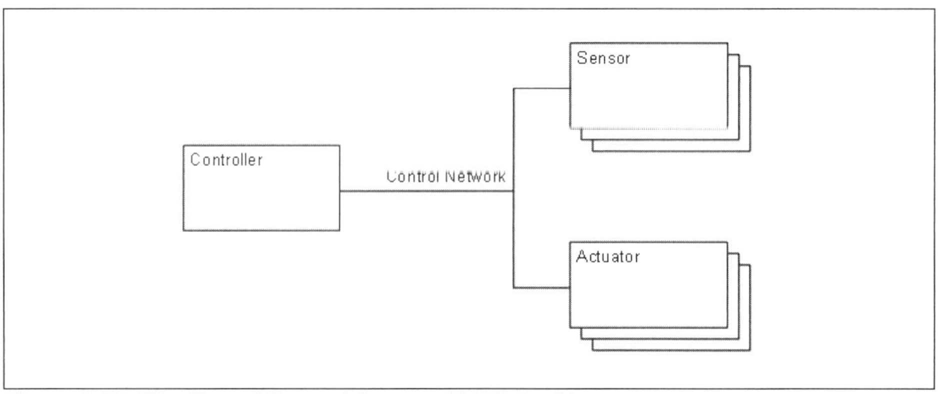

Figure 9-25. Distributed Control System (DCS) Architecture

for them proactively has been shown to greatly improve control performance [2, 3]. This section includes a significant portion of a paper published by ISA [3].

The Standard PID Algorithm

PID is the most widely used control algorithm in industrial process control [25]. As shown in Figure 9-26, the controller compares the process variable (PV) with a reference set point (SP). The error is then processed to calculate a new output to bring the PV back to its desired SP [26].

PID stands for "proportional, integral and derivative" components of the algorithm. Each of the three components performs a different task and has a different effect on the functioning of a system. Their outputs are summed to produce the system output.

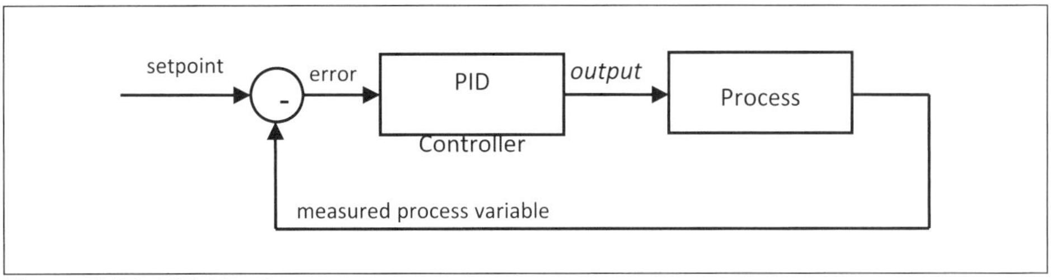

Figure 9-26. PID Block

Though there are many variations of PID algorithms, in its non-interacting form without rate limiting and all actions based on error, the equation for the standard PID algorithm is

$$Output = K_P\left[e(t) + K_I\int e(t)d(t) + K_D\frac{de(t)}{dt}\right]$$

K_P, K_I and K_D are the proportional, integral, and derivative gains, respectively.

In its digital form, the software implementation of the PID algorithm is based on the sampled data for its PV being provided on a periodic basis.

When there is no communication loss, PID, once configured for the process it controls, keeps the process in a steady state. Figure 9-27(a) shows the PID reaction to a process disturbance.

Before time t_0 the PID output ("out") is kept at a constant value to maintain PV at the SP value. At t_0 a drop of PV is observed due to process disturbances. To correct the drop, the PID increases its output. At time t_2, PV comes back at SP, and the out is stabilized at some value that is slightly bigger than its original value to compensate for the disturbance. As is shown in Figure 9-27(a), the out is the sum of three parts: P, I, and D.

Input Communication Lost

Now consider what would happen to each component of PID output if the communication from the sensor input is lost between times t_1 and t_2. For

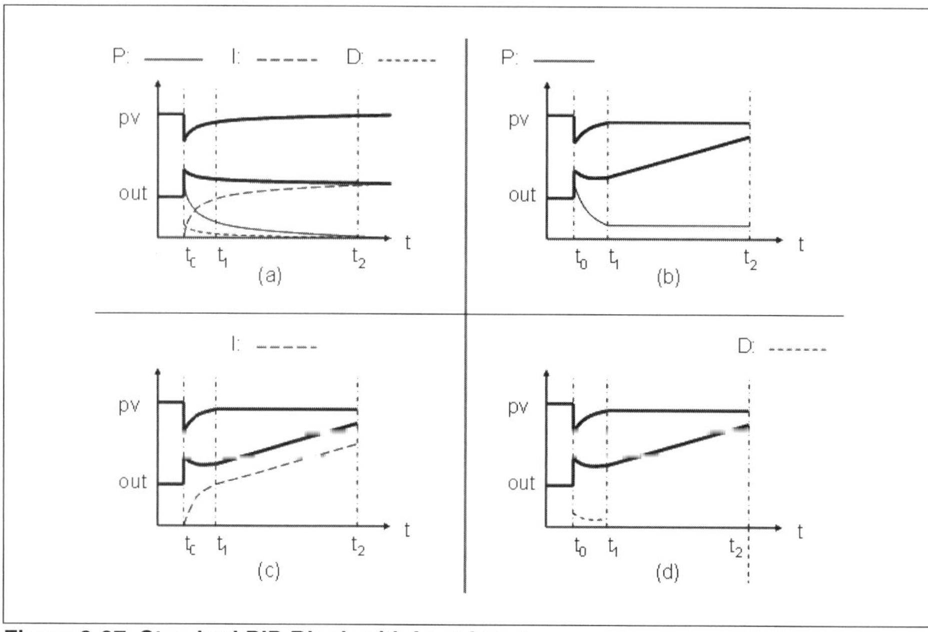

Figure 9-27. Standard PID Block with Lost Input

the PID block, the measured PV remains the same as at time t_1 during this period, as shown in Figure 9-27b, c, and d.

Figure 9-27(b) shows the proportional gain P. Since the measured PV and SP remain the same, the proportional gain is constant from t_1 to t_2. Figure 9-27(c) shows the integral part I. Since PV and SP remain the same, the error remains the same, so the integral part is a linear increasing line from t_1 to t_2. Figure 9-27(d) shows the derivative part D. Since PV and SP remain the same, the error remains the same, so the derivative stays 0 from t_1 to t_2. As a result, the out of the PID block is a linear increasing line from t_1 to t_2, shown in Figure 9-27b, c, and d. This destabilizes the process. The longer the communication is lost, the bigger the deviation between PV and SP.

Once the communication is reestablished at t_2, PID is back to normal. However, since the derivative part calculated at time t_2 is based on the difference of measured PVs between t_2 and one period before t_2, we shall expect a spike for the derivative part. This is because the PV at t_1 is used as the PV at one period before t_2. PV at t_2 could be significantly different from the PV at t_1. Since the PV changes between time t_1 and t_2, we expect the derivative spike to be even bigger. Due to sudden changes of the proportional and derivative parts, the value of the out will have a big impulse before and after t_2.

Output Communication Lost

We continue to analyze how standard PID behaves if output communication is lost between t_1 and t_2. Here we assume there are no other disturbances and input communication is OK.

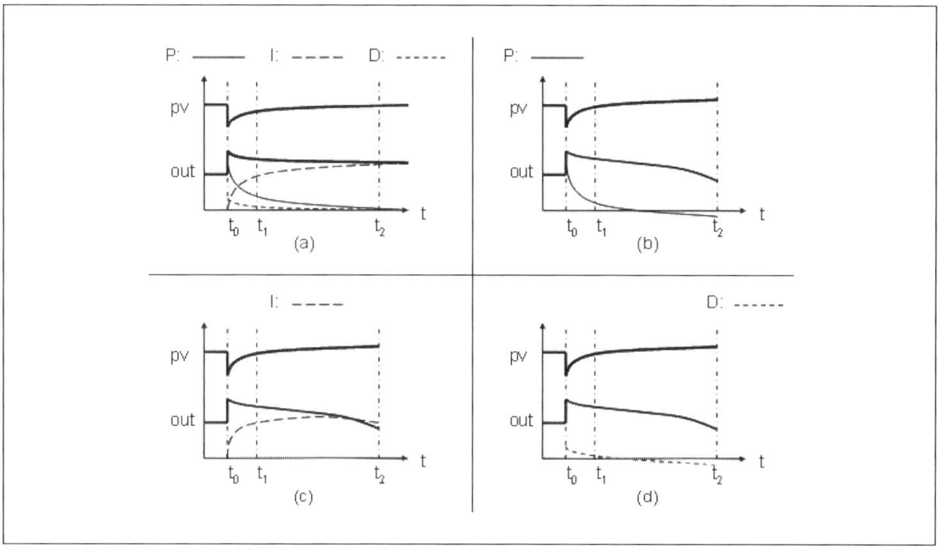

Figure 9-28. Standard PID Block with Output Input

From t_1 the actuator will stay with the out value of t_1 until t_2 when a new out value is received from the PID. This will cause the measured PV to eventually reach SP and then overshoot a little bit, shown in Figure 9-28(b,c,d). The P, I, and D components are all calculated based on the current PV, shown in Figure 9-28 (b,c,d). This is as good as we could expect from PID. The only drawback is that the actuator gets a bump in the out value, from the one calculated at t_1 to the one calculated at t_2.

Both Input and Output Communication Lost

When both input and output communications are lost, the PID behaves the same as when only input communication is lost, as in Figure 9-27. The only difference is that when communication is reestablished at t_2, the actual PV is different. In this case, the PV strays less as the actuator output stays constant. Similarly, the actuator receives a bump in the out value.

The Enhanced PID Algorithm

The underlying assumption in the digital implementation of the PID algorithm above is that the algorithm is executed on a periodic basis. When the input containing the measurement is lost, the calculated reset action may not be appropriate [14]. When later on a new measurement gets through, the calculated derivative action may produce a spike in the output. If a PID block continues to execute using the last process variable, the output will continue to move based on the reset tuning and error between the last measured process variable and the set point. If the control block is only executed when a new measurement is communicated, then this could delay control response to set point changes and feedforward action on measured disturbances. Also, when control is executed, calculating the reset contribution based on the scheduled period of execution or on the time since the last reset contribution may result in changes that increase process variability [14].

In [14], an enhanced PI algorithm was proposed to reduce wireless communications between sensors and controllers without significantly impacting control performance. Based on that work, further improvements to the derivative part of the equations were made and applied [3].

To provide the best control when measurements are not updated on a periodic basis, the PID may be restructured to reflect the reset and derivative contributions for the expected process response since the last measurement update. One means of doing this is illustrated in Figure 9-29.

As shown in Figure 9-29, the reset/rate contributions (integral/derivative parts of the PID) are determined based on the use of a new value flag from the communications stack, the same idea as in [14]. To account for the process response, the filter output is calculated in the following manner when a new measurement is received:

$$F_N = F_{N-1} + (O_{N-1} - F_{N-1}) * \left(1 - e^{\frac{-\Delta T}{T_{Reset}}}\right)$$

where F_N = *New filter output*

F_{N-1} = *Filter output for last execution*

O_{N-1} = *Controller output for last execution*

ΔT = *Elapsed time since a new value was communicated*

Since the last communicated actuator position as reflected in the feedback of actuator position is used in the integral calculation, this automatically compensates for any loss in the output communicated to the downstream element.

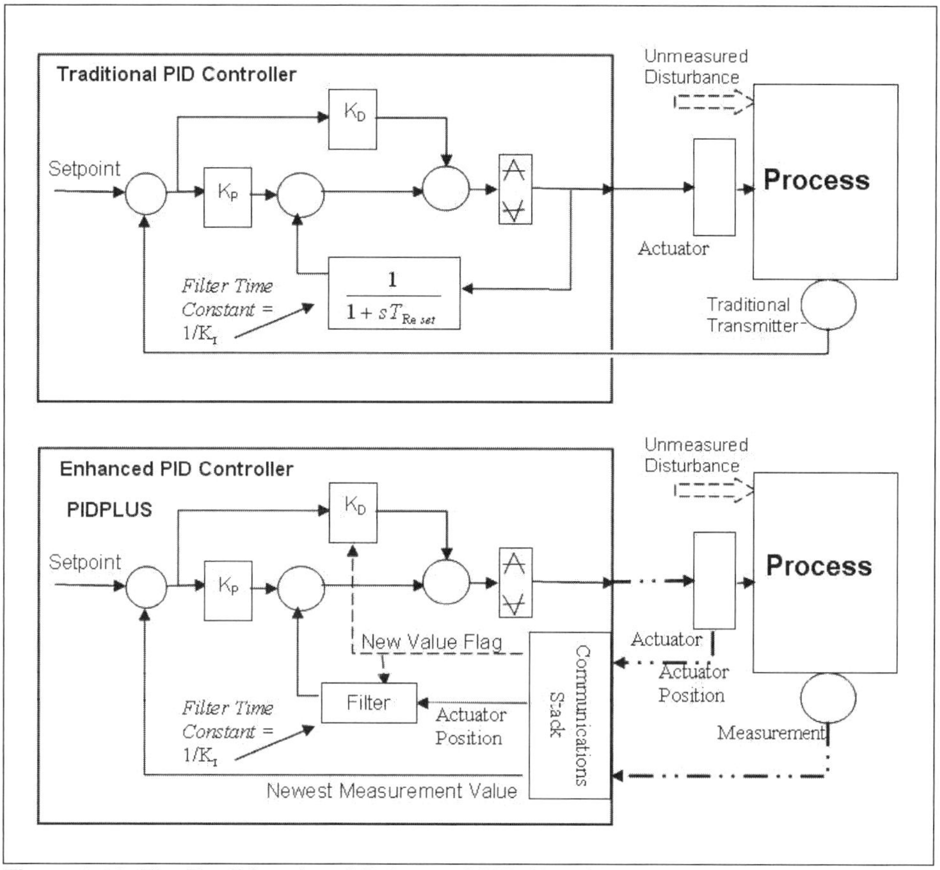

Figure 9-29. The Traditional and Enhanced PID Algorithms

The derivative part for this example (rate limiting not applied) is determined by the following equation:

$$O_D = K_D \bullet \frac{e_N - e_{N-1}}{\Delta T}$$

where e_N = *current error*

e_{N-1} = *last error*

ΔT = *elapsed time since a new value was communicated*

O_D = *controller derivative term*

Consider the contribution of the derivative part when the inputs are lost for several periods. When the communication is reestablished, $e_N - e_{N-1}$ in the equation above would be the same for the original and modified algorithms. However, for the standard PID algorithm, the divisor in the derivative part would be the period, while that in the new algorithm is the elapsed time between two successfully received measurements. It is obvious that the modified algorithm would produce smaller derivative action than the standard PID algorithm.

There are two major problems with the standard PID algorithm when dealing with communication losses: continued execution during communication loss and sudden output change when communication is reestablished. The enhanced PID algorithm solves these problems by only computing the integral and derivative components when communication is established and incorporating actuator feedback into the reset calculation.

To summarize, the enhanced PID algorithm has the following beneficial features:

- PID integral mode is restructured to provide integral action to match the process response in the elapsed time (reset time is set equal to process time constant)
- PID derivative mode is modified to compute a rate of change over the elapsed time from the last new measurement value
- PID reset and rate action are only computed when there is a new value

Experiments and Results

Several experiments were carried out to validate this algorithm. The setup for the test is shown in Figure 9-30. PROC_1 and PROC_2 are two identical processes, each of which consists of a second order process with a delay of one second and time constants of six seconds and three seconds.

The modified PID algorithm is implemented in PIDPLUS, and PID2 is a standard PID block. The parameters for PID2 are determined by testing it with a tuning application, which suggests a gain of 0.85, a reset of 10.71, and a rate of 1.71. Then the tuning parameters of PIDPLUS are set the same as PID2. PIDPLUS is configured to utilize the BKCAL_IN value for the reset components.

The process variable communication is simulated by the COM_IN_1 and COM_IN_2 blocks, which are controlled by COM_STATUS_IN. If COM_STATUS_IN is set to 1, blocks COM_IN_1 and COM_IN_2 relay measurements accurately. Otherwise, the two blocks drop the measurements. The same logic is applied to the output using COM_OUT_1, COM_OUT_2 and COM_STATUS_OUT.

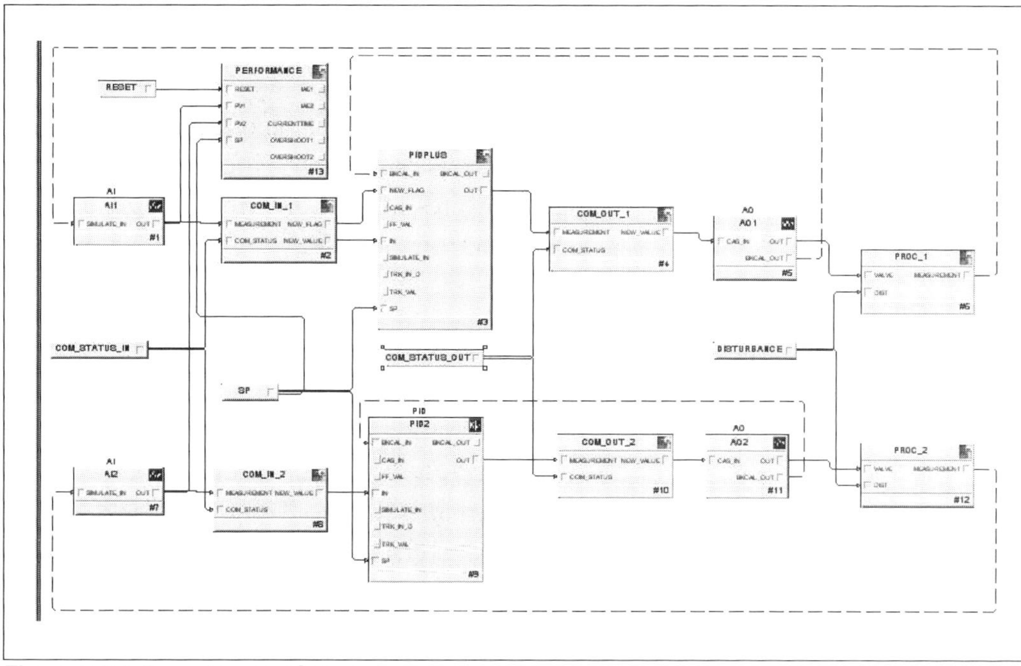

Figure 9-30. Experimental Setup

By changing the external set point and introducing some disturbances that impact each PID and associated process equally, the performance of the two PID blocks can be evaluated. The performance of the two PID blocks is collected in the PERFORMANCE block. The metric used is Integral Absolute Error (IAE). The scan rate for all blocks is set to 0.2 second. Initially, the uncontrolled disturbance (DISTURBANCE) to the processes is 20.

Reliable Communications

To simulate the case of reliable communications, both COM_STATUS_IN and COM_STATUS_OUT were set to 1. SP was changed from 50 to 60. The result is shown in the left part of Figure 9-31 (from time 11:10 to 11:11).

The curve of AI1/OUT.CV matches nicely with that of AI2/OUT.CV, and AO1/SP.CV matches nicely with AO2/SP.CV. Therefore, we conclude that the two PID blocks work in the same way when communication is reliable.

Unreliable Communications

To study the effect of unreliable communications on the two PID blocks, both unreliable input and unreliable output are considered.

Unreliable Input

During the period of lost inputs, the last communicated process variable is maintained and used in the PID blocks. As a first experiment set points were used to introduce changes. The result is shown in the right part of Figure 9-31, where SP was decreased from 60 to 50.

When the input channel is shut down, the error between the set point and process variable fed to PID blocks is a constant. For the standard PID block PID2, the integral part would keep integrating, which explains the linear decreasing of AI2/OUT and AO2/SP. However, as it has a flag for missed communications, PIDPLUS would simply freeze the reset component during loss of communication, which explains the level-off in AO1/SP during lost communications. Since AO1 is given a constant value, the process variable of PROC_1 approaches the set point gradually, as shown by AI1/OUT.

When communications are re-established, the input process variables to PIDPLUS/PID2 reflect the true measurement provided by AI1/AI2, respectively. For PID2, AI2/OUT is very low compared to the set point, which causes a sharp spike in AO2 by the derivative part of PID2 at the moment communications are re-established. For PIDPLUS, AI1/OUT is close to the set point, and that small deviation is further evened out by the divisor used in the derivative part of the new algorithm. Thus both AI1 and AO1 transit to their steady states smoothly.

The different behaviors of the two PID bocks are further proved by the performance data. In a duration of 121 seconds, the IAE for PIDPLUS is 169, while that for PID2 is 372.

Not shown are the effects of using a process disturbance instead of set-point change. The results were very close to what was observed for the set-point change [3].

Unreliable Output

In this case, the behaviors of the two PID blocks are examined with unreliable output communications. To test out the PID enhancements, the set

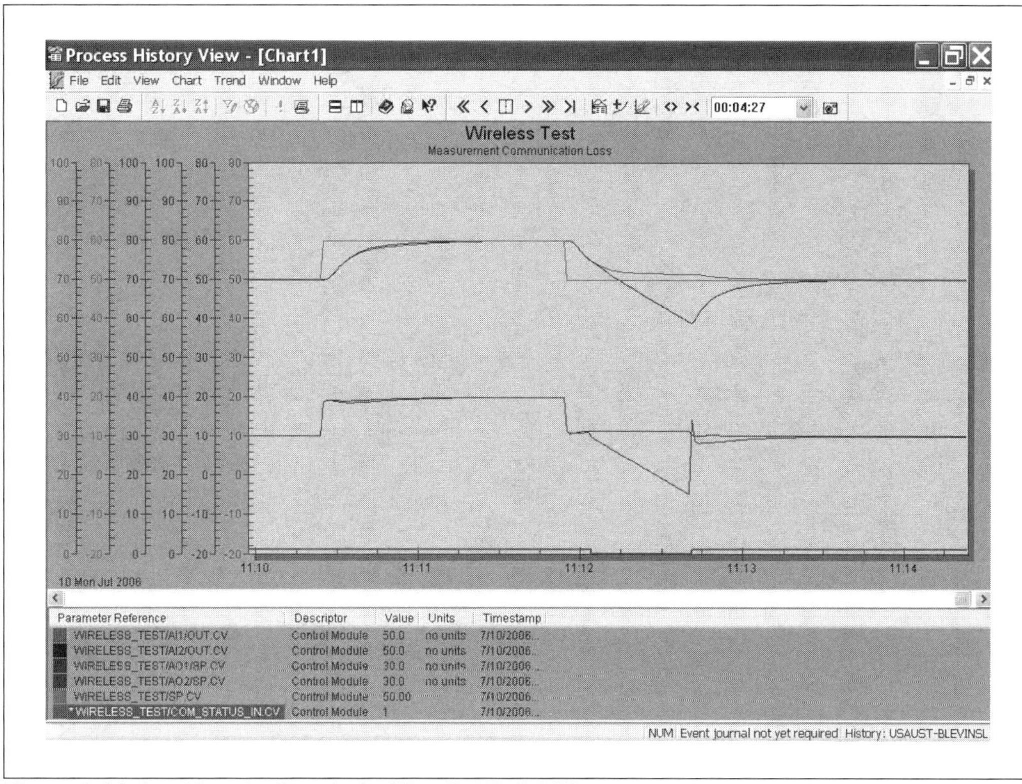

Figure 9-31. Lost Inputs Coupled with a Set Point Change

point is changed from 50 to 60 when the communication is reliable. Then, after the processes settle at the set point, the set point is changed back to 50, and the output channels are cut off. Figure 9-32 shows the resulting curves.

When communication is lost, the outputs of PIDPLUS/PID2 are equal. Since the two controlled processes are the same, the values of AI1.OUT and AI2.OUT follow the same curve. When communication is reestablished, the input errors for PIDPLUS and PID2 are the same. However, the divisor in the derivative part of PIDPLUS is much bigger than that in PID2, which explains the sharper spike in the curve for AO2/OUT. During the transition period, the IAE for PIDPLUS is 190, while that for PID2 is 196.

9-7. Installation of a Wireless Network

Installing a wireless network includes installing wireless devices, a gateway (in some cases both a gateway and one or more access points), and a connection to a host or control system. Once the gateway, devices, and

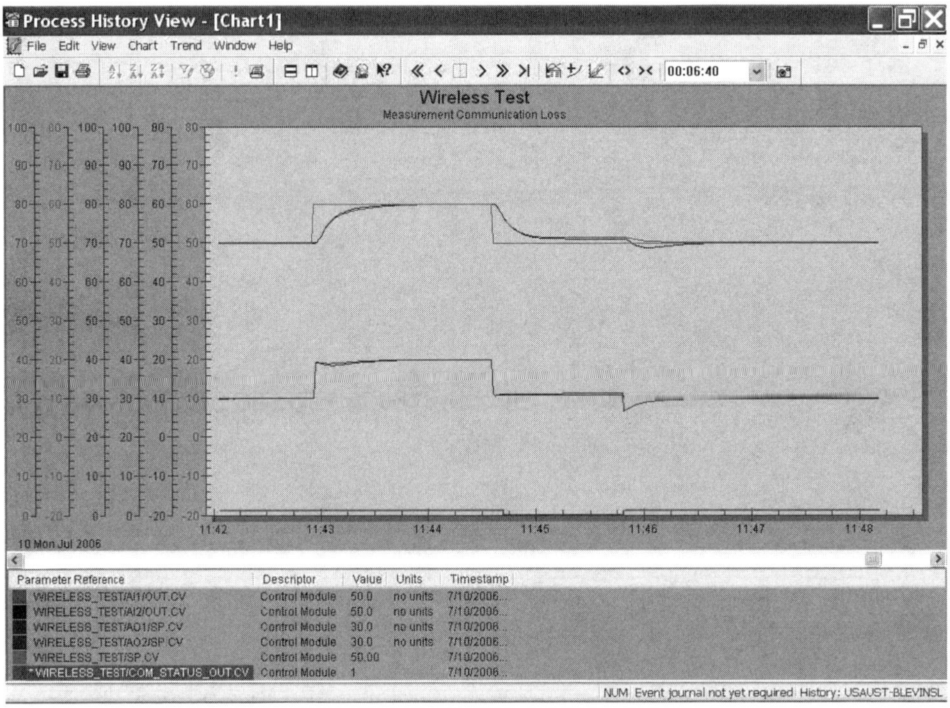

Figure 9-32. Lost Inputs Coupled with a Set-Point Change

control system are configured the wireless mesh forms and communications begin.

Devices

As has been discussed, before devices can join the wireless mesh network they must be given a Network ID and a join key. The device uses the Network ID to determine which advertisement message to join through (there could be more than one wireless network within range). Once the device has heard one or more neighbors, it uses the join key to send a join request to the network manager. The network manager in turn authenticates the device and incorporates the device into the network.

Gateways

The gateway provides the connection between host applications and control systems and the devices that make up the wireless network. The gateway is responsible for collecting and maintaining cached response messages from all devices in the network. These cached response messages originate from burst mode commands, event notifications, and commonly communicated HART commands. These cached response messages are returned as responses to host-based application requests. The gateway has additional built-in functionality to support adapters [5] that allow transparent access to the devices connected to an adapter.

In many cases the gateway connects to the mesh network through multiple access points (Figure 9-22). When multiple access points are in use the network manager will schedule communication traffic through all of them. If one of these access points fails then the network manager will adjust the schedule by spreading traffic across the remaining access points.

An example of a gateway is the Rosemount 1420 Wireless Gateway (Figure 9-33) [21][27].

Integration with Control Systems

Figure 9-33 shows how one or more wireless gateways are connected to the control system.

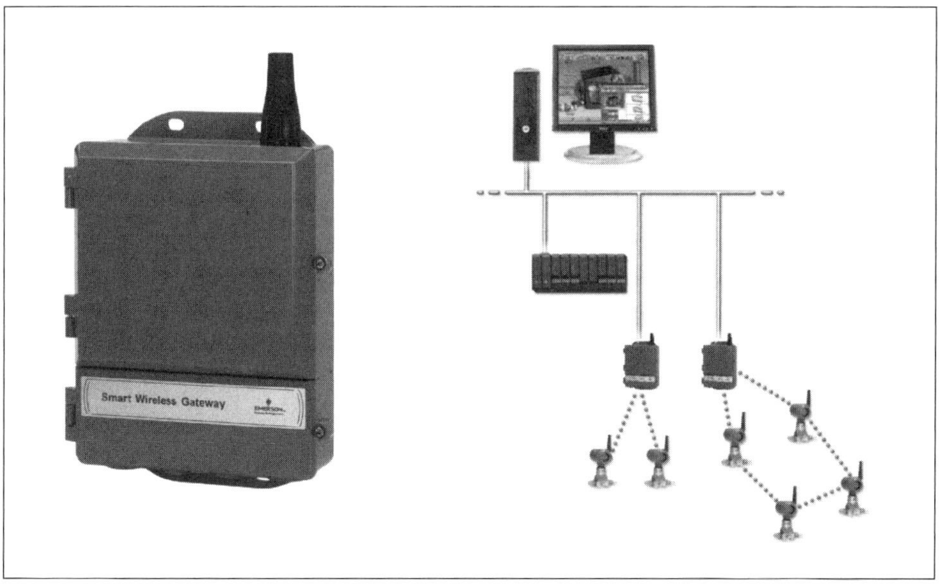

Figure 9-33. Connecting the Gateway to the Control System

In many situations the gateway connects to the control system through native interfaces and shows up as a part of the overall configuration system (Figure 9-34). In other cases, industry standard mechanisms (e.g., Modbus and OPC) can be used.

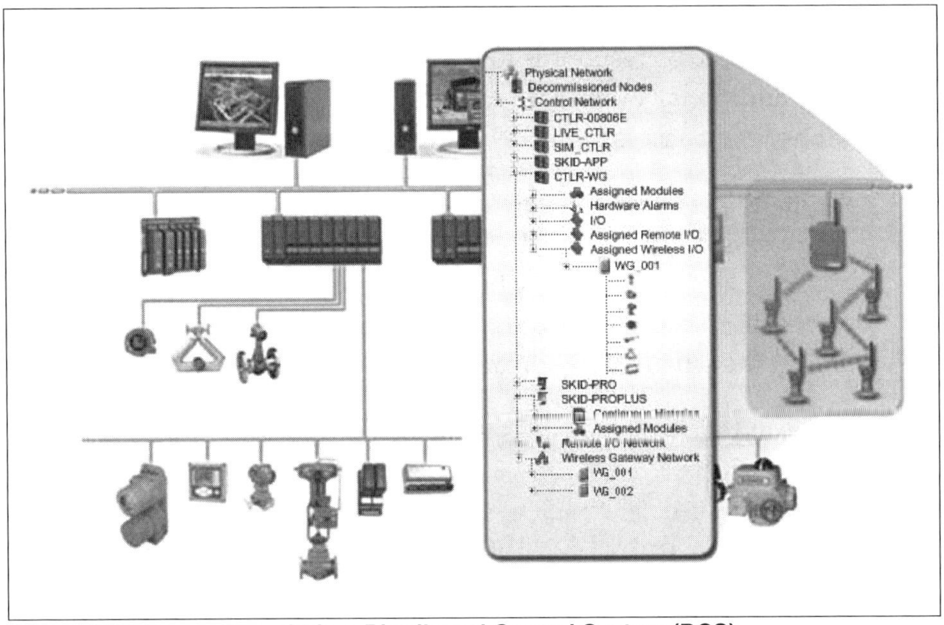

Figure 9-34. Integration within a Distributed Control System (DCS)

9-8. Questions

1. What are key features of smart devices, and how do these features improve operations in the plant?
2. What are key advantages of wireless networks?
3. What is a mesh topology, and why is it used in industrial settings?
4. In wireless device networks, what are several key features required for improved measurement and control? How do these differ from wired networks?
5. What is TDMA, and how are communications scheduled in a WirelessHART network?
6. How is high reliability achieved in a wireless network?
7. How are communications secured in a WirelessHART network?
8. A side effect of equipment trip is a sudden increase in the number of communications. How does a wireless network handle this surge in communications?
9. How can oversampling be reduced in a wireless network?
10. How do wired communication systems address variation in communication times? How do control strategies handle missed communications?

11. How can wireless communication systems address this same variation in communication times?
12. What is involved in installing a wireless network? What must devices be given in order to join a wireless network?

9-9. References

1. Nixon, M., Chen, D., Blevins, T., and Mok, A. K. "Meeting Control Performance over a Wireless Mesh Network." Presented at IEEE CASE Technical Conference, 2008.
2. McMillan, G. K. "Is Wireless Process Control Ready for Prime Time?" *Control*, May 2009.
3. Song, J., Mok, A. K., Chen, D., Nixon, M., Blevins, T., and Wojsznis, W. "Improving PID Control with Unreliable Communications." Presented at ISA EXPO Technical Conference, 2006.
4. Blevins, T., McMillan, G. K., Wojsznis, W., and Brown, M. "Advanced Control Unleashed: Plant Performance Management for Optimum Benefit." Research Triangle Park: ISA, 2002.
5. HART Communication Foundation (HCF) WirelessHART Standard, Sep. 2007.
6. IEEE 802.15.4 Standard, 2006.
7. Song, J., Han, S., Mok, A. K., Chen, D., Lucas, M., and Nixon, M. "A Study of Process Data Transmission Scheduling in Wireless Mesh Networks." ISA EXPO Technical Conference, Oct. 2007.
8. Song, J., Han, S., Mok, A. K., Chen, D., Lucas, M., Nixon, M, and Pratt, W. "WirelessHART: Applying Wireless Technology in Real-Time Industrial Process Control." 14th IEEE Real-Time and Embedded Technology and Applications Symposium, 2008.
9. Dworkin, M. "Recommendation for Block Cipher Modes of Operation: The CCM Mode for Authentication and Confidentiality." *NIST Special Publication 800-38C*, May 2004.
10. *FIPS Publication 197, Advanced Encryption Standard (AES)*. U.S. DoC/NIST, Nov. 26, 2001
11. Diffie, W. and Hellman, M. "Privacy and Authentication: An Introduction to Cryptography." *Proceedings of the IEEE*, Vol. 67 No. 3, Mar. 1979, pp. 397-427.

12. ISO 7498-1 Information Processing Systems – OSI Reference Model – The Basic Model.

13. Halsall, F. *Data Communications, Computer Networks and Open System, Third Edition.* Addison Wesley, 1992.

14. Chen, D., Nixon, M., Aneweer, T., Shepard, R., Blevins, T., McMillan, G. K., and Mok, A. K. "Similarity-based Traffic Reduction to Increase Battery Life in a Wireless Process Control Network." Chicago: ISA EXPO 2005.

15. Foundation Fieldbus standard, http://www.fieldbus.org/.

16. International Society of Automation, http://www.isa.org/.

17. Profibus standard, http://www.profibus.org/.

18. Wireless Industrial Networking Alliance, http://www.wina.org/.

19. ZigBee Alliance, http://www.zigbee.org/.

20. Chipcon Products, http://www.chipcon.com/.

21. DeltaV digital control system, http://www.easydeltav.com/.

22. Caro, D. "Wireless Networks for Industrial Automation." Research Triangle Park: ISA, 2004.

23. IEC 61804-3 standard, Electronic Device Description Language (EDDL).

24. Control Information from John Shaw, http://learncontrol.com/pid.

25. "PID Controller." http://en.wikipedia.org/wiki/PID_controller.

26. http://www.emersonprocess.com/rosemount/document/pds/00813-0100-4420.pdf.

Appendix A – Answers to Chapter Questions

Chapter 1 – Field Measurement Fundamentals

1. The installed accuracy of the 1980s and earlier DP transmitters was about 2% due to non-ideal effects of installation, process, and ambient conditions. A smart DP transmitter with advanced sensor technology maintains its bench-top accuracy of 0.025% when used in the process because of the compensation of non-ideal effects of the application.
2. If thermowells are installed at numerous locations in the column, an integral temperature transmitter and sensor assembly can be moved from thermowell to thermowell and the results historized and analyzed in the DCS. The optimum location is the one that shows the largest change in temperature of about the same magnitude in both directions for a change in reflux or steam flow, whichever is being manipulated for temperature control. This location provides the greatest sensitivity for temperature as an inference of composition and provides a process gain that is about the same for temperatures above and below set point.
3. The lower installation cost and increased mobility of wireless transmitters can result in more process variables and more locations in the process measured.
4. For pharmaceutical processes, the process and automation developed in the laboratory and pilot plant determine what will be used in the industrial plant; therefore, improvements in measurement and control should be demonstrated as early as possible in the research and development phase of the product.
5. An industrial automation system consisting of a DCS with integrated advanced control, data analytics, and simulation capability offer a state-of-the-art implementation of lab exercises. The students learn the

tools and nomenclature that are consistent with industry standards that will help them to be more productive immediately on a job in industry. Professors get to do research and development in an environment well understood by companies supporting the university. The advanced algorithms can also be readily demonstrated in industrial applications.

6. Sensing lines create significantly higher installation and maintenance costs. The lines often need to be insulated and heated to prevent freezing. Also, plugging, polymerization, vaporization, and condensation can introduce errors.
7. The design, installation, and maintenance of wiring are a significant initial and ongoing maintenance cost. Wiring systems pose obstacles to process control improvement by the relocation or addition of measurements. Wiring systems are susceptible from EMI noise.
8. HART and Fieldbus transmitter signals provide additional process variables, alerts, and diagnostic information.
9. Flow is the most frequently measured process variable in the process industry.
10. A control loop ultimately affects the process by manipulating a flow in nearly all applications in the process industry.
11. If the vessel discharge flow is non-zero during normal operation, the process is continuous. In a batch process, the principal discharge flow is normally only opened at the end of the batch cycle. The level in continuous processes are normally constant whereas is batch processes the level is generally increasing when feeds are being added.
12. The temperature and composition open loop response (controller in manual) in continuous operations is normally self-regulating and will reach a steady state after disturbances die out. In batch operations, the open loop response is usually integrating where the response continues to ramp.
13. The total deadtime in a control loop is the summation of all the pure time delays and the effective time delays from time lags smaller than the largest time constant. Digital device scan and execution times and process transportation and sample delays are common sources of pure time delays. Control valve lags, sensor lags, secondary process time constants (e.g., thermal lags), transmitter damping, and signal filters

are common sources of time lags that contribute to the total loop deadtime.

14. Fed-batch processes tend to have more flow control loops than their chemical counterpart. Cell culture bioreactors tend to have more composition measurements than chemical reactors. Batch processes generally have more automated sequencing and product grade management than continuous processes.

15. In pure batch control, variability in the feeds is transferred to process outputs. In fed-batch variability in process outputs is transferred to feeds. In profile control, variability in the batch profile is transferred to process outputs that in turn transfer variability to the feeds.

16. Industrial processes rarely operate at a steady state or at the stream conditions defined on a process flow diagram.

17. Field pressure and temperature switches tend to mechanical in design and provide no indication of the process variable or the switches performance.

18. A discrete 2oo3 voting logic will not vote for a trip if a single path from measurement to voting fails. Analog middle signal selection will not indicate a trip condition on a measurement failure of any type. Middle selection also prevents the disabling of a control loop that can cause an abnormal process condition and true trip condition to develop.

19. The startup of a process is usually the most dangerous and difficult operation due to transients and unbalances. Many plants take a long time to startup and get to the point where product quality acceptable making the loss in production capacity and efficiency an incredible hit on the bottom line.

20. Unmeasured disturbances that are fast and large or at frequency close to the loop frequency are most disruptive.

21. Online analyzers with sensors mounted in the process provide the fastest response. At-line analyzers introduce significant sample transportation and processing time delays. Off-line analyzers create huge variable and often unknown time delays between the time the sample was taken and the lab analysis was entered into the control system.

22. Heat recovery operations and undersized cooling towers, refrigeration units, boilers, and headers are major sources of disturbances to temperature loops.

23. Systematic errors since they are determined by scientific principles can be reduced. For example, signal linearization can compensate for measurement nonlinearities (e.g., pH), set points can be shifted for bias errors, and calibration procedures can reduce manufacturing tolerance errors. Random errors are unpredictable.

24. An error given in percent of reading is smaller than an error in percent of span at any transmitter output less than 100% of scale. Errors in percent of span limit the turndown (rangeability) of measurements. For a differential head flowmeter, the effect of a DP error in percent of span severely limits the rangeability of the measurement.

25. Since a control loop is trying to compensate for changes in the process, resolution and sensitivity errors are the most limiting because the loop does not see the change. Sensitivity and resolution go hand in hand. The resolution must be better than the sensitivity or you won't see the sensitivity. Resolution can also create an apparent separation between the trend of the true and measured process variable as illustrated in Figure 1-23. Repeatability errors are next in terms of concern because the measured change in process varies for a given true change in the process leading to inconsistent loop performance and erroneous corrections to the process. Repeatability errors can be averaged out to some extent by back mixed volumes.

26. EDDL allows the supplier to develop customized user-friendly interfaces to smart instrumentation with interoperability across multiple hosts, devices, and technologies.

27. An asset management system offers better visualization of information by screen graphics and faster navigation by tabs.

28. If the measured change is larger than the true change in the process variable, the slope calibration adjustment needs to be decreased.

29. The principal problem with making online calibration adjustments at process conditions is the availability and accuracy of a measurement of the true value of the process variability. For pH samples, procedures must be enforced to prevent changes in composition and temperature and provide accurate time stamping of the sample.

30. The direct mounting of differential pressure transmitters for level measurement eliminates the need for sensing lines, equalization lines, and capillary systems.

Chapter 2 – Temperature Measurement

1. Non-graybodies and unequal changes in radiation emitted at different wavelengths for changes in process and target conditions introduce considerable errors into even ratio pyrometers.
2. A Platinum RTD with a thermowell-mounted transmitter with "Transmitter-Sensor Matching" is needed to provide the resolution, sensitivity, repeatability, and nonlinearity for tight temperature control at the right set point.
3. Commercial software is available that take into account the process characteristics of the gas or fluid being measured to compute the wake and natural frequencies. The thermowell supplier should be able to make this calculation.
4. Type K (chromel-alumel) thermocouples are commonly used because they are relatively inexpensive and offer a wide temperature range.
5. The faster response of the thermocouple element is of no practical importance in nearly all industrial processes because the variation in secondary process lags and the thermowell or protection tube lags are larger than the element lag.
6. Careful selection of the sensor and thermowell to the application requirements is the first step in making a reliable measurement. Dual-element sensors coupled with multi-input transmitters can provide automatic backup in the case one sensor should fail. These multi-input transmitters can also be attached to sensors of different types enabling a thermocouple to provide backup to an RTD.

Chapter 3 – Pressure Measurement

1. Pressure measurements are important for reducing emissions, preventing contamination, and optimizing unit operations.
2. Gas phase reactions, absorption, and distillation require tight control for best performance (e.g., 0.1% control band).
3. DP transmitters are also used for level and flow measurements.

4. Wireless direct-mounted pressure transmitters can be moved between spare pressure connections to measure pressure profiles for monitoring fouling, frosting, plugging, and tracking down disturbances.

5. The direct mounting of the pressure or DP transmitter on the process connection eliminates the plugging and leakage problems and response lag time of impulse lines (sensing lines).

6. A Coplanar™ transmitter reduces the number of leak points and improves the accuracy of the measurement.

7. A smart DP can detect plugged impulse lines, bubbles, and solids.

Chapter 4 – Flow Measurement

1. Flow measurements are important for closing the material balance, computing process metrics, cascade control, feedforward (ratio) control, data analytics, and process modeling.

2. The heat capacity, viscosity, and thermal conductivity of the process fluid affect the correction factor for the heat input in Equation 4-2c.

3. Differential head orifice flowmeters require the longest straight runs upstream (e.g., 44 to 145 pipe diameters).

4. Multi-point thermal mass flowmeters can measure gas flows in large ducts when accuracy is not critical and changes in the gas composition have a negligible effect on thermal conductivity and heat capacity of the stream.

5. Pitot tubes can be used in large pipes even though the composition is changing.

6. Vortex meters can be used. A pure monomer probably has too low of an electrical conductivity for a magnetic flowmeter.

7. A venturi tube can be used with a direct-mounted DP or with diaphragm seals to eliminate the plugging of impulse lines. A magnetic flowmeter is also possible but in this case the extra rangeability and accuracy is not needed.

8. Magnetic flowmeters may be the best choice because of the rangeability requirement. Since the stream has process water, the conductivity should be significantly above the low limit. The line size is too big for Coriolis meters.

9. Only a Coriolis flowmeter can meet these accuracy, rangeability and drift requirements.

Chapter 5 – Level Measurement

1. Level measurements are important for ensuring vessels have a level that meets operation requirements. Since level has an integrating response as described in Chapter 1, there is no steady state. Level will ramp off to a high or low physical limit unless level is measured and controlled. Some of the possible adverse consequences of hitting a high limit are vessel overpressure, environmental releases, and contamination of an overhead system. The undesirable of approaching the low limit is insufficient liquid head and starvation of downstream users of the material in the vessel. Precise level measurement is particularly important for inventory, material balance, and residence time control in certain types of vessels (e.g., storage tanks, column receivers, crystallizers, and reactors).
2. Capillary systems or direct mounting of the DP on the nozzle can be used to eliminate sensing line (impulse and equalization line) problems. However, the use of a capillary system slows down the response and adds a liquid head so asymmetrical systems with a direct-mounted DP on the bottom of the vessel and the use of a capillary system to replace the equalization line are used. Bubblers also eliminate sensing line problems but introduce other problems such as measurement noise due to purge gas pressure and flow fluctuations. Dual direct-mounted DP transmitters can be used for both the total and vessel pressure measurement if the liquid head is a large enough portion of the total pressure measurement at the bottom (see Chapter 1 DP level installation examples for a discussion bubbler and dual DP level measurements).
3. Electromagnetic energy is emitted from all radar devices. When the emitted signal reaches a point where there is a change in DC, usually the media surface, some of the signal is reflected back to the transmitter. The amount of energy that is reflected back to the transmitter is proportional to the DC of the media. A rule-of-thumb is that the value of the dielectric constant represents the percentage of energy that is reflected. Thus a DC of eight means that eight percent of the emitted energy is reflected back to the transmitter. Fundamentally, media with a higher DC provide stronger return signals and are therefore easier to measure.

4. Yes. The emitted signal is less than three percent of maximum leakage allowed from a microwave oven. Radar waves are just as safe as the constant radio, cellular, and other communication waves that surround us every day. Furthermore, the transmitter is normally placed in a metallic tank that acts as a Faraday's cage and therefore the radar waves are isolated within the tank.

5. The effects of foam on a radar measurement can be difficult to predict. In some applications, the foam may dampen out the signal completely while other types of foam may be transparent to the transmitter. The thickness, density, and the dielectric constant are factors that need to be considered when evaluating an application with foam. On dry foam the microwaves typically pass through and detect the liquid surface below. On medium type foam, the signal can be absorbed or scattered and the results are therefore hard to predict. If the foam is wet, the microwaves are often reflected from the foam surface and thereby the foam surface level is measured. The frequency at which the radar operates also affects how foam is measured. Low frequency radar in general penetrates foam to a larger extent than high frequency radar. Guided Wave Radar is in general better suited to measure on applications where foam is present since the radar uses a lower frequency pulse.

6. The signals from two or more transmitters in one tank will not blend and therefore the radar units will not conflict with each other. If it concerns two or more guided wave radar units, the rules for nearby objects would apply to the probes as they do to other metallic objects nearby. For that reason, the probes need to be installed a certain distance away from each other.

7. Ultrasonic transmitters can be used up to its stated range limits, typically 25 to 40 feet and thus should be chosen based on the tank height and installation mounting requirements. The only maintenance needed is to occasionally check the sensor surface for cleanliness. Most ultrasonic transmitters are non-repairable; thus, if failure occurs, they must be replaced.

Chapter 6 – pH Measurement

1. pH measurements are important for protecting the environment and meeting water and product quality specifications. The pH measurement has extraordinary sensitivity and rangeability that goes way

beyond the capability of any other common industrial measurement. The degree of composition control made possible by pH measurements is incredible.

2. Process temperature and ionic strength can change the indicated pH despite a constant acid and base concentration by changes in the dissociation constants and activity coefficients, respectively. Non-aqueous solvents (e.g., alcohols) change the equivalence point and the activity coefficients.

3. The combination non-flowing junction electrode eliminates the need to install, troubleshoot, and replace separate measurement and reference electrodes. The non-flowing junction eliminates the need to pressurize and refill the reference electrolyte to ensure a positive electrolyte flow through the reference junction into the process. This type of electrode promotes the throwaway concept. If there is a problem, the whole sensor is pitched rather than selectively diagnosing, servicing, and replacing the measurement and reference electrodes. A flowing junction reference electrode would provide a more consistent reference potential and be less susceptible to plugging and poisoning, but these advantages are not well recognized as being important enough except in some high accuracy applications.

4. Electrical continuity is achieved by ensuring that the reference electrolyte is in contact with the process fluid by the use of a porous junction or small holes.

5. Plugging is the trapping of process material or silver precipitate from the electrolyte in the reference junction that causes a restriction to electrolyte movement and contact with the process which increases the electrical resistance of the junction. Poisoning occurs when the process material gets inside the reference's innermost chamber and changes the composition of the electrolyte or the internal electrode. Reference electrodes have tortuous paths or multiple junctions to prolong the time it takes for the contamination to reach the innermost section of the reference where its internal electrode resides.

6. The Nernst equation seen in the literature for pH assumes the activity of the inner and outer gel surfaces are identical. The exposure of the outer surface to harsh operating conditions or low water concentrations can change the hydration and composition of the glass in contact with the process.

7. A decrease in the resistance (R_{11}) due to a non-ideal electrode structure creates a short (low resistance) path in parallel (shunt) with the high resistance of the glass that will decrease the change in milliVolts with pH measured by the transmitter.

8. Wireless transmitters are particularly advantageous for pH measurements, because pH signals are particularly susceptible to wiring problems, ground loops, and electromagnetic interference. Also, transmitter mobility offers the opportunity for developing applications in the lab and moving them to the optimum location in the field. Even if the lab route is bypassed and pH measurement needs to be tried out in the field to see if the electrode will work in a tough application, a wireless transmitter eliminates the need to install wiring and have an intelligent input card for the trial installation. Wireless transmitters also offer the latest improvements in diagnostics and unsolicited alerts.

9. The performance and life of electrodes subjected to high process temperatures and steam sterilization can be improved by the use of new glass and electrode structure designs to maintain the integrity of the glass surface and reference junction and maximize the shunt resistance (R_{11}).

10. Poisoning of the reference electrode can be mitigated in increasing order of capability by the use of triple junction, solid, and flowing junction reference electrodes. For gel and liquid reference electrodes, it may be necessary to change the type of electrolyte and internal electrode. This is particularly important for waste streams with cyanides, bromides, iodides, sulfides, or nitrates that can cause precipitation of the silver.

11. Higher velocities (e.g., 5 to 7 fps) reduce glass measurement electrode coating and response time, but increase abrasion. The optimum velocity depends on whether abrasion or coating is decreasing electrode life. The orientation and design of the protective shroud should prevent impingement of the process stream directly onto the bulb if abrasion is a major concern. To minimize coating, it is desirable to have the process stream sweep across the bulb. Most of the benefit of flat glass electrodes in reducing coating was due to stream velocity and direction.

12. For pure water (low conductivity) applications to insure a stable reference junction and minimize noise, special assemblies should be used to keep sample flow rate low and constant.

Chapter 7 – Final Element Fundamentals

1. Most control loops end up manipulating a flow in the process by means of a final element to deal with disturbances or reach a new set point.

2. Temperature and pressure regulators do not provide the indication and historization of the set point, process variable, or controller output like there is for a DCS loop. Consequently, performance problems or impending failures largely go undetected. The offset from proportional-only control is largely unknown and set by regulator construction and operating conditions. There are no options for improving the controller tuning or structure like there is in a PID block in a DCS. Finally, a control valve in a loop has many more options for materials of construction than regulators.

3. Piping valves were originally designed for on-off service and are generally not suitable for responding to small changes in desired position normally demanded of throttling valves. Piping valves tend to have too much stiction (stick-slip) and backlash (deadband).

4. The sliding stem valve (globe valve) with a diaphragm actuator, equal percentage trim, and a digital positioner or the modulating solenoid valve generally offers the best sensitivity. The size, materials of construction, and application experience in the process industry is much greater for the sliding stem valve than the modulating solenoid valve.

5. For small flows, the roller diaphragm valve offers an extraordinary rangeability. There are no pockets or obstructions in the flow path so plugging at low flows is less of an issue. The flow is forced to be laminar so there is no erratic response from operation in the transition region from turbulent to laminar at low flows.

6. For moderate flows, the sliding stem valve (globe valve) with a diaphragm actuator, equal percentage trim, and a digital positioner or the modulating solenoid valve generally offers the best rangeability. The size, materials of construction, and application experience in the process industry is much greater for the sliding stem valve than for the modulating solenoid valve.

7. For large flows, the contoured control disk type of butterfly valve with a diaphragm actuator and digital positioner offers the best rangeability.

8. For slurries, the segmented V-notch ball valve provides a straight obstruction free path through the body of the valve. The V-notch helps to shred and prevent pulp and bark from getting hung up in the ball face.

9. Rotary valves and dampers tend to have the largest backlash because of the transmission from linear to rotary motion and the play in connections and linkages. High seal friction and long closure component shafts cause shaft windup that causes both deadband and stick-slip. Piping valves tend to be the worse because they were originally designed for on-off instead of throttling service.

10. The use of integral action in two or more controllers will cause a limit cycle. For a single loop in the DCS, the use of integral action in the process loop and the digital positioner will cause a limit cycle from deadband. For a cascade control loop in the DCS, the use of integral action in the primary and secondary process loops will cause a limit cycle for deadband even if the positioner is not using integral action.

11. The use of integral action in just one controller or in the positioner will cause a limit cycle from a resolution limit.

12. A three-way solenoid valve driven by a discrete signal is installed on each output of the positioner to force the valve to a fixed position. The controller output that provides the signal to the positioner should set to match the fixed position for coordination of the loop and positioner.

13. A volume booster with a bypass is installed on each output of the positioner to make a valve with a large pneumatic actuator respond faster. It is critical that the bypass be opened enough to prevent a limit cycle.

14. The principal limit to the turndown of a variable speed drive is static head. With modern variable speed drives, the turndown is limited to where pump head approaches the static head, which causes an erratic flow, an excessive process gain, and a possible reversal of flow.

15. The actual energy savings from a variable speed drive is often overestimated by not adequately evaluating static head and loss in efficiency at low flow and the amount time spent at low flow (duty cycle).

16. There are usually no decision points in a project for the use of a variable speed drive (VSD) instead of a control valve, users don't like the complexity of a VSD, someone needs to do the right calculations that includes duty cycle, and the potential improvements in loop performance speed from the elimination of valve stick-slip, backlash, and

shaft windup is not understood or undermined by a poor resolution input card, the introduction of a signal deadband, or poor drive setup.

17. For a well-designed and properly setup VSD and control valve, the VSD and valve will generally respond faster to small and large changes, respectively.

18. The equal percentage inherent flow characteristic is the most forgiving for errors in sizing, has the most linear installed flow characteristic for a piping system with significant friction losses, and offers the best rangeability because of the decrease in flow coefficient as the valve approaches the closed position where stiction is greatest.

19. For a large step (e.g., > 10%), the response has a short delay due to a pre-stroke deadtime followed by a ramp from the stroking time of the valve. The delay and ramp rate (slewing rate) are set by the actuator volume and positioner or booster flow capacity. For a small step (e.g., < 1%), the response has a delay due pres-stroke deadtime, secondary time constant, and the sensitivity of the positioner followed an exponential response due to a primary time constant. The actuator and positioner type and dynamics largely determine this response.

20. For a ramp in the valve signal, there is a large time delay in the deadband or resolution limit divided by the ramp rate and a lag due to the primary and secondary time constants affected by actuator and positioner type and dynamics.

Chapter 8 – Control Valves

1. Two methods of converting linear actuator output to rotary input for a rotary valve is a lever or <u>rack-and-pinion.</u>

2. Two common sealing methods for valve stems and valve shafts against process fluid are <u>packing</u> and <u>O-rings.</u>

3. For flow-splitting applications, which valve is an obvious choice?

4. Rotary valve

5. Globe valve

6. Three-way valve

7. Plug valve

8. The globe valve exhibits a desired gain over a narrow range of travel. <u>False</u>

9. Typical pneumatic signal ranges are 3 to 15 psig (0.2 to 1.0 bar) and 6 to 30 psig (0.4 to 2.0 bar).

10. Advantages of a digital valve controller include:

    a. Communication with the final control element
    b. Auto calibration
    c. Diagnostic capabilities
    d. All of the above

11. Common valve flow characteristics are quick opening, linear, and equal percentage.

12. Setting the spring compression in a spring-and-diaphragm actuator is called "initial wind in" or "bench set."

13. Installing a control valve assembly often includes the step of setting the zero and span before putting the unit in operation.

14. Monitoring equipment and process operation in order to anticipate and prevent equipment failures is called predictive maintenance.

15. A study found that as many as 80% of control loops did not do an adequate job of reducing process variability.

16. Two of the most common causes of deadband in process control equipment are friction and backlash.

17. The recommended range of gain in a control valve is:

    a. 30% to 60%
    b. 1 to 4
    c. The higher the better
    d. 0.5 to 2.0

18. The angle-pattern valve is a variant of the globe valve.

Chapter 9 – Wireless

1. Smart devices incorporate features such as status on measured values, time stamps, event latching and confirmation, and block data transfer. The availability of additional variables, such as density and temperature for Coriolis mass flowmeters, provides additional information useful for understanding changes in the process. Using these smart devices results in fewer trips into the plant, a much better controlled process, and significantly less downtime.

2. Wireless networks can result in reduced installation and commissioning costs, elimination of wiring noise and connection problems, and a reduction in the overall life cycle.

3. In a mesh topology, each device is connected to several other devices and all devices are capable of routing communications traffic. Mesh networks are self-healing: the network can still operate even when a device breaks down or a connection goes bad. As a result a mesh topology is very reliable.

4. Several key features for improved measurement and control with wireless networks are values reported with status, exception reporting, and time stamps. Status is important to the control strategy, exception reporting significantly reduces the number of communications, and time stamps can be used to measure latency and jitter. Although all of these same features are also important for wired communications, in wireless communications exception reporting improves battery life and time stamps can be used to adjust control calculations for variations.

5. TDMA stands for Time Division Multiple Access. TDMA communications are scheduled by a centralized network manager. The network manager is responsible for the creation of this schedule and the associated connections. It is also responsible for the distribution of this schedule to the individual network devices.

6. The wireless network achieves very high reliability through the use of several mechanisms: multiple paths to devices, multiple RF channels, and multiple communication tries. If improved reliability is required, more paths can be inserted by the network manager. As additional devices are added to the mesh (network) the strength of the overall mesh gets better and better. Additional devices improve path diversity. Additional access points, and devices in general, increase throughput, reduce latency, and can be used to route around areas of potential interference.

7. Communications are secured by ensuring that only devices approved to join the network join, that communications are not tampered with, and that bad guys cannot spoof the network through the use of replay attacks (a replay attack occurs when the same packets are resent on the same network). Communications are made reliable by retrying on different channels, retrying at different points in time, and communicating on different paths.

8. When there is contention for network resources, wireless networks use message priority to determine which packets are sent first. Message priority helps mitigate situations where devices need to perform flow control. Priorities are set so that network management commands are the highest priority (this allows the network manager to temporarily increase communication resources, re-route, etc.), followed by real-time published messages, followed by alarms and alerts. In the case of alarms and alerts, since they are time-stamped at the device, no critical information is lost if an alarm or alert is delayed a few time slots.

9. With wireless systems, because of the inherent time-synchronized nature of the mesh network, it is possible to reduce or eliminate this oversampling. Sampling can be scheduled to occur just before communications are scheduled.

10. Current control strategies address variation in communication by oversampling the I/O system. In standard process control systems, a missed I/O communication is considered an error. Usually, a control loop is configured with a maximum number of lost I/O communications, after which the loop declares failure and the values are set according to the fail-safe configuration.

11. Wireless communication systems can address variation in communications by modifying the calculation of the integral and derivative components of PID algorithms. The modification detects missed communications and compensates for the variation in communication times.

12. Installing a wireless network includes installing wireless devices, a gateway and one or more access points, and a connection to a host or control system. Once the gateway, devices, and control system are configured, the wireless mesh forms and communications begin. Before devices can join the wireless mesh network, they must be given a Network ID and a join key. The device uses the Network ID to determine which advertisement message to join through.

Appendix B – Top Ten Things to Consider Before Selecting Your VFD

1. **What will be the torque demands of the loads or processes in your planned system? Will any of the loads be hard to start?** Keep in mind, VFDs have limited over-current capacity. Hard-to-start loads may require an over-sized unit to cover higher current demands.

2. **How many motors will the drive have to control? If more than one, will they start up sequentially or simultaneously?** You should calculate the total peak currents of all motor loads under the worst operating conditions your planned system will see. Size your VFD according to this maximum current requirement.

3. **Will your applications require a quick start or an emergency stop of the load?** If so, high currents will be demanded of the VFD. Over-sizing of the drive may be necessary.

4. **Is motor overheating a potential concern for any of your planned VFD applications?** It may be, for reduced speed, constant torque applications.

5. **What range of motor sizes will your process or processes require the VFD to handle?** Keep in mind that smaller motors are not as efficient as larger ones, so improvements due to the VFD will likely be readily apparent. However, since large motors consume much more power, even small increases in efficiency can, over the life of the motor, lead to appreciable savings.

6. **Will the VFD system be operating in an environment containing volatiles, airborne particulate matter, high ambient temperatures?** In the case of volatiles, be sure all materials in use are resistant to chemical attack and are properly grounded; for an atmosphere full of particulates, ensure proper sealing; for high temps, allow for cooling.

7. **Do you need equipment or drive protection features that will ensure a continuity in your processing?** Be aware that you can have a drive that trips instantly in an over-current condition, or one that maintains constant motor torque and reduces motor speed to maintain current required.

8. **Do you want or need a lot of diagnostic capability in your unit?** How critical is it to get a handle on load downtimes and be able to do a detailed fault analysis?

9. **Are power factor correction capacitors present on the motor loads your VFD system will have to handle?** When these are switched, they typically generate power disturbances, and VFDs can be negatively impacted. Isolation transformers or line reactors may be necessary.

10. **Is it possible the power source for your planned system will occasionally be switched while the VFD is operating?** This might happen, for example, when loads are switched to standby generators during a power outage. Some drives can handle a brief power outage, others can't.

(Courtesy of Belden Inc., "Building a Reliable VFD System," VFDWP, 2009)

Appendix C – Effect of Sample Time on Process Control

There is considerable confusion as to when sample times affect the ability of a control system to compensate for unmeasured disturbances. The purpose of this appendix is to provide concepts and examples to sort out fact from fiction and offer some guidance.

Setting appropriate sample times and minimizing the effect of sample times on process control is increasingly important for the following reasons:

1. Since we live in a digital world, sampled data is the norm. Just from the volume of applications, the opportunity for setting and dealing with sample times is large.
2. There are no clear guidelines for various types of process control applications.
3. In some applications, conventional sample times can cause severe safety and performance issues.
4. In most cases, the tuning of the controller dictates that sample times could be significantly slower. If distributed control system (DCS) module execution times and wireless communication time intervals could be increased, controller loading is reduced and wireless battery life is prolonged, respectively.
5. If we want more at-line analyzers to provide measurements of stream compositions that tell us what is really going on in the process and offer the opportunity for a more advanced level of control, we need to understand and address sample processing and analyzer cycle times.
6. If we want to move to more wireless measurement that give us the flexibility and intelligence for process control improvement, we need to understand and address wireless communication intervals.

This appendix uses the term "sample time" as the time between updates in sampled data from digital measurements and controllers and from analyzers in the broadest sense. The following discussion should be useful for determining whether DCS scan or module execution times, wireless communication time intervals, model predictive control execution time, and at-line analyzer cycle time will affect control system performance.

Table C-1 provides a guideline for sample times that will generally have negligible effect on control loop performance. The **practical** sample time is for loops with conservative (moderately slow) tuning. The **ultimate** sample time is for loops with aggressive (fast) tuning. Note that the sample times are for commercial production size equipment. The sample times needed for pilot plants and research facilities are considerably faster.

Table C-1 suggests **practical** and **ultimate** sample times for different types of processes with typical dynamics in loops with excellent final elements, clean sensors, and minimal transmitter damping settings. The **ultimate** limit is set to be less than $1/10^{th}$ of the total of the smallest deadtime plus the smallest time constant, with some consideration as to maximum practical controller gains to reduce valve cycling and noise amplification. For any loop with a control valve, the minimum loop deadtime is about 1 second for an unmeasured disturbance so the **ultimate** limit on sample time is about 0.1 second. The **practical** limit reflects current tuning practices. For integrating processes, the process time constant shown is the inverse of the integrating process gain (denoted by single exclamation point). The double exclamation point denotes a runaway (positive feedback) process time constant. Consultants say it is impossible to generalize, but some guidance is helpful to the user with the realization there are always exceptions and the actual process dynamic and tuning should be identified by automated online tuners and adaptive controllers. I didn't consider **ultimate** sample times slower than 60 sec. Note that slower sample times will affect the deadtime identified.

The loop should be closely monitored and tested for any increase in sample time.

Table C-1. Rough Guide of DCS Sample and Wireless Communication Times for Industrial Process Control Loops

Type of Process Loop	Process Deadtime	Process Time Constant	Practical Sample Time	Ultimate Sample Time
Liquid Flow	0.05 - 0.5 sec	0.5 - 5 sec	2 sec	0.1 sec
Gas Flow	0.1 - 0.5 sec	1 - 10 sec	1 sec	0.1 sec
Liquid Pressure*	0.05 - 0.5 sec	0.2 - 1 sec	0.1 sec	0.02 sec
Column Pressure!	1 - 10 sec	10 - 100 sec	10 sec	2 sec
Furnace Pressure*	0.1 - 0.5 sec	0.2 - 20 sec	0.1 sec	0.02 sec
Vessel Pressure!	0.2 - 1 sec	10 - 100 sec	10 sec	1 sec
Surge Control	0.05 - 0.5 sec	0.2 - 10 sec	0.1 sec	0.02 sec
Liquid Level!	0.05 - 0.5 sec	10 - 100000 min	300 sec	60 sec
Exchanger Temperature	0.2 - 2 min	0.5 - 5 min	10 sec	2 sec
Batch Temperature!	1 - 10 min	5 - 100000 min	150 sec	30 sec
Runaway Temperature!	0.5 - 5 min	1 - 100 min	10 sec	5 sec
Column Temperature	2 - 100 min	10 - 1000 min	300 sec	60 sec
Furnace Temperature	0.2 - 2 min	0.5 - 5 min	10 sec	2 sec
Vessel Temperature	1 - 10 min	5 - 50 min	150 sec	30 sec
Column Composition	1 - 50 min	10 - 1000 min	300 sec	60 sec
Furnace Oxygen	0.2 - 1 min	0.2 - 1 min	10 sec	2 sec
Vessel Composition	0.5 - 5 min	5 - 50 min	150 sec	30 sec
Inline (Static Mixer) pH	2 - 10 sec	2 - 10 sec	2 sec	0.5 sec
Vessel pH	0.5 - 5 min	1 - 50 min	30 sec	5 sec

Table Notes

* - denotes loop uses a variable speed drive with a negligible deadtime, deadband, and resolution limit as the final element. If a control valve or damper is used for these loops, you can multiply the sample times for asterisked items by a factor of 5.

! - denotes an integrating response whose integrating process gain is the inverse of the process time constant shown.

!! - denotes a runaway response that can accelerate and reach a point of no return.

For surge control, it assumed that a volume booster has been added to each of the positioner outputs to reduce the pre-stroke dead time to less than 0.2 seconds. A valve with excessive stiction and backlash will add significant deadtime to the response to unmeasured disturbances that deteriorates the ultimate limit to possible performance.

For inline (static mixer) pH control, the largest time constant comes from the sensor lag or the process variable filter time with a nominal value of 5 seconds.

For the vessel pH control, it is assumed the mixing time is less than 30 sec and the reagent delivery time delay is negligible by injection of the reagent into a recirculation line just before it enters the vessel. The lower value for the time constant is for a set point on a steep titration curve that causes the pH to move much faster than for a linear response. The response can look like a runaway as the pH accelerates through the neutral region.

For level control set point changes, the deadtime observed is usually about 10 times larger than the actual process deadtime due to level measurement sensitivity limits and noise. For unmeasured disturbances, the deadtime observed is often about 20 times larger than the actual process deadtime because of the amount of time it takes the controller output to work through the resolution limit and deadband of the control valve.

Theoretical Basis and Background

The performance of a control loop depends on the tuning. Specifically, the peak and integrated errors are inversely proportional to the controller gain. The peak error is not affected much by the integral time setting. However, the integrated error is proportional to the integral time. Thus, a loop with good dynamics can be made to perform as poorly as a process with bad dynamics by sluggish tuning. The effect of slow sample times is hidden by large integral times or small controller gains. Thus, it is critical for any comparison, that tuning criteria be specified. In fact, there is an implied deadtime as a result of the tuning of the loop [1]. The tuning of the controller puts a **practical** limit on how fast the sample time must be for the effect to be negligible.

If a controller is tuned for maximum performance, the peak error is proportional to the loop deadtime to process time constant ratio. The integrated error is proportional to the deadtime squared. These statements are strictly true only when the process time constant is large compared to the loop deadtime. The loop deadtime is the sum of final element deadtime (e.g., valve pre-stroke time delay, deadband, and stiction), process deadtime (e.g., mixing, thermal, and transportation), automation deadtime (e.g., sensor lag, transmitter damping, and sample times), and small process time constants. All of the time constants smaller than the largest time constant become effectively deadtime in the first order plus dead time approximation used in industry. Process and automation system dynamics places an **ultimate** limit on loop performance. There is a corresponding **ultimate** limit on the sample time.

The effect of sample times can be accessed in terms of **practical** and **ultimate** limits on performance. Critical loops where peak errors can cause destruction or environmental releases, such as compressor surge control, furnace pressure control, exothermic reactor temperature control, and RCRA pH control, the tuning is necessarily aggressive. As a result the **practical** limit is much closer to the **ultimate** limit. For a discussion of cases where exceptionally fast sample times are needed, see reference [2].

For many digital devices the update is available near the beginning of the sample time (latency is negligible), which means the average deadtime from the sample time is about half the sample period. For at-line analyzers (field analyzers with automated sample systems), the result is not available until the end of the sample processing and analyzer cycle time, which translates to an average effective deadtime that is about 1.5 times the time interval between updates in the analyzer output signal.

The detrimental effect of sample time is greater than deadtime in that for continuous sources of deadtime, such as process transportation and mixing time delays and small process time constants, there is a continuous train of updates. For sampled data, there are no intervening values. Consequently, the effects can be worse. For example, there is aliasing of oscillations where the indicated amplitude is smaller and the period is larger than actual. There can be jitter due to variations in latency and lack of synchronization of digital data that introduce variable time delays and noise for rapidly changing signals.

References

1. McMillan, G. K., "Effect of Sample Delay on Standard PID Loop Tuning and Performance," Advanced Application Note 005, Aug. 28, 2008, http://www.modelingandcontrol.com/application-notes/.

2. McMillan, G. K., "Analog Control Holdouts", Apr. 2, 2007, http://www.modelingandcontrol.com/.

Index